Fungal Biology

Series Editors

Vijai Kumar Gupta
ERA Chair of Green Chemistry
Department of Chemistry and Biotechnology
School of Science
Tallinn University of Technology
Tallinn
Estonia

Maria G. Tuohy
Head of the Molecular Glycobiotechnology Group
Biochemistry
School of Natural Sciences
National University of Ireland Galway
Galway
Ireland

About the Series

Fungal biology has an integral role to play in the development of the biotechnology and biomedical sectors. It has become a subject of increasing importance as new fungi and their associated biomolecules are identified. The interaction between fungi and their environment is central to many natural processes that occur in the biosphere. The hosts and habitats of these eukaryotic microorganisms are very diverse; fungi are present in every ecosystem on Earth. The fungal kingdom is equally diverse, consisting of seven different known phyla. Yet detailed knowledge is limited to relatively few species. The relationship between fungi and humans has been characterized by the juxtaposed viewpoints of fungi as infectious agents of much dread and their exploitation as highly versatile systems for a range of economically important biotechnological applications. Understanding the biology of different fungi in diverse ecosystems as well as their interactions with living and non-living is essential to underpin effective and innovative technological developments. This series will provide a detailed compendium of methods and information used to investigate different aspects of mycology, including fungal biology and biochemistry, genetics, phylogenetics, genomics, proteomics, molecular enzymology, and biotechnological applications in a manner that reflects the many recent developments of relevance to researchers and scientists investigating the Kingdom Fungi. Rapid screening techniques based on screening specific regions in the DNA of fungi have been used in species comparison and identification, and are now being extended across fungal phyla. The majorities of fungi are multicellular eukaryotic systems and therefore may be excellent model systems by which to answer fundamental biological questions. A greater understanding of the cell biology of these versatile eukaryotes will underpin efforts to engineer certain fungal species to provide novel cell factories for production of proteins for pharmaceutical applications. Renewed interest in all aspects of the biology and biotechnology of fungi may also enable the development of "one pot" microbial cell factories to meet consumer energy needs in the 21st century. To realize this potential and to truly understand the diversity and biology of these eukaryotes, continued development of scientific tools and techniques is essential. As a professional reference, this series will be very helpful to all people who work with fungi and should be useful both to academic institutions and research teams, as well as to teachers, and graduate and postgraduate students with its information on the continuous developments in fungal biology with the publication of each volume.

More information about this series at http://www.springer.com/series/11224

*This volume is dedicated to my father **Late Shri Ram Prasad**, who motivated and supported me at every stage of my life*

Late Shri Ram Prasad (1938–2011)

Foreword

Endophytes are the group of microorganisms which reside in the internal tissues (roots, leaves, stems, bark, flowers, etc.) of plants in a symbiotic manner without causing any disease symptoms. Endophytes play an important role in helping plants to fight against biotic and abiotic stresses and enhance plant growth. Among them, endophytic fungi are one of the ubiquitous groups present in all land plants. Endophytic fungi obtained from medicinal plants are considered as an alternative to produce the bioactive compounds to fight against several human diseases. Research data suggested that during the last two decades, extensive research has been carried out of endophytic fungi and several biologically active compounds have been isolated from endophytic fungi. The anticancer drug, Taxol, has been reported from more than 15 genera of endophytic fungi having the potential to produce paclitaxel and its analogues. Similarly, the endophytic fungi showed a potential to fight against the multiple drug resistance pathogens which is increasing in an alarming rate throughout the world. Fungal endophytes like *Phaeosphaeria avenaria*, *Leptosphaeria* sp., *Fusarium* sp., and *P. chrysanthemicola* showed significant antimicrobial potential against human bacterial and fungal pathogens.

Advances in Endophytic Fungal Research: Present Status and Future Challenges, the volume published by the Springer Nature, is an important volume, and I strongly believe that it will attract the readers working in the field. The present volume has 15 chapters contributed by academicians and scientists working on endophytic fungal research throughout the world. I congratulate the editor for bringing out this volume with excellent contributions from scientists working on endophytic fungi and their application in health and industry.

K. R. S. Sambasiva Rao
Mizoram University (a Central University)
Aizawl, Mizoram, India

Preface

Endophytes are the group of microorganism (bacteria, Actinobacteria, or fungi) that reside in the internal tissues of the plants in a symbiotic association without causing any disease symptoms. Among the endophytes, endophytic fungi are one of the important ingredients of plant micro-ecosystems having significant effect on the growth and development of host plant. Endophytic fungi have been well documented and showed beneficial effects to the host plant either by preventing pathogenic organisms from colonization or enriching the rhizospheric soil by enhancing the nutrients for the plants to uptake. Majorly, the endophytic fungi have been categorized into two main groups (clavicipitaceous and non-clavicipitaceous) based on the differences in the evolution, taxonomy, their host, and ecological roles. Our knowledge about the relationship between the endophytic fungi and the host plant is still very limited. Though several secondary metabolites having biological activities are reported from endophytic fungi associated with plants, understanding the relationship between the fungi and plant will facilitate the more potential molecules by manipulating the growth of plants as well as endophytic fungi associated with them like the inoculation of particular endophytic fungi having secondary metabolite production ability to the plants to improve the drug quality and quantity.

Endophytic fungi associated with traditional medicinal plants collected from protected forest areas have been given a special attention by thinking that the medicinal plants share the potential of synthesizing the bioactive compounds with the endophytic fungi associated with them. The bioactive compounds produced by the endophytic fungi are originated by using different metabolic biosynthetic pathways and fall into several groups like terpenoids, steroids, quinones, phenols, etc. Therefore, the endophytic fungi represent a chemical reservoir for the production of new bioactive compounds having several bioactivities like anticancer, immunomodulatory, antioxidant, insecticidal, etc. for use in the pharmaceutical and agricultural industries.

The objective of this volume on *Advances in Endophytic Fungal Research: Present Status and Future Challenges* is to keep the readers informed about the recent developments that took place in the endophytic fungal research and the challenges for the researchers to look in the upcoming years. It is very important to exploit the plants especially the medicinally important plants for the isolation of endophytic fungi as an alternative tool for the production of secondary bioactive compounds before the plants get extinct. This is very much required due to the rapid increase of multiple drug resistance (MDR) against available drugs. The volume consists of 15 chapters contributed by the author(s) having vast experience in the field of endophytic fungal biology. Book chapters cover the wide applications of endophytic fungi obtained from different ecosystems and the methods of identification and characterization of the endophytic fungi. The main focus of the book is to look for an alternative method for the sustainable production of bioactive secondary metabolites from endophytic fungi having potential in pharmaceutical and agricultural industries. Recent developments given in the book will give more ideas to the researchers, students, and eminent scientists working on endophytic fungal research.

I express my sincere gratitude to all the contributors for their valuable contributions and support throughout. I extend my sincere thanks to the research scholar working in Molecular Microbiology and Systematics Laboratory, Department of Biotechnology, Mizoram University, for their hard work.

Mizoram, India Bhim Pratap Singh

Acknowledgments

My sincere thanks are extended to all the academicians and scientists who have contributed chapters and happily agreed to share their work on endophytic fungal research in this volume. At the same time, I also express my deepest gratitude to my family members especially my wife (Dr. Garima Singh) and my daughter (Aadita Singh) for their kind support which has prompted me to complete the assignment on time. I am also thankful to the Department of Biotechnology (DBT), New Delhi, Government of India, for supporting us financially in the form of several externally funded projects time to time. I am equally thankful to the Springer Publishing for their full cooperation during the production of the volume. In particular, I am thankful to the series editors, Dr. Vijai Kumar Gupta and Prof. Maria G. Tuohy, for accepting our proposal and providing their full support and encouragements. I am also thankful to the production team of Springer Nature for all their efforts for publishing the volume on time. I admit that it is quite possible to have some mistakes in the text inadvertently, and I take responsibility for the mistakes, and please feel free to inform me the same.

I am thankful to Prof. KRS Sambasiva Rao, Vice-Chancellor, Mizoram University, for his endeavor and motivations at all stages of the progress. I am also thankful to the Department of Biotechnology, New Delhi, for the establishment of DBT Bioinformatics Center at Mizoram University which was quite useful during the compilation of the book.

Mizoram, India　　　　Bhim Pratap Singh

Editor
Bhim Pratap Singh
Molecular Microbiology and Systematics
Laboratory
Department of Biotechnology
Mizoram University
Aizawl, Mizoram, India

ISSN 2198-7777 ISSN 2198-7785 (electronic)
Fungal Biology
ISBN 978-3-030-03588-4 ISBN 978-3-030-03589-1 (eBook)
https://doi.org/10.1007/978-3-030-03589-1

Library of Congress Control Number: 2019930125

This Springer imprint is published by the registered company Springer Nature Switzerland AG
The registered company address is: Gewerbestrasse 11, 6330 Cham, Switzerland

Bhim Pratap Singh
Editor

Advances in Endophytic Fungal Research

Present Status and Future Challenges

Springer

Contents

Contributors

Soad A. Abdelgalil Institute of Bioproduct Development, Universiti Teknologi Malaysia, Johor Bahru, Malaysia

City of Scientific Research and Technological Applications, New Burg Al Arab, Alexandria, Egypt

S. Anju Bhavan's Vivekananda College of Science, Humanities and Commerce, Secunderabad, India

Manash Pratim Barkataki Department of Biosciences and Bioengineering, Indian Institute of Technology Guwahati, Guwahati, Assam, India

Chanda V. Berde Department of Microbiology, Gogate Jogalekar College, Ratnagiri, Maharashtra, India

Biju Bharali Centre for the Environment, Indian Institute of Technology Guwahati, Guwahati, Assam, India

Utpal Bora Centre for the Environment, Indian Institute of Technology Guwahati, Guwahati, Assam, India

Department of Biosciences and Bioengineering, Indian Institute of Technology Guwahati, Guwahati, Assam, India

P. V. Bramhachari Department of Biotechnology, Krishna University, Machilipatnam, Andhra Pradesh, India

Hasnahana Chetia Department of Biosciences and Bioengineering, Indian Institute of Technology Guwahati, Guwahati, Assam, India

Manabendra Dutta Choudhury Department of Life Science and Bioinformatics, Assam University, Silchar, Assam, India

Srinivas Chowdappa Fungal Metabolite Research Laboratory, Department of Microbiology and Biotechnology, Bangalore University, Jnanabharathi Campus, Bangalore, Karnataka, India

Biplab Dash Department of Agril. Microbiology, Indira Gandhi Krishi Vishwavidhyalaya, Raipur, Chhattisgarh, India

Camila Rodrigues de Carvalho Laboratório de Química de Produtos Naturais Bioativos, Fundação Oswaldo Cruz-Fiocruz, Instituto René Rachou, Belo Horizonte, MG, Brazil

Harcharan Singh Dhaliwal Department of Biotechnology, Akal College of Agriculture, Eternal University, Baru Sahib, Himachal Pradesh, India

Hesham A. El Enshasy Institute of Bioproduct Development, Universiti Teknologi Malaysia, Johor Bahru, Malaysia

Faculty of Chemical Engineering and Energy, Universiti Teknologi Malaysia, Johor Bahru, Malaysia

City of Scientific Research and Technological Applications, New Burg Al Arab, Alexandria, Egypt

Mariana Costa Ferreira Department of Microbiology, Federal University of Minas Gerais, Belo Horizonte, MG, Brazil

Reeta Goel Department of Microbiology, GB Pant University of Agriculture and Technology, Pantnagar, Uttarakhand, India

Siti Zulaiha Hanapi Institute of Bioproduct Development, Universiti Teknologi Malaysia, Johor Bahru, Malaysia

Sairengpuii Hnamte Department of Microbiology, School of Life Sciences, Pondicherry University, Kalapet, Puducherry, India

Lata Jain ICAR-National Institute of Biotic Stress Management, Raipur, Chhattisgarh, India

Debajyoti Kabiraj Department of Biosciences and Bioengineering, Indian Institute of Technology Guwahati, Guwahati, Assam, India

T. Raja Komaraiah Department of Microbiology, Kakatiya University, Warangal, Telangana, India

Divjot Kour Department of Biotechnology, Akal College of Agriculture, Eternal University, Baru Sahib, Himachal Pradesh, India

Vinay Kumar ICAR-National Institute of Biotic Stress Management, Raipur, Chhattisgarh, India

Vinod Kumar Department of Biotechnology, Akal College of Agriculture, Eternal University, Baru Sahib, Himachal Pradesh, India

J. S. Kushveer Department of Biotechnology, Pondicherry University, Kalapet, Puducherry, India

Ong Mei Leng Harita Go Green Sdn. Bhd., Johor Bahru, Johor, Malaysia

Sarangam Majumdar Dipartimento di Ingegneria Scienze Informatiche e Matematica, Università degli Studi di L'Aquila, L'Aquila, Italy

Roslinda Abd Malek Institute of Bioproduct Development, Universiti Teknologi Malaysia, Johor Bahru, Malaysia

Himani Meena Department of Microbiology, School of Life Sciences, Pondicherry University, Kalapet, Puducherry, India

Preeti Mehta Centre of Advance Bioenergy Research, Research & Development Centre, Indian Oil Corporation Limited, Faridabad, India

Rashmi Mishra Department of Biotechnology, Pondicherry University, Kalapet, Puducherry, India

Vineet Kumar Mishra Molecular Microbiology and Systematic Laboratory, Department of Biotechnology, Mizoram University, Aizawl, Mizoram, India

Chakrabhavi D. Mohan Department of Studies in Molecular Biology, University of Mysore, Manasagangotri, Mysore, Karnataka, India

Ganugula Mohana Sheela Department of Biotechnology, Vignan University, Guntur, Andhra Pradesh, India

Subhoshmita Mondal Department of Chemical Engineering, Jadavpur University, Kolkata, India

Ponnala Vimal Mosahari Centre for the Environment, Indian Institute of Technology Guwahati, Guwahati, Assam, India

Deepa Nath Department of Botany and Biotechnology, Karimganj College, Karimganj, Assam, India

Sunita Ojha Department of Biosciences and Bioengineering, Indian Institute of Technology Guwahati, Guwahati, Assam, India

Parasuraman Paramanantham Department of Microbiology, School of Life Sciences, Pondicherry University, Puducherry, India

Ajit Kumar Passari Molecular Microbiology and Systematic Laboratory, Department of Biotechnology, Mizoram University, Aizawl, Mizoram, India

Subhaswaraj Pattnaik Department of Microbiology, School of Life Sciences, Pondicherry University, Puducherry, India

A. M. V. N. Prathyusha Department of Biotechnology, Krishna University, Machilipatnam, Andhra Pradesh, India

Chayanika Putatunda Department of Microbiology, DAV University, Jalandhar, India

Kusam Lata Rana Department of Biotechnology, Akal College of Agriculture, Eternal University, Baru Sahib, Himachal Pradesh, India

P. Revanthbabu Department of Biotechnology, Pondicherry University, Kalapet, Puducherry, India

Luiz Henrique Rosa Department of Microbiology, Federal University of Minas Gerais, Belo Horizonte, MG, Brazil

Priyanka Saha Department of Life Science and Bioinformatics, Assam University, Silchar, Assam, India

Dharitri Saikia Centre for the Environment, Indian Institute of Technology Guwahati, Guwahati, Assam, India

V. Venkateswara Sarma Department of Biotechnology, Pondicherry University, Kalapet, Puducherry, India

Anil Kumar Saxena ICAR-National Bureau of Agriculturally Important Microorganisms, Mau, Uttar Pradesh, India

Pragya Sharma Department of Bioengineering and Technology, Gauhati University Institute of Science and Technology (GUIST), Gauhati University, Guwahati, Assam, India

Rashmi Sharma Department of Microbiology, DAV University, Jalandhar, India

Imran Sheikh Department of Biotechnology, Akal College of Agriculture, Eternal University, Baru Sahib, Himachal Pradesh, India

Chandra N. Siddaiah Department of Studies in Biotechnology, University of Mysore, Manasagangotri, Mysore, Karnataka, India

Busi Siddhardha Department of Microbiology, School of Life Sciences, Pondicherry University, Kalapet, Puducherry, India

Bhim Pratap Singh Molecular Microbiology and Systematic Laboratory, Department of Biotechnology, Mizoram University, Aizawl, Mizoram, India

Tinka Singh Centre for the Environment, Indian Institute of Technology Guwahati, Guwahati, Assam, India

Ravindra Soni Department of Agril. Microbiology, Indira Gandhi Krishi Vishwavidhyalaya, Raipur, Chhattisgarh, India

Anupam Das Talukdar Department of Life Science and Bioinformatics, Assam University, Silchar, Assam, India

Lalrokimi Tochhawng Molecular Microbiology and Systematic Laboratory, Department of Biotechnology, Mizoram University, Aizawl, Mizoram, India

Fazilath Uzma Fungal Metabolite Research Laboratory, Department of Microbiology and Biotechnology, Bangalore University, Jnanabharathi Campus, Bangalore, Karnataka, India

Devanaboyina Venkataiah Department of Microbiology, Kakatiya University, Warangal, Telangana, India

Abhishek Walia Department of Microbiology, DAV University, Jalandhar, India

Ajar Nath Yadav Department of Biotechnology, Akal College of Agriculture, Eternal University, Baru Sahib, Himachal Pradesh, India

Neelam Yadav Gopi Nath P.G. College, VBSP University, Ghazipur, Uttar Pradesh, India

Carlos Leomar Zani Laboratório de Química de Produtos Naturais Bioativos, Fundação Oswaldo Cruz-Fiocruz, Instituto René Rachou, Belo Horizonte, MG, Brazil

Chapter 1
Endophytic Fungi: Role in Dye Decolorization

Lalrokimi Tochhawng, Vineet Kumar Mishra, Ajit Kumar Passari, and Bhim Pratap Singh

1.1 Introduction

Synthetic dyes have become an important part of life as they have several advantages over natural dyes; for example, they are very stable as a result of the formation of covalent bonds with fiber, whereas natural dyes are less permanent and difficult to apply. There are approximate 10,000 synthetic dyes available on market with annual production of 7×10^5 metric tons (Campos et al. 2001). Significant amounts of synthetic dyes are used by the textile and dye industries; 15–20% of the dyes used in the dye industry are unable to bind to fabric and are released as effluent and contaminate nearby water sources (Husain 2010). Water is a natural resource that is required in almost all household and industrial uses. Industrial effluents that contain mainly dye compounds are among the serious causes of water pollution, making water unfit for aquatic life (Hassani et al. 2008; Chanyal and Agrawal 2017). Many synthetic dyes are also carcinogenic and toxic to human and aquatic life as they are made up of compounds like benzidine and aromatic compounds. Cleavage of some of the most widely used commercial dyes, azo dyes, resulted in the formation of amine, which is mutagenic to humans (Asgher et al. 2006). Direct discharge of contaminants from several industries to the environment is the main cause of water pollution, so the removal of these compounds from the environment is essential for sustainable development (Aksu and Donmez 2005; Balaji et al. 2012).

The methods used for dye treatment are classified into three categories: physical, chemical, and biological methods (Robinson et al. 2001). The physical and chemical methods are quite expensive not very effective for the treatment of dyes from wastewater (Si et al. 2013). Because of their disadvantages, like their high cost, associated waste disposal problems, and lower adaptability, these methods are not

L. Tochhawng · V. K. Mishra · A. K. Passari · B. P. Singh (✉)
Molecular Microbiology and Systematic Laboratory, Department of Biotechnology, Mizoram University, Aizawl, Mizoram, India
e-mail: bhimpratap@mzu.edu.in

B. P. Singh (ed.), *Advances in Endophytic Fungal Research*, Fungal Biology,
https://doi.org/10.1007/978-3-030-03589-1_1

well suited for the treatment of dye-contaminated water (Asgher et al. 2008). Hence, the production of cost-effective, environmentally friendly, and efficient biological methods is necessary for effluent treatment.

Bioremediation-based methods using microorganisms for the treatment of effluent could be an efficient method as several bacteria and especially fungi and their enzymes are reported to have potential in decolorizing textile dyes (Forgacs et al. 2004). Bioremediation as a technology enjoys wide public support owing to its low cost and environmental friendliness. Among microorganisms, endophytes that reside in plants grown in contaminated sites may be a promising biological organism for the elimination of dye or for the decolorization of dye produced by several industries. Other important functions of endophytes include the ability to augment the thermal and salinity tolerance of host plants, increasing their ability to live in extreme environmental conditions (Redman et al. 2002). Rodriguez et al. (2004) reported that some tropical trees can increase their resistance by hosting endophytic microorganisms. Hence, bioremediation, which involves the use of microorganisms in the degradation or absorption of pollutants, is one of the most effective alternatives for the removal or reduction of industrial waste (Mcmullan et al. 2001; Singh et al. 2014). Most microorganisms, especially fungi and their enzymes, have the potential to degrade dyes by producing enzymes like laccases (Forgacs et al. 2004). Several researchers have reported that laccases produced by endophytic fungi have the ability to degrade azo dyes (Chanyal and Agrawal 2017; Ngieng et al. 2013; Novotny et al. 2004). Ngieng et al. (2013) reported on 20 endophytic fungi from *Melastoma malabathricum* with the ability to decolorize 5 different azo dyes to varying degrees. This chapter provides an overview of various forms of dye degradation using endophytic fungi, presents a biotechnological approach to using these fungi, and discusses their future prospects.

1.2 Endophytes

The groups of microorganisms that inhabit different parts of plant tissues without causing any disease symptoms are known as endophytes (Gouda et al. 2016; Schulz and Boyle 2005; Specian et al. 2012). There are no apparent symptoms of disease found in plants hosting endophytes. Hence, endophytes exhibit strong positive associations with their host plant (Nair and Padmavathy 2014).

Endophytes are well known to produce a wide variety of bioactive compounds, including antibiotics, antitumor and immunosuppressive agents, plant growth hormones, and biological substances like enzymes, alkaloids, and vitamins, which can play a vital role in the pharmaceutical and agriculture industries (Uzma et al. 2018; Mishra et al. 2017; Wu et al. 2010). Endophytes are reported to produce compounds having the ability to inhibit bacterial and fungal pathogens and could help in protecting plants against phytopathogens (Mishra et al. 2016; Rya et al. 2007; Staniek et al. 2008; Rai et al. 2014). Interestingly, few endophytes share the compounds produced by plants as well (Kumara et al. 2014; Shweta et al. 2014). These findings demonstrate that endophytes are present within plant tissues and subsist as reservoirs of bioactive metabolites (Tan and Zou 2001; Priti et al. 2009).

Endophytes are categorized into bacteria and fungi. More than 200 bacterial genera have been identified as endophytes distributed among the phyla Actinobacteria, Proteobacteria, and Firmicutes (Golinska et al. 2015). Among these, the genus *Streptomyces* from the phylum Actinobacteria is the most dominant of genera that have been isolated in large volume from medicinal plants as endophytes (Zothanpuia et al. 2018; Zhao et al. 2011; Golinska et al. 2015). Endophytic fungi are classified into clavicipitaceous and nonclavicipitaceous. Clavicipitaceous fungi are those that infect certain grasses present in cool regions, and nonclavicipitaceous fungi are isolated from healthy tissues of nonvascular plants like ferns and their allies, conifers, and angiosperms. Ascomycota and Basidiomycota represent the two main groups of nonclavicipitaceous fungi (Jalgaonwala et al. 2011; Bhardwaj and Agrawal 2014).

1.3 Dyes

Dyes are complex colored organic compounds mainly used in the textile, leather, paper, and food industries (Harvey and Keith 1983). Since dyes are the major component used in these industries, the effluents discharged are mainly composed of dye compounds that act as the main source of organic pollutants released into the environment that disturb normal biotic as well as abiotic systems (Muthezhilan et al. 2014). Dyes can exist in many different formed, but they include a minimum of one azo (N=N) bond. For instance, dyes having one N=N bond are known as azo and monoazo dyes, whereas dyes having two and three N=N bonds are known as diazos and triazos, respectively. Azo groups linked with naphthalene and benzene rings give dyes their color and make various shades and intensities possible (Zollinger 1991). Dyes are broadly divided into synthetic and natural dyes, and do not have a definite chemical structure. Chromophores give fibers their color, whereas auxochromes fixed colors on fibers (Banat et al. 1996; Welham 2000). Dye processing occurs largely in three steps: preparation, dyeing, and finishing. Different types of dye processing are carried out based on the nature of the fiber and the properties of the dyes, like the chemical structure, fixing property compatibility with the materials to be used, classification, and pigments used (Guaratini and Zanoni 2000). Large amounts of chemical compounds are used in each step of dye processing (Moore and Ausley 2004).

Dyes cause serious problems related to the release of industrial effluents and the removal of dyes and other chemicals pollutants (Balaji et al. 2012). Most dyes can affect human beings and animals by causing allergies, cardiovascular problems, gastrointestinal problems, DNA damage, and cancer (Harvey and Keith 1983; Mittal et al. 2006). To address these problems, many researchers are trying to develop a simple, eco-friendly, and cost-effective techniques for dye degradation, but this remains a very difficult and complex challenge (Pant et al. 2008).

1.3.1 *Types of Dyes*

1.3.1.1 Textile Dyes and Their Importance

Over 100,000 colors of synthetic dyes are produced commercially, and over 7×10^5 tons of dyes are produced every year worldwide (Zollinger 1987). Throughout the world, dyes are heavily utilized in the textile, cosmetic, paper, pharmaceutical, and food industries along with additives in the petroleum industry (Husain 2010). Synthetic dyes are broadly used in the textile and dyeing industries. When dyes are processed, approximately 15–20% of the dyes does not bind with fibers and as a result is lost in effluent (Husain 2010). Maas and Chaudhari (2005) suggested that around the world, approximately 280,000 tons of textile dyes are released from industrial sewage every year. The released textile dyes are transferred in rivers or lakes, which can cause water pollution at very low concentrations of 1 ppm (O'Neill et al. 1999). The accumulation of dye in water can reduce sunlight penetration, which in turn can reduce photosynthetic activity. Moreover, accumulated dyes unfavorably influence biological oxygen demand (BOD), chemical oxygen demand (COD), and dissolved oxygen, which creates a toxic environment for aquatic organisms (Khan and Bhawana 2013; Sen et al. 2016). Most dyes have a hard structure and are of synthetic origin, so bleaching dyes is complicated.

Natural Dyes

Natural dyes are obtained from natural resources like different parts of plants, animals, insects, and minerals without application of any chemical treatment. Indigo dye from the plant *Indigofera tinctoria*, lawsone obtained from *Lawsonia mermis*, carajurin obtained from *Bignonia chica*, and carotenoids are examples of natural dyes (Vankar 2000). Synthetic dyes are widely used in various industries, but they are also known mutagens, allergens, and carcinogens, whereas natural dyes do not have any toxic effects. Dyes obtained from natural resources like plants, animals, or minerals are very easy to wash and clean, and for this reason natural dyes are very eco-friendly (Vankar 2000). In addition to their dyeing properties, natural dyes also have a broad range of medicinal properties. These days, there is growing awareness of the use of natural dyes and dye producer plants. Owing to their nontoxic nature, medicinal values, and less harmful side effects, natural dyes are now applied commonly in everyday food products and in the pharmaceutical industry (Chengaiah et al. 2010; Shahid et al. 2013).

In India, over 400 plants can produce dye. For instance, a brilliant, naturally occurring yellow dye is produced from turmeric, which also has strong antiseptic activity to cure skin ailments (Siva 2007). Natural dyes are commonly applied in food coloring, leather, natural protein fibers like wool, silk, and cotton textiles, and even in drugs and cosmetic products owing to their nontoxicity. Due to the evolution of synthetic dyes, the use of natural dyes worldwide in textiles has been limited to artisan or craftsman uses, small-scale or cottage-level dyers, and printers, and some producers manufacture environmental friendly dyes for textile production (Chavan 1995). Natural dyes require a compound known as a mordant used to set dyes on fabric, which prevents the dye from bleeding or being washed out easily. Mordants facilitate chemical reactions for the absorption of dyes between dyes and fibers (Siva 2007).

Most natural dyes are found from plants that produce various colors. Most plant parts like the seeds, leaves, bark, roots, flowers, fruits, and so forth produce dyes. Interestingly, 2000 pigments are made by different plant parts. Of these, only 150 have been commercially sold on the market. In India, approx. 450 taxa are known to produce dyes, of which the 50 best-known plants produce natural dyes. Briefly, Siva (2007) reported that most dyes are obtained from roots (n = 10), followed by wood (n = 8), flowers and fruits (n = 7), leaves (n = 5), bark (n = 4), seeds, and gum and resins (n = 3). To date, few dyes have been synthesized from natural sources; for instance, few plants, like *Lithospermum erythrorhizon* and *Bixa Orellana*, can be used to derive dyes for lipstick and indigo for eye shadow. Siva (2007) demonstrated that the occurrence of dyes in plants varies depending on the age of the plant and the season. Many plants, such as *Punica granatum* L., which have high antimicrobial potential due to the occurrence of high amounts of tannins, can be used to extract natural dyes. Moreover, several other plant dyes, such as lawsone, commonly known as henna obtained from *Lawsonia inermis* L., Juglone from walnut, and lapachol from alkanet, demonstrated antibacterial and antifungal properties (Siva 2007). Another advantage of natural dyes is that some have antimicrobial potential against pathogens. Singh et al. (2005) reported on the antimicrobial activity of five natural dyes (*Acacia catechu, Kerria lacca, Quercus infectoria, Rubia cordifolia*, and *Rumex maritimus*) against *Klebsiella pneumonie, Escherichia coli, Proteus vulgaris*, and *Pseudomonas aeruginosa*. Of these, *Quercus infectoria*, a natural dye, showed the highest antimicrobial activity (Singh et al. 2005).

Synthetic Dyes

Synthetic dyes are widely used in various industries such as textiles, leather, beauty care products, food, pharmaceuticals, and paper printing. The most toxic compounds that create environmental pollution include azo, antraquinone, heterocyclic, triphenylmethane, and phthalocyanine dyes. Interestingly, 10–15% of dyes are removed in wastewater in these industries while dyeing and washing textiles (Husain 2010). This wastewater can cause several diseases around the world. Moreover, the release of wastewater can pollute both ground and surface water, which may lead to various health issues in both humans and animals because they are considered very toxic, mutagenic, and carcinogenic. Therefore, it is immensely important to search for alternative methods to degrade dyes easily from wastewater treatment systems. The degradation of dyes using physical and chemical methods is not very effective owing to the high costs and time and because the method could be methodologically demanding (Ben Younes et al. 2012). Nowadays, most researchers are trying to use microorganisms to degrade dyes (Eichlerova et al. 2005).

Of the 100,000 commercially available dyes, at least 10–15% of the used dye traps in the environment in the form of wastes become major environmental pollutants (Zollinger 1987; O'Neill et al. 1999; Robinson et al. 2001). Among synthetic dyes, azo reactive dyes are the most prevalent types that are soluble in water. Azo dyes have the greatest range of color structures and colors among the synthetic dyes and are usually resistant to biodegradation by aerobic methods. Most of these dyes become toxic only after being released into aquatic environments, where they may be converted into harmful carcinogenic amines (Spadaro et al. 1992; Chung and Stevens 1993). Hence, they are considered xenobiotic compounds due to their resistance to natural microbiological degradation (Stolz 2001). There are more than 3000 azo dyes, of which Maxilon Blue GRL, Astrazon Red GTLN, and Sandolan Yellow are widely used in certain industries like textiles, leather, cosmetics, food coloring, and paper production (Elbanna et al. 2010). It was estimated that approximately 10% of dyes do set in fiber during the dyeing process and are released directly into the environment, causing major harm (Asad et al. 2007) and complications like mutagenicity, genotoxicity, and carcinogenicity to living beings (Puvaneswari et al. 2006).

White rot fungi are commonly used as decomposers of lignin and have the potential to degrade various organic pollutants, including pesticides, polycyclic aromatic hydrocarbons (PAHs), polychlorinated biphenyls and synthetic dyes (Bezalel et al. 1997; Sack et al. 1997; Novotny et al. 2000; Pointing 2001). They are commonly used in nonspecific, free-radical-mediated processes that utilize enzymes to degrade lignin and other structurally similar compounds (Chagas and Durrant 2001). Many researchers have reported on the decolorization and degradation of synthetic dyes from various ligninolytic fungi such as *Phanerochaete chrysosporium* (Couto et al. 2000; Podgornik et al. 2001; Conneely et al. 2002; Moldes et al. 2003), *Trametes versicolor* (Swamy and Ramsay 1999a, b; Borchert and Libra 2001; Martins et al. 2003), *Pleurotus ostreatus* (Shin et al. 1997; Rodrigues et al. 1999; Novotny et al. 2001), and *Bjerkandera adusta* (Heinfling et al. 1998; Heinfling-Weidtmann et al. 2001; Jarosz-Wilkołazka et al. 2002). In total, 20 fungal endophytes have been obtained from *Melastoma malabathricum* and examined for their potential to decolorize azo dyes such as congo red, methyl red, orange G, and an anthraquinone dye, Remazol Brilliant Blue R (RBBR). Of these, the isolated *Marasmius cladophyllus* strain MS8 has decolorized RBBR, methyl red, congo red, and orange G in percentages of 97%, 56%, 48%, and 33%, respectively (Ngieng et al. 2013).

1.4 Biosorption and Bioaccumulation

Among all the methods for the treatment of wastewater, adsorption has been regarded as an efficient and low-cost process for the treatment of dyes mixed with effluents released by industry. The removal of textile dyes from wastewater using fungal biomass is an attractive option as it might reduce the overall cost of the treatment process (Saraf and Vaidya 2015). Biosorption is the combination of biomass solutes that cannot use any metabolic energy or transport in the binding process.

This binding process may occur suddenly where live biomass is used. However, both living and dead biomass can be used in biosorption (Tobin et al. 1994).

The degradation of dyes has been mostly performed by microbial isolates originating from soils polluted with dyes, effluents from industry, or marine sources (Saratale et al. 2006; Abedin 2008; Gou et al. 2009). For instance, Kabbout and Taha (2014) studied the biosorption of methylene blue by the dead fungal biomass of *Aspergillus fumigates* and optimized the conditions for better absorption. *Aspergillus niger* and *Rhizopus stolonifer* showed the ability to remove dyes like Congo red and bromophenol blue respectively by biosorption (Fu and Viraraghavan 2001; Zeroual et al. 2006). Bayramoglu and Arica (2007) studied the native and heat-treated fungal biomass of *Trametes versicolor* for the removal of two benzidine dyes, direct red 128 and direct blue −1 using different parameters. The biosorption activity of the heat-treated and native biomass of *T. versicolor* was 152.3 and 101.1 mg/g for direct blue −1 and 225.4 and 189.7 mg/g for the removal of direct red. Iqbal and Saeed (2007) reported that the uptake efficiency of RBBR by immobilized *Phanerochaete chrysosporium* biomass in loofa sponge was much better than that of free fungal biomass with enhanced uptake efficiency of RBBR. The value of loofa sponge–immobilized biomass increased (18.6%) compared to free fungal biomass. Marcharchand and Ting (2017) showed that *Trichoderma asperellum* growing on fewer nutrients can retain its dye-decolorizing efficiency. *Trichoderma asperellum* has been grown in synthetic medium with different concentrations such as 20%, 50%, 75%, and 100%, and shown an ability to perform biosorption. This is also a good cost-effective strategy to use on fungi for dye removal; this method can help to lower the cost of cultivating the biomass of Ta for dye removal activities isolated at lower concentrations (20%, 50%, and 75%). This result demonstrated that lower nutrient levels might be useful in cultivating the biomass of Ta for dye removal activities (Marcharchand and Ting 2017) (Table 1.1).

However, endophytes associated with plants have been much less explored for their biosorbent activity. Ting et al. (2016) recovered an endophytic *Diaporthe* sp. fungus from *Portulaca* weed and explored it for its potential activity in biodegradation and biosorption in triphenylmethane (TPM) dyes. The live cells of *Diaporthe* sp. showed stronger decolorizing activity against malachite green, crystal violet, and methyl violet with decolorizing efficiencies (%) of 87.80%, 84.87%, and 78.81%, respectively. Additionally, Ting et al. (2016) reported that the decolorization of live cells is far better than the decolorizing efficiency (18.82%, 39.88%, 48.32%) of dead cells.

Bioaccumulation is explained as the accumulation of pollutants or any xenobiotic substance by actively growing cells by metabolism (Aksu and Donmez 2005). Aksu and Donmez (2005) suggested that some fungi, like *Candida tropicalis*, have the ability to remove different dyes like Remazol Black B and Remazol Blue using a bioaccumulation process. Similarly, *Saccharomyces cerevisiae* can remove dyes like Remazol Blue, Remazol Black B, and Remazol Red RB by bioaccumulation (Aksu 2003). Taskin and Erdel (2010) also showed the efficiency of fungi soilborne *Aspergillus niger* in decolorizing textile dye Reactive Black 5 through bioaccumulation.

Table 1.1 Biodegradation of various synthetic dyes by endophytic fungi

Source	Name of organism	Type of dye	Mechanism	Reference
M. malabathricum	*Marasmius cladophyllus*	Remazol Brilliant Blue R (RBBR), Orange G, Congo red, and Methyl red	–	Ngieng et al. (2013)
P. hispidum Sw	*Phlebia* spp.	Reactive Blue 19 and Reactive Black 5	Absorption	Bulla et al. (2017)
H. annuus L	*P. formosus*	Reactive Blue 19 and Reactive Black 5	Absorption	Bulla et al. (2017)
Canavalia rosea, Ipomoea pescaprae and Spinifex spp.	*Fusarium* spp.	Yellow MR, Blue M2R, Black-B, Red BSID, Manenta MP, Blue MR, Orange M2R, Orange 3R and Brown GR	–	Muthezhilan et al. (2014)
Pigieon pea	*Myrothecium verrucaria*	Congo red, Methyl orange, Methyl red, and Crystal violet	–	Sun et al. (2017)
Decayed wood	*Ganoderma cupreum*	Reactive violet 1	Biodegradation	Gahlout et al. (2013)
Fagus sylvatica	*Fomes fomentarius*	Remazol Brilliant Blue R (RBBR)	–	Eichlerova et al. (2005)
Fagus sylvatica	*Oudemansiella mucida*	Remazol Brilliant Blue R (RBBR) and Orange G	–	Eichlerova et al. (2005)
Liquidambar styraciflua and *Quercus nuttallii*	*Pleurotus cystidiosus*	Remazol Brilliant Blue R (RBBR)	–	Eichlerova et al. (2005)
Acer pseudosieboldianum	*Trichaptum abietinum*	Remazol Brilliant Blue R (RBBR) and Orange G	–	Eichlerova et al. (2005)

1.5 Enzymes Involved in Biodegradation of Dyes

Lignin is easily degraded by most fungi owing to their extracellular secretion of nonspecific and nonselective enzymes. These enzymes are laccases, manganese peroxidases, and lignin peroxidases, which can work together to produce H_2O_2 and secondary metabolites. (Kirk and Farrell 1987; Levin et al. 2004). Similarly, due to their nonspecific mechanisms, these fungi are able to degrade lignin and various other pollutants such as PAHs, polychlorinated biphenyls, dioxins, chlorinated phenols, explosives, dyes, and pesticides (Pointing 2001). Some researchers have reported on the ability of laccases, manganese peroxidases, and lignin peroxidases to decolorize dyes (Cripps et al. 1990; Podgornik et al. 2001; Pointing 2001). Cha et al. (2001) reported that most fungi have cytochrome P-450 monooxygenase, which is involved in organic pollutant degradation.

Laccases are multicopper enzymes with the ability to oxidize phenolic and non-phenolic lignin-related compounds. Moreover, laccases have applications in bioremediation as well as other biotechnological applications (Yang et al. 2017). They are broadly distributed in fungi, bacteria, insects, and plants. They can catalyze the oxidation of a broad range of substrates, including mono-, di-, and polyphenols, ascorbate, and methoxyphenols through the concurrent reduction of oxygen to water. Most widely studied laccases are of fungal origin. They are primarily produced by ascomycetes, basidiomycetes, and deuteromycetes, of which white-rot basidiomycetes are regarded as the most proficient laccase producers (Arora and Sharma 2010; Si et al. 2013; Yang et al. 2017; Zhuo et al. 2017). Laccases are also commonly used in bleaching, pulping, dye decolorization, biosensing, food technology, and wastewater treatment (Campos et al. 2016). Endophytic fungi also have the potential to decolorize dyes discharged in various industries.

Laccases are among the major enzymes produced by endophytic fungi for decolorizing dye. Muthezhilan et al. (2014) reported on the fungal endophyte *Fusarium* sp. strain AEF17 isolated from the *Ipomoea pescaprae* plant showing laccase enzyme production. The laccase purified from this strain had the potential to decolorize nine different textile dyes. The maximum decolorization activity was found in blue M2R (BM2R), orange M2R(OM2R), and black-B (BB), followed by limited decolorization activity by red BSID (RBSID), yellow MR(YMR), blue MR (BMR), magenta MP (MMP), brown GR (BGR), and orange 3R (O3R) and decolorization. In another study, a fungal endophyte known as *Myrothecium verrucaria* was isolated from pigeon pea that has the potential to produce laccase enzyme. In the presence of ABTS (2,2′-azino-bis(3-ethylbenzothiazoline-6-sulfonate), the isolate *Myrothecium verrucaria* exhibited significant decolorization potential toward methyl orange, Congo red, crystal violet, and methyl red) (Sun et al. 2017). Endophytic fungi *Phlebia* spp. isolated from *P. hispidum* Sw. and *P. formosus*, obtained from *H. annuus* L., were active in biodegradation of textile dyes. They displayed the ability to degrade Reactive Black 5 and Reactive Blue 19 textile dyes. After 30 days of treatment, isolated *Phlebia* spp. and *P. formosus* showed 90% and 70% degradation of black and blue dye, respectively. *Phlebia* spp. showed the highest production of extracellular laccase, which suggests a role of the enzyme in the decolorization of the blue and black textile dyes (Bulla et al. 2017). The strain *Ganoderma cupreum* AG-1 isolated from decayed wood showed the potential to produce lignolytic enzymes and the capacity to decolorize around 19 different azo dyes in a pH range of 4.5–6. The *Ganoderma cupreum* strain AG-1 was also found to produce laccase and manganese peroxidase. *G. cupreum* AG-1 showed a capacity to produce large amounts of laccase, which indicates its role in decolorizing dyes (Gahlout et al. 2013). A white-rot fungis from Argentina, *Coriolus versicolor* f. *antarcticus*, showed potential for the production of various extracellular enzymes like laccase, lignin peroxidase, and manganese peroxidase on solid medium, which have the ability to decolorize different dyes such as malachite green, azure B, poly R478, anthraquinone blue, Congo red, and xylidine (Levin et al. 2004). Levin et al. (2004) suggested that the strain *Coriolus versicolor* f. *antarcticus* could be used as a candidate in biodecoloration processes. Fungal endophytes associated with *Melastoma malabathricum* exhibited the ability to decolorize three azo dyes: methyl red, Congo red, and orange G. Among all the endophytic isolates recovered from *M. malabathricum*, a basidiomycete fungus, *Marasmius cladophyllus*, decolorized 97% of RBBR dye, followed by 56% decolorization of methyl red (Table 1.1).

1.6 Conclusion

Fungus-mediated biodegradation or removal of dye from wastewater involves either the use of pure enzymes or biosorption. This chapter discussed the importance of endophytic fungi in both biosorption and enzyme-mediated removal of dyes using endophytic fungi.

Laccases from endophytic fungi have demonstrated their potential in degrading various types of dyes. Additionally, many studies on fungal biosorbents have suggested that biosorbents represent an emerging and promising solution to conventional practices. Fungal endophytes have been much less explored for their ability to remove dyes using a biosorption approach. However, many studies have reported on purified laccases from endophytic fungi for their potent activity in degrading dyes. Nevertheless, it is essential to search for endophytic fungal strains that are very eco-friendly and nontoxic that can degrade various dyes easily.

Acknowledgments BPS is thankful to the Department of Biotechnology, Government of India, New Delhi, for financial support under DBT's Unit of Excellence Programme for NE (102/IFD/SAN/4290-4291/2016-2017).

References

Abedin RMA (2008) Decolorization and biodegradation of crystal violet and malachite green by *Fusarium solani* (Martius) Saccardo. A comparative study on biosorption of dyes by the dead fungal biomass. AEJB 1:17–31

Aksu Z (2003) Reactive dye bioaccumulation by *Saccharomyces cerevisiae*. Process Biochem 38:1437–1444

Aksu Z, Donmez G (2005) Combined effects of molasses sucrose and reactive dye on the growth and dye bioaccumulation properties of *Candida tropicalis*. Process Biochem 40:2443–2454

Arora DS, Sharma RK (2010) Ligninolytic fungal laccases and their biotechnological applications. Appl Biochem Biotechnol 160(6):1760–1788

Asad S, Amoozegar MA, Pourbabaee AA, Sarbolouki MN, Dastgheib SM (2007) Decolorization of textile dyes by newly isolated halophilic and halotolerant bacteria. Bioresour Technol 98:2082–2088

Asgher M, Shah SAH, Ali M, Legge RL (2006) Decolorization of some reactive textile dyes by white rot fungi isolated in Pakistan. World J Microbiol Biotechnol 22:89–93

Asgher M, Bhatti HN, Ashraf M, Legge RL (2008) Recent developments in biodegradation of industrial pollutants by white rot fungi and their enzyme system. Biodegradation 19:771–783

Balaji V, Vinayagamoorthi D, Palanisamy A, Anbalagan S (2012) Degradation of reactive Red HE7B and Yellow FN2R dyes by fungal isolates. J Acad Indus Res 1(3):132–136

Banat IM, Nigam P, Singh D, Marchant R (1996) Microbial decolorization of textile-dye-containing effluents: a review. Bioresour Technol 61:103–103

Bayramoglu G, Arica MY (2007) Biosorption of benzidine based textile dyes "Direct Blue 1 and Direct Red 128" using native and heat-treated biomass of *Trametes versicolor*. J Hazard Mater 143:135–143

Ben Younes S, Bouallagui Z, Sayadi S (2012) Catalytic behaviour and detoxifying ability of an atypical homotrimeric laccase from the thermophilic strain *Scytalidium thermophilum* on selected Azo and Triarylmethane dyes. J Mol Catal B Enzym 79:41–48

Bezalel L, Hadar Y, Cerniglia CE (1997) Enzymatic mechanisms involved in phenantrene degradation by white-rot fungus *Pleurotus ostreatus*. Appl Environ Microbiol 63:2495–2501

Bhardwaj A, Agrawal P (2014) A review fungal endophytes: as a store house of bioactive compound. World J Pharm Pharm Sci 3:228–237

Borchert M, Libra JA (2001) Decolorization of reactive dyes by the white rot fungus *Trametes versicolor* in sequencing batch reactors. Biotechnol Bioeng 75:313–321

Bulla LMC, Polonio JC, Portela-Castro ALB, Kava V, Azevedo JL, Pamphile JA (2017) Activity of the endophytic fungi *Phlebia sp.* and *Paecilomyces formosus* in decolourisation and the reduction o reactive dyes' cytotoxicity in fish erythrocytes. Environ Monit Assess 189(88):1–11

Campos R, Kandelbauer A, Robra KH, Cavaco-Paulo A, Gubitz GM (2001) Indigo degradation with purified laccases from *Trametes hirsuta* and *Sclerotium rolfsii*. J Biotechnol 89:131–139

Campos PA, Levin LN, Wirth SA (2016) Heterologous production, characterization and dye decolorization ability of a novel thermostable laccase isoenzyme from *Trametes trogii* BAFC 463. Process Biochem 51:895–903

Cha CJ, Doerge DR, Cerniglia CE (2001) Biotransformation of Malachite Green by the fungus *Cunninghamella elegans*. Appl Environ Microbiol 67:4358–4360

Chagas EP, Durrant LR (2001) Decolorization of azo dyes by *Phanerochaete chrysosporium* and *Pleurotus sajorcaju*. Enzym Microb Technol 29:473–477

Chanyal S, Agrawal PK (2017) Decolorization of textile by laccase from newly isolated endophytic fungus *Daldinia* sp. Kavaka 48(1):33–41

Chavan RB (1995) Revival of natural dyes—a word of caution to environmentalists. In: Symposium proceedings eco-friendly textile processing. India and Japan, IIT, Delhi, p 83–186

Chengaiah B, Rao KM, Kumar KM, Alagusundaram M, Chetty CM (2010) Medicinal importance of natural dyes—a review. Int J PharmTech Res 2(1):144–154

Chung KT, Stevens SE (1993) Decolorization of azo dyes environmental microorganisms and helminthes. Environ Toxicol Chem 12:2121–2132

Conneely A, Smyth WF, McMullan G (2002) Study of the white-rot fungal degradation of selected phthalocyanine dyes by capillary electrophoresis and liquid chromatography. Anal Chim Acta 451:259–270

Couto SR, Rivela I, Munoz MR, Sanroma'n A (2000) Stimulation of ligninolytic enzyme production and the ability to decolorize Poly R-478 in semi-solid-state cultures of *Phanerochaete chrysosporium*. Bioresour Technol 74:159–164

Cripps C, Bumpus JA, Aust SD (1990) Biodegradation of azo and heterocyclic dyes by *Phanerochaete chrysosporium*. Appl Environ Microbiol 56:1114–1148

Eichlerova I, Homolka L, Lisa L, Nerud F (2005) Orange G and Remazol Brilliant Blue R decolorization by white rot fungi *Dichomitus squalens*, *Ischnoderma resinosum* and *Pleurotus*. Chemosphere 60:398–404

Elbanna K, Hassan G, Khider M, Mandour R (2010) Safe biodegradation of textile azo dyes by newly isolated lactic acid bacteria and detection of plasmids associated with degradation. J Bioremed Biodegr 1:1–6

Forgacs E, Cserhati T, Oros G (2004) Removal of synthetic dyes from wastewaters: a review. Environ Int 30:953–971

Fu Y, Viraraghavan T (2001) Fungal decolourization of dye wastewaters: a review. Bioresour Technol 79:251–262

Gahlout M, Gupte S, Gupte A (2013) Optimization of culture condition for enhanced decolorization and degradation of azo dye reactive violet 1 with concomitant production of ligninolytic enzymes by *Ganoderma cupreum* AG-1. 3 Biotech 3(2):143–152

Golinska P, Wypij M, Agarkar G, Rathod D, Dahm H, Rai M (2015) Endophytic actinobacteria of medicinal plants: diversity and bioactivity. Anton Van Leeuwen 108:267–289

Gou M, Qu Y, Zhou J, Ma F, Tan L (2009) Azo dye decolorization by a new fungal isolate *Penicillium* sp. QQ and fungal-bacterial cocultures. J Hazard Mater 170:314–319

Gouda S, Das G, Sen SK, Shin HS, Patra JK (2016) Endophytes: a treasure house of bioactive compounds of medicinal importance. Front Microbiol 7:1538

Guaratini CCI, Zanoni MVB (2000) Textile dyes. Química Nova 23(1):71–78

Harvey JW, Keith AS (1983) Studies of the efficacy and potential hazards of methylene blue therapy in aniline-induced haemoglobinaemia. Br J Haematol 54(1):29–41

Hassani AH, Mirzayee R, Nasseri S, Borghei M, Gholami M, Torabifar B (2008) Nanofiltration process on dye removal from simulated textile wastewater. Int J Environ Sci Technol 5(3):401–408

Heinfling A, Martı'nez MJ, Martínez AT, Bergbauer M, Szewzyk U (1998) Transformation of industrial dyes by manganese peroxidases from *Bjerkandera adusta* and *Pleurotus eringii* in a manganese-independent reaction. Appl Environ Microbiol 64:2788–2793

Heinfling-Weidtmann A, Reemtsma T, Storm T, Szewzyk U (2001) Sulfophthalimide as major metabolite formed from sulfonated phthalocyanine dyes by the white-rot fungus *Bjerkandera adusta*. FEMS Microbiol Lett 203:179–183

Husain Q (2010) Peroxidase mediated decolorization and remediation of wastewater containing industrial dyes: a review. Rev Environ Sci Biotechnol 9(2):117–140

Iqbal M, Saeed A (2007) Biosorption of reactive dye by loofa sponge-immobilized fungal biomass of *Phanerochaete chrysosporium*. Process Biochem 42:1160–1164

Jalgaonwala RE, Mohite BV, Mahajan RT (2011) Natural products from plant associated endophytic fungi. J Microbiol Biotechnol Res 1:21–32

Jarosz-Wilkołazka A, Rdest-Kochman˜ ska J, Malarczyk E, Wardas W, Leonowicz A (2002) Fungi and their ability to decolorize azo and antraquinonic dyes. Enzym Microb Technol 30:566–572

Kabbout R, Taha S (2014) Biodecolorization of textile dye effluent by biosorption on fungal biomass materials. Phys Procedia 55:437–444

Khan RP, Bhawana FMH (2013) Microbial decolorization and degradation of synthetic dyes: a review. Rev Environ Sci Biotechnol 12(1):75–97

Kirk TK, Farrell RL (1987) Enzymatic "combustion": the microbial degradation of lignin. Annu Rev Microbiol 41:465–505

Kumara MP, Soujanya KN, Ravikanth G, Vasudeva R, Gane- shaiah KN, Uma Shaanker R (2014) Rohitukine, a chromone alkaloid and a precursor of flavopiridol, is produced by endophytic fungi isolated from *Dysoxylum binectariferum Hook.f* and *Amoora rohituka* (Roxb). Phytomedicine 21:541–546

Levin L, Papinutti L, Forchiassin F (2004) Evaluation of Argentinean white-rot fungi for their ability to produce lignin-modifying enzymes and decolorize industrial dyes. Bioresour Technol 94:169–176

Maas R, Chaudhari S (2005) Adsorption and biological decolourization of azo dye Reactive Red 2 in semicontinuous anaerobic reactors. Process Biochem 40(2):699–705

Marcharchand S, Ting ASY (2017) *Trichoderma asperellum* cultured in reduced concentrations of synthetic medium retained dye decolourization efficacy. J Environ Manag 203(1):542–549

Martins MAM, Lima N, Silvestre AJD, Queiroz MJ (2003) Comparative studies of fungal degradation of single or mixed bioaccessible reactive azo dyes. Chemosphere 52:967–973

Mcmullan G, Mehan C, Conneely A, Kirby N, Robinson T, Nigam P, Banat IM, Marchant R, Smyth WF (2001) Microbial decolourisation and degradation of textile dyes. Appl Microbiol Biotechnol 56(1–2):81–87

Mishra VK, Passari AK, Singh BP (2016) In vitro antimycotic and biosynthetic potential of fungal endophytes associated with *Schima Wallichii*. In: Kumar P et al (eds) Current trends in disease diagnostics. Springer International Publishing, Basel, pp 367–381

Mishra VK, Passari AK, Chandra P, Leo VV, Kumar B, Gupta VK, Singh BP (2017) Determination and production of antimicrobial compounds by *Aspergillus clavatonanicus* strain MJ31, an endophytic fungus from *Mirabilis jalapa* L. using UPLC-ESI-MS/MS and TD GC-MS. PLoS One 12(10):1–24. https://doi.org/10.1371/journal.pone.0186234

Mittal A, Mittal J, Kurup L, Singh A (2006) Process development for the removal and recovery of hazardous dye erythrosine from wastewater by waste materials – bottom ash and de-oiled soya as adsorbents. J Hazard Mater 138(1):95–105

Moldes D, Couto SR, Cameselle C, MA S'n (2003) Study of the degradation of dyes by MnP of *Phanerochaete chrysosporium* produced in a fixed-bed bioreactor. Chemosphere 51:295–303

Moore SB, Ausley LW (2004) Systems thinking and green chemistry in the textile industry: concepts, technologies and benefits. J Clean Prod 12:585–601

Muthezhilan R, Vinoth S, Gopi K, Jaffar Hussain A (2014) A dye degrading potential of immobilized laccase from endophytic fungi of coastal sand dune plants. Int J Chem Tech Res 6(9):4154–4160

Nair DN, Padmavathy S (2014) Impact of endophytic microorganisms on plants, environment and humans. Sci World J 2014:11

Ngieng NS, Zulkharnain A, Roslan HA, Husaini A (2013) Decolorization of synthetic dyes by endophytic fungal flora isolated from Senduduk Plant (*Melastoma malabathricum*). Biotechnology 2013:260730

Novotny C, Erbanova P, Cajthaml T, Rothschild N, Dosoretz C, Sasek V (2000) *Irpex lacteus*, a white rot fungus applicable to water and soil bioremediation. Appl Microbiol Biotechnol 54:850–853

Novotny C, Rawal B, Bhatt M, Patel M, Sasek V, Molitoris HP (2001) Capacity of *Irpex lacteus* and *Pleurotus ostreatus* for decolorization of chemically different dyes. J Biotechnol 89:113–122

Novotny C, Svobodova K, Kasinath A, Erbanova P (2004) Biodegradation of synthetic dyes by *Irpex lacteus* under various growth conditions. Int Biodeter Biodegr 54:215–223

O'Neill C, Hawkes F, Hawkes D, Lourenco N, Pinheiro H, Delee W (1999) Colour in textile effluents-sources, measurement, discharge consents and simulation: a review. J Chem Technol Biotechnol 74:1009–1018

Pant D, Singh A, Satyawali Y, Gupta RK (2008) Effect of carbon and nitrogen source amendment on synthetic dyes decolourizing efficiency of white-rot fungus, *Phanerochaete chrysosporium*. J Environ Biol 29:79–84

Podgornik H, Poljansek I, Perdih A (2001) Transformation of Indigo carmine by *Phanerochaete chrysosporium* ligninolytic enzymes. Enzym Microb Technol 29:166–172

Pointing SB (2001) Feasibility of bioremediation by white-rot fungi. Appl Microbiol Biotechnol 57:20–33

Priti V, Ramesha BT, Singh S, Ravikanth G, Ganeshaiah KN, Suryanaraynan TS, Shaanker U (2009) How promising are endophytic fungi as alternative sources of plant secondary metabolites? Curr Sci 97(4):477–478

Puvaneswari N, Muthukrishnan J, Gunasekaran P (2006) Toxicity assessment and microbial degradation of azo dyes. Indian J Exp Biol 44:618 626

Rai M, Agarkar G, Rathod D (2014) Multiple applications of endophytic *Colletotrichum* species occurring in medicinal plants, in novel plant bioresources: applications in food, medicine and cosmetics. In: Gurib-Fakim A (ed) Novel plant bioresources. Wiley, Chichester, pp 227–236

Redman RS, Sheehan KB, Stout RG, Rodriguez RJ, Henson JM (2002) Thermotolerance conferred to plant host and fungal endophyte during mutualistic symbiosis. Science 298:1581

Robinson T, McMullan G, Marchant R, Nigam P (2001) Remediation of dyes in textile effluent a critical review on current treatment technologies with a proposed alternative. Bioresour Technol 77:247–255

Rodrigues E, Pickard A, Vazquez-Duhalt R (1999) Industrial dye decolorization by laccases from ligninolytic fungi. Curr Microbiol 38:27–32

Rodriguez RJ, Redman RS, Henson JM (2004) The role of fungal symbioses in the adaptation of plants to high stress environments. Mitig Adapt Strat Global Change 9:261–272

Rya RPK, Germaine A, Franks DJ, Ryan DN (2007) Bacterial endophytes: recent developments and applications. FEMS Microbiol Lett 278:1–9

Sack U, Heinze TM, Deck J, Cerniglia CE, Martens R, Zadrazil F, Fritsche W (1997) Comparison of phenantrene and pyrene degradation by different wood-decaying fungi. Appl Environ Microbiol 63:3919–3925

Saraf S, Vaidya VK (2015) Comparative study of biosorption of textile dyes using fungal biosorbents. Int J Curr Microbiol App Sci 2:357–365
Saratale GD, Kalme SD, Govindwar SP (2006) Decolorisation of textile dyes by *Aspergillus ochraceus* (NCIM-1146). IJBT 5:407–410
Schulz B, Boyle C (2005) The endophytic continuum. Mycol Res 109:661–686
Sen SK, Raut S, Bandopadhyay P, Raut S (2016) Fungal decolouration and degradation of azo dyes: a review. Fungal Biol Rev 30(3):112–133
Shahid M, Mohammad F, Islam S (2013) Recent advancements in natural dye applications: a review. J Clean Prod 53:310–331
Shin KS, Oh IK, Kim CHJ (1997) Production and purification of Remazol brilliant blue decolorizing peroxidase from the culture filtrate of *Pleurotus ostreatus*. Appl Environ Microbiol 63:1744–1748
Shweta S, Shivanna MB, Gurumurthy BR, Shaanker U, Santhosh Kumar TR, Ravikanth G (2014) Inhibition of fungal endophytes by camptothecine produced by their host plant, *Nothapodytes nimmoniana* (Grahm) Mabb. (Icacinaceae). Curr Sci 107:994–1000
Si J, Peng F, Cui BK (2013) Purification, biochemical characterization and dye decolorization capacity of an alkali-resistant and metal tolerant laccase from *Trametes pubescens*. Bioresour Technol 128:49–57
Singh R, Jain A, Panwar S, Gupta D, Khare SK (2005) Antimicrobial activity of some natural dyes. Dyes Pigments 66(2):99–102
Singh R, Singh P, Sharma R (2014) Microorganism as a tool of bioremediation technology for cleaning environment: a review. Proc Int Acad Ecol Environ Sci 4(1):1–6
Siva R (2007) Status of natural dyes and dye yielding plants in India. Curr Sci 92(7):916–925
Spadaro JT, Gold MH, Renganathan V (1992) Degradation of azo dyes by the lignin degrading fungus *P chrysosporium*. Appl Environ Microb 58:2397–2340
Specian V, Sarragiotto MH, Pamphile JA, Clemente E (2012) Chemical characterization of bioactive compounds from the endophytic fungus *Diaporthe helianthi* isolated from *Luehea divaricata*. Braz J Microbiol 43:1174–1182
Staniek A, Woerdenbag HJ, Kayser O (2008) Endophytes: exploiting biodiversity for the improvement of natural product-based drug discovery. J Plant Interact 3:75–98
Stolz A (2001) Basic and applied aspects in the microbial degradation of azo dyes. Appl Microbiol Biotechnol 56:69–80
Sun J, Guo N, Niu LL, Wang QF, Zang YP, Zu YG et al (2017) Production of laccase by a new *Myrothecium verrucaria* MD-R-16 isolated from Pigeon Pea [*Cajanus cajan* (L.) Millsp.] and its application on dye decolorization. Molecules 22:673
Swamy J, Ramsay JA (1999a) The evaluation of white rot fungi in the decoloration of textile dyes. Enzym Microb Technol 24:130–137
Swamy J, Ramsay JA (1999b) Effects of Mn2+ and NHþ 4 concentrations on laccase and manganese peroxidase production and Amaranth decoloration by *Trametes versicolor*. Appl Microbiol Biotechnol 51:391–396
Tan RX, Zou WX (2001) Endophytes: a rich source of functional metabolites. Nat Prod Rep 18:448–459
Taskin M, Erdel S (2010) Reactive dye bioaccumulation by fungus *Aspergillus niger* isolated from the effluent of sugar fabric-contaminated soil. Toxicol Ind Health 26(4):239–247
Ting ASY, Lee MVJ, Chow YY, Cheong SL (2016) Novel exploration of endophytic *Diaporthe* sp. for the biosorption and biodegradation of triphenylmethane dyes. Water Air Soil Pollut 227:109
Tobin JM, White C, Gadd GM (1994) Metal accumulation by fungi: applications in environment biotechnology. J Ind Microbiol 13:126–130
Uzma F, Hashem A, Murthy N, Mohan HD, Kamath PV, Singh BP, Venkataramana M, Gupta VK, Siddaiah CN, Chowdappa S, Alqaeawi AA, Abd-Allah EF (2018) Endophytic fungi—alternative sources of cytotoxic compounds: a review. Front Pharmacol 9(309):1–37. https://doi.org/10.3389/fphar.2018.00309
Vankar PS (2000) Chemistry of natural dyes. Resonance 5(10):73–80
Welham A (2000) The theory of dyeing (and the secret of life). J Soc Dye Colour 116:140–143

Wu X, Monchy S, Taghavi S, Zhu W, Ramos J, Van der Lelie D (2010) Comparative genomics and functional analysis of niche specific adaptation in *Pseudomonas putida*. FEMS Microbiol Rev 35(2):299–323

Yang J, Li W, Ng TB, Deng X, Lin J, Ye X (2017) Laccases: production, expression, regulation, and applications in pharmaceutical biodegradation. Front Microbiol 8:832

Zeroual Y, Kim BS, Kim CS, Blaghen M, Lee KM (2006) Biosorption of Bromophenol blue from aqueous solutions by *Rhizopus stolonifer* biomass. Water Air Soil Pollut 177:135–146

Zhao J, Shan T, Mou Y, Zhou L (2011) Plant-derived bioactive compounds produced by endophytic fungi. Mini Rev Med Chem 11:159–168

Zhuo R, Yuan P, Yang Y, Zhang S, Ma F, Zhang X (2017) Induction of laccase by metal ions and aromatic compounds in *Pleurotus ostreatus* HAUCC 162 and decolorization of different synthetic dyes by the extracellular laccase. Biochem Eng J 117:62–72

Zollinger H (1987) Colour chemistry: synthesis, properties of organic dyes and pigments. VCH Publishers, New York, pp 92–100

Zollinger H (1991) Color chemistry: synthesis, properties and application of organic dyes and pigments. Angew Chem Int Ed 496:456–980

Zothanpuia PAK, Leo VV, Kumar B, Chnadra P, Nayak C, Hashem A, Abd Allah EF, Alqarawi AA, Singh BP (2018) Bioprospection of actinobacteria derived from freshwater sediments for their potential to produce antimicrobial compounds. Microb Cell Fact 17(1):68. https://doi.org/10.1093/chromsci/bmy050

Chapter 2
Antimycobacterial and Antiplasmodial Compounds Produced by Endophytic Fungi: An Overview

Mariana Costa Ferreira, Camila Rodrigues de Carvalho, Carlos Leomar Zani, and Luiz Henrique Rosa

Abbreviations

ACT	Artemisinin-based combination therapy
AIDS	Acquired immunodeficiency syndrome
EC_{50}	Half maximal effective concentration
IC_{50}	Half maximal inhibitory concentration
MDR	Multi drug-resistant
MIC	Minimum inhibitory concentration
mL	Milliliters
WHO	World Health Organization
μg	Microgram
μM	Micromolar

2.1 Introduction

Endophytic microbes reside asymptomatically within plant tissues for at least one period of their life cycle without showing any harmful effects on the ecology of the host plant, its healthiness, and its evolution (Brundrett 2006). In addition, several studies suggest that the diversity of biologically active compounds produced by endophytes is due to the close biological association between endophytes and their host plants (Uzma et al. 2018; Strobel 2003). Among the endophytes, fungi play

M. C. Ferreira · L. H. Rosa (✉)
Department of Microbiology, Federal University of Minas Gerais, Belo Horizonte, MG, Brazil
e-mail: lhrosa@icb.ufmg.br

C. R. de Carvalho · C. L. Zani
Laboratório de Química de Produtos Naturais Bioativos, Fundação Oswaldo Cruz-Fiocruz, Instituto René Rachou, Belo Horizonte, MG, Brazil

B. P. Singh (ed.), *Advances in Endophytic Fungal Research*, Fungal Biology,
https://doi.org/10.1007/978-3-030-03589-1_2

different important ecological and biotechnological roles. Endophytic fungi include taxa of Chytridiomycota, Zygomycota, Ascomycota, and Basidiomycota that occur in several environments of the planet and produce different bioactive metabolites (Rosa et al. 2011).

The search for potential pharmaceuticals from natural sources is of interest due to low production costs, structural diversity, and multiple uses of active compounds to treat various diseases. Studies aimed at exploring natural products from endophytes increases the chances of discovering novel bioactive compounds from several sources (Demain 2000; Mishra et al. 2017a, b; Passari et al. 2017). The need for new chemical compounds to treat human diseases is ever increasing mainly due to the rapid development of drug-resistant microbes, the discovery of new cases of life-threatening infections, and the constant recurrence of diseases, which have served to highlight the need for investment in research that spurs new drug discovery (Zothanpuia et al. 2018; Demain 2000; Strobel et al. 2004). Among the important human diseases, many studies are being carried out continuously to combat those caused by *Mycobacterium* (tuberculosis, leprae) species and protozoa of the genus *Plasmodium* (malaria).

2.2 Mycobacterial Infections

Mycobacterium is a genus of Actinobacteria, family Mycobacteriaceae, which includes about 190 species (King et al. 2017). The genus includes pathogens known to cause grave diseases in mammals, including tuberculosis (*M. tuberculosis*) and leprosy (*M. leprae*) in humans (Ryan and Ray 2004). In addition, other *Mycobacterium* species cause infections called atypical mycobacterial infections; they are not “typical” because they do not cause tuberculosis, but they can still harm people, especially people with other problems that affect their immunity, such as AIDS (Griffith et al. 2007).

According to Koul et al. (2011), one third of the world’s current population have been infected with *M. tuberculosis*. Especially in countries with large unregulated private sectors and weak health care systems, the diagnosis of tuberculosis cases continues to be a challenge. Of the estimated 10.4 million new cases, only 6.3 million were detected and officially declared in 2016. It is believed that India, Indonesia, and Nigeria accounted for almost half of this global gap of the 4.1 million undetected cases (WHO 2017a). While many mycobacteria cause human disease, the complexes of *M. tuberculosis*, *Mycobacterium avium*, *Mycobacterium intracellulare*, *Mycobacterium ulcerans*, and *M. leprae* are by far the most important pathogens in global health. Mycobacterial diseases are difficult to treat, and the emergence of resistant strains is a major challenge facing researchers (Gillespie 2002).

2.3 Malaria

Malaria is an infectious disease caused by protozoa, *Plasmodium* spp., and transmitted through the bite of an infected female *Anopheles* mosquito (WHO 2014). The disease is widespread in tropical and subtropical regions around the world, and in sub-Saharan Africa, Asia, and Latin America (White 2004). The disease affects approximately 215 million people, and its highest incidence is in Africa (90%). The World Health Organization (WHO) (2014) warns that this disease remains a global health problem, and in 2016, 429,000 deaths were reported due to malaria alone.

Plasmodium falciparum, among the four species of *Plasmodium* that infect humans, accounts for the majority of cases with high mortality and morbidity rate (White 2004) and is now highly resistant to chloroquine in most malaria-affected areas (Nosten et al. 2000). According to Nosten et al. (2000), resistance to sulfadoxine/pyrimethamine is also widespread and has developed much more rapidly. Resistance to mefloquine is confined only to those areas where it has been used on a large scale (Vietnam, Cambodia, and Thailand) but has arisen within 6 years of systematic deployment. The epidemiology of resistance in *Plasmodium vivax* is less well studied; chloroquine resistance is serious only in parts of Papua New Guinea, Indonesia, and adjacent areas. Sulfadoxine/pyrimethamine resistance in *P. vivax* is more widespread (White 2004). For these reasons, the search for new drugs has become extremely important, aimed at obtaining more efficient and less toxic drugs and solving the resistance problem. In this endeavor, natural products derived from endophytic fungi are promising.

2.4 Antimycobacterial Potential Associated with Endophytic Fungi

Compounds considered natural products have shown to be of great importance in the treatment of mycobacterial infections, mainly tuberculosis (Alvin et al. 2014). Between 2003 and 2005, around 300 new antituberculosis drugs were characterized as natural products (Copp and Pearce 2007). In addition, between 2008 and 2012, 28 new secondary metabolites were isolated from microbial sources, and, according to Salomon and Schmidt (2012), an additional 450 novel compounds were identified and reported during the period 2006–2009 (Alvin et al. 2014).

According to Singh and Pelaez (2008), biological diversity serves as a bridge for the study of metabolic diversity. Thus, choosing suitable source material becomes essential for obtaining novel bioactive compounds. Plants have been recognized as a promising source of drugs, either as pure active principles or as preparations (Farnsworth et al. 1985). Additionally, whereas each plant hosts a variety of endophytes, it has been beneficial exploring these microorganisms to discover new compounds, especially from plants located in unique locations (Uzma et al. 2018; Mishra et al. 2017a, b; Strobel 2003). Since each microorganism can produce a wide variety of compounds and few are considered bioactive, it is necessary to use screening, purification, and identification techniques to characterize the potential active compounds (Alvin et al. 2014). The most important secondary metabolites from endophytic fungi shown to present antimycobacterial activity are listed in Table 2.1.

Cheng et al. (2012) analyzed secondary metabolites from the endophytic fungus *Biscogniauxia formosan*, which was recovered from the bark of a medicinal plant, *Cinnamomum* spp., present in the tropical forest of Taiwan. From *B. formosan* 2 new azaphilone derivatives, biscogniazaphilones A (**2**) and B (**3**), and 10 known compounds were identified; of these, 7 presented antimycobacterial activity against *M. tuberculosis* strain H37Rv. Biscogniazaphilones A (**2**) and B (**3**) exhibited minimum inhibitory concentration (MIC) values of 13.24 and 6.10 μM, respectively.

Investigation of secondary metabolites produced by mangrove endophytes revealed that fusaric acid was the dominant constituent in a bioactive extract of *Fusarium* spp. In view of the increasing prevalence of multidrug-resistant (MDR) *M. tuberculosis* and, consequently, the need for drugs acting through novel mechanisms of action, the potential use of natural products complexed to metals aimed at improving their antimycobacterial activity was evaluated (Pan et al. 2011). Thus, the authors prepared different metal complexes of fusaric acid, which were screened to detect the antimycobacterial activity. Cadmium (II) and copper (II) complexes (**10-11**) exhibited potent inhibitory activity against the *M. bovis* BCG strain with a MIC value of 8.49 μM and the *M. tuberculosis* H37Rv strain (MIC = 23.69 μM); these two compounds displayed the same inhibition degree as that of ethambutol used as the control.

The compounds 4-deoxybostrycin (**21**) and nigrosporin (**22**), isolated from the mangrove endophytic *Nigrospora* sp. collected from South China, were screened against different mycobacteria species (*M. bovis, M. avium, M. intracellulare, M. tuberculosis*, and clinical MDR *M. tuberculosis* strains), demonstrated MICs ranging from 15.61 to 187.33 μM. The 4-deoxybostrycin exhibited promising inhibition against some clinical MDR *M. tuberculosis* strains, better than that of first-line antituberculosis drugs. In addition, it was demonstrated that 4-deoxybostrycin could affect the expression of *M. tuberculosis* H37Rv genes involved in nucleotide, lipid, energy, coenzyme, carbohydrate metabolism, and information storage, making 4-deoxybostrycin a good candidate for the development of new agents to treat tuberculosis (Wang et al. 2013).

Verma et al. (2011) obtained the compound piperine (**12**) (5-(3,4-methylenedioxyphenyl)-1-piperidinopent-2,4-dien-1-one) from the endophytic fungus *Periconia* sp., which was the first report of an alternative source of this chemical other than its host, the plant *Piper longum*. The highly functionalized fungus-derived piperine exhibited strong activity against *M. tuberculosis* and *Mycobacterium smegmetis* with MICs of 6.09 and 9.18 μM, respectively. This finding is of significance as piperine is a potential cancer preventative agent, reaffirming that endophytic microorganisms can produce important pharmaceuticals, and these compounds appear to be mimetic to their host origin, or that they can metabolize their host natural products.

Table 2.1 Antimycobacterial compounds reported from endophytic fungi

Fungal endophyte	Host plant/tissue	Compounds isolated	Species of *Mycobacterium*	MIC	Reference
Phomopsis sp. PSU-D15	*Garcinia dulcis*/ leaves	**1.** Phomoenamide ($C_{14}H_{24}N_2O_4$)	*M. tuberculosis*	21.98 μM	Rukachaisirikul et al (2008)
Biscogniauxia formosana	*Cinnamomum* sp./ bark	**2.** Biscogniazaphilone A ($C_{24}H_{34}O_4$)	*M. tuberculosis*	13.24 μM	Cheng et al. (2012)
		3. Biscogniazaphilone B ($C_{25}H_{32}O_5$)		6.10 μM	
		4. N-trans-feruloy-3-O-methyldopamine ($C_{18}H_{19}NO_5$)		37.95 μM	
		5. 5-hydroxy-3,7,40-trimethoxyflavone ($C_{18}H_{16}O_6$)		76.14 μM	
		6. 4-methoxycinnamaldehyde ($C_{10}H_{10}O_2$)		259.58 μM	
		7. methyl 3,4-methylenedioxycinnamate ($C_{11}H_{10}O_4$)		282.26 μM	
		8. 4-methoxy-trans-cinnamic acid ($C_{10}H_{10}O_3$)		280.6 μM	
Nodulisporium sp.	*Antidesma ghaesembilla*/fresh twigs	**9.** (-)-cis-(7R*,8S*)-4,7,8-Trihydroxy-3,6,7,8-tetrahydronaphtho (2,3-c)furan-5(1H)-one ($C_{12}H_{12}O_5$)	*M. tuberculosis*	13.22 μM	Prabpai et al. (2015)
Fusarium sp. DZ-27	*Kandelia candel*/ bark	**10.** Fusaric acid—Cd($C_{10}H_{13}NO_2$)$_2$	*M. bovis*	8.49 μM	Pan et al. (2011)
		11. Fusaric acid—Cu($C_{10}H_{13}NO_2$)$_2$	*M. tuberculosis*	23.69 μM	
Periconia sp.	*Piper longum*/ leaves	**12.** Piperine ($C_{17}H_{19}NO_3$)	*M. tuberculosis*; *M. smegmetis*	6.09 μM 9.18 μM	Verma et al. (2011)
Fungal endophyte PSU-N24	*Garcinia nigrolineata*/ branch	**13.** Hydronaphthalenone derivative ($C_{16}H_{20}O_8$)	*M. tuberculosis*	36.73 μM	Sommart et al. (2008)
Phomopsis sp. USIA5	*Urobotrya siamensis*/leaves	**14.** 3-Nitropropionic Acid ($C_3H_5O_4N$-H)	*M. tuberculosis*	27.48 μM	Chomcheon et al. (2005)

(continued)

Table 2.1 (continued)

Fungal endophyte	Host plant/tissue	Compounds isolated	Species of *Mycobacterium*	MIC	Reference
Dothideomycete sp. LRUB20	*Leea rubra*/stem	**15.** 2,4-dinitrophenylhydrazone derivative ($C_{13}H_{16}O_5N_4$)	*M. tuberculosis*	648.73 μM	Chomcheon et al. (2006)
Phoma sp.NRRL46751	*Saurauia scaberrinae*/lower crown	**16.** Phomapyrrolidone B ($C_{34}H_{42}NO_4$)	*M. tuberculosis*	11.15 μM	Wijeratne et al. (2013)
		17. Phomapyrrolidone C ($C_{34}H_{41}NO_5$)		9.56 μM	
Coniothyrium cereale	*Marine green alga Enteromorpha sp.*	**18.** (−)-Trypethelone ($C_{16}H_{15}NO_5$)	*M. phlei*	20 μg/disk zones of inhibition of 18 mm	Elsebai et al. (2011)
Chaetomium globosum IFB-E036	*Cynodon dactylon*/ root	**19.** Chaetoglocins A ($C_{11}H_{15}O_5$)	*M. smegmatis*	70.41 μM	Ge et al. (2011)
		20. Chaetoglocins B ($C_{11}H_{15}O_5$)		70.41 μM	
Nigrospora sp.	Plant collected from South China Sea	**21.** 4-deoxybostrycin ($C_{16}H_{16}O_7$) **22.** Nigrosporin ($C_{16}H_{16}O_6$)	*M. bovis*; *M. avium;* *M. intracellulare;* *M. tuberculosis;* Clinical multi drug-resistant(MDR) *M. tuberculosis* strains	<15.61 to >187.33 μM	Wang et al. (2013)
Diaporthe sp. P133	*Pandanus amaryllifolius*/ leaves	**23.** Diaportheone A ($C_{12}H_{10}O_4$)	*M. tuberculosis*	462.39 μM	Bungihan et al. (2011)
		24. Diaportheone B ($C_{12}H_{12}O_4$)		15.89 μM	
Streptomyces sp. Y3111	*Heracleum soulieil*/ stem	**25.** Heraclemycins A ($C_{24}H_{20}O_5$)	*M. bovis*	257.45 μM	Liu et al. (2014)
		26. Heraclemycins B ($C_{24}H_{22}O_5$)		>256.12 μM	
		27. Heraclemycins C ($C_{22}H_{22}O_5$)		16.98 μM	
		28. Heraclemycins D ($C_{22}H_{20}O_5$)		68.6 μM	
Fusarium solani	*Glycyrrhiza glabra*	**29.** 3,6,9-trihydroxy-7-methoxy-4,4-dimethyl-3,4-dihydro-1H-benzo(g) isochromene-5,10-dione ($C_{16}H_{16}H$)	*M. tuberculosis*	1.22 mM	Shah et al. (2017)

Fungal endophyte	Host plant/tissue	Compounds isolated	Species of *Mycobacterium*	MIC	Reference
		30. Fusarubin ($C_{15}H_{14}O_7H$)		26.03 μM	
		31. 3-O-methylfusarubin ($C_{16}H_{16}O_7H$)		199.19 μM	
		32. Javanicin ($C_{15}H_{14}O_6H$)		109.85 μM	
Chaetomium globosum	*Avena sativa*	**33.** Chetomin ($C_{31}H_{30}N_6O_6S_4$)	*M. tuberculosis*	1.09 μM	Marmouzi et al. (2017)
Penicillium sp.	*Garcinia nobilis*/ leaves	**34.** Penialidin A ($C_{14}H_{12}O_8$)	*M. smegmatis*	202.76 μM	Jouda et al. (2016)
		35. Penialidin B ($C_{15}H_{14}O_8$)		193.93 μM	
		36. Penialidin C ($C_{14}H_{10}O_8$)		50.94 μM	
		37. Citromycetin ($C_{14}H_{10}O_7$)		107.50 μM	
		38. p-Hydroxy-phenyl-glyoxalaldoxime ($C_8H_8O_3N$)		376.16 μM	
		39. Brefeldin A ($C_{16}H_{24}O_4$)		891.71 μM	

The endophyte *Fusarium solani* isolated from the medicinal plant *Glycyrrhiza glabra*, which occurs in the Kashmir Himalayas, was highly active against mycobacteria and nonmycobacteria pathogenic strains (Shah et al. 2017). The fractionation of the *F. solani* extract afforded four compounds: 3,6,9-trihydroxy-7-methoxy-4,4-dimethyl-3,4-dihydro-1H-benzo(g)isochromene-5,10-dione (**29**), fusarubin (**30**), 3-O-methylfusarubin (**31**), and javanicin (**32**). When tested against *M. tuberculosis* H37Rv, they exhibited MIC values of 1.22 mM, 26.03, 199.19, and 109.85 μM, respectively.

Marmouzi et al. (2017) evaluated the secondary metabolites produced by *Chaetomium globosum*, an endophyte of *Avena sativa* growing in Morocco, for their antitubercular activities. Among the compounds isolated, chetomin (**33**) presented strong inhibition against *M. tuberculosis* H37Rv, with MIC values of 1.09 μM.

Garcinia species are considered promising hosts of endophytes able to produce new antimycobacterial compounds. *Phomopsis* sp. PSU-D15, recovered from *G. dulcis*, produced three new metabolites identified as phomoenamide, phomonitroester, and deacetylphomoxanthone B. Phomoenamide (**1**) was active against *M. tuberculosis* H37Ra and exhibited a MIC value of 21.98 μM (Rukachaisirikul et al. 2008). Ten compounds isolated from the endophytic fungi PSU-N24, obtained from *Garcinia nigrolineata*, showed antimicrobial, antiplasmodial, and antimycobacterial activities. Hydronaphthalenone (**13**) showed a MIC value of 36.73 μM against *M. tuberculosis* (Sommart et al. 2008). From a crude extract of *Penicillium* sp., endophyte of *Garcinia nobilis*, was isolated six metabolites with activity against *M. smegmatis*: penialidin A (**34**), B (**35**), and C (**36**), citromycetin (**37**), p-Hydroxyphenyl-glyoxalaldoxime (**38**), and brefeldin A (**39**). The metabolites exhibited MIC values between 193.93 and 891.71 μM (Jouda et al. 2016).

There is a constant need to discover new drugs against pathogenic microorganisms that are able to rapidly develop drug-resistant strains in a persistent battle to survive. According to Alvin et al. (2014), when it comes to discovering new bioactive metabolites, it is extremely important to select the most appropriate source material. The objective of bioprospecting new compounds is to isolate compounds that are safe and effective for human use, and the development of efficient screening methods is essential for finding potential bioactive compounds (Mishra et al. 2016, 2017a, b; Alvin et al. 2014; Gordon 2007).

2.5 Antimalarial (In Vivo) Antiplasmodial (In Vitro) Compounds Produced by Endophytic Fungi

According to WHO (2017a, b), one of the world's great public health achievements is the global response to malaria over the course of many years. In 2016, 91 countries reported an increase of 5 million cases of malaria over the 2015 level (216 million cases) (WHO 2017b), emphasizing the intensive search for new drugs (Kaushik et al. 2014). The most effective intervention to guarantee that a mild case of malaria does not develop into severe disease and death is the diagnosis and treatment of malaria patients (Kaushik et al. 2014). Artemisinin-based combination therapy (ACT) has been integral to the recent success of global malaria control, so a global health priority is protecting the efficacy of the treatment of the disease (WHO 2017b). The major advantage of that therapy is that artemisinin quickly leads to a reduction in the number of malaria parasites and the partner drug clears the remaining ones, although the efficacy of artemisinin-based combination therapy is threatened by the emergence of both partner drug and artemisinin resistance.

Partial artemisinin resistance causes delayed parasite clearance following treatment with ACT (WHO 2017b). Such resistance does not usually lead to treatment failure, although if the artemisinin component is less effective, the partner drug must clear a greater parasite number, jeopardizing the future efficacy of the partner drug (WHO 2017b). One of the threats to malaria control is the drug resistance, which has important implications for global public health (WHO 2017b). According to Cowman and Duraisingh (2001), there is an urgent need to discover new antimalarial agents because of the spread of drug-resistant malarial strains.

Some research groups have been investigating fungi for antiplasmodial metabolite production (Fatima et al. 2017). Wiyakrutta et al. (2004) studied the antiplasmodial activity of several extracts obtained from cultures of endophytic fungi recovered from the stems and leaves of Thai medicinal plants and found six isolates able to produce extracts with IC_{50} in the range of 1.2–9.1 μg mL^{-1}. Phongpaichit et al. (2007) studied extracts from endophytic fungi isolated from plants of the genus *Garcinia* against the MDR strain of the *P. falciparum* K1. The extracts showed activity with IC_{50} values of 1.94–7.87 μg mL^{-1}. The extracts obtained from the fungal isolates A13 and N41 recovered from *G. atroviridis* and *G. nigrolineata*, respectively, showed antiplasmodial activity with IC_{50} values of 7.87 and 3.97 μg mL^{-1}, respectively.

Kaushik et al. (2014) recovered 5 endophytic fungi associated with seagrasses, 36 isolates from marine algae (both from the coast of Rameswaram, India), and 43 isolates from the bark or leaves of trees growing in the forests of Western Ghats, India. Sixteen endophytic fungi (representing about 19% of the 84 fungi screened), members of the genera *Alternaria*, *Aspergillus*, *Chaetomium*, *Curvularia*, *Corynespora*, *Fusarium*, *Nigrospora*, *Paecilomyces*, *Penicillium*, *Phomopsis*, *Trichoderma*, and *Xylaria*, showed high ($IC_{50} < 10$ μg mL^{-1}) antiplasmodial activity. The IC_{50} potency profiles of the remaining extracts were 11–25 μg mL^{-1} (18% isolates), 26–50 μg mL^{-1} (21%), and 51–100 μg mL^{-1} (19%).

The major bioactive metabolites obtained from endophytic fungi presenting antiplasmodial activity are described in Table 2.2. Three novel antiplasmodial dihydroisocoumarins were identified by Kongsaeree et al. (2003) from the endophytic fungus *Geotrichum* sp. recovered from fresh stems of *Crassocephalum crepidioides*. The compounds 3-(R)-7-butyl-6,8-dihydroxy-3-pent-11-enylisochroman-1-one (**40**), 7-butyl-6,8-dihydroxy-3-pentylisochroman-1-one (**41**), and 7-but-15-enyl 6,8-dihydroxy-3-pent-11-enylisochroman-1-one (**42**) were tested against *P. falciparum* (K1), but only the compounds **40** and **41** showed antiplasmodial action with IC_{50} values of 15.44 and 8.59 μM, respectively.

Table 2.2 Antiplasmodial compounds reported from endophytic fungi

Fungal endophyte	Host plant/tissue	Compounds isolated	Strains of *Plasmodium*	IC_{50}	Reference
Geotrichum sp.	*Crassocephalum crepidioides*/stem	**40.** 3-(R)-7-butyl-6,8-dihydroxy-3-pent-11-enylisochroman-1-one ($C_{18}H_{24}O_4$)	*Plasmodium falciparum* (K1)	15.44 μM	Kongsaeree et al. (2003)
		41. 7-butyl6,8-dihydroxy-3-pentylisochroman-1-one ($C_{18}H_{22}O_4$)		8.59 μM	
Xylaria sp. PBR-30	*Sandoricum koetjape*/leaf	**43.** 2-chloro-5-methoxy-3-methylcyclohexa-2,5-diene-1,4-dione ($C_8H_7O_3Cl$)	*Plasmodium falciparum* (K1)	1.84 μM	Tansuwan et al. (2007)
		44. xylariaquinone A ($C_{15}H_{12}O_5$)		6.68 μM	
Xylaria sp.	*Siparuna* sp./leaf	**45.** (+) phomalactone ($C_8H_{10}O_3$)	*Plasmodium falciparum* (chloroquine resistant)	84.32 μM	Jiménez-Romero et al. (2008)
		46. 6-(1-propenyl)-3,4,5,6-tetrahydro-5-hydroxy-4H-pyran-2-one ($C_8H_{12}O_4$)		>29.03 μM	
		47. 5-hydroxymellein ($C_{10}H_{10}O_4$)		97.84 μM	
Exserohilum rostratum	*Stemona* sp./leaf	**48.** 11(*R*)-hydroxymonocerin ($C_{16}H_{20}H_7$)	*Plasmodium falciparum* (K1)	0.68 μM	Sappapan et al. (2008)
		49. monocerin ($C_{16}H_{20}O_6$)		7.70 μM	
Phomopsis sp. BCC 1323	*Tectona grandis*/leaf	**51.** phomoxanthone A ($C_{38}H_{38}O_{16}$)	*Plasmodium falciparum* (K1)	146.52 nM	Isaka et al. (2001)
		52. phomoxanthone B ($C_{38}H_{38}O_{16}$)		439.58 nM	
Unidentified endophytic fungus PSU-N24	*Garcinia nigrolineata*	**55.** 9α-hydroxyhalorosellinia A ($C_{16}H_{20}O_8$)	*Plasmodium falciparum*	23.33 μM	Sommart et al. (2008)
		62. 3-(2-hydroxypropyl)benzene-1,2-diol		6.68 μg/mL	
		63. desoxybostrycin		≤10 μg/mL	
Chalara alabamensis	*Asterogyne martiana*	**67.** viridiol ($C_{20}H_{18}O_6$)	*Plasmodium falciparum*	15.52 μM	Cao et al. (2010)
Pestalotiopsis sp.	*Melaleuca quinquenervia*/stem	**69.** pestalactam A ($C_{10}H_{12}ClNO_4$)	*Plasmodium falciparum* (chloroquine sensitive/chloroquine resistant)	16.2/41.3 μM	Davis et al. (2010)
		70. pestalactam B ($C_{10}H_{13}NO_4Na$)		20.7/36.3 μM	

Fungal endophyte	Host plant/tissue	Compounds isolated	Strains of *Plasmodium*	IC_{50}	Reference
Unidentified endophytic fungus	*Tinaspora crispa*	**72.** 7-hydroxy-3,4,5-trimethyl-6-on-2,3,4,6-tetrahydroisoquinoline-8-carboxylic acid ($C_{13}H_{15}O_3N$)	*Plasmodium falciparum*	0.129 µM	Elfita et al. (2011)
		73. 2,5-dihydroxy-1-(hydroxymethyl) pyridin-4-on		0.127 µM	
Phomopsis archeri	*Vanilla albidial* cortex stem	**74.** phomoarcherin B ($C_{23}H_{28}O_5$)	*Plasmodium falciparum*	2.05 µM	Hemtasin et al. (2011)
Phomatospora bellaminuta	–	**78.** pestalopyrone ($C_{10}H_{12}O_3$)	*Plasmodium falciparum* (Dd2)	37 µM	Cao & Clardy (2011)
Diaporthe sp.	–	**79.** dicerandrol D ($C_{36}H_{37}O_{15}$)	*Plasmodium falciparum* (3D7, a drug sensitive strain)	600 nM	Calcul et al. (2013)
Codinaeopsis gonytrichoides	*Vochysia guatemalensis*	**80.** codinaeopsin ($C_{32}H_{40}N_2O_3$)	*Plasmodium falciparum* (3D7)	4.7 µM	Kontnik & Clardy (2008)
Pullularia sp. BCC 8613	*Culophyllum* sp./leaf	**81.** pullularin A($C_{42}H_{57}N_5O_9$) **82.** pullularin B($C_{43}H_{59}N_5O_9$)	*Plasmodium falciparum*	4.63 µM 4.17 µM	Isaka et al. (2007)
Diaporthe miriciae	*Vellozia gigantea*	**85.** epoxycytochalasin H ($C_{30}H_{39}NO_5$)	*P. falciparum* chloroquine-sensitive/chloroquine-resistant strains	105.34/79 nM	Ferreira et al. (2017)

K1 = *Plasmodium falciparum* multi-drug-resistant strain

Tansuwan et al. (2007) identified two novel benzoquinone metabolites: 2-chloro-5-methoxy-3-methylcyclohexa-2, 5-diene-1,4-dione (**43**) and xylariaquinone A (**44**), from endophytic *Xylaria* sp. PBR-30 isolated from Thai medicinal plants. The compounds exhibited antiplasmodial activity against *P. falciparum* (K1) with IC_{50} values of 1.84 and 6.68 μM, respectively. Additionally, the cytotoxic activity of both compounds against African green monkey kidney fibroblasts (Vero cells) was examined using a colorimetric method. The compound **43** exhibited cytotoxicity with IC_{50} values of 1.35 μM, and the elipticine used as positive control showed an IC_{50} value of 2.03 μM.

Jiménez-Romero et al. (2008) evaluated the ethyl acetate crude extract of *Xylaria* sp. recovered from the leaf of *Siparuna* sp. (Altos Campanas National Park, Panama), which showed weak activity (IC_{50} 30 μg mL^{-1}) against a chloroquine-resistant strain of *P. falciparum*. Three compounds, (+)-phomalactone (**45**), 6-(1-propenyl)-3,4,5,6-tetrahydro-5-hydroxy-4H-pyran-2-one (**46**), and 5-hydroxymellein (**47**), were isolated from the crude extract and also showed weak antiplasmodial activity when tested against the target with IC_{50} values of 84.32, >29.03, and 97.84 μM, respectively (Jiménez-Romero et al. 2008). The ethyl acetate extract of the culture of *Exserohilum rostratum*, an endophytic isolated from leaves of a *Stemona* sp., were active against *P. falciparum* (K1) in a study conducted by Sappapan et al. (2008). The authors obtained a new isocoumarin derivative, 11(*R*)-hydroxymonocerin (**48**), together with known monocerin (**49**) and 12(*R*)-hydroxymonocerin (**50**). The compounds **48** and **49** exhibited antiplasmodial activity with IC_{50} values of 0.68 and 7.70 μM, respectively. Activity-guided fractionation of the extract obtained from the endophytic *Phomopsis* sp. BCC 1323 isolated from a teak leaf of *Tectona grandis* (Mee Rim district, Thailand) led to the isolation and identification of two novel xanthone dimers, phomoxanthones A (**51**) and B (**52**), which exhibited significant activity against *P. falciparum* (K1) with IC_{50} values of 146.52 and 439.58 nM, respectively (Isaka et al. 2001).

Sommart et al. (2008) demonstrated that the extract of the unidentified endophytic fungus PSU-N24, isolated from *Garcinia nigrolineata*, showed interesting antimicrobial, antimalarial, and antimycobacterial activities. Further investigation of the extract led to the isolation and structural elucidation of three new hydronaphthalenone derivatives (**53-55**) and one new dihydroramulosin derivative: 6β-hydroxy-8-dihydroramulosin (**56**) together with eight known compounds: (*R*)-mullein (**57**), cis-4-hydroxymellein (**58**), 8-dihydroramulosin (**59**), 6-hydroxyramulosin (**60**), griseofulvin (**61**), 3-(2-hydroxypropyl)benzene-1,2-diol (**62**), desoxybostrycin (**63**), and austrocortinin (**64**). All metabolites were evaluated against *P. falciparum* since compound **62** showed the best antiplasmodial activity, with an IC_{50} value of 6.68 μg mL^{-1}, while **55** A and **63** were less active, with IC_{50} values of 23.33 μM and ≤10 μg mL^{-1}, respectively.

Cao et al. (2010) showed that the dicholoromethane extract of *Chalara alabamensis*, an endophytic fungus isolated from *Asterogyne martiana* collected in Costa Rica, displayed activity against *P. falciparum*, with an EC_{50} of 24 μg mL^{-1}. The fractionation led to the isolation of asterogynins A (**65**) and B (**66**), viridiol (**67**), and viridian (**68**). All four compounds were tested against *P. falciparum*, but only viridiol was active, with an EC_{50} value of 15.52 μM. The ethyl acetate extract of endophytic fungus *Pestalotiopsis* sp. isolated from the stem of *Melaleuca quinquenervia* (Toohey Forest, Australia) was purified to yield pestalactams A, B, and C (**69-71**). The pestalactams A and B showed activity against chloroquine-sensitive (IC_{50} 16.2 and 20.7 μM, respectively) and chloroquine-resistant (IC_{50} 41.3 and 36.3 μM, respectively) cell lines of *P. falciparum.* According to Davis et al. (2010) pestalactams A and C are the first examples of natural products containing halogenated caprolactam ring. Elfita et al. (2011) isolated the alkaloids 7-hydroxy-3,4,5-trimethyl-6-on-2,3,4,6-tetrahydroisoquinoline-8-carboxylic acid (**72**) and 2,5-dihydroxy-1-(hydroxymethyl)pyridin-4-on (**73**) from an endophytic fungus associated with *Tinaspora crispa*, a plant used traditionally for the treatment of malaria. Both compounds showed in vitro activity against *P. falciparum* 3D7, with IC_{50} values 0.129 and 0.127 μM respectively.

Three new sesquiterpenes and four known compounds were isolated from the endophytic fungus *Phomopsis archeri* recovered from the stem cortex of *Vanilla albidia* (Thailand). Phomoarcherin B (**74**), a new pentacyclic aromatic sesquiterpene, showed activity against *P. falciparum* with an IC_{50} value of 2.05 μM (Hemtasin et al. 2011). Cao and Clardy (2011) reported that extracts obtained from the endophytic fungus *Delitzchia winteri*, isolated in Costa Rica, afforded two new naphthoquinones, delitzchianones A (**75**) and B (**76**). From *Phomatospora bellaminuta* the same authors obtained a new 8-acetoxy pestalopyrone (**77**) and the known compound pestalopyrone (**78**). All four compounds were evaluated against *P. falciparum* (Dd2), and only pestalopyrone (**77**) exhibited activity against Dd2 with an IC_{50} value of 37 μM.

Almost 6000 fungi were isolated from bark and leaf tissues of *Avicennia marina*, *Kandelia obovate*, and *Lumnitzera racemosa* by Calcul et al. (2013). All fungal isolates were cultivated and their extracts screened against *P. falciparum* (3D7), revealing 34 extracts active and 266 extracts partially active. Of these, 58 endophytic extracts were scaled up and subjected to medium-pressure liquid chromatography to yield 18 compounds, identified either through dereplication or structure analysis (eight cytochalasins, two trichothecenes, two lipids, and six polyketides). Among these compounds, dicerandrol D (**79**), produced by *Diaporthe* sp., displayed nanomolar antiplasmodial activity (600 nM) and a low cytotoxicity, with a selectivity index of 13. A new tryptophan-polyketide hybrid named codinaeopsin (**80**) was isolated from the extract of *Codinaeopsis gonytrichoides*, an endophytic fungus recovered from *Vochysia guatemalensis* in Costa Rica. Codinaeopsin was active against *P. falciparum* (3D7), showing an IC_{50} of 4.7 μM (2.3 μg mL^{-1}) (Kontnik and Clardy 2008). Isaka et al. (2007) recovered *Pullularia* sp. BCC 8613 from a leaf of *Culophyllum* sp. collected in Thailand. This fungus produced the cyclohexadepsipeptides pullularins A–D (**81-84**). Pullularin A (**81**) and B (**82**) exhibited antiplasmodial activity with IC_{50} values of 4.63 and 4.17 μM, respectively. Ferreira et al. (2017) demonstrated that crude extracts of the endophytic fungus *Diaporthe miriciae*, isolated from *Vellozia gigantea* from Brazil, showed antifungal, antibacterial, and antiplasmodial activities. The extract of *D. miriciae* yielded the compound epoxycytochalasin H (**85**), which showed high activity against both chloroquine-sensitive and chloroquine-resistant strains of *P. falciparum*, with IC_{50} values of 105.34 and 79 nM, respectively, without any cytotoxicity toward mammalian kidney (Vero) cells.

2.6 Conclusion

An endophytic fungus includes taxa with high metabolic activity, versatility, and an ability to produce different bioactive compounds. This chemical diversity is especially important for the discovery of new compounds with potential in the treatment or prevention of neglected diseases that have a large negative impact on poor and developing countries. Because the global diversity of endophytic fungi is far from being accessed, mainly in tropical forests, these fungi are considered potential metabolic factories capable of producing bioactive natural products with chemical, pharmacological, and toxicological profiles that allow them to be considered prototype molecules for the development of new antimycobacterial and antimalarial agents. Further studies of the taxonomy, diversity, and bioprospection of bioactive compounds produced by endophytes must be conducted to discover more antimycobacterial and antimalarial compounds, especially those active against drug-resistant strains of *Mycobacterium* and *Plasmodium* species.

References

Alvin A, Miller KI, Neilan BA (2014) Exploring the potential of endophytes from medicinal plants as sources of antimycobacterial compounds. Microbiol Res 169(7):483–495

Brundrett MC (2006) Understanding the roles of multifunctional mycorrhizal and endophytic fungi. In: BJE S, CJC B, Sieber TN (eds) Microbial root endophytes. Springer-Verlag, Berlin, Germany, pp 281–293

Bungihan ME, Tan MA, Kitajima M, Kogure N, Franzblau SG, dela Cruz TEE, Takayama H, Nonato MG (2011) Bioactive metabolites of Diaporthe sp. P133, an endophytic fungus isolated from Pandanus amaryllifolius. J Nat Med 65(3-4):606–609

Calcul L, Waterman C, Ma WS, Lebar MD, Harter C, Mutka T et al (2013) Screening mangrove endophytic fungi for antimalarial natural products. Mar Drugs 11:5036–5050

Cao S, Clardy J (2011) New naphthoquinones and a new δ-lactone produced by endophytic fungi from Costa Rica. Tetrahedron Lett 52(17):2206–2208

Cao S, Ross L, Tamayo G, Clardy J (2010) Asterogynins: secondary metabolites from a Costa Rican endophytic fungus. Org Lett 12(20):4661–4663

Cheng MJ, Wu MD, Yanai H, Su YS, Chen IS, Yuang GF et al (2012) Secondary metabolites from the endophytic fungus *Biscogniauxia formosana* and their antimycobacterial activity. Phytochem Lett 5:467–472

Chomcheon P, Sriubolmas N, Wiyakrutta S, Ngamrojanavanich N, Chaichit N, Mahidol C, Ruchirawat S, Kittakoop P (2006) Cyclopentenones, scaffolds for organic syntheses produced by the endophytic fungus mitosporic dothideomycete sp. LRUB20. J Nat Prod 69(9):1351–1353

Chomcheon P, Wiyakrutta S, Sriubolmas N, Ngamrojanavanich N, Isarangkul D, Kittakoop P (2005) 3-Nitropropionic acid (3-NPA), a potent antimycobacterial agent from endophytic fungi: is 3-NPA in some plants produced by endophytes? J Nat Prod 68:1103–1105

Copp BR, Pearce AN (2007) Natural product growth inhibitors of *Mycobacterium tuberculosis*. J Nat Prod 24:278–297

Cowman AF, Duraisingh MT (2001) An old enemy, a new battle plan: perspectives on combating drug resistance malaria. EMBO Rep 2(2):77–79

Davis RA, Carroll AR, Andrews KT, Boyle GM, Tran TL, Healy PC et al (2010) Pestalactams A–C: novel caprolactams from the endophytic fungus *Pestalotiopsis* sp. Org Biomol Chem 8(8):1785–1790

Demain AL (2000) Microbial natural products: a past with a future. In: Wrigley SK, Hayes MA, Thomas R, Chrystal EJT, Nicholson N (eds) Biodiversity: new leads for pharmaceutical and agrochemical industries. Royal Society of Chemistry, Cambridge, pp 3–16

Elfita E, Muharni M, Munawar M, Legasari L, Darwati D (2011) Antimalarial compounds from endophytic fungi of brotowall (*Tinaspora crispa* L). Indones J Chem 11(1):53–58

Elsebai MF, Natesan L, Kehraus S, Mohamed IE, Schnakenburg G, Sasse F, Shaaban S, Gütschow M, König GM (2011) HLE-inhibitory alkaloids with a polyketide skeleton from the marine-derived fungus. J Nat Prod 74(10):2282–2285

Farnsworth NR, Akerele O, Bingel AS, Soejarto DD, Guo Z (1985) Medicinal plants in therapy. Bull World Health Organ 63:965–981

Fatima N, Muhammad SA, Qazi MA, Shah ZU, Khan AK, Maalik A et al (2017) Fungal metabolites and antimalarial drug discovery; a review. Acta Pol Pharm 74(5):1327–1341

Ferreira MC, Cantrell CL, Wedge DE, Gonçalves VN, Jacob MR, Khan S et al (2017) Antimycobacterial and antimalarial activities of endophytic fungi associated with the ancient and narrowly endemic neotropical plant *Vellozia gigantea* from Brazil. Mem Inst Oswaldo Cruz 112(10):692–697

Ge HM, Zhang Q, Xu SH, Guo ZK, Song YC, Huang WY, Tan RX (2011) Chaetoglocins A–D, four new metabolites from the endophytic fungus Chaetomium globosum. Planta Med 77(03):277–280

Gillespie SH (2002) Evolution of drug resistance in *Mycobacterium tuberculosis*: clinical and molecular perspective. Antimicrob Agents Chemother 46:267–274

Gordon B (2007) Jeweller of the jungle. Americas 59:38–45

Griffith DE, Aksamit T, Brown-Elliott BA, Catanzaro A, Daley C, Gordin F, Iseman M et al (2007) An official ATS/IDSA statement: diagnosis, treatment, and prevention of nontuberculous mycobacterial diseases. Am J Respir Crit Care Med 175(4):367–416

Hemtasin C, Kanokmedhakul S, Kanokmedhakul K, Hahnvajanawong C, Soytong K, Prabpai S et al (2011) Cytotoxic pentacyclic and tetracyclic aromatic sesquiterpenes from *Phomopsis archeri*. J Nat Prod 74:609–613

Isaka M, Jaturapat A, Rukseree K, Danwisetkanjana K, Tanticharoen M, Thebtaranonth Y (2001) Phomoxanthones A and B, novel xanthone dimers from the endophytic fungus *Phomopsis* species. J Nat Prod 64:1015–1018

Isaka M, Berkaew P, Intereya K, Komwijit S, Sathitkunanon T (2007) Antiplasmodial and antiviral cyclohexadepsipeptides from the endophytic fungus *Pullularia* sp. BCC 8613. Tetrahedron 63(29):6855–6860

Jiménez-Romero C, Ortega-Barría E, Arnold AE, Cubilla-Rios L (2008) Activity against *Plasmodium falciparum* of lactones isolated from the endophytic fungus *Xylaria* sp. Pharm Biol 46(10-11):1–4

Jouda JB, Mawabo IK, Notedji A, Mbazoa CD, Nkenfou J, Wandji J, Nkenfou CN (2016) Antimycobacterial activity of polyketides from *Penicillium* sp. endophyte isolated from *Garcinia nobilis* against *Mycobacterium smegmatis*. Int J Mycobacteriol 5(2):192–196

Kaushik NK, Murali TS, Sahal D, Suryanarayanan TS (2014) A search for antiplasmodial metabolites among fungal endophytes of terrestrial and marine plants of southern India. Acta Parasitol 59(4):745–757

King HC, Khera-Butler T, James P, Oakley BB, Erenso G, Aseffa A, Knight R, Wellington EM, Courtenay O (2017) Environmental reservoirs of pathogenic mycobacteria across the Ethiopian biogeographical landscape. PLoS One 12(3):e0173811

Kongsaeree P, Prabpai S, Sriubolmas N, Vongvein C, Wiyakrutta S (2003) Antimalarial dihydroisocoumarins produced by *Geotrichum* sp., an endophytic fungus of *Crassocephalum crepidioides*. J Nat Prod 66:709–711

Kontnik R, Clardy J (2008) Codinaeopsin, an antimalarial fungal polyketide. Org Lett 10(18):4149–4151

Koul A, Arnoult E, Lounis N, Guillemont J, Andries K (2011) The challenge of new drug discovery for tuberculosis. Nature 469:483–490

Liu M, Abdel-Mageed WM, Ren B, He W, Huang P, Li X et al (2014) Endophytic *Streptomyces* sp. Y3111 from traditional Chinese medicine produced antitubercular pluramycins. Appl Microbiol Biotechnol 98(3):1077–1085

Marmouzi I, Faouzi MEA, Saidi N, Cherrah Y, Rehberg N, Ebada SS et al (2017) Bioactive secondary metabolites from *Chaetomium globosum*, an endophyte from the moroccan plant *Avena sativa*. Chem Nat Compd 53(6):1208–1211

Mishra VK, Singh G, Passari AK, Yadav MK, Gupta VK, Singh BP (2016) Distribution and antimicrobial potential of endophytic fungi associated with ethnomedicinal plant *Melastoma malabathricum* L. J Environ Biol 37(2):229–237

Mishra VK, Passari AK, Chandra P, Leo VV, Kumar B, Gupta VK, Singh BP (2017a) Determination and production of antimicrobial compounds by *Aspergillus clavatonanicus* strain MJ31, an endophytic fungus from *Mirabilis jalapa* L. using UPLC-ESI-MS/MS and TD GC-MS. PLoS One 12(10):1–24. https://doi.org/10.1371/journal.pone.0186234

Mishra VK, Passari AK, Leo VV, Singh BP (2017b) Molecular diversity and detection of endophytic fungi based on their antimicrobial biosynthetic genes. In: Singh BP, Gupta VK (eds) Molecular markers in mycology, fungal biology. Springer International Publishing, Switzerland, pp 1–35. https://doi.org/10.1007/978-3-319-34106-4_1

Nosten F, Van Vugt M, Price R, Luxemburger C, Thway KL, Brockman A et al (2000) Effects of artesunate-mefloquine combination on incidence of *Plasmodium falciparum* malaria and mefloquine resistance in western Thailand: a prospective study. Lancet 356(9226):297–302

Pan JH, Chen Y, Huang YH, Tao YW, Wang J, Li Y et al (2011) Antimycobacterial activity of fusaric acid from a mangrove endophyte and its metal complexes. Arch Pharm Res 34(7):1177–1181

Passari AK, Mishra VK, Singh G, Singh P, Kumar B, Gupta VK, Sharma RK, Saikia R, Donovan A, Singh BP (2017) Insights into the functionality of endophytic actinobacteria with a focus on their biosynthetic potential and secondary metabolites production. Sci Rep 7(11809):1–17. https://doi.org/10.1038/s41598-017-12235-4

Phongpaichit S, Nikom J, Rungjindamai N, Sakayaroj J, Hutadilok-Towatana N, Rukachaisirikul V et al (2007) Biological activities of extracts from endophytic fungi isolated from *Garcinia* plants. FEMS Immunol Med Microbiol 51:517–525

Prabpai S, Wiyakrutta S, Sriubolmas N, Kongsaeree P (2015) Antimycobacterial dihydronaphthalenone from the endophytic fungus Nodulisporium sp. of Antidesma ghaesembilla. Phytochem Lett 13:375–378

Rosa LH, Vieira MLA, Cota BB, Johann S (2011) Endophytic fungi of tropical forests: a promising source of bioactive prototype molecules for the treatment of neglected diseases. In: Ekinci D (ed) Drug development—a case study based insight into modern strategies. Intech, Croatia, pp 1–18

Rukachaisirikul V, Sommart U, Phongpaichit S, Sakayaroj J, Kirtikara K (2008) Metabolites from the endophytic fungus *Phomopsis* sp. PSU-D15. Phytochemistry 69(3):783–787

Ryan KJ, Ray CG (2004) Sherris medical microbiology, 4th edn. McGraw Hill, New York

Salomon CE, Schmidt LE (2012) Natural products as leads for tuberculosis drug development. Curr Top Med Chem 12:735–765

Sappapan R, Sommit D, Ngamrojanavanich R, Pengpreecha S, Wiyakrutta C, Sriubolmas N et al (2008) 11-Hydroxymonocerin from the Plant Endophytic Fungus *Exserohilum rostratum*. J Nat Prod 71:1657–1659

Shah A, Rather MA, Hassan QP, Aga MA, Mushtaq S, Shah AM et al (2017) Discovery of antimicrobial and antitubercular molecules from *Fusarium solani*: an endophyte of *Glycyrrhiza glabra*. J Appl Microbiol 122(5):1168–1176

Singh SB, Pelaez F (2008) Biodiversity, chemical diversity and drug discovery. In: Natural compounds as drugs, vol I. Birkhäuser, Basel, pp 141–174

Sommart U, Rukachaisirikul V, Sukpondma Y, Phongpaichit S, Sakayaroj J, Kirtikara K (2008) Hydronaphthalenones and a dihydroramulosin from the endophytic fungus PSU-N24. Chem Pharm Bull 56(12):1687–1690

Strobel GA (2003) Endophytes as sources of bioactive products. Microbes Infect 5:535–544

Strobel G, Daisy B, Castillo U, Harper J (2004) Natural products from endophytic microorganisms. J Nat Prod 67:257–268

Tansuwan S, Pornpakakul S, Roengsumran S, Petsom A, Muangsin N, Sihanonta P et al (2007) Antimalarial benzoquinones from an endophytic fungus, *Xylaria* sp. J Nat Prod 70:1620–1623

Uzma F, Hashem A, Murthy N, Mohan HD, Kamath PV, Singh BP, Venkataramana M, Gupta VK, Siddaiah CN, Chowdappa S, Alqaeawi AA, Abd-Allah EF (2018) Endophytic fungi—alternative sources of cytotoxic compounds: a review. Front Pharmacol 9(309):1–37. https://doi.org/10.3389/fphar.2018.00309

Verma VC, Lobkovsky E, Gange AC, Singh SK, Prakash S (2011) Piperine production by endophytic fungus *Periconia* sp. isolated from *Piper longum* L. J Antibiot 64(6):427–431

Wang C, Wang J, Huang Y, Chen H, Li Y, Zhong L et al (2013) Antimycobacterial activity of marine fungus-derived 4-deoxybostrycinand nigrosporin. Molecules 18:1728–1740

White NJ (2004) Antimalarial drug resistance. J Clin Invest 113(8):1084

WHO (2014) Malaria fact sheet N°94. Archived from the original on 3 September 2014. Retrieved 28 Aug 2014

WHO (2017a) Global tuberculosis control. World Health Organization, Geneva

WHO (2017b) World malaria report 2017. World Health Organization, Geneva

Wiyakrutta S, Sriubolmas N, Panphut W, Thongon N, Danwisetkanjana K, Ruangrungsi N et al (2004) Endophytic fungi with anti-microbial, anti-cancer and anti-malarial activities isolated from Thai medicinal plants. World J Microbiol Biotechnol 20:265–272

Wijeratne EM, He H, Franzblau SG, Hoffman AM, Gunatilaka AA (2013) Phomapyrrolidones A-C, antitubercular alkaloids from the endophytic fungus Phoma sp. NRRL46751. J Nat Prod 76:1860–1865

Zothanpuia, Passari AK, Leo VV, Kumar B, Chnadra P, Nayak C, Hashem A, Abd_Allah EF, Alqarawi AA, Singh BP (2018) Bioprospection of actinobacteria derived from freshwater sediments for their potential to produce antimicrobial compounds. Microb Cell Fact 17(1):68. https://doi.org/10.1093/chromsci/bmy050

Chapter 3
Bioprospecting for Fungal-Endophyte-Derived Natural Products for Drug Discovery

Priyanka Saha, Anupam Das Talukdar, Manabendra Dutta Choudhury, and Deepa Nath

3.1 Introduction

Fungi were recognized to be a loaded depository of clinically potent imperative compounds since the time penicillin was discovered. Presently, fungus-derived drugs range from regular antibiotics to anticholesterol agents. Whereas plants still continue to be the chief source of potent molecules and drug leads, some clinically noteworthy plants have been reported to be endangered or facing extinction due to their exploitation. For this reason, the focus has shifted toward endophytic fungi that inhabit these medicinally potent plants and thus may have inherited their medicinal abilities. The marker in this field of endophyte bioprospecting was certainly the sighting of the first, *Taxomyces andreanae* (Stierle et al. 1993).

3.1.1 Definition of Endophyte

An endophytic fungus lives in the form of mycelium temporararily in biological association with a specific plant system. Hence, a fungus is termed as an endophyte when the demonstration of its hyphae is clear in living tissues (Kaul et al. 2016). Throughout the last 30 years, this well-known terminology of endophyte and endophytic fungus has frequently appeared in the literature of mycology to explain the interior mycota of plants. Although pathogens generally influence plant activity negatively, endophytes may not necessarily be symptomatic in their host colonization

---

P. Saha · A. D. Talukdar (✉) · M. D. Choudhury
Department of Life Science and Bioinformatics, Assam University, Silchar, Assam, India
e-mail: anupam@bioinfoaus.ac.in

D. Nath
Department of Botany and Biotechnology, Karimganj College, Karimganj, Assam, India

B. P. Singh (ed.), *Advances in Endophytic Fungal Research*, Fungal Biology,
https://doi.org/10.1007/978-3-030-03589-1_3

(Schulz and Boyle 2005). Endophytes are usually fungi or bacteria that exert favorable effects upon their host under certain circumstances of metabolic interaction, for example, abiotic or biotic stress conditions (Vandenkoornhuyse et al. 2015). Note that a plant may contain several different types and classes of endophytes (Cao et al. 2004). This physiological condition is further investigated both in hosts and in endophytes for the synthesis of novel metabolites (Kusari et al. 2012).

3.1.2 Background of Endophyte–Plant Interactions

Fungal endophytes have been defined as microbes that invade almost all tissues of host plants (Wilson 1995). Recent findings demonstrated them as pervasive microorganisms residing in all plants of the world (Mishra et al. 2017a; Strobel and Daisy 2003). They independently inhabit the internal tissues of living plants (Huang et al. 2007; Porras-Alfaro and Bayman 2011) and are involved in constant metabolic interactions with their host (Aly et al. 2013). The relationship between a host plant and an endophyte lies on a symbiotic continuum, expressing itself as mutualism, commensalism, or parasitism (Kogel et al. 2006). This intimate association and collaboration give rise to complex plant microbiome interactions, bringing forth an exciting frontier to be explored (Yuan et al. 2009). In addition, the coevolution of endophytes and their particular host plants in adapting to environmental adversities and altering endophyte–host communication is believed to be linked to the appearance of natural endophyte products (Gunatilaka 2006; Aly et al. 2011). Endophytic fungus colonization has been documented to contribute to host plant defense in circumstances of biotic and abiotic stress (Redman et al. 2002; Waller et al. 2005; Akello et al. 2007; Bae et al. 2009; Rodriguez et al. 2009). Though endophytes were first elaborated from the Darnel (*Lolium temulentum*) back in 1904, they attracted increased attention following the discovery of the paclitaxel-producing endophytic fungus *Taxomyces andreanae*, which was derived from the parent source of this vital anticancer drug i.e, from Taxus brevifolia (Stierle et al. 1993). This evoked a shift toward endophytes as new sources of therapeutic agents (Aly et al. 2010; Debbab et al. 2012) and led to more comprehensive studies on yielding endophytes that produce pharmacologically main natural products previously only known to be obtainable from plants. Pharmaceutically important fungal-endophytes derived phytochemicals have been reported with unraveling bioactivity. Moreover, research on the fungal biosynthesis of plant-associated metabolites are in process, and new approaches to improve the production of commercially essential plant-derived compounds of endophytic fungi origin are described.

Endophytic fungi are generally categorized into two distinct forms: meiosporic and mitosporic. The mitosporic form dwells independently in the innermost layers of plants underneath the epidermal cell layer. Such fungi initiate a specific infection known as quiescent infections by localizing themselves inside healthy juvenile tissues (Bacon and White 2000; Rodriguez et al. 2009). A massive diversity of the biology of endophytic fungi is evident in tropical forests and in temperate areas of vegetation. These potentially active fungi account for over a million endophytic fungal species that may reside in their hosts alone or in association with other groups of endophytes living in a symbiotic association. An endophytic fungus belongs to a polyphyletic class of microbes that can survive in different parts of plants, whether above or below the soil. It can live in roots, stems, or even in the leaves of the plant system (Faeth and Fagan 2002; Vesterlund et al. 2011).

3.1.3 Classification of Endophytic Fungi

As described by Schultz, active endophytic fungi can be subdivided into three major classes (Faeth and Fagan 2002): mycorrizal endophytes, pasture endophytic fungi, and non-pasture forms. Bioactive isolates derived from this potentially active group of fungi play a tremendous role in boosting the immune systems of both endophytic fungi and their respective host species. This immune tolerance is attributed to their severe adverse tolerances to abiotic and biotic circumstances. Additionally, these compounds can trigger novel active secondary metabolite production (Zhang et al. 2006; Firáková et al. 2007) that can serve as useful and important medicinal resources. From such specific behavioral activity it is clear that the colonization process of these classes of fungi involves a series of chemical interactions produced by host plants and fungi. These interactions also produce a wide range of secondary metabolites, the majority of which are prime secondary metabolites such as essential oils and saponins. They are created as a result of complex biochemical mechanisms between pathogens and plants and most probably the presence of endophytic fungi. The produced secondary metabolites in turn exhibit a negative feedback loop that hinders the colonization of such fungi. Endophytic fungi defend themselves in another set of mechanisms where they produce various types of detoxification enzymes to decompose the secondary metabolites by host plant species. This allows them to easily invade and enter the host plant part where they are to reside. Upon entry into the host system, the associative fungi inherent a dormant state in which they either dwell for the entire lifespan of the host or for a particular period. These modes of association are called antagonism or mutualism, and it continues to be in this mode till they occur favorable condition for endophytic fungi (Helander et al. 2007) (Fig. 3.1).

3.2 Natural Products and Traditional Approaches in Medicine

The term *natural product* in general can be defined as a natural-based metabolite obtained from animals or plants (Strobel and Daisy 2003). Its usage has been into use to human for an era with plants as a source of natural compounds. To date, the Chinese are the most populous country to use traditional medicines with not less than 5000

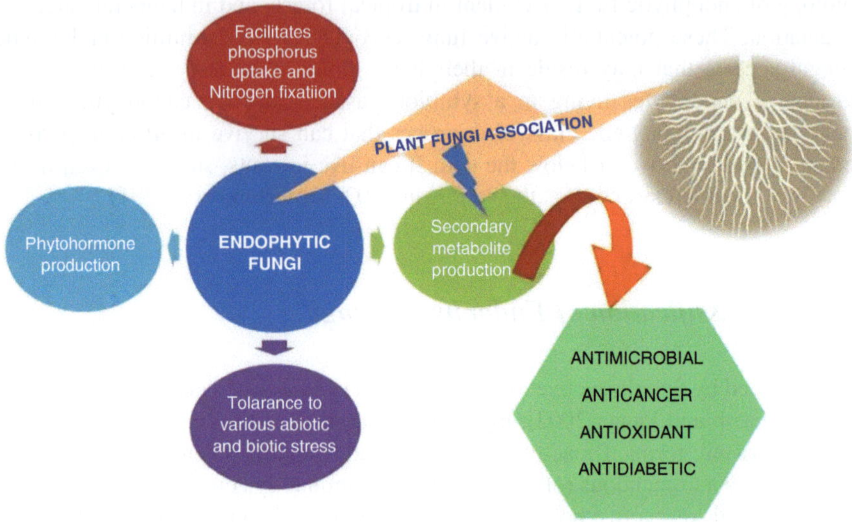

Fig. 3.1 Various contributions of endophytic fungi in nature

plants and their byproducts in their pharmacopoeia (Gurib-Fakim 2006). Aspirin, an effective medicine recognized worldwide, has its natural origins in a group of phytochemicals based glycoside, more specifically salicin. The plant producing it belongs to genera the *Populus* and *Salix*. Such bioactive phytochemicals can be used in numerous ways. For instance, indigenous people from the Amazon region and the Aborigines of Australia identified plant groups that are capable of providing relief from mild headaches and colds to severe intestinal and internal anomalies (London 2012). Other evidence indicates that now-extinct civilizations were aware of the benefits of many important medicinal plants. Interestingly, almost 3000 years ago, the ancient Mayans used fungi on roasted green corn to cure intestinal ailments (London 2012). The Benedictine monks applied *Papaver somniferum* as an analgesic and anesthetic, which the Greeks had done for years (Schiff Jr 2002). In ancient times, many people realized the potency of various parts of the plant as decoctions for curing various ailments. Such plant byproducts promisingly improved their value, diminished anxiety due to severe pain, and provided relief. Though there is plethora of bioactive compounds isolated from endophytic fungi, a clear and concise knowledge of understanding the chemical nature of such bioactive molecules still remained unknown.

3.2.1 Importance of Natural Compounds in Pharmaceutical Industries

Natural product research has attracted significant research attention over the years in almost all pharmaceutical sectors with the advent of the concept of combinatorial chemistry and information technology. Combinatorial chemistry generally involves further synthesis of a drug lead molecule through structural modification with either side chain modifications or enhanced group alterations of structurally related small molecules (Strobel and Daisy 2003). Presently, a majority of pharmaceutical investors are inclined toward natural products because of the promise of large profits. These profits represent amounts much greater than those from ordinary antibiotics of chemical origin. Natural compounds can be rightly defined as potent phytomolecules that bestow benefits for a range of clinical ailments. It is also evident that phytochemistry and biochemistry are gaining importance in connection with traditional molecular structures and now provide newer molecules to the pharmaceutical industry.

The progress and success of various pharmaceutical-based industries worldwide directly depend on the availability of free plant-based bioactive molecules that are responsible for the effectiveness of drugs in the treatment of diseases. This strategy in turn favors and helps in the initiation of various methodologies for the isolation of single, pure compounds in natural-product discovery. Furthermore, combinatorial chemistry also aids in the study of drug-based discovery by proposing effective medicines in the treatment of clinical ailments (Lederberg and Harrison 1998). Plant-based products normally serve as the initial lead molecule where its potentialities can be multiplied with the assistance of synthetic chemistry. Natural products are thus pioneer entities offering a countless range of chemical structures related by the application of combinatorial large databases.

3.2.2 *Endophytes Produce Bioactive Compounds*

Endophytes have attracted much attention within these last ten decades for their potential to produce many bioactive potentialities (Tan and Zou 2001; Ludwig-Müller 2015). The bioactive molecules isolated from endophytes are generally categorized as steroids, xanthones, terpenoids, isocoumarins, phenols, tetralones, benzopyranones and enniatines (Schulz et al. 2002). Well known compounds such as ergosterol, or indole-3-acetic acid are also found to some extent (Yu et al. 2010). These phytocompounds are essential for the endophyte to successfully compete with its neighboring competitors in the host system (Schulz et al. 2002), which is a likely explanation for why majority of these compounds exhibit antimicrobial activities (Mishra et al. 2017b). An effective plant defense system against its pest is counted on the amount of its ability of secondary metabolites production (Orole and Adejumo 2009; Waqas et al. 2015).

Endophyte isolation amidst hosts close environments is a sure shot methodology or an effective protocol for the identification and understanding of a novel organism. This isolation is generally followed by characterization of such molecule. This mechanism thus helps immensely in discovering new bioactive compounds of great importance (Casella et al. 2013). Nevertheless, there are studies to facilitate and understand the antimicrobial efficacy without any complementary recognition of the bioactive plant derived phyto isolates of the endophytes (Liu et al. 2001), so amongst such reports there is yet great prospective for new molecule.

3.3 Endophytic Fungi and Plant Secondary Metabolite Production

Extensive studies have confirmed the efficacy of endophytic fungi in synthesizing several bioactive secondary metabolites, which, until now, was only reported to be attributed to plants. This presents the potential activities of endophytes as substitute for a traditional source of metabolites (Priti et al. 2009). Endophytes are thought to directly provide these metabolites to their host, thus imparting to their chemical defense. It is also thought that endophytes mediate resistance by transferring the related genes to their host genome (Wink 2008). Paclitaxel, a taxol group of compounds derived through the endophytic fungus *Taxomyces andreanae*, is a model that are produced from the plant species *Taxus brevifolia* (Strobel and Daisy 2003), is an vital anticancer drug which promotes tubular development and stabilization, which promotes mitosis disruption. This initial breakthrough enlightened researchers to conduct research with various other genera of *Taxus* species for taxol production. Many other endophytic fungi, like *S tepuiense, S nepalense, Tubercularia sp.* strain TF5 were then investigated and reported to produce the same Paclitaxel (Bashyal et al. 1999). Paclitaxel was also found to be present in endophytes inhabiting hosts in species other than *Taxus*: *Bartalinia robillardoides, Periconia sp.*; *Colletotrichum gloeosporioides* and *Justicia gendarussa* respectively (Gangadevi and Muthumary 2008). The fungal culture for paclitaxel production yielded a small amount of product, which did not validate this method for commercial production. For the large production of paclitaxel production was thus initiated a engineered condition among endophytic fungi which is not confined to a few species of the endophytes. Subsequently, detailed research in this area identified medicinal plants apart from *T. brevifolia*, such as species like *Taxus baccata* (Hamayun et al. 2009) and *Taxus cuspidate* (Santoyo et al. 2016).

Vincristine, a potent and popular anticancer drug, is obtained from *Catharanthus roseus.* Interestingly, it was seen that the same phytomolecule was reported in endophytic fungi *Fusarium oxysporum* cultures isolated from the *Catharanthus roseus* (Aly et al. 2011). The mode of action of this anticancer compound is mediated by arresting mitosis by similar tubulin dimers fastening and tubular arrangement assembly.

Podophyllum is the particular species capable of synthesizing the end product podophyllotoxin. They are considered an important antecedent of anticancer potions, teniposide and etoposide. They mediated its efficacy topoisomerase II inhibition, blocking the step of ligation in cell cycle, which leads to necrosis and ultimately cell death. Topoisomerases care an important enzyme class which catalyzes and directs the process of DNA uncoiling and further promotes DNA transcription.

Another potent toxin, Podophyllotoxin was reported from the plant *Phialocephala fortinii* and *Trametes hirsute*, which are simply the endophytic fungi isolated from and *P peltatum* and *P hexandrum* respectively (Eyberger et al. 2006). The production of podophyllotoxin from the species *P. fortinii* was not reported to be satisfactory (Puri et al. 2006). So, more research and findings conferred the presence of Podophyllotoxin extracted from endophytic fungi *F.* dwelling in *Juniperus recurva*, the host plant that produces also produce the same toxin Podophyllotoxin alike to its residing endophyte (Kour et al. 2008). This toxin was later found to not be solely produced by the above fungi. Another important fungus, *Aspergillus fumigatus*, which is an endophyte of *Juniperus communis,* also produce the same (Kusari et al. 2009).

An innate topoisomerase I Camptothecin from the plant species *Camptotheca acuminate* has been well studied. This compound was later studied in various plant families of different sources like Apocynaceae (*Ervatamia heyneana*), Icacinaceae, Rubiaceae (Ophiorrhiza mungos, *Ophiorrhiza pumila*) (Wink 2008). The compound, when studied extensively, was found to be a product of endophytic infection that subsequently led to its gene transfer (Wink 2008). In an experimental study, Camptothecin was confirmed in *Entrophospora infrequens* cultures residing in the host *N. foetida,* which yield about 4.96 ± 0.73 and 0.575 ± 0.036 mg/100 g of dry cell mass respectively (Puri et al. 2006). Later on other research reported the production camptothecin from *F solani*, secluded from bark of *C. acuminate* (Kusari et al. 2009).

The Poaceae family is another illustrated example of the endophytic defense activity. Grasses belonging to this family are reported to produce a particular alkaloid when infection by endophytic fungi. This association of grass and endophytic fungi extends a defense mechanism by such alkaloid production. This alkaloid is effective against various insects feeding on them. This broad-spectrum alkaloids were simultaneously studied in *F. pratensis*, a typical grass, and in roots of *R. serotinus* species (Lehtonen et al. 2005). The host plant *F. pratensis* bears another important yet common fungi, *Neotyphodium uncinatum*, which possesses biosynthetic potentialities for almost all the major alkaloids of the plant (Blankenship et al. 2001) and displays a wide range of potent isolates to protect the grasslands against the hemiparasitic group and thus check destructions (Lehtonen et al. 2005). Endophytic fungi produce varieties of secondary metabolites effective against such insectivores (Table 3.1). Previous reports suggest that such production of secondary metabolites is not the task of the fungi alone; rather it is a complex interaction between the host and the fungal in association with the development of infection. The complex bio pathway may involve horizontal gene transfer during the process of co-evolution, thus importing the biochemical pathways from fungi into the host plant (Kusari et al. 2012). It will be of great interest to determine the extent of the role of endophytic biosynthetic pathways in the secondary metabolite profiles of plants as it will explain the supplementary the irregular distribution of natural phytoisolates viz; glycosides, phenols, flavones, alkaloids and various other products possibly anthraquinones in the plant species (Wink 2008) (Fig. 3.2).

Table 3.1 Bioactivity of various secondary metabolites derived from various classes of endophytic fungi

Endophytic fungi	Host	Phyto isolates (secondary metabolite)	Bioactivity of phytoisolates	References
Fusarium solani	*Camptotheca acuminata*	Camptothecin	Antitumor	Shweta et al. (2010) and Kusari et al. (2011)
Diaporthe sp., Arthrinium sp., Schizophyllum sp., Penicillium sp., Phomopsis sp.	*Cinchonal edgeriana*	Alkaloids of Chincona	Antipyretic and analgesic anti-inflammatory antimalarial	Nicoletti and Fiorentino (2015) and Ludwig-Müller (2015)
Fusarium oxysporum	*Catharanthus roseus*	Vinblastine, vincristine	Antitumor	Kumar et al. (2013)
Phyllostictas pinarum	*Cupressus sp.*	Paclitaxel	Antitumor	Senthil Kumaran et al. (2008)
Alternaria sp.	*Phellodendron amurense*	Berberine	Antibiotic	Zhou et al. (2009)
Phyllostictacitri carpa	*Citrus medica*	Paclitaxel	Antitumor	Senthil Kumaran et al. (2008)
Fusarium solani	*Taxus celebica*	Paclitaxel	Antitumor	Chakravarthi et al. (2008)
Cladosporium cladosporio	*Taxus media*	Paclitaxel	Antitumor	Zhang et al. (2015)
Taxomyces sp.	*Taxus yunnanensis*	Paclitaxel	Antitumor	Zhao et al. (2011) and Flores-Bustamante et al. (2010)
Rhizopus oryzae	*Iris germanica*	a-Irone, b-Irone	Anti-inflammatory	Elumalai et al. (2009)
Monilia sp.	*Dysosma veitchii*	Penicillium implication Podophyllotoxin	Antitumor	Zhao et al. (2010) and Zhao et al. (2011)
Penicillium sp. Phialocephala fortinii, Trameteshirsuta	*Sinopodophyllum hexandrum*	Podophyllotoxin	Antitumor	Pandey et al. (2014) and Prakash (2015)
Chaetomium globosum	*Hypericum perforatum*	Hypericin	Anti-depressant	Shah et al. (2016)
Sordariomycete sp.	*Eucommiaul moides*	Caffeyl quninic acid, Chlorogenic acid	Antimicrobial Antitumor	Deshmukh et al. (2015) and Palombo (2012)
Fusarium oxysporum	*Ginkgo biloba*	Ginkgolide B	Antishock antiseptic and anti-inflammatory	Marchart (2011)

(continued)

Table 3.1 (continued)

Endophytic fungi	Host	Phyto isolates (secondary metabolite)	Bioactivity of phytoisolates	References
Alternaria neesex	*Melia azedarach*	Toosendanin	Contact toxicity and anti-feeding	Caboni et al. (2012)
Fusarium redolens	*Fritillaria unibracteata*	Peimisine and Imperialine glucoside	Sputum Relaxer Whooping cough and antitumor	Pan et al. (2015)
Colletotrichum gloeosporioides	*Piper nigrum*	Piperine	Antimicrobial, antidepressant, anti-inflammatory and anticancer	Chithra et al. (2014) and Arora and Ramawat (2017)
Chaetomium globosum	*Hypericum perforatum*	Hypericin	Anti-depressant	Kusari et al. (2012)
Botryodiplodia theobroma	*Taxus baccata*	Paclitaxel	Antitumor	Omeje et al. (2017)
Neurospora sp.	*Nothapodytes foetida*	Camptotheci	Antitumor	Rehman et al. (2008)

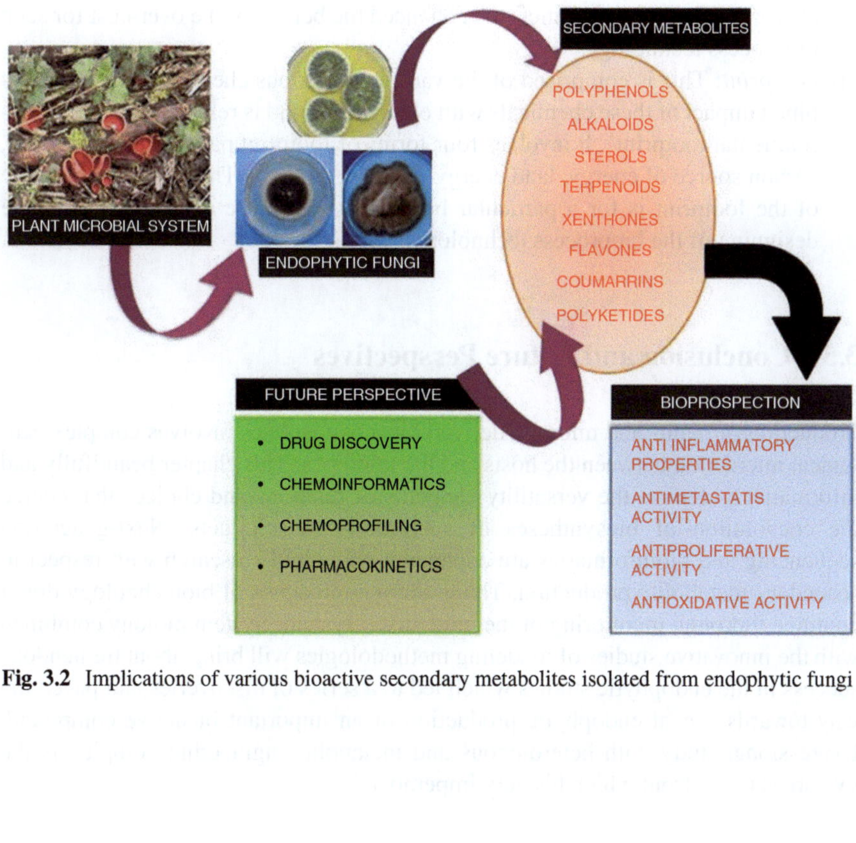

Fig. 3.2 Implications of various bioactive secondary metabolites isolated from endophytic fungi

3.4 Bioprocessing of Endophytic Fungi

It is essential to expand upon the bioprocess understanding for fungal endophytes used in sustainable production of pharmaceutically important compounds. This is necessary for enhancing quality sustenance protocols, increasing productivity and strategizing large-scale industrial production. A few of them re-discussed below:

(a) *Bioprocess knowledge*: This is the total set of techniques and methods required for the completion of a bioprocess (e.g. assembly of a secondary metabolite using an endophytic group).
(b) *Bioprocess design*: This aspect of bioprocessing technology involves the applications for calculating modules or equipments to fabricate innovative and new entity of biopharma based products (medicinally important) obtained or rather harvested from fungal endophytes. It includes process synthesis and process investigation.
(c) *Production practice economics*: This portion of the technology generally highlights the economics and financial strategies involved in the bioprocess technology. They may be bringing forth some scrupulous molecules that are fungal endophytes in origin that are directly dependable over the entire structure of the bioprocess technology, continuance, and setup stability and maintenance. The closer the process economics, the enhanced the benefit of the over-cost for such bioprocess technology.
(d) *Footprint*: This is comprised of the various hazardous chemicals and the combined impact of these chemicals with other wasts, and is referred to as the 'environmental footprint'. It involves four forms of footprint patterns: fossil energy, carbon source of energy, land energy and water energy. The lesser the measure of the footprint is for a particular bioprocess, the more appropriate goes the designing of the bioprocess technology.

3.5 Conclusion and Future Perspectives

Production of plant- and microbe-derived bioactive isolates involves complex ecological interaction between the hosts and the microbes. This chapter beautifully and informatively depicts the versatility elements of relations and choices that coerce the coevolution of biosyntheses of such bioactive products. Next-generation sequencing and bioinformatics are enhancing endophytic research with respect to secondary metabolite production. The modern approaches of biotechnology donot continue thorough monitoring of the mutualistic system. System biology combined with the innovative studies of modeling methodologies will bring about tremendous success in the endophytic studies which led to a series of discoveries and paved the way towards fungal endophytic production of an important bioactive compound. Expressional study with heterologous and metabolic engineering complex is the two areas to confront which his very imperative.

Elicitation, the process of production of both primary and secondary plant metabolites by means of microbial culture system (Ali et al. 2014) or microbes induced enhancers (Lingfei et al. 2005), has been described in many cases. Plant biosynthetic pathways and their contributions are the alternatives. In spite of extensive and elaborative research in the applications of endophytes, there are more exhilarating frontiers to be explored regarding the host-fungi relationship. Molecular biology and gene expression studies further add to this field of research and will possibly bring about insights for better understanding the underlying complex biochemical interaction and signaling in the regulation of such specialized fungal metabolisms. Moreover, the endosymbiont concept in the study of fungal host relationship is another noteworthy area to have a clear knowledge of such interaction. An in-depth understanding at multiple levels, including genetic as well as molecular aspects, gene network study, gene cluster analysis and environmental effect studies with vivid gene expression culture will be supportive for production of endophytic fungi-derived phyto isolates. Further, it is anticipated that this methodology will pave the way in increasing the production of biopotent drugs and consequently validated drug molecule bearing promising future.

Acknowledgement The authors are grateful to DBT (Govt. of India) Sponsored Bioinformatics Infrastructure Facility (BIF) of Assam University and Delcon's e-Journal Access Facility.

Conflict of interest: The authors declare that there is no conflict of interests regarding the publication of this book chapter.

References

Akello J, Dubois T, Gold CS, Coyne D, Nakavuma J, Paparu P (2007) Beauveria bassiana (Balsamo) Vuillemin as an endophyte in tissue culture banana (Musa spp.). J Invertebr Pathol 96(1):34–42

Ali S, Charles TC, Glick BR (2014) Amelioration of high salinity stress damage by plant growth-promoting bacterial endophytes that contain ACC deaminase. Plant Physiol Biochem 80:160–167

Aly AH, Debbab A, Kjer J, Proksch P (2010) Fungal endophytes from higher plants: a prolific source of phytochemicals and other bioactive natural products. Fungal Divers 41(1):1–16

Aly AH, Debbab A, Proksch P (2011) Fungal endophytes: unique plant inhabitants with great promises. Appl Microbiol Biotechnol 90(6):1829–1845

Aly A, Debbab A, Proksch P (2013) Fungal endophytes—secret producers of bioactive plant metabolites. Pharmazie 68(7):499–505

Arora J, Ramawat K (2017) An introduction to endophytes. In: Endophytes: biology and biotechnology. Springer, pp 1–23

Bacon CW, White J (2000) Microbial endophytes. CRC Press, Boca Raton

Bae H, Sicher RC, Kim MS, Kim S-H, Strem MD, Melnick RL, Bailey BA (2009) The beneficial endophyte Trichoderma hamatum isolate DIS 219b promotes growth and delays the onset of the drought response in Theobroma cacao. J Exp Bot 60(11):3279–3295

Bashyal B, Li J, Strobel G, Hess W, Sidhu R (1999) Seimatoantlerium nepalense, an endophytic taxol producing coelomycete from Himalayan yew (Taxus wallachiana). Mycotaxon 72:33–42

Blankenship JD, Spiering MJ, Wilkinson HH, Fannin FF, Bush LP, Schardl CL (2001) Production of loline alkaloids by the grass endophyte, Neotyphodium uncinatum, in defined media. Phytochemistry 58(3):395–401

Caboni P, Ntalli N, Bueno C, Alche L (2012) Isolation and chemical characterization of components with biological activity extracted from Azadirachta Indica and Melia Azedarach. In: Emerging trends in dietary components for preventing and combating disease. ACS Publications, pp 51–77

Cao L, Qiu Z, You J, Tan H, Zhou S (2004) Isolation and characterization of endophytic Streptomyces strains from surface-sterilized tomato (Lycopersicon esculentum) roots. Lett Appl Microbiol 39(5):425–430

Casella TM, Eparvier V, Mandavid H, Bendelac A, Odonne G, Dayan L, Duplais C, Espindola LS, Stien D (2013) Antimicrobial and cytotoxic secondary metabolites from tropical leaf endophytes: isolation of antibacterial agent pyrrocidine C from Lewia infectoria SNB-GTC2402. Phytochemistry 96:370–377

Chakravarthi B, Das P, Surendranath K, Karande AA, Jayabaskaran C (2008) Production of paclitaxel by Fusarium solani isolated from Taxus celebica. J Biosci 33(2):259

Chithra S, Jasim B, Sachidanandan P, Jyothis M, Radhakrishnan E (2014) Piperine production by endophytic fungus Colletotrichum gloeosporioides isolated from Piper nigrum. Phytomedicine 21(4):534–540

Debbab A, Aly AH, Proksch P (2012) Endophytes and associated marine derived fungi—ecological and chemical perspectives. Fungal Divers 57(1):45–83

Deshmukh SK, Verekar SA, Bhave SV (2015) Endophytic fungi: a reservoir of antibacterials. Front Microbiol 5:715

Elumalai E, Chandrasekaran N, Thirumalai T, Sivakumar C, Therasa SV, David E (2009) Achyranthes aspera leaf extracts inhibited fungal growth. Int J Pharm Tech Res 1(4):1576–1579

Eyberger AL, Dondapati R, Porter JR (2006) Endophyte fungal isolates from Podophyllum peltatum produce podophyllotoxin. J Nat Prod 69(8):1121–1124

Faeth SH, Fagan WF (2002) Fungal endophytes: common host plant symbionts but uncommon mutualists. Integr Comp Biol 42(2):360–368

Firáková S, Šturdíková M, Múčková M (2007) Bioactive secondary metabolites produced by microorganisms associated with plants. Biologia 62(3):251–257

Flores-Bustamante ZR, Rivera-Orduña FN, Martínez-Cárdenas A, Flores-Cotera LB (2010) Microbial paclitaxel: advances and perspectives. J Antibiot 63(8):460

Gangadevi V, Muthumary J (2008) A simple and rapid method for the determination of taxol produced by fungal endophytes from medicinal plants using high performance thin layer chromatography. Chin J Chromatogr 26(1):50–55

Gunatilaka AL (2006) Natural products from plant-associated microorganisms: distribution, structural diversity, bioactivity, and implications of their occurrence. J Nat Prod 69(3):509–526

Gurib-Fakim A (2006) Medicinal plants: traditions of yesterday and drugs of tomorrow. Mol Asp Med 27(1):1–93

Hamayun M, Khan SA, Ahmad N, Tang D-S, Kang S-M, Na C-I, Sohn E-Y, Hwang Y-H, Shin D-H, Lee B-H (2009) Cladosporium sphaerospermum as a new plant growth-promoting endophyte from the roots of Glycine max (L.) Merr. World J Microbiol Biotechnol 25(4):627–632

Helander M, Ahlholm J, Sieber T, Hinneri S, Saikkonen K (2007) Fragmented environment affects birch leaf endophytes. New Phytol 175(3):547–553

Huang W-Y, Cai Y-Z, Hyde KD, Corke H, Sun M (2007) Endophytic fungi from Nerium oleander L (Apocynaceae): main constituents and antioxidant activity. World J Microbiol Biotechnol 23(9):1253–1263

Kaul S, Sharma T, Dhar MK (2016) "Omics" Tools for Better Understanding the Plant–Endophyte Interactions. Front Plant Sci 7:955

Kogel K-H, Franken P, Hückelhoven R (2006) Endophyte or parasite–what decides? Curr Opin Plant Biol 9(4):358–363

Kour A, Shawl AS, Rehman S, Sultan P, Qazi PH, Suden P, Khajuria RK, Verma V (2008) Isolation and identification of an endophytic strain of Fusarium oxysporum producing podophyllotoxin from Juniperus recurva. World J Microbiol Biotechnol 24(7):1115–1121
Kumar A, Patil D, Rajamohanan PR, Ahmad A (2013) Isolation, purification and characterization of vinblastine and vincristine from endophytic fungus Fusarium oxysporum isolated from Catharanthus roseus. PLoS One 8(9):e71805
Kusari S, Lamshöft M, Spiteller M (2009) Aspergillus fumigatus Fresenius, an endophytic fungus from Juniperus communis L. Horstmann as a novel source of the anticancer pro-drug deoxypodophyllotoxin. J Appl Microbiol 107(3):1019–1030
Kusari S, Zuhlke S, Spiteller M (2011) Effect of artificial reconstitution of the interaction between the plant Camptotheca acuminata and the fungal endophyte Fusarium solani on camptothecin biosynthesis. J Nat Prod 74(4):764–775
Kusari S, Hertweck C, Spiteller M (2012) Chemical ecology of endophytic fungi: origins of secondary metabolites. Chem Biol 19(7):792–798
Lederberg J, Harrison PF (1998) Antimicrobial resistance: issues and options. National Academies Press, Washington, DC
Lehtonen P, Helander M, Saikkonen K (2005) Are endophyte-mediated effects on herbivores conditional on soil nutrients? Oecologia 142(1):38–45
Lingfei L, Anna Y, Zhiwei Z (2005) Seasonality of arbuscular mycorrhizal symbiosis and dark septate endophytes in a grassland site in southwest China. FEMS Microbiol Ecol 54(3):367–373
Liu CH, Zou WX, Lu H, Tan RX (2001) Antifungal activity of Artemisia annua endophyte cultures against phytopathogenic fungi. J Biotechnol 88(3):277–282
London DS (2012) Diet as a double-edged sword: the pharmacological properties of food among the Waorani Hunter-Gatherers of Amazonian Ecuador. Arizona State University
Ludwig-Müller J (2015) Plants and endophytes: equal partners in secondary metabolite production? Biotechnol Lett 37(7):1325–1334
Marchart K (2011) Qualitative and quantitative analysis of Chinese medical herbs for the prevention and therapy of acute myocardial infarction and diabetes. Uniwien
Mishra VK, Passari AK, Chandra P, Leo VV, Kumar B, Gupta VK, Singh BP (2017a) Determination and production of antimicrobial compounds by *Aspergillus clavatonanicus* strain MJ31, an endophytic fungus from Mirabilis jalapa L. using UPLC-ESI-MS/MS and TD GC-MS. PLoS One 12(10):1–24. https://doi.org/10.1371/journal.pone.0186234
Mishra VK, Passari AK, Leo VV, Singh BP (2017b) Molecular diversity and detection of endophytic fungi based on their antimicrobial biosynthetic genes. In: Singh BP, Gupta VK (eds) Molecular markers in mycology, fungal biology. Springer, Cham, Switzerland, pp 1–35. https://doi.org/10.1007/978-3-319-34106-4_1
Nicoletti R, Fiorentino A (2015) Plant bioactive metabolites and drugs produced by endophytic fungi of Spermatophyta. Agriculture 5(4):918–970
Omeje EO, Ahomafor JE, Onyekaba TU, Monioro PO, Nneka I, Onyeloni S, Chime C, Eboka JC (2017) Endophytic fungi as alternative and reliable sources for potent anticancer agents. In: Natural products and cancer drug discovery. InTech
Orole O, Adejumo T (2009) Activity of fungal endophytes against four maize wilt pathogens. Afr J Microbiol Res 3(12):969–973
Palombo EA (2012) Endophytes from Medicinal Plants as Novel Sources of Bioactive Compounds. In: Medicinal plants: biodiversity and drugs. pp 355–411
Pan F, Su X, Hu B, Yang N, Chen Q, Wu W (2015) Fusarium redolens 6WBY3, an endophytic fungus isolated from Fritillaria unibracteata var. wabuensis, produces peimisine and imperialine-3β-d-glucoside. Fitoterapia 103:213–221
Pandey PK, Singh S, Yadav RNS, Singh AK, Singh MCK (2014) Fungal endophytes: promising tools for pharmaceutical science. Int J Pharm Sci Rev Res 25(2):128–138
Porras-Alfaro A, Bayman P (2011) Hidden fungi, emergent properties: endophytes and microbiomes. Annu Rev Phytopathol 49:291–315

Prakash V (2015) Endophytic fungi as resource of bioactive compounds. Int J Pharm Bio Sci 6:887–898

Priti V, Ramesha B, Singh S, Ravikanth G, Ganeshaiah K, Suryanarayanan T, Uma Shaanker R (2009) How promising are endophytic fungi as alternative sources of plant secondary metabolites? Curr Sci 97(4):477–478

Puri SC, Nazir A, Chawla R, Arora R, Riyaz-ul-Hasan S, Amna T, Ahmed B, Verma V, Singh S, Sagar R (2006) The endophytic fungus Trametes hirsuta as a novel alternative source of podophyllotoxin and related aryl tetralin lignans. J Biotechnol 122(4):494–510

Redman RS, Sheehan KB, Stout RG, Rodriguez RJ, Henson JM (2002) Thermotolerance generated by plant/fungal symbiosis. Science 298(5598):1581–1581

Rehman S, Shawl A, Kour A, Andrabi R, Sudan P, Sultan P, Verma V, Qazi G (2008) An endophytic Neurospora sp. from Nothapodytes foetida producing camptothecin. Appl Biochem Microbiol 44(2):203–209

Rodriguez R, White J Jr, Arnold A, Redman RS (2009) Fungal endophytes: diversity and functional roles. New Phytol 182(2):314–330

Santoyo G, Moreno-Hagelsieb G, Orozco-Mosqueda Mdel C, Glick BR (2016) Plant growth-promoting bacterial endophytes. Microbiol Res 183:92–99

Schiff PL Jr (2002) Opium and its alkaloids. Am J Pharm Educ 66(2):186

Schulz B, Boyle C (2005) The endophytic continuum. Mycol Res 109(6):661–686

Schulz B, Boyle C, Draeger S, Römmert A-K, Krohn K (2002) Endophytic fungi: a source of novel biologically active secondary metabolites. Mycol Res 106(9):996–1004

Senthil Kumaran R, Muthumary J, Hur B (2008) Production of Taxol from Phyllosticta spinarum, an endophytic fungus of Cupressus sp. Eng Life Sci 8(4):438–446

Shah A, John Dar N, Parvaiz Hassan Q, Ahmad M (2016) Endophytes and Neurodegenerative diseases: a hope in desperation. CNS Neurol Disord Drug Targets 15(10):1231–1239

Shweta S, Zuehlke S, Ramesha B, Priti V, Kumar PM, Ravikanth G, Spiteller M, Vasudeva R, Shaanker RU (2010) Endophytic fungal strains of Fusarium solani, from Apodytes dimidiata E. Mey. ex Arn (Icacinaceae) produce camptothecin, 10-hydroxycamptothecin and 9-methoxycamptothecin. Phytochemistry 71(1):117–122

Stierle A, Strobel G, Stierle D (1993) Taxol and taxane production by Taxomyces andreanae, an endophytic fungus of Pacific yew. Science 260(5105):214–216

Strobel G, Daisy B (2003) Bioprospecting for microbial endophytes and their natural products. Microbiol Mol Biol Rev 67(4):491–502

Tan RX, Zou WX (2001) Endophytes: a rich source of functional metabolites. Nat Prod Rep 18(4):448–459

Vandenkoornhuyse P, Quaiser A, Duhamel M, Le Van A, Dufresne A (2015) The importance of the microbiome of the plant holobiont. New Phytol 206(4):1196–1206

Vesterlund S-R, Helander M, Faeth SH, Hyvönen T, Saikkonen K (2011) Environmental conditions and host plant origin override endophyte effects on invertebrate communities. Fungal Divers 47(1):109–118

Waller F, Achatz B, Baltruschat H, Fodor J, Becker K, Fischer M, Heier T, Hückelhoven R, Neumann C, von Wettstein D (2005) The endophytic fungus Piriformospora indica reprograms barley to salt-stress tolerance, disease resistance, and higher yield. Proc Natl Acad Sci U S A 102(38):13386–13391

Waqas M, Khan AL, Hamayun M, Shahzad R, Kang S-M, Kim J-G, Lee I-J (2015) Endophytic fungi promote plant growth and mitigate the adverse effects of stem rot: an example of Penicillium citrinum and Aspergillus terreus. J Plant Interact 10(1):280–287

Wilson D (1995) Endophyte: the evolution of a term, and clarification of its use and definition. Oikos 73:274–276

Wink M (2008) Plant secondary metabolism: diversity, function and its evolution. Nat Prod Commun 3(8):1205–1216

Yu H, Zhang L, Li L, Zheng C, Guo L, Li W, Sun P, Qin L (2010) Recent developments and future prospects of antimicrobial metabolites produced by endophytes. Microbiol Res 165(6):437–449

Yuan Z-L, Chen Y-C, Yang Y (2009) Diverse non-mycorrhizal fungal endophytes inhabiting an epiphytic, medicinal orchid (Dendrobium nobile): estimation and characterization. World J Microbiol Biotechnol 25(2):295
Zhang HW, Song YC, Tan RX (2006) Biology and chemistry of endophytes. Nat Prod Rep 23(5):753–771
Zhang F, Wang M, Zheng Y, Liu H, Zhang X, Wu S (2015) Isolation and characterzation of endophytic Huperzine A-producing fungi from Phlegmariurus phlegmaria. Microbiology 84(5):701–709
Zhao J, Zhou L, Wang J, Shan T, Zhong L, Liu X, Gao X (2010) Endophytic fungi for producing bioactive compounds originally from their host plants. Curr Res, Technol Educ Trop Appl Microbiol Microbial Biotechnol 1:567–576
Zhao J, Shan T, Mou Y, Zhou L (2011) Plant-derived bioactive compounds produced by endophytic fungi. Mini Rev Med Chem 11(2):159–168
Zhou H, Wu S, Joo JY, Zhu S, Han DW, Lin T, Trauger S, Bien G, Yao S, Zhu Y (2009) Generation of induced pluripotent stem cells using recombinant proteins. Cell Stem Cell 4(5):381–384

Chapter 4
Exploring the Benefits of Endophytic Fungi via Omics

Hasnahana Chetia, Debajyoti Kabiraj, Biju Bharali, Sunita Ojha, Manash Pratim Barkataki, Dharitri Saikia, Tinka Singh, Ponnala Vimal Mosahari, Pragya Sharma, and Utpal Bora

4.1 Introduction

4.1.1 What Are Endophytic Fungi?

The term "endophyte" stems from the two Greek words—*endon* = within and *phyte* = plant. Thus, an endophyte is an organism surviving within a host plant without deleterious consequences. In 1809, a German botanist, Johann Heinrich Friedrich Link termed them as "Entophytae" which constituted of moderately parasitic species living within plants in 1809. In 1991, Orlando Petrini re-termed these organisms as "endophytes". However, modern studies have shown that endophytes colonize plant tissues without detriment to the host and at most times, rather acts as a benefactor. An endophyte may be bacteria, fungi, algae or oomycetes. Here, we will focus on endophytic fungi only.

H. Chetia · D. Kabiraj · S. Ojha · M. P. Barkataki
Department of Biosciences and Bioengineering, Indian Institute of Technology Guwahati, Guwahati, Assam, India

B. Bharali · D. Saikia · T. Singh · P. V. Mosahari
Centre for the Environment, Indian Institute of Technology Guwahati, Guwahati, Assam, India

P. Sharma
Department of Bioengineering and Technology, Gauhati University Institute of Science and Technology (GUIST), Gauhati University, Guwahati, Assam, India

U. Bora (✉)
Centre for the Environment, Indian Institute of Technology Guwahati, Guwahati, Assam, India

Department of Biosciences and Bioengineering, Indian Institute of Technology Guwahati, Guwahati, Assam, India
e-mail: ubora@iitg.ac.in

B. P. Singh (ed.), *Advances in Endophytic Fungal Research*, Fungal Biology,
https://doi.org/10.1007/978-3-030-03589-1_4

Endophytic fungi are a diverse group of fungi living a primarily asymptomatic life within almost every terrestrial plant lineage available in natural and anthropogenic habitats. Fossil records indicate that most of the plants, if not all, have been associated with endophytic fungi for >400 myr (million years). They are capable of surviving in intra- or inter-cellular anatomical sections of a plant found either above ground (e.g. stem) or underground (e.g. roots). Growing evidences suggest that they have played an important role in driving the evolution, adaptation and ecological fitness of plant life on land and at the same time, shaped the structure and diversity of the community of associated organisms like bacteria, insects, etc. (Passari et al. 2017; Rodriguez et al. 2009).

Based on the lifestyle, functional diversity, biology and mode of transmission followed by an endophytic fungi, they are classified as systemic/true endophytes (mutualistic for the entirety of its lifespan) or transient/non-systemic endophytes (adapting transient modes of lifestyles, mutualistic and parasitic being the two most common modes, during different stages of life cycle) (Wani et al. 2015). Another classification for endophytic fungi based on their phylogeny and life history traits, divides them into clavicipitaceous and non-clavicipitaceous classes (Rodriguez et al. 2009). Both sexual and asexual modes of reproduction have been reported in these endophytes. Asexual endophytes transmit genetic material via hyphae-to-seed transformation while sexual endophytes adopts spore-based transmission in addition to the strategy of asexual endophytes (Saikkonen et al. 2016).

Currently, there are ~4648 publications (Scopus) and ~1166 patents (Google Patents) on the search term "endophytic fungi" which signifies the gradual growth of scientific and/or industrial interest on these groups of organisms (Fig. 4.1).

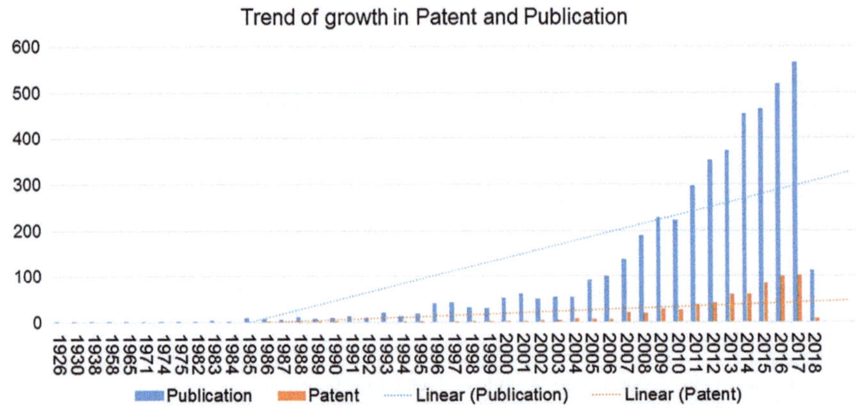

Fig. 4.1 Trend of growth in the number of patents and publications per year (1926 to present) on endophytic fungi. This graph was generated on the basis of data generated by Google Patents and Scopus (Publications) using the search term "endophytic fungi". Linear trend lines were used to depict the increase in the patents and publications every year

In this chapter, we review the benefits of endophytic fungi towards plants, humans and how emerging omics-based studies have widened the periphery of existing knowledge on this group of mutualistic biotrophs.

4.1.2 Endophytic Fungi-Host Plant Association Benefits the Host

Endophytes are generally found to be in mutualistic relationship with host plants; they symbiotically exchange favours with the plant without causing any apparent trauma. Treading past the phenotypical manifestations, these bi-taxa associations are now known to have profound beneficial impact of fungal endophytes on plants communities like improvement of plant fitness, tolerance to biotic and abiotic stress, increase in biomass, lowering water consumption, etc. In return, the host plant provides the fungal endophytes with refuge, nutrients and carves the way to their life within the next generation of host plants. The discovery of Paclitaxel of endophytic fungal origin has tremendously boosted overall interest and studies like bioprospecting and metabolite screening studies on them. The role of fungal endophytes as effectors during different forms of stress in plants is discussed below.

4.1.2.1 Role of Endophytic Fungi in Biotic Stress Tolerance

The spectrum of biotic stress faced by a plant includes challenges from herbivores, competing plant species, pathogens and parasites. Herbivorous challenges to a plant mediated by vertebrate and invertebrate pests, affects its fitness due to extensive defoliation. This is mitigated by the plant via photosynthetic enhancement, increased foliar and meristematic growth rate, delay in senescence and so forth (Redondo-Gómez 2013). Sometimes, a plant is affected by allelochemicals released by the neighbouring plants. These allelopathic challenges are mitigated by strategies like detoxification and morphological changes. Plant immune responses to pathogen/parasites also vary from one plant to another. In this section, we shall discuss the role of resident fungal endophytes during these biotic stresses to host plants.

Defence Against Herbivores

There are multiple studies which suggest that presence of endophytic fungi does aid plants in mitigating herbivorous challenges. Comparison between plants with and without endophytic fungi has shown that plants without endophytes are more vulnerable to herbivore-mediated damage than the former. Most widely studied plants in this regard are grasses. One of the earliest observations of herbivore deterrence by endophyte-containing plant was reported by Bacon et al., when their study proved the existence of a positive correlation between *Epichloë typhina*-hosting fescue grass (*Festuca arundinacea* Schreb.) and toxicity-mediated disorders in grazing cattle (Bacon et al. 1977). Another *Epichloë* species, *E. festucae* var. *lolii* which grows in its host grass, *Lolium perenne*, harms both vertebrate and invertebrate herbivores by production of defensive alkaloids, viz. peramine, ergovaline and lolitrem B (Müller and Krauss 2005; Cheplick et al. 1989). Peramine deters or intoxicates invertebrates while the other two alkaloids act as neurotoxins against vertebrates (Philippe 2016).

Other reports on benefits of endophytic fungus against herbivorous species were *Phomopsis oblonga* which protects elm trees against the beetle, *Physocnemum brevilineum* and perennial ryegrass *Lolium perenne* L. against the sod webworm (Clay 1987; Webber 1981). Enhanced resistance of perennial ryegrass and tall fescue to insects such as the Argentine stem weevil (*Listronotus bonariensis* Kuschel) imparted by endophyte infection *Acremonium lolii* Latch were also reported (Prestidge 1993). Thus, endophyte–host plant relationship is a defensive mutualistic relationship marked by production of a diversity of bioactive chemicals produced by the endophytes and involves multifaceted signalling and chemical crosstalk in both the species. These findings have led to conceptualizing usage of endophytic fungi as potential biocontrol agents that can be further probed via field censuses, in vitro studies and greenhouse experiments.

Defence Against Allelopathic Challenges

Plants release allelochemicals for inhibiting or stimulating desired effects on growth, development and survival of its neighbouring plants. Its co-inhabitant microbial communities including endophytic fungi also partakes in these chemical interventions by providing a spectrum of novel, unique bioactive compounds not produced by their hosts. Bioactive compounds are generally identified via assays for antimicrobial, cytotoxicity or insecticidal properties. However, their role in allelopathic challenges has remained restricted to potted or greenhouse experiments with low emphasis on their genomic dynamics.

The allelopathic effect of hosted endophytes on neighbouring plants has been studied lesser than other related aspects. One study showed that a tall fescue species *L. rundina* with endophyte *Neotyphodium coenophialum* can impair the growth of seedlings by releasing inhibitory allelochemicals (Orr et al. 2005; Rudgers and Orr 2009). Another study reported under controlled environment of greenhouse, the allelopathic impact of an invasive for *Centaurea stoebe* on a native bunchgrass *Koeleria macrantha* to be twice in presence of the endophyte *Alternaria alternata* (Aschehoug et al. 2014). Similarly, experimentation on potted *Achnatherum inebrians* Keng (host plant)–*Neotyphodium gansuense* (endophytic fungi) against multiple species of turfgrass, viz. *Lolium perenne, L. Festuca arundinacea* Schreb. and *Poa pratensis* L. showed significant inhibitory effects on turfgrass seed germination and growth rates (Yang 2010). Presence of alkaloid toxins has been co-related with the allelopathic potential of grasses associated with endophytes (Clay and Holah 1999).

The aforementioned studies have shown that endophytic fungi are not just some asymptomatic species living inside a plant and the co-adaptation of both species is usuallybeneficial to the host. However, it remains to be deduced as to how the allelochemical type and their expression changes in these interactions, both in vitro and in vivo conditions. Microbial interactions with host plants must also be investigated to comprehend the roles of allelochemicals formed by fungal endophytes in host plants.

Defence from Parasites and Pathogens

Endophytic fungi confers resistance against plant pathogens through the enhanced production of plant secondary metabolites, endophyte-specific metabolite synthesis, induction of systemic or local immune response or through competitive exclusion of the pathogen (Arnold et al. 2003). The intimate interaction between the endophyte and the host may lead to the development of a plethora of substances that work in a multi-functional way.

The relationship between production of a known fungicide and anti-cancer drug, Taxol in Yew trees (*Taxus* spp.) and its inhabitant endophytic fungus, *Paraconiothyrium* sp. provides an interesting example to understand beneficial role of endophytes. *Paraconiothyrium* sp. grows at the points of bark-cracking, which are most vulnerable to infection, during branch formation in Yew trees. On reception of chemical stimulus triggered by secondary metabolites of invading wood decaying fungi, *Paraconiothyrium* enhances taxol synthesis and its sequestration in extracellular hydrophobic bodies to create a barrier of anti-fungal protection in the stem's vascular system, thereby acting in a way beneficial to both host and endophyte itself (Soliman et al. 2015). These observations also prove that an endophytic fungal species is capable of occupying the same niche as mobile host immune cells (Talbot 2015).

Wide range of endophytic fungal strains with antibacterial or anti-fungal compounds have been isolated which can be used, either singly or in composition, as a potential biocontrol agent (Mishra et al. 2017a; Soliman et al. 2015; Talbot 2015; Strobel et al. 2002; Gunatilaka 2006). Application of endophytic fungi can resist many devastating plant pathogens such as *Phytophthora, Puccinia, Fusarium* and *Colletotrichum.* Blasticidin-S is a commercialized fungal metabolite that inhibits the growth of various pathogenic fungal and bacterial species (Mishra et al. 2017b; Takeuchi et al. 1958). Validamycin, pyrrolnitrin, etc. are used as a lead for developing agrochemicals for sustainable agriculture system.

Plant pathogenic viruses are obligate intracellular parasites that cause huge loss in crop plants. Application of endophytes for preventing or tackling these challenges is still experimental. A study showed that virulence of Barley yellow dwarf (BYD) virus is reduced when aphids-infected with BYD are applied on endophyte-hosting plants in comparison to that of endophyte-free plants (Lehtonen et al. 2006). Nematocidal activity of plant exudates infected by endophytic fungi has also been studied. 3-hydroxypropionic acid is one of the principle compounds found in several endophytic fungi that shows selective nematocidal effect against plant parasitic nematode (Schwarz et al. 2004).

Another study involving isolation of endophytic fungi from vegetables to assess anti-oomycetic potential identified 152 species of endophytes where a *Fusarium oxysporum* isolate from roots of red pepper was shown to be the most effective against three pathogens, namely, *Phytophthora infestans* (tomato late blight disease), *P. capsica* (vegetable blight and rot disease) and *Pythium ultimum* (vegetables root rot disease) (Kim et al. 2007). This study along with the previous ones paves the way for utilization of fungal endophytes as potent bio-control agents with multiple targets. Table 4.1 lists some of the anti-pathogenic/parasitic compounds identified from endophytic fungi.

Table 4.1 Activity-wise distribution of bioactive compounds reported from endophytic fungi

Anti-cancer (cytotoxic/anti-proliferative) activity				
Host plant	Fungi	Compound/s	Activity	Reference
Mimusops elengi	*Ascomycetes* sp.	Ergoflavin	Induction of cytotoxicity	Deshmukh et al. (2009)
Tabebuia argentea	*Alternaria alternata, Alternaria* spp.*, Aspergillus niger, Penicillium* spp.	Lapachol	Interference of topoisomerase enzyme activity	Channabasava and Govindappa (2014)
Crotalaria pallida	*Alternaria* sp.	Coumarin	Caspase-dependent apoptosis	Umashankar et al. (2015)
Terminalia arjuna	*Pestalotiopsis terminaliae*	Paclitaxel	Apoptosis induction	Gangadevi and Muthumary (2009)
Morinda citrifolia Linn.	*Botryodiplodia theobromae*	Paclitaxel	Downregulation of COX-2	Pandi et al. (2010)
Eugenia jambolana Lam.	*Cephalotheca faveolata*	Sclerotiorin	Activation of apoptotic and downregulation of anti-apoptotic genes	Giridharan et al. (2012)

Immunomodulation (Immuno-suppressive and anti-inflammatory) activity				
Host plant	Fungi	Compound/s	Activity	Reference
Tripterygium wilfordii	*Fusarium subglutinans*	Sub-glutinol A and B	Immuno-suppressive activity	Strobel and Pliam (1995)
Mimusops elengi	*Ascomycetes* sp.	Ergoflavin	Anti-inflammatory	Deshmukh et al. (2009)
Aloe vera	*Talaromyces wortmannii*	Component C	Anti-inflammatory	Pretsch et al. (2014)

(continued)

Table 4.1 (continued)

Antioxidant activities				
Host plant	Fungi	Compound/s	Activity	Reference
Morinda citrifolia Linn.	*Botryodiplodia theobromae*	Paclitaxel	Restoring the anti-peroxidative enzyme activity	Pandi et al. (2010)
Terminalia morobensis	*Pestalotiopsis microspora*	Isopestacin	Scavenge superoxide O_2 and OH radicals	Strobel et al. (2002)

Antidiabetic activity				
Host plant	Fungi	Compounds	Activity	Reference
Salvadora oleoides	*Aspergillus* sp. *JPY1 and 2, Phoma* sp.	2, 6-di-tert-butyl-p-cresol; Phenol, 2, 6-bis (1,1-dimethylethyl)-4-methyl	Regeneration of pancreatic cells and increases insulin secretion from beta cells	Dhankhar et al. (2013)
Adathoda beddomei	*Syncephalastrum* sp.	NA	α-amylase inhibition	Prabavathy and Nachiyar (2013)
Viscum album	*Alternaria* sp.	Lectin (N-acetyl galactosamine)	Inhibition of α-amylase, α-glucosidase and sucrase; stimulation of ductal stem cells	Govindappa et al. (2015)
Taxus sumatrana	NA	NA	Alpha-glucosidase inhibition	Artanti et al. (2011)

Anti-parasitic activity				
Host plant	Fungi	Compounds	Target pathogen	Reference
Garcinia atroviridis, G. mangostana, G. dulcis, G. nigrolineata	*Muscodor* sp., *Botryosphaeria rhodina, Aspergillus aculeatus, Penicillium sclerotiorum*	NA	*Plasmodium falciparum K1*	Phongpaichit et al. (2007)
Vitex pinnata	*Lasiodiplodia theobromae*	Cladospirone *B*, Desmethyl-lasiodiplodin, R-(−)- mullein	*Trypanosoma* spp.	Kamal et al. (2016)

Antiviral compounds				
Host plant	Fungi	Compounds	Against pathogen	Reference
Quercus sp.	*Cytonaema* sp.	Cytonic acid	Hepatitis virus, human cytomegalovirus	Guo et al. (2000)
Garcinia atroviridis, G. mangostana	*Guignardia mangiferae, Penicillium sclerotiorum, Botryosphaeria rhodina*	NA	Herpes simplex virus type 1 (HSV-1 ATCC VR-260)	Phongpaichit et al. (2007)

(continued)

Table 4.1 (continued)

Antiviral compounds

Host plant	Fungi	Compounds	Against pathogen	Reference
Aegiceras corniculatum	*Emericella* sp.	Emerimidine, emeriphenolicins, asper-nidine, austin, austinol, de-hydroaustin and acetoxydehydroaustin	Influenza A (H1N1)	Zhang et al. (2011)
Quercus emoryi	*Alternaria tenuissima*	Altertoxin I, II, III and V	HIV-1 virus	Bashyal et al. (2014)

Anti-fungal activity

Host plant	Fungi	Compounds	Against pathogen	Reference
Xylopia aromatica	*Periconia atropurpurea*	2,4-dihydroxy-6-[(1′*E*,3′*E*)-penta-1′, 3′-dienyl]-benzaldehyde	*Cladosporium sphaerospermum, C. cladosporioides*	Teles et al. (2006)
Astragalus membranaceus	*Aspergillus fumigatus*	Fumitremorgin B, C, Cyclotryprostatin B, Verruculogen, Cyclotryprostatin C, Gliotoxin	*C. albicans, F. solani, P. chrysogenum*	Zhang et al. (2017a, 2017b)
Terminalia morobensis	*Pestalotiopsis microspora*	Isopestacin	*Pythium ultimum, Sclerotinia sclerotioru, Rhizoctonia solani*	Strobel et al. (2002)
Paris polyphylla var. *yunnanensis*	*Pichia guilliermondii*	5α,8α-epidioxyergosta-6,22-dien-3β-ol and helvolic acid	*M. oryzae*	Zhao et al. (2010)
Sesbania grandiflora (L.) Pers.	*Fusarium* sp., *Phaeoacremonium* sp., *Acremonium* sp., *Cladosporium* sp.	NA	*Candida albicans, Cryptococcus neoformans, Rhizopus* spp., *Mucor* spp., *Aspergillus* spp., *Alternaria* spp., *Curvularia* spp., *Penicillium* spp.	Powthong et al. (2013)

Nematocidal activity

Host plant	Fungi	Compounds	Against nematode	Reference
Tropical tree (Species not specified)	*Phomopsis phaseoli*	3-Hydroxy propionic acid	*Meloidogyne incognita*	Schwarz et al. (2004)
Betula pendula, B. pubescens	*Melanconium betulinum*			
Sargassum thunbergii	*Eurotium cristatum* EN-220	Indole diketopiperazine alkaloids	*Panagrellus redivivus*	Du et al. (2017)

(continued)

Table 4.1 (continued)

Nematocidal activity				
Host plant	Fungi	Compounds	Against nematode	Reference
Menegazzia sp.	*Xylaria grammica*	Grammicin	*Meloidogyne incognita*	Kim et al. (2018)

Insecticidal activity				
Host plant	Fungi	Compounds	Against insect	Reference
Limonia acidissima L.	*Penicillium oxalicum* LA-1	Hamisonine	*Culex quinquefasciatus*	Seetharaman et al. (2017)
Acrostichum speciosum	*Aspergillus fumigatus* JRJ111048	Anhydride-derivative aspergide	*Spodoptera litura*	Guo et al. (2017)
Dioscorea zingiberensis	*Berkleasmium* sp.	Spirobisnaphthalenes	*Aedes albopictus*	Tian et al. (2016)

Antibacterial activity				
Host plant	Fungi	Compounds	Against pathogen	Reference
Aquilaria sinensis	*Fusarium equiseti, Fusarium solani, Fusarium avenaceum, Leptosphaerulina chartarum, Fusarium oxysporum, Phaeoacremonium rubrigenum, Coniothyrium nitidae*	NA	*B. subtilis, Staphylococcus aureus, Escherichia coli*	Cui et al. (2011)
Astragalus membranaceus	*Aspergillus fumigatus*	Fumitremorgin B and C, cyclotryprostatin B, verruculogen, cyclotryprostatin C, gliotoxin, methylthiogliotoxin, fumiquinazoline C, D and J	*B. subtilis, S. aureus, E. coli, P. aeruginosa*	Zhang et al. (2017a, b)
Adathoda beddomei	*Syncephalastrum* sp.	Furandione, phenyl esters, anthracene methanol, diethyl phthalate, pentadecanoic acid	*Escherichia coli, Candida albicans, Pseudomonas aeruginosa, Klebsiella pneumoniae, Staphylococcus aureus*	Prabavathy and Nachiyar (2013)

(continued)

Table 4.1 (continued)

Antibacterial activity				
Host plant	Fungi	Compounds	Against pathogen	Reference
Garcinia atroviridis, G. mangostana, G. dulcis, G. nigrolineata, G. scortechinii	*Guignardia mangiferae, Penicillium sclerotiorum, Phomopsis* sp., *Aspergillus aculeatus, Guignardia mangiferae, Curvularia* sp., *Fusicoccum* sp., *Muscodor* sp., *Pestalotiopsis* sp., *Botryosphaeria rhodina*	NA	*Mycobacterium tuberculosis*	Phongpaichit et al. (2007)
Aloe vera	*Talaromyces wortmannii*	Component C	*Propionibacterium acnes*	Pretsch et al. (2014)
Smallanthus sonchifolius	*Papulaspora immersa, Apiospora montagnei* Sacc.	NA	*S. aureus, Kocuria rhizophila, P. aeruginosa, E. coli*	Ramos et al. (2010)
Paris polyphylla var. *yunnanensis*	*Pichia guilliermondii*	5α,8α-epidioxyergosta-6,22-dien-3β-ol and helvolic acid	*A. tumefaciens, E. coli, P. lachrymans, R. solanacearum, X. vesicatoria, B. subtilis, S. aureus, S. haemolyticus*	Zhao et al. (2010)
Sesbania grandiflora (L.) Pers.	*Fusarium* sp, *Phaeoacremonium* sp., *Acremonium* sp., *Cladosporium* sp.	NA	*Bacillus subtilis, Staphylococcus aureus, Pseudomonas aeruginosa* and *Escherichia coli*	Powthong et al. (2013)
Vitex negundo Linn.	*Phomopsis* sp., *Aspergillus flavus* gr.	NA	*S. typhimurium, E. coli, B. subtilis, B. cereus, K. pneumoniae* and *S. aureus*	Desale and Bodhankar (2013)
Ricinus communis	*A. fumigates, A. japonicas, A. niger, Fusarium semitectum, Curvularia pallescens, Phoma hedericola, Alternaria tenuissima, F, solani, Drechslera australien and A. repens*	NA	*B. subtilis, Enterococcus* sp., *K, pneumoniae, E, coli, S. typhimurium* and *S. aureus*	Sandhu et al. (2014)

NA information not available/not stated

4.1.2.2 Role of Endophytic Fungi in Abiotic Stress Tolerance

Plants are subjected to various abiotic stresses in their life cycle such as drought stress, temperature stress, salinity stress and heavy metal toxicity stress (Kaur and Asthir 2015). In addition to their intrinsic mechanism of adaptation, endophytes also enhance alleviation of the negative effects of stress conditions faced by host plants. These types of mutualistic relation confer tolerance to host plants' abiotic stress through two mechanisms: (a) by activating stress response after exposure in host, allowing plants to avoid or alleviate the effects of stress and (b) by synthesis of anti-stress biochemical by endophyte (Lata et al. 2018). There are reports wherein endophytes in cool season grass orchestrate avoidance or tolerance of droughts as well as enhanced recovery via morphological, biochemical and physiological adaptations (Singh et al. 2011). These mechanisms also vary according to conditions including host genotypes or environmental factors and accordingly adapt to different symbiotic lifestyle (Johnson et al. 2003). Using analyses such as comparative genomics, studying the transcriptomic profile, metabolic profile and next-generation sequencing, plant–endophyte relationship and their capability for stress tolerance can be better understood (Kaul et al. 2016).

Drought Stress Tolerance

Plants adapt to the stress to drought conditions via avoidance, tolerance and recovery. In temperate regions, drought conditions usually result from water deficiency and high temperature (Malinowski and Belesky 2000). A seed-transmissible non-pathogenic fungal endophyte (symbiont), *Neotyphodium* stimulates drought tolerance, dry matter production, efficient utilization of soil nitrogen, and deters insects in tall fescue plants (Bayat et al. 2009). During drought, length of root hair increases and diameter decreases in endophyte-infected tall fescue plants resulting in enhanced surface area for water and mineral absorption (Malinowski et al. 1999). Endophytes also precondition the host to the drought stress so that plant exhibits drought response sooner. For example, leaf stomatal resistance arising due to endophytic compounds can initiate root to shoot signals in plants. Early decrease in transpiration and cell elongation builds up the soluble carbohydrates near leaf sheaths and blades. Severe water deficit results in hydrolysis of carbohydrates providing solutes for osmotic pressure. Water content of endophyte-infected plants are higher due to an increase in solute concentrations in tissues (Bacon et al. 2018).

Another mechanism of tackling drought in plants has increased production of osmolytes. Plants adapt to water stress by accumulating solutes so that osmotic potential is lowered and turgor pressure is maintained during water deficiency. This mechanism is known as osmotic adjustment. Symbiotic plants increase the accumulation of fructose and glucose in the leaf sheath and glucose in blades which contributes partially to the solute pool for osmotic potential adjustment. Biochemically, the endophytes secrete sugar alcohols such as mannitol and arabitol which protect host enzymes and membranes from desiccation (Richardson et al. 1992). Accumulation of amino acids due to nitrogen fertilization in the endophyte-infected leaf sheaths and blades were reported, indicating increased amino acid synthesis by the plant (Lyons et al. 1990). Tall fescue plants infected with endophytes produce loline alkaloids which also act as osmo-regulators in water deficit conditions (Malinowski and Belesky 2000).

Cell size of plants also changes according to the water availability, and this is possible due to cellular elasticity. During water-stressed conditions, endophyte infection in tall fescue plants results in less rigid cell wall and increased turgid weight:dry weight ratio which indicates greater water intake capacity and higher water absorbance ability of the cells. This results in less damage to the cells in stressed conditions. *Acremonium*-infected cool season grasses undergo dormancy during summer for plant survival (Bacon et al. 2018).

Overall evidences suggest that endophyte-associated plants can use water more efficiently in comparison to non-symbiotic plants. During recovery, endophyte-infected plants rapidly uptake water which restores their latent physiological functions (Malinowski and Belesky 2000).

Salinity Stress

Plant tolerance mechanism to salinity stress involves alleviation of antioxidant enzymes like catalases (CAT) or superoxide dismutases (SOD). ROS (reactive oxygen species) scavengers such as ascorbate, glutathione and tocopherol act as antioxidants (Rouhier et al. 2008). These scavengers remove ROS in salt-stressed plants directly or indirectly by generation of glutathione or ascorbate in the cells. Salt tolerance due to increasing levels of antioxidants induced by *Piriformospora indica*, an endophyte of barley was also reported (Baltruschat et al. 2008). In *Penicillium minio-luteum*-treated soybean plants, less damage due to salinity stress was observed compared to untreated plants (Khan et al. 2011).

Temperature Stress

Enhanced survival rate and increased number of leaves has been reported in *Capsicum annuum* plants with resident *Penicillium resedanum* in comparison to the control un-infected plants under heat stress conditions (Khan et al. 2013). The plants inoculated with *P. resedanum* showed reduced electrolyte leakage and lipid peroxidation and had higher nutrient content in terms of proline accumulation. Another example is the tripartite relationship between *Dichanthelium lanuginosum* (endophyte), *Curvularia protuberata* and a *Curvularia* thermal tolerance virus (CThTV). The virus-infected fungus confers tolerance of elevated temperatures to its native host and additionally, to *Solanum lycopersicum,* thus implying a mechanism of temperature tolerance between, i.e. scavenging the damaging ROS generated by the plant defence mechanisms in response to environmental stress (Marquez et al. 2007).

Accumulation of flavonoids which helps to resist the negative effects of heat stress has been observed in endophyte-hosting plants. A study observed such phenomena in cucumber plants associated with *Exophiala* sp. LHL08, a gibberellin-producing strain (Khan et al. 2012). The heat-stressed plants with *Exophiala* had higher plant growth, less pronounced heat induced oxidative stress with enhanced levels of total polyphenol including flavonoids and reduced scavenging glutathione and lipid peroxidation activities.

Heavy Metal Stress

The survival of *Clethra barbinervis* associated with root endophytic fungi *Phialocephala fortinii*, *Rhizodermea veluwensis* and *Rhizoscyphus* sp. in mining sites with high levels of heavy metal concentrations has been reported (Yamaji et al. 2016). These fungi are known to enhance the host plant growth by modulation of metal ion uptake and storage in roots and shoots. However, growth of the same plant was very difficult without endophyte (Yamaji et al. 2016).

The association of a plant with endophytic fungi is, therefore, a supremely beneficial relationship for both the partners, especially, the plant. Plant, being a sessile organism, cannot move away from a stressful environment and tackles the myriad of biotic and abiotic stress conditions via its immune system. Endophytic fungi, acting as a faithful guest, don't just reside inside the plant tissues asymptomatically, but also helps their host to maintain their growth and physiological functions through various mechanisms discussed above. However, these actions are not altruistic, and are rather, an act of protecting its hide for long-term existence.

4.1.2.3 Endophytic Fungi as a Source of Therapeutic Compounds

Endophytic fungi are an enormously diverse, polyphyletic group and their diversity is coupled with the ability to synthesize bioactive compounds. This fact has triggered numerous investigations on perceiving their biosynthetic potential. The landmark study on *N. coenophialum* causing "fescue toxicosis,", in grazing livestock divulged a plethora of toxic alkaloids that acted as the fescue's secondary defence against herbivores (Bacon et al. 1977; Schardl et al. 2004).

Also, the discovery of anti-cancer potential of a bioactive compound, paclitaxel (Taxol) from *Taxomyces andreanae* in 1993, kick started the studies on fungal endophytes for hunting novel, biologically active compounds (Stierle et al. 1993). Current knowledge boasts on its capabilities to synthesize different bioactive metabolites that could be used as drugs for treatment of various diseases and other potential applications in food technology, agriculture and other related industries (Uzma et al. 2018; Godstime et al. 2014; Jalgaonwala et al. 2011; Shukla et al. 2014; Strobel and Daisy 2003). The metabolites produced by these endophytes are classified into different functional categories, for example, benzopyranones, phenolic acids, flavonoids, alkaloids, quinones, chinones, tetralones, xanthones, steroids, saponins, terpenoids, tannins and many others (Kharwar et al. 2011; Kusari and Spiteller 2012; Staniek et al. 2008).

According to one study, bioactive compounds derived from medicinal plants and their endophytes contribute to a major percentage of natural drugs in the market (Passari et al. 2017; Singh and Dubey 2015). Several such secondary metabolites with therapeutic value or potential have been discovered till date; for instance, paclitaxel (also known as Taxol) (Stierle et al. 1993), azadirachtin (Kusari et al. 2012), camptothecin and structural analogues (Puri et al. 2005; Kusari et al. 2009a, b, c, 2011; Shweta et al. 2010), deoxypodophyllotoxin (Kusari et al. 2009a, b, c), podophyllotoxin (Eyberger et al. 2006; Puri et al. 2006), hypericin and emodin (Kusari et al. 2008, 2009a, b, c). Taxol is an alkaloid formed by *Metarhizium anisopliae* residing in Taxus tree and is considered as one of the most potent anti-cancerous compounds reported till date (Jalgaonwala et al. 2011; Zhang et al. 2009; Visalakchi and Muthumary 2010). Lignans extracted from *Podophyllum hexandrum* have been reported with anti-cancerous activity (Lignanların et al. 1995). Resins like etoposide and teniposide isolated from *P. emodi* have also shown strong activity against cancer cells (Lignanların et al. 1995). Fungal-derived terpenoids have antibacterial, antineoplastic, antiviral and gastric stimulating effects (Godstime et al. 2014; Jalgaonwala et al. 2011). Camptothecin derived from *Nothapodytes foetida* shows cytotoxic activity and acts as anti-fungal agent (Joseph and Priya 2011; Zhang et al. 2012). Huperzine A purified from *Huperzia serrata* can act as inhibit cholinesterases (Nair and Padmavathy 2014). Endophyte-derived compounds like catechin, gallic acid, oxacillin, ampicillin and cephalexin can kill bacteria (Akiyama et al. 2001). Helvolic acid isolated from *Cytonaema* sp., produces a triterpenoid with robust antibacterial activity (Kumar et al. 2014).

As discussed above, endophytic fungus can act as a warehouse of novel bioactive compounds serving as a source of drugs with antimicrobial, antidiabetic, anti-arthritic, anti-cancer, immunosuppressant and insect-repellent activities (Mishra et al. 2016a, b; Godstime et al. 2014; Jalgaonwala et al. 2011). Their metabolomic diversity can contribute in the rising war against drug-resistant strains of pathogens (Godstime et al. 2014). However, very few plants have been probed for their whole endophytic diversity. Bioactive metabolites of endophytic-origin, like paclitaxel, hypericin, podophyllotoxin, etc. have both medicinal and agricultural significance (Joseph and Priya 2011; Zhao et al. 2011). Some of these are analogues of phytohormones and essential oils as well (Molina et al. 2012; Nicoletti and Fiorentino 2015; Zhao et al. 2010). A more detailed table on bioactive compounds derived from endophytic fungi with their respective has been provided below in Table 4.1.

4.1.3 Application of Omics-Approaches in Endophytic Fungi

While majority of the discoveries and observations about endophytic fungi-plant mutualism has been based on traditional scientific approaches, integration of omics to this field is recent. The traditional approaches involved field or controlled environment (greenhouse or pot trials), culture, in vitro and in vivo assays, toxicity and bioactivity assessments. While these approaches have contributed to majority of the findings in this field, the relatively new methods are rapidly emerging in omics, especially the high-throughput approaches like whole genome sequencing, RNA-Seq (transcriptomics), metagenomics, two-dimensional fluorescence difference gel electrophoresis (2D-DIGE) for proteomics and microfluidics-coupled mass spectrometry devices for metabolomics. The omics-based insights can be advantageous in various aspects and can be used to create multi-resistant plants which are economically important (Nicolás et al. 2014).

4.1.3.1 Genomic Studies on Endophytic Fungi

There has a sprout of growth in the whole genome sequencing studies since high-throughput methods of genome sequencing have been developed. NCBI started archiving raw sequence data from massively parallel sequencing platforms in SRA (short read archive) database in 2007 and according to a 2012 report, total sequence data stored in SRA has surpassed 100 Terabases of which ~61% is data on human genome (Kodama et al. 2012). Comparatively, the number of genomic studies on fungal endophytes is meagre.

One such species is *Pestalotiopsis fici*, an endophyte of *Camellia sinensis* (tea), which produces unique secondary metabolites like pestaloficiol, pestalofones, pestalodiols, chloropupukeananin, chloropestolides, chloropupukeanone and chloropupukeanolides, implicated in inhibition of HIV-1 replication, tumour cytotoxicity and anti-fungal effect (Wang et al. 2015a, b). Its genome analysis reported the abundance of carbohydrate-active enzymes, especially pectinases and a rich set of secondary metabolite synthesis genes, which depicted its high potential of natural product synthesis and indicated that these may have aided it in adapting to endophytic lifestyle (Wang et al. 2015a, b). Another study used a combinatorial approach of genomics, transcriptomics and metabolomics to understand *A. sarcoides* (endophyte of *Picea mariana)* biology with respect to production of C8 volatiles, a group of uncommon secondary metabolites (Gianoulis et al. 2012). Whole genome sequencing and analysis of *Penicillium aurantiogriseum* enabled identification of genes for paclitaxel biosynthesis which probably evolved independently of the host plant biosynthetic pathway Yang et al. 2014). *L. maculans* (bitrophic-endophyte of Brassica crops) adapts to life within its host plant, with multiple alternating lifestyles—parasitic, saprotrophic, necrotrophic and endophytic. Analysis revealed its genome has an AT-rich blocks which contains effector genes and transposable element families interspersed with point mutations, a genomic defence measure specific to fungal genomes, enabling rapid sequence diversification and adaptation to host-directed constraints (Rouxel et al. 2011). Similar studies on *E. nigrum* genome has shown its capability to sugarcane root system biomass increase and production of metabolites against the host plant pathogens—*Xanthomonas albilineans, Fusarium verticillioides*, *Ceratocystis paradoxa* and *Colletotrichum falcatum*, demonstrating its potential as a biocontrol agent (de Fávaro et al. 2012). The novel draft genome sequence of *Gaeumannomyces* sp., an endophytic fungus isolated from a reed plant (*Phragmites communis*) revealed that it enables production of a number of secondary metabolites which have significant anti-inflammatory activity (Kim et al. 2017;

Lee et al. 2017). Whole genome analysis of some other fungal endophytes of economically important plants including *Piriformospora indica* (Zuccaro et al. 2011), *Rhodotorula graminis* (Firrincieli et al. 2015), *Harpophora oryzae* (Xu et al. 2015) and *Xylona heveae* (Gazis et al. 2016) have provided information of the genes required for their nutrient uptake, colonization, biotic and abiotic stress tolerance, etc. *P. indica* genome was compared with other fungi with different lifestyle to get insights into endophyte-associated bio- and saphro-trophism (Zuccaro et al. 2011). A similar study characterized putative HOG genes from *P. indica* and demonstrated its role in stress tolerance in rice plants via comparative genomics combined with phylogenetic and expression analysis of HOG pathway (Jogawat et al. 2016).

Host genomes with minor genomic differences have also been found to have impact on the results of symbiotic interaction (Rodriguez and Redman 2008). Stress tolerance of the host plant may be enhanced, reduced, or has no effect based on various genotype-specific interactions (Cheplick et al. 1989). Assembly of genome and transcriptome of the perennial ryegrass has provided evidence for ancient horizontal transfer of a gene for β-1,6-glucanase enzyme of fungal origin which encodes fungal cell wall degrading capability (Shinozuka et al. 2017). This enzyme has been implied to confer biotic and abiotic stress tolerance to *Epichloë* species. Orthologues of this gene has also been observed in other plant families like Loliinae and Dactylidinae (Shinozuka et al. 2017). Identification of these type of key endophytic enzymes associated with stress tolerance has proven that such detailed genomic investigations can contribute significantly towards uncovering traits with agronomic importance.

Other than whole genome studies, several metagenomics studies on endophytic fungi have also been reported. The earliest study to report the complete mitochondrial genome (mtgenome) of an endophytic fungal species was that of *E. festucae var. Lolii* (GenBank accession—KF906135) whose role in plant stress has been discussed above (Ekanayake et al. 2013). The second mtgenome report was on *Penicillium polonicum* isolated from *Huperzia serrata* (toothed clubmoss), which produces huperzine-A, a drug used to treat diseases like Alzheimer's and myasthenia gravis (Kang et al. 2016). *P. fici* mtgenome was reported in 2017 (its genomic and agronomic aspects have been discussed previously) (Zhang et al. 2017a, b).

All these studies used a combination of next-generation sequencing and advanced *in silico* tools for assembly, annotation and analysis. Generally, fungal mtgenome display significant variations in gene order, genome size and other commonly studied factors. As these genomes are found to be more than suitable for phylogenetic studies, more emphasis on sequencing these myriads of fungal endophytes associated with plants of therapeutic or industrial relevance must be applied. These data and corresponding analyses can open new avenues to explore ecological niches with extraordinary metabolomic diversity and dynamics of their synthesis in natural and engineered scenarios.

4.1.3.2 Transcriptome Studies on Endophytic Fungi

Large-scale transcriptomic studies on an organism can reveal novel information on the modulations in gene expression associated with fungal-plant mutualism. Transcriptomes can either be studied on the basis of existing whole genome sequences or can be assembled de novo; both approaches are known to capture a significantly accurate snapshot of the gene expression status. Both kind of transcriptomic studies exist on endophytic fungi (Fig. 4.1).

Several experiments were aimed at generating information about the endophyte biology and its relatedness to plant health. The transcriptomic report *Epulorhiza* sp. isolated from roots of an orchid, *Anoectochilus roxburghii,* was one of the earliest reports and observed upregulation of genes involved in pyrimidine metabolism (Li et al. 2012). Another study published in 2012 identified an anti-fungal gene, which was significantly differentially expressed in *E. festucae*-associated with *Festuca rubra* and was unique to the Epichloë genus (Ambrose and Belanger 2012). They hypothesized that the corresponding gene product which was secretary in nature might have a role in conferring "disease-resistance", a characteristic trait of endophyte-infected fescue.

Other than biology, identification of genes involved in secondary metabolite production with potential industrial applications was also probed via transcriptomics. A study identified the genes involved in cellulose biodegradation and possibly, in bio-fuel production in *Ascocoryne sarcoides* via combined genomics, transcriptomics and metabolomics, constituting one of the most thoroughly studied endophytic fungal species (Gianoulis et al. 2012). Another study on *P. fici* identifying several potential gene clusters involved in endophyte-derived bioactive compound biosynthesis has been discussed in the previous section (Wang et al. 2015a, b).

Several other studies have focused on understanding how endophytic fungi-plant stress tolerance might trigger transcriptional changes in the endophyte genome. Multiple glutathione-S-transferase genes were identified in response to different heavy metal exposures from a metal-tolerant endophyte, *Exophiala pisciphila*'s transcriptome (Shen et al. 2015).

Overall, these reports of transcriptomic analysis, generated using microarray, SAGE (serial analysis of gene expression) and more recent forms of second-generation sequencing, have provided us valuable sequence information with deduced putative functions inferred from homology, allowing us to predict traits and metabolic pathways important for the endophytic lifestyle ranging from nutrient acquisition and storage, quorum sensing, reactive species detoxification and biosynthesis of novel compounds.

4.1.3.3 Proteome Studies on Endophytic Fungi

Other than the transcriptomic generated putative protein sequence data, stand-alone experiments on proteomic profiling on fungal endophytes are also available in literature but minimally. There have been two proteome-based studies on *P. indica* from Sebacinales order, a ubiquitous endophyte colonizing roots of almost all known terrestrial plants. This species is known to promote plant growth during normal conditions as well as under duress. The effect of presence of this endophyte during different levels of drought stress in barley plants has been demonstrated (Ghabooli et al. 2013). *P. indica* promoted biomass increase in roots and shoots under both water-deficit and well-watered conditions and subsequent mass spectrometry analysis identified differentially accumulated proteins involved in various pathways like photosynthesis, reactive species defence and energy transport which were hypothesized to be co-related to *P. indica* mediated systemic drought resistance. Another study by published a standardized method on efficient cellular protein extraction from endophytic fungi for generating two-dimensional gel electrophoresis. They applied this methodology to obtain quantitative proteome data on *P. indica* (Yadava et al. 2015, 130). The study was also able to compare protein fingerprints isolated from their study for unveiling the role of endophytes under biotic stress. There are barely any high-throughput studies on transporter proteins in plant–endophytic fungi interactions. One study reportedly emphasized on the role of phosphorus transporter protein in *P. indica* involved in phosphorus homeostasis in maize plants (Yadav et al. 2010).

These studies contribute to the spectrum of limited information on plant-endophyte relationships and further combinatorial studies involving a four-way approach (genome, transcriptome, proteome and metabolome) will yield even more clear insights into the same.

4.1.3.4 Metabolome Studies on Endophytic Fungi

Current research has been focusing mainly on the endophytic fungi to exploit its unique capability of producing rare and therapeutic metabolites. Most of the large-scale proteomic studies are performed using LC (liquid chromatography) or GC (gas chromatography)-MS (mass spectrometry) based multivariate analysis. Comparative metabolomic study on *Aspergillus terreus,* an endophyte of *Opuntia ficus indica*, cultured under 11 different culture conditions was reported (Adpressa and Loesgen 2016). They identified a novel compound 7-desmethylcitreoviridin in addition to 16 other known fungal metabolites and demonstrated that their LC-MS based analytical method can be conveniently used to track environment-modulated variations in metabolite production and thereby, for discovery of novel bioactive compounds. Another study reported vincamine, a nootropic drug and its analogues from the fungal endophyte *Geomyces* sp. of *Nerium indicum* via LC-MS based metabolomics, establishing the endophyte as an alternative source of these compounds (Na et al. 2016).

Some metabolomic studies have added new directions to the dynamics of plant–fungi interactions. Global metabolomic analysis via GC-MS analysis was utilized to show that reactive species generation in host plant is repressed by *P. indica* to promote more favourable conditions of symbiotic interaction (Hua et al. 2017). Presence of the endophyte may prime the host plant for biotic or abiotic stress by significantly inducing the expression of gamma-aminobutyric acid (GABA) at its roots, as was observed in Chinese cabbage during the experiment.

With the advantages provided by the multiple modes of biological analyses, it will be possible to establish the evolutionary origins and diversity of mutualistic endophytic fungi and predict different outcomes of plant–fungal interactions based on genomic or environmental differences. Benefits from this information can be translated for benefit of agricultural practices, especially, tackling stressed conditions in an environment-friendly way, without resorting to chemical means.

4.1.3.5 Metagenomic Studies on Endophytic Fungi

Analysis of microbial communities through high-throughput metagenomic technologies has enabled better understanding of structure, functions and community dynamics. However, these techniques have been applied for fungal endophyte screening since late 2013 only.

Early studies on fungal species were mostly culture-based and involved ITS (internal transcribed spacer) region sequencing for limited number of samples. This approach was applied for molecular identification of endophytic fungi residing in a medicinal tree, *Aquilaria malaccensis* and majority of the species discovered on it, belonged to the phylum Ascomycota (Premalatha and Kalra 2013). A modified protocol for ARISA (automated ribosomal intergenic spacer analysis) for diagnostic assessment of myco-diversity in environmental samples was developed by targeting polymorphisms in fungal ITS1 rDNA length (Weig et al. 2013).

A study published in late 2013 reported for the first time, the use of next-generation sequencing for large-scale analysis of ITS region of resident endophytic species in *Eucalyptus grandis* from South Africa (Kemler et al. 2013). Another study applied Illumina sequencing of fungi associated with roots of a monodominant forest to reveal a network representing co-occurrence patterns of symbiotic associations (Toju et al. 2016).

A close observation of the publications on endophytic fungi from the earliest available literature till March 2018 showed that the number of traditional approach-based studies on endophytic fungi is roughly ten folds greater than that of the omics-based studies. Most of these studies comprise of *in situ* observations, field and potted experimentations and novel fungal endophyte reports (Fig. 4.2).

All the approaches discussed above were applied to peruse the biology of endophytic fungi and how they promote plant health during normal as well as stress conditions. These next-generation tools with the juxtaposition of systems biology have allowed exploration of these endophytic residents and simultaneously explore their interactions with the host. At the same time, identification of biosynthetic pathways for novel metabolites from data mining has become very easy. High-throughput studies have facilitated unbiased identification of metabolites across genome and transcriptomes of newly discovered endophytes as well as acquisition of novel source of already commercialized natural bioactive metabolites. Genome reports of filamentous fungi like *Aspergillus* spp. has revealed that they have a greater share of secondary metabolite biosynthetic gene clusters than anticipated (Osbourn 2010).

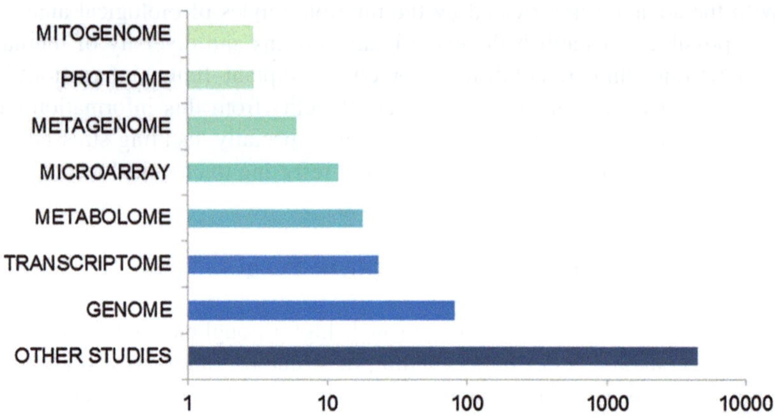

Fig. 4.2 Distribution of total publications for the term "endophytic fungi" in Scopus (March 2018) based on the "type of study" reported. All the titles of the "type of studies" are self-explanatory other than "other studies" which reflects the remaining set of publications other than the ones listed above, and it mostly includes *in situ*/greenhouse experiments or reviews

A. nidulans can produce multiple polyketides, non-ribosomal peptides and indole alkaloids (Brakhage et al. 2008; Rank et al. 2010). While these genes are silent in test conditions but can be stimulated using approaches like knock out, promoter exchange or even simulation of existing ecological niche (Vasundhara et al. 2016).

4.1.4 Web-Resources on Endophytic Fungi

The voluminous amount of genomic, mito- and meta-genomic, transcriptomic, proteomic and metabolomic data generated by the studies discussed above are hosted in multiple web-servers and databases of the cyber space. Most of the data are readily available in the various dedicated databases of NCBI, Ensembl and UniProt. The following table lists the resources other than these three popular databases (Table 4.2).

4.1.5 Conclusion and Perspectives

Multi-disciplinary approaches have yielded a deeper understanding of the mutualistic relation of endophytic fungi with plants. However, a careful analysis of the studies discussed above, reveals an uneven distribution of the sampled species that are currently being probed. Most of the hosts for endophytes are grass and as more endophytic species associated with other plant species have been reported, a shift of focus towards these species and their interaction dynamics with their respective host plants are necessary.

Table 4.2 Databases or web portals hosting data on endophytic fungi

Database name	Data type	Description	Link	Inception year	Reference
BCCM/IHEM: MALDI-TOF MS database for filamentous fungi	MALDI-TOF MS reference spectra	An online web application providing access to MALDI-TOF MS reference spectra database for filamentous fungi	http://bccm.belspo.be/about-us/bccm-ihem	2017	Not published
Fungal Barcoding Database	DNA Barcode	Provides updated information on fungal barcoding	http://www.fungalbarcoding.org	Not available	Not published
FungiDB	Genomic sequence	Integrates sequence and annotation data on whole genome from experimental and environmental isolates	http://fungidb.org	2012	Stajich et al. (2012)
Fungi From India	Taxonomic information	Collection of taxonomic databases of Indian fungi	http://www.fungifromindia.com/	2012	Ranadive et al. (2017)
Fungi Phosphorylation Database (FPD)	Phosphoproteins and phosphosites	Comprises high-confidence in vivo phosphosites identified by MS-based proteomics in fungi species	http://bis.zju.edu.cn/FPD/	2017	Bai et al. (2017)
Genome projects at University of Kentucky	Whole genome sequences	Collection of whole genome projects carried out on endophytic fungi species at the University of Kentucky	http://www.endophyte.uky.edu/	2007	Not published
Maarj*AM*	DNA sequence	Contains glomeromycota DNA sequence data that originate from ecological studies based on "environmental samples" or taxonomic investigations based on cultured arbuscular mycorrhizal fungi	https://maarjam.botany.ut.ee/	2010	Õpik et al. (2010)
Mycobank	Mycological nomenclatural novelties and associated data	Pairwise sequence alignments and polyphasic identifications of fungi and yeasts against curated references databases	http://www.mycobank.org	2004	Crous et al. (2004)

(continued)

Table 4.2 (continued)

Database name	Data type	Description	Link	Inception year	Reference
Mycology Collections Portal (MyCoPortal)	Geo-referenced species checklists, distribution maps, interactive identification keys and digital images	A suite of user-friendly, web-based data access to diverse information on fungal diversity	http://mycoportal.org	2011	Miller and Bates (2017)
MycoDB	Hierarchically arranged metadata	Ecological meta-analysis database	http://www.mycodb.fr/	2016	Chaudhary et al. (2016)
New Endophyte Database	Morphological characteristics	Portal on sequence and species identification based on morphotypes	https://www.mpgranch.com/dispatches/new-endophyte-database	2013	Not published
Q-bank Fungi database	DNA sequence data	Contains barcodes and morphological, phenotypical and ecological data of more than 725 species that are of relevance to mycological phytopathology	http://www.q-bank.eu/fungi/	2013	Bonants et al. (2013)

Next-generation sequencing technologies have undergone drastic advancement in terms of data accuracy as the third generation of sequencers can provide more accurate sequences with better length. While some excellent examples of multi-dimensional studies on fungal endophytes exist, our general observation is that a few genomes or transcriptomics have low coverage with moderate sequencing depths. These assemblies can be improved via re-sequencing and analyses.

Also, more focused and combinatorial databases can be developed using the complementary information generated from the omics-based studies discussed above. Associating these data with systems biology approaches can be applied for predictive modelling of endophyte-mediated beneficial processes and complexities of interactions with host plants. Similar approaches should also be applied to study the dynamics of fungal endophytes with the plant microbes.

Finally, the knowledge garnered till date on plant-associated fungal endophytes can be utilized in developing plant probiotics, a more environment-friendly and sustainable approach to ensure plant health without jeopardizing the quality of the food products generated from them or other downstream processes related to it. The scope for exploitation of these species is vast, especially in the fields of drug discovery and agriculture.

References

Adpressa DA, Loesgen S (2016) Bioprospecting chemical diversity and bioactivity in a marine derived *Aspergillus terreus*. Chem Biodivers 13(2):253–259

Akiyama H, Fujii K, Yamasaki O, Oono T, Iwatsuki K (2001) Antibacterial action of several tannins against *Staphylococcus aureus*. J Antimicrob Chemother 48(4):487–491

Ambrose KV, Belanger FC (2012) SOLiD-SAGE of endophyte-infected red fescue reveals numerous effects on host transcriptome and an abundance of highly expressed fungal secreted proteins. PLoS One 7(12):e53214

Arnold AE, Mejía LC, Kyllo D, Rojas EI, Maynard Z, Robbins N et al (2003) Fungal endophytes limit pathogen damage in a tropical tree. Proc Natl Acad Sci U S A 100(26):15649–15654. http://www.ncbi.nlm.nih.gov/pubmed/14671327

Artanti N, Tachibana S, Kardono L, Sukiman H (2011) Screening of endophytic fungi having ability for antioxidative and α-glucosidase inhibitor activities isolated from *Taxus sumatrana*. Pak J Biol Sci 14(22):1019–1023

Aschehoug ET, Callaway RM, Newcombe G, Tharayil N, Chen S (2014) Fungal endophyte increases the allelopathic effects of an invasive forb. Oecologia 175(1):285–291. https://doi.org/10.1007/s00442-014-2891-0

Bacon CW, Porter JK, Robbins JD, Luttrell ES (1977) *Epichloë typhina* from toxic tall fescue grasses. Appl Environ Microbiol 34(5):576–581

Bacon CW, White JF, James F (2018) Biotechnology of endophytic fungi of grasses. CRC, Boca Raton

Bai Y, Chen B, Li M, Zhou Y, Ren S, Xu Q et al (2017) FPD: a comprehensive phosphorylation database in fungi. Fungal Biol 121(10):869–875

Baltruschat H, Fodor J, Harrach BD, Niemczyk E, Barna B, Gullner G et al (2008) Salt tolerance of barley induced by the root endophyte *Piriformospora indica* is associated with a strong increase in antioxidants. New Phytol 180(2):501–510. http://www.ncbi.nlm.nih.gov/pubmed/18681935

Bashyal BP, Wellensiek BP, Ramakrishnan R, Faeth SH, Ahmad N, Gunatilaka AAL (2014) Altertoxins with potent anti-HIV activity from *Alternaria tenuissima* QUE1Se, a fungal endophyte of *Quercus emoryi*. Bioorg Med Chem 22(21):6112–6116

Bayat F, Mirlohi A, Khodambashi M (2009) Effects of endophytic fungi on some drought tolerance mechanisms of tall fescue in a hydroponics culture. Russ J Plant Physiol 56(4):510–516. https://doi.org/10.1134/S1021443709040104

Bonants P, Edema M, Robert V (2013) Q-bank, a database with information for identification of plant quarantine plant pest and diseases. EPPO Bull 43(2):211–215

Brakhage AA, Schuemann J, Bergmann S, Scherlach K, Schroeckh V, Hertweck C (2008) Activation of fungal silent gene clusters: a new avenue to drug discovery. In: Petersen F, Amstutz R (eds) Natural compounds as drugs. Progress in drug research, vol 66. Birkhäuser, Basel, pp 1–12

Channabasava, Govindappa M (2014) First report of anticancer agent, lapachol producing endophyte, *Aspergillus niger* of *Tabebuia argentea* and its *in vitro* cytotoxicity assays. Bangladesh J Pharmacol 9(1):129–139

Chaudhary VB, Rúa MA, Antoninka A, Bever JD, Cannon J, Craig A et al (2016) MycoDB, a global database of plant response to mycorrhizal fungi. Sci Data 3:160028

Cheplick GP, Clay K, Marks S (1989) Interactions between infection by endophytic fungi and nutrient limitation in the grasses *Lolium perenne* and *Festuca arundinacea*. New Phytol 111(1):89–97. https://doi.org/10.1111/j.1469-8137.1989.tb04222.x

Clay K (1987) Effects of fungal endophytes on the seed and seedling biology of *Lolium perenne* and *Festuca arundinacea*. Oecologia 73(3):358–362. https://doi.org/10.1007/BF00385251

Clay K, Holah H (1999) Fungal endophyte symbiosis and plant diversity in successional fields. Forensic Sci Int 285(5434):1742–1745. http://www.ncbi.nlm.nih.gov/pubmed/10481011

Crous PW, Gams W, Stalpers JA, Robert V, Stegehuis G (2004) MycoBank: an online initiative to launch mycology into the 21st century. Stud Mycol 50:19–22

Cui J, Guo S, Xiao P (2011) Antitumor and antimicrobial activities of endophytic fungi from medicinal parts of *Aquilaria sinensis*. J Zhejiang Univ Sci B 12(5):385–392

de Fávaro LCL, de Sebastianes FLS, Araújo WL (2012) *Epicoccum nigrum* P16, a sugarcane endophyte, produces antifungal compounds and induces root growth. PLoS One 7(6):e36826

Desale MG, Bodhankar MG (2013) Antimicrobial activity of endophytic fungi isolated from V*itex negundo* Linn. Int J Curr Microbiol App Sci 2(12):389–395

Deshmukh SK, Mishra PD, Kulkarni-Almeida A, Verekar S, Sahoo MR, Periyasamy G et al (2009) Anti-inflammatory and anticancer activity of ergoflavin isolated from an endophytic fungus. Chem Biodivers 6(5):784–789

Dhankhar S, Dhankhar S, Parkash YJ (2013) Investigations towards new antidiabetic drugs from fungal endophytes associated with salvadora oleoides decne. Med Chem (Los Angeles) 9:624–632

Du F-Y, Li X, Li X-M, Zhu L-W, Wang B-G (2017) Indolediketopiperazine alkaloids from *Eurotium cristatum* en-220, an endophytic fungus isolated from the marine alga *Sargassum thunbergii*. Mar Drugs 15(12):24

Ekanayake PN, Rabinovich M, Guthridge KM, Spangenberg GC, Forster JW, Sawbridge TI (2013) Phylogenomics of fescue grass-derived fungal endophytes based on selected nuclear genes and the mitochondrial gene complement. BMC Evol Biol 13(1):270

Eyberger AL, Dondapati R, Porter JR (2006) Endophyte fungal isolates from *Podophyllumpeltatum* produce podophyllotoxin. J Nat Prod 69(8):1121–1124

Firrincieli A, Otillar R, Salamov A, Schmutz J, Khan Z, Redman RS et al (2015) Genome sequence of the plant growth promoting endophytic yeast *Rhodotorula graminis* WP1. Front Microbiol 6:978

Gangadevi V, Muthumary J (2009) Taxol production by *Pestalotiopsis terminaliae*, an endophytic fungus of *Terminalia arjuna* (arjun tree). Biotechnol Appl Biochem 52(1):9–15

Gazis R, Kuo A, Riley R, LaButti K, Lipzen A, Lin J et al (2016) The genome of *Xylona heveae* provides a window into fungal endophytism. Fungal Biol 120(1):26–42

Ghabooli M, Khatabi B, Ahmadi FS, Sepehri M, Mirzaei M, Amirkhani A et al (2013) Proteomics study reveals the molecular mechanisms underlying water stress tolerance induced by *Piriformospora indica* in barley. J Proteome 94:289–301

Gianoulis TA, Griffin MA, Spakowicz DJ, Dunican BF, Alpha CJ, Sboner A et al (2012) Genomic analysis of the hydrocarbon-producing, cellulolytic, endophytic fungus *Ascocoryne sarcoides*. PLoS Genet 8(3):e1002558

Giridharan P, Verekar SA, Khanna A, Mishra PD, Deshmukh SK (2012) Anticancer activity of sclerotiorin, isolated from an endophytic fungus *Cephalotheca faveolata* Yaguchi, Nishim. and Udagawa. Indian J Exp Biol 50:464–468

Godstime O, Felix E, Jewo Augustina O, Christopher E (2014) Mechanisms of antimicrobial actions of phytochemicals against enteric pathogens – a review. J Pharm Chem Biol Sci 2(22):77–85

Govindappa M, Sadananda TS, Channabasava, Ramachandra YL, Chandrappa CP, Padmalatha RS et al (2015) *In vitro* and *in vivo* antidiabetic activity of lectin (N-acetylgalactosamine, 64 kDa) isolated from endophytic fungi, *Alternaria* species from *Viscum album* on alloxan induced diabetic rats. Integr Obes Diabetes 1(1):11–19

Gunatilaka AAL (2006) Natural products from plant-associated microorganisms: distribution, structural diversity, bioactivity, and implications of their occurrence. J Nat Prod 69(3):509–526. http://www.ncbi.nlm.nih.gov/pubmed/16562864

Guo B, Dai J-R, Ng S, Huang Y, Leong C, Ong W et al (2000) Cytonic acids A and B: novel tridepside inhibitors of hCMV protease from the endophytic fungus *Cytonaema* species. J Nat Prod 63(5):602–604

Guo Z, Gai C, Cai C, Chen L, Liu S, Zeng Y et al (2017) Metabolites with insecticidal activity from *Aspergillus fumigatus* JRJ111048 isolated from mangrove plant acrostichum specioum endemic to hainan island. Mar Drugs 15(12):381

Hua MDS, Senthil Kumar R, Shyur LF, Cheng Y Bin, Tian Z, Oelmüller R, et al (2017) Metabolomic compounds identified in *Piriformospora indica*-colonized chinese cabbage roots delineate symbiotic functions of the interaction. Sci Rep 7(1)

Jalgaonwala RE, Mohite BV, Mahajan RT (2011) A review: natural products from plant associated endophytic fungi. J Microbiol Biotechnol Res 1(2):21–32

Jogawat A, Vadassery J, Verma N, Oelmüller R, Dua M, Nevo E et al (2016) PiHOG1, a stress regulator MAP kinase from the root endophyte fungus *Piriformospora indica*, confers salinity stress tolerance in rice plants. Sci Rep 6(1):36765

Johnson LJ, Johnson RD, Schardl CL, Panaccione DG (2003) Identification of differentially expressed genes in the mutualistic association of tall fescue with *Neotyphodium coenophialum*. Physiol Mol Plant Pathol 63(6):305–317. https://www.sciencedirect.com/science/article/pii/S0885576504000384

Joseph B, Priya RM (2011) Bioactive compounds from endophytes and their potential in pharmaceutical effect: a review. Am J Biochem Mol Biol 1(3):291–309

Kamal N, Viegelmann C, Clements C, Edrada-Ebel R (2016) Metabolomics-guided isolation of anti-trypanosomal metabolites from the endophytic fungus *Lasiodiplodia theobromae*. Planta Med 83(6):565–573

Kang X, Liu C, Liu D, Zeng L, Shi Q, Qian K et al (2016) The complete mitochondrial genome of huperzine A-producing endophytic fungus *Penicillium polonicum*. Mitochondrial DNA B 1(1):202–203

Kaul S, Sharma T, Dhar MK (2016) Omics tools for better understanding the plant-endophyte interactions. Front Plant Sci 7:955. http://www.ncbi.nlm.nih.gov/pubmed/27446181

Kaur G, Asthir B (2015) Proline: a key player in plant abiotic stress tolerance. Biol Plant [Internet] 59(4):609–619. https://doi.org/10.1007/s10535-015-0549-3

Kemler M, Garnas J, Wingfield MJ, Gryzenhout M, Pillay K-A, Slippers B (2013) Ion torrent PGM as tool for fungal community analysis: a case study of endophytes in *Eucalyptus grandis* reveals high taxonomic diversity. PLoS One 8(12):e81718

Khan AL, Hamayun M, Ahmad N, Hussain J, Kang S-M, Kim Y-H et al (2011) Salinity stress resistance offered by endophytic fungal interaction between *Penicillium minioluteum* LHL09 and *Glycine max.* L. J Microbiol Biotechnol 21(9):893–902. http://www.ncbi.nlm.nih.gov/pubmed/21952365

Khan AL, Hamayun M, Kang S-M, Kim Y-H, Jung H-Y, Lee J-H et al (2012) Endophytic fungal association via gibberellins and indole acetic acid can improve plant growth under abiotic stress: an example of *Paecilomyces formosus* LHL10. BMC Microbiol 12:3. http://www.ncbi.nlm.nih.gov/pubmed/22235902

Khan AL, Waqas M, Hamayun M, Al-Harrasi A, Al-Rawahi A, Lee I-J (2013) Co-synergism of endophyte *Penicillium resedanum* LK6 with salicylic acid helped *Capsicum annuum* in biomass recovery and osmotic stress mitigation. BMC Microbiol 13(1):51. http://www.ncbi.nlm.nih.gov/pubmed/23452409

Kharwar RN, Mishra A, Gond SK, Stierle A, Stierle D (2011) Anticancer compounds derived from fungal endophytes: their importance and future challenges. Nat Prod Rep 28(7):1208

Kim H-Y, Choi GJ, Lee HB, Lee S-W, Lim HK, Jang KS et al (2007) Some fungal endophytes from vegetable crops and their anti-oomycete activities against tomato late blight. Lett Appl Microbiol 44(3):332–337. http://www.ncbi.nlm.nih.gov/pubmed/17309513

Kim JA, Jeon J, Kim K-T, Choi G, Park S-Y, Lee H-J et al (2017) Draft genome sequence of an endophytic fungus,Gaeumannomyces sp. strain JS-464, isolated from a reed plant, *Phragmites communis*. Genome Announc 5(31):e00734-17

Kim TY, Jang JY, Yu NH, Chi WJ, Bae C-H, Yeo JH et al (2018) Nematicidal activity of grammicin produced by *Xylaria grammica* KCTC 13121BP against *Meloidogyne incognita*. Pest Manag Sci 74(2):384–391

Kodama Y, Shumway M, Leinonen R, International Nucleotide Sequence Database Collaboration (2012) The sequence read archive: Explosive growth of sequencing data. Nucleic Acids Res 40(D1):D54–D56

Kumar S, Aharwal RP, Shukla H, Rajak RC, Sandhu SS (2014) Endophytic fungi: as a source of antimicrobials bioactive compounds. World J Pharm Pharm Sci 3(2):1179–1197

Kusari S, Spiteller M (2012) Metabolomics of endophytic fungi producing associated plant secondary metabolites: progress, challenges and opportunities. In: Roessner U (ed) Metabolomics. InTech, London

Kusari S, Lamshöft M, Zühlke S, Spiteller M (2008) An endophytic fungus from *Hypericum perforatum* that produces hypericin. J Nat Prod 71(2):159–162

Kusari S, Lamshöft M, Spiteller M (2009a) *Aspergillus fumigatus* Fresenius, an endophytic fungus from *Juniperus communis* L. Horstmann as a novel source of the anticancer pro-drug deoxypodophyllotoxin. J Appl Microbiol 107(3):1019–1030

Kusari S, Zühlke S, Košuth J, Čellárová E, Spiteller M (2009b) Light-independent metabolomics of endophytic *Thielavia subthermophila* provides insight into microbial hypericin biosynthesis. J Nat Prod 72(10):1825–1835

Kusari S, Zühlke S, Spiteller M (2009c) An endophytic fungus from *Camptotheca acuminata* that produces camptothecin and analogues. J Nat Prod 72(1):2–7

Kusari S, Zühlke S, Spiteller M (2011) Effect of artificial reconstitution of the interaction between the plant *Camptotheca acuminata* and the fungal endophyte *Fusarium solani* on camptothecin biosynthesis. J Nat Prod 74(4):764–775

Kusari S, Verma VC, Lamshoeft M, Spiteller M (2012) An endophytic fungus from *Azadirachta indica* A. Juss. that produces azadirachtin. World J Microbiol Biotechnol 28(3):1287–1294

Lata R, Chowdhury S, Gond SK, White JF (2018) Induction of abiotic stress tolerance in plants by endophytic microbes. Lett Appl Microbiol 66(4):268–276. https://doi.org/10.1111/lam.12855

Lee C, Kim S, Li W, Bang S, Lee H, Lee HJ et al (2017) Bioactive secondary metabolites produced by an endophytic fungus *Gaeumannomyces* sp. JS0464 from a maritime halophyte *Phragmites communis*. J Antibiot (Tokyo) 70(6):737–742

Lehtonen PT, Helander M, Siddiqui SA, Lehto K, Saikkonen K (2006) Endophytic fungus decreases plant virus infections in meadow ryegrass *(Lolium pratense)*. Biol Lett 2(4):620–623. http://www.ncbi.nlm.nih.gov/pubmed/17148304

Li B, Tang M, Tang K, Zhao L, Guo S (2012) Screening for differentially expressed genes in *Anoectochilus roxburghii* (Orchidaceae) during symbiosis with the mycorrhizal fungus *Epulorhiza* sp. Sci China Life Sci 55(2):164–171

Lignanların A, Ve Bitkiler Ö, Dağılımı A, Konuklugil B (1995) The importance of aryltetralın (podophyllum) lignans and their distribution in the plant kingdom. J Fac Pharm Ankara 2422(13):301–302

Lyons PC, Evans JJ, Bacon CW (1990) Effects of the fungal endophyte *Acremonium coenophialum* on nitrogen accumulation and metabolism in tall fescue. Plant Physiol 92(3):726–732. http://www.ncbi.nlm.nih.gov/pubmed/16667341

Malinowski DP, Belesky DP (2000) Adaptations of endophyte-infected cool-season grasses to environmental stresses. Crop Sci 40(4):923. https://www.crops.org/publications/cs/abstracts/40/4/923

Malinowski DP, Brauer DK, Belesky DP (1999) The endophyte *Neotyphodium coenophialum* affects root morphology of tall fescue grown under phosphorus deficiency. J Agron Crop Sci 183(1):53–60. https://doi.org/10.1046/j.1439-037x.1999.00321.x

Marquez LM, Redman RS, Rodriguez RJ, Roossinck MJ (2007) A virus in a fungus in a plant: three-way symbiosis required for thermal tolerance. Science. 315(5811):513–515. http://www.ncbi.nlm.nih.gov/pubmed/17255511

Miller AN, Bates ST (2017) The Mycology Collections Portal (MyCoPortal). IMA Fungus 8(2):65–66

Mishra VK, Singh G, Passari AK, Yadav MK, Gupta VK, Singh BP (2016a) Distribution and antimicrobial potential of endophytic fungi associated with ethnomedicinal plant Melastoma malabathricum L. J Environ Biol 37(2):229–237

Mishra VK, Passari AK, Singh BP (2016b) In vitro antimycotic and biosynthetic potential of fungal endophytes associated with Schima Wallichii. In: Kumar P et al (eds) Current trends in disease diagnostics. Springer International, Basel, pp 367–381

Mishra VK, Passari AK, Chandra P, Leo VV, Kumar B, Gupta VK, Singh BP (2017a) Determination and production of antimicrobial compounds by Aspergillus clavatonanicus strain MJ31, an endophytic fungus from Mirabilis jalapa L. using UPLC-ESI-MS/MS and TD GC-MS. PLoS One 12(10):1–24. https://doi.org/10.1371/journal.pone.0186234

Mishra VK, Passari AK, Leo VV, Singh BP (2017b) Molecular diversity and detection of endophytic fungi based on their antimicrobial biosynthetic genes. In: Singh BP, Gupta VK (eds) Molecular markers in mycology, fungal biology. Springer International, Basel, pp 1–35. https://doi.org/10.1007/978-3-319-34106-4_1

Molina G, Pimentel MR, Bertucci TCP, Pastore GM (2012) Application of fungal endophytes in biotechnological processes. Chem Eng Trans 27:289–294

Müller CB, Krauss J (2005) Symbiosis between grasses and asexual fungal endophytes. Curr Opin Plant Biol 8(4):450–456. https://www.sciencedirect.com/science/article/pii/S1369526605000683

Na R, Jiajia L, Dongliang Y, Yingzi P, Juan H, Xiong L et al (2016) Indentification of vincamine indole alkaloids producing endophytic fungi isolated from *Nerium indicum,* Apocynaceae. Microbiol Res 192:114–121

Nair DN, Padmavathy S (2014) Impact of endophytic microorganisms on plants, environment and humans. Sci World J 2014:250693

Nicolás C, Hermosa R, Rubio B, Mukherjee PK, Monte E (2014) Trichoderma genes in plants for stress tolerance- status and prospects. Plant Sci 228:71–78

Nicoletti R, Fiorentino A (2015) Plant bioactive metabolites and drugs produced by endophytic fungi of spermatophyta. Agriculture 5(4):918–970

Öpik M, Vanatoa A, Vanatoa E, Moora M, Davison J, Kalwij JM et al (2010) The online database MaarjAM reveals global and ecosystemic distribution patterns in arbuscular mycorrhizal fungi (Glomeromycota). New Phytol 188(1):223–241

Orr SP, Rudgers JA, Clay K (2005) Invasive plants can inhibit native tree seedlings: testing potential allelopathic mechanisms. Plant Ecol 181(2):153–165. https://doi.org/10.1007/s11258-005-5698-6

Osbourn A (2010) Secondary metabolic gene clusters: evolutionary toolkits for chemical innovation. Trends Genet 26(10):449–457

Pandi M, Manikandan R, Muthumary J (2010) Anticancer activity of fungal taxol derived from *Botryodiplodia theobromae* Pat., an endophytic fungus, against 7, 12 dimethyl benz(a)anthracene (DMBA)-induced mammary gland carcinogenesis in Sprague dawley rats. Biomed Pharmacother 64(1):48–53

Passari AK, Mishra VK, Singh G, Singh P, Kumar B, Gupta VK, Sharma RK, Saikia R, Donovan A, Singh BP (2017) Insights into the functionality of endophytic actinobacteria with a focus on their biosynthetic potential and secondary metabolites production. Sci Rep 7(11809):1–17. https://doi.org/10.1038/s41598-017-12235-4

Philippe G (2016) Lolitrem B and indole diterpene alkaloids produced by endophytic fungi of the genus Epichloë and their toxic effects in livestock. Toxins (Basel) 8(2):47. http://www.mdpi.com/2072-6651/8/2/47

Phongpaichit S, Nikom J, Rungjindamai N, Sakayaroj J, Hutadilok-Towatana N, Rukachaisirikul V et al (2007) Biological activities of extracts from endophytic fungi isolated from *Garcinia* plants. FEMS Immunol Med Microbiol 51(3):517–525

Powthong P, Jantrapanukorn B, Thongmee A, Suntornthiticharoen P (2013) Screening of antimicrobial activities of the endophytic fungi isolated from *Sesbania grandiflora* (L.) Pers. J Agric Sci Technol 15:1513–1522

Prabavathy D, Valli Nachiyar C (2013) Antimicrobial and antidiabetic activity of an endophytic fungi isolated from *Adathoda beddomei*. Int J Pharm Pharm Sci 5(3):780–783

Premalatha K, Kalra A (2013) Molecular phylogenetic identification of endophytic fungi isolated from resinous and healthy wood of *Aquilaria malaccensis*, a red listed and highly exploited medicinal tree. Fungal Ecol 6(3):205–211

Prestidge RA (1993) Causes and control of perennial ryegrass staggers in New Zealand. Agric Ecosyst Environ 44(1–4):283–300. https://www.sciencedirect.com/science/article/pii/016788099390051P

Pretsch A, Nagl M, Schwendinger K, Kreiseder B, Wiederstein M, Pretsch D et al (2014) Antimicrobial and anti-inflammatory activities of endophytic fungi *Talaromyces wortmannii* extracts against acne-inducing bacteria. PLoS One 9(6):e97929

Puri SC, Verma V, Amna T, Qazi GN, Spiteller M (2005) An endophytic fungus from *Nothapodytes foetida* that produces camptothecin. J Nat Prod 68(12):1717–1719

Puri SC, Nazir A, Chawla R, Arora R, Riyaz-ul-Hasan S, Amna T et al (2006) The endophytic fungus *Trametes hirsuta* as a novel alternative source of podophyllotoxin and related aryl tetralin lignans. J Biotechnol 122(4):494–510

Ramos HP, Braun GH, Pupo MT, Said S (2010) Antimicrobial activity from endophytic fungi Arthrinium state of *Apiospora montagnei* Sacc. and *Papulaspora immersa*. Braz Arch Biol Technol 53(3):629–632

Ranadive K, Jagtap N, Khare H (2017) Fungifromindia: the first online initiative to document fungi from India. IMA Fungus 8(2):67–69. http://www.imafungus.org/Issue/82/MycoLens.pdf

Rank C, Larsen TO, Frisvad JC (2010) Functional systems biology of *Aspergillus*. In: Aspergillus molecular biology and genomics. p 173–198

Redondo-Gómez S (2013) Abiotic and biotic stress tolerance in plants. In: Rout GR, Das AB (eds) Molecular stress physiology of plants. Springer India, New Delhi, pp 1–20. https://doi.org/10.1007/978-81-322-0807-5_1

Richardson MD, Chapman GW, Hoveland CS, Bacon CW (1992) Sugar alcohols in endophyte-infected tall fescue under drought. Crop Sci 32(4):1060. https://www.crops.org/publications/cs/abstracts/32/4/CS0320041060

Rodriguez R, Redman R (2008) More than 400 million years of evolution and some plants still can't make it on their own: plant stress tolerance via fungal symbiosis. J Exp Bot 59(5):1109–1114

Rodriguez RJ, White Jr JF, Arnold AE, Redman RS (2009). Fungal endophytes: diversity and functional roles. New Phytol 182(2):314–330. https://doi.org/10.1111/j.1469-8137.2009.02773.x

Rouhier N, Koh CS, Gelhaye E, Corbier C, Favier F, Didierjean C et al (2008) Redox based anti-oxidant systems in plants: Biochemical and structural analyses. Biochim Biophys Acta 1780(11):1249–1260. https://www.sciencedirect.com/science/article/pii/S0304416507002991

Rouxel T, Grandaubert J, Hane JK, Hoede C, Van De Wouw AP, Couloux A et al (2011) Effector diversification within compartments of the *Leptosphaeria maculans* genome affected by repeat-induced point mutations. Nat Commun 2(1)

Rudgers JA, Orr S (2009) Non-native grass alters growth of native tree species via leaf and soil microbes. J Ecol 97(2):247–255. https://doi.org/10.1111/j.1365-2745.2008.01478.x

Saikkonen K, Young CA, Helander M, Schardl CL (2016) Endophytic Epichloë species and their grass hosts: from evolution to applications. Plant Mol Biol 90(6):665–675

Sandhu SS, Kumar S, Aharwal RP (2014) Isolation and identification of endophytic fungi from *Ricinus communis* linn. and their antibacterial activity. Int J Res Pharm Chem 4(3):611–618

Schardl CL, Leuchtmann A, Spiering MJ (2004) Symbioses of grasses with seedborne fungal endophytes. Annu Rev Plant Biol 55(1):315–340

Schwarz M, Köpcke B, Weber RWS, Sterner O, Anke H (2004) 3-Hydroxypropionic acid as a nematicidal principle in endophytic fungi. Phytochemistry 65(15):2239–2245

Seetharaman P, Gnanasekar S, Chandrasekaran R, Chandrakasan G, Syed A, Hodhod MS et al (2017) Isolation of limonoid compound (Hamisonine) from endophytic fungi *Penicillium oxalicum* LA-1 (KX622790) of *Limonia acidissima* L. for its larvicidal efficacy against LF vector, *Culex quinquefasciatus* (Diptera: Culicidae). Environ Sci Pollut Res 24(26):21272–21282

Shen M, Zhao DK, Qiao Q, Liu L, Wang JL, Cao GH et al (2015) Identification of glutathione S-transferase (GST) genes from a dark septate endophytic fungus (*Exophiala pisciphila*) and their expression patterns under varied metals stress. PLoS One 10(4):e0123418

Shinozuka H, Hettiarachchige IK, Shinozuka M, Cogan NOI, Spangenberg GC, Cocks BG et al (2017) Horizontal transfer of a ß-1,6-glucanase gene from an ancestral species of fungal endophyte to a cool-season grass host. Sci Rep 7(1):9024

Shukla ST, Habbu PV, Kulkarni VH, Jagadish KS, Pandey AR, Sutariya VN (2014) Endophytic microbes: a novel source for biologically/pharmacologically active secondary metabolites. Asian J Pharmacol Toxicol 2(3):1–16

Shweta S, Zuehlke S, Ramesha BT, Priti V, Mohana Kumar P, Ravikanth G et al (2010) Endophytic fungal strains of *Fusarium solani*, from *Apodytes dimidiata* E. Mey. ex Arn (Icacinaceae) produce camptothecin, 10-hydroxycamptothecin and 9-methoxycamptothecin. Phytochemistry 71(1):117–122

Singh R, Dubey AK (2015) Endophytic actinomycetes as emerging source for therapeutic compounds. Indo Glob J Pharm Sci 5(2):106–116

Singh LP, Gill SS, Tuteja N (2011) Unraveling the role of fungal symbionts in plant abiotic stress tolerance. Plant Signal Behav 6(2):175–191. http://www.ncbi.nlm.nih.gov/pubmed/21512319

Soliman SSM, Greenwood JS, Bombarely A, Mueller LA, Tsao R, Mosser DD et al (2015) An endophyte constructs fungicide-containing extracellular barriers for its host plant. Curr Biol 25(19):2570–2576. http://www.ncbi.nlm.nih.gov/pubmed/26412131

Stajich JE, Harris T, Brunk BP, Brestelli J, Fischer S, Harb OS et al (2012) FungiDB: an integrated functional genomics database for fungi. Nucleic Acids Res 40(Database issue):D675–D681

Staniek A, Woerdenbag HJ, Kayser O (2008) Endophytes: exploiting biodiversity for the improvement of natural product-based drug discovery. J Plant Interact 3(2):75–93

Stierle A, Strobel G, Stierle D (1993) Taxol and taxane production by *Taxomyces andreanae*, an endophytic fungus of Pacific yew. Science 260(5105):214–216

Strobel GA, Pliam NB (1995) Immunosuppressant diterpene compound, p 24

Strobel G, Daisy B (2003) Bioprospecting for microbial endophytes and their natural products. Microbiol Mol Biol Rev 67(4):491–502

Strobel G, Ford E, Worapong J, Harper JK, Arif AM, Grant DM et al (2002) Isopestacin, an isobenzofuranone from *Pestalotiopsis microspora*, possessing antifungal and antioxidant activities. Phytochemistry 60(2):179–183

Takeuchi S, Hirayama K, Ueda K, Sakai H, Yonehara H (1958) Blasticidin S, a new antibiotic. J Antibiot (Tokyo) 11(1):1–5. http://www.ncbi.nlm.nih.gov/pubmed/13525246

Talbot NJ (2015) Plant immunity: a little help from fungal friends. Curr Biol 25(22):R1074–R1076. http://www.ncbi.nlm.nih.gov/pubmed/26583896

Teles HL, Sordi R, Silva GH, Castro-Gamboa I, Bolzani Vda S, Pfenning LH et al (2006) Aromatic compounds produced by *Periconia atropurpurea*, an endophytic fungus associated with *Xylopia aromatica*. Phytochemistry 67(24):2686–2690

Tian J, Liu XC, Liu ZL, Lai D, Zhou L (2016) Larvicidal spirobisnaphthalenes from the endophytic fungus *Berkleasmium* sp. against *Aedes albopictus*. Pest Manag Sci 72(5):961–965

Toju H, Tanabe AS, Ishii HS (2016) Ericaceous plant-fungus network in a harsh alpine-subalpine environment. Mol Ecol 25(13):3242–3257

Umashankar T, Govindappa M, Yarappa Lakshmikantha R, Padmalatha Channabasava, RS (2015). Isolation and characterization of coumarin isolated from endophyte, Alternaria species-1 of Crotalaria pallida and its apoptotic action on HeLa cancer cell line. Metabolomics, 5(4), 1–8.

Uzma F, Hashem A, Murthy N, Mohan HD, Kamath PV, Singh BP, Venkataramana M, Gupta VK, Siddaiah CN, Chowdappa S, Alqaeawi AA, Abd Allah EF (2018) Endophytic fungi—alternative sources of cytotoxic compounds: a review. Front Pharmacol 9(309):1–37. https://doi.org/10.3389/fphar.2018.00309

Vasundhara M, Kumar A, Reddy MS (2016) Molecular approaches to screen bioactive compounds from endophytic fungi. Front Microbiol 7:1774

Visalakchi S, Muthumary J (2010) Taxol (anticancer drug) producing endophytic fungi: an overview. Int J Pharma Bio Sci 1(3):1–9

Wang X, Zhang X, Liu L, Xiang M, Wang W, Sun X et al (2015a) Genomic and transcriptomic analysis of the endophytic fungus *Pestalotiopsis fici* reveals its lifestyle and high potential for synthesis of natural products. BMC Genomics 16(1):28

Wang X, Zhang X, Liu L, Xiang M, Wang W, Sun X et al (2015b) Genomic and transcriptomic analysis of the endophytic fungus reveals its lifestyle and high potential for synthesis of natural products. BMC Genomics 16:28

Wani ZA, Ashraf N, Mohiuddin T, Riyaz-Ul-Hassan S (2015) Plant-endophyte symbiosis, an ecological perspective. Appl Microbiol Biotechnol 99(7):2955–2965. https://doi.org/10.1007/s00253-015-6487-3

Webber J (1981) A natural biological control of Dutch elm disease. Nature 292(5822):449–451. https://doi.org/10.1038/292449a0

Weig AR, Peršoh D, Werner S, Betzlbacher A, Rambold G (2013) Diagnostic assessment of mycodiversity in environmental samples by fungal ITS1 rDNA length polymorphism. Mycol Prog 12(4):719–725

Xu X-H, Su Z-Z, Wang C, Kubicek CP, Feng X-X, Mao L-J et al (2015) The rice endophyte *Harpophora oryzae* genome reveals evolution from a pathogen to a mutualistic endophyte. Sci Rep 4(1):5783

Yadav V, Kumar M, Deep AK, Kumar H, Sharma R, Tripathi T et al (2010) A phosphate transporter from the root endophytic fungus *Piriformospora indica* plays a role in phosphate transport to the host plant. J Biol Chem 285(34):26532–26544

Yadava P, Bhuyan SK, Bandyopadhyay P, Yadava PK (2015) Extraction of proteins for two-dimensional gel electrophoresis and proteomic analysis from an endophytic fungus. Protoc Exch. https://doi.org/10.1038/protex.2015.084

Yamaji K, Watanabe Y, Masuya H, Shigeto A, Yui H, Haruma T (2016) Root fungal endophytes enhance heavy-metal stress tolerance of clethra barbinervis growing naturally at mining sites via growth enhancement, promotion of nutrient uptake and decrease of heavy-metal concentration. PLoS One 11(12):e0169089. https://doi.org/10.1371/journal.pone.0169089

Yang S (2010) Allelopathic effects of endophytic fungi of *Achnatherum inebrians* Keng on the seed and seedling of three turf grasses. Acta Agrestia Sinica. http://en.cnki.com.cn/Article_en/CJFDTOTAL-CDXU201001017.htm

Yang Y, Zhao H, Barrero RA, Zhang B, Sun G, Wilson IW et al (2014) Genome sequencing and analysis of the paclitaxel-producing endophytic fungus *Penicillium aurantiogriseum* NRRL 62431. BMC Genomics 15(1):69

Zhang P, Zhou P-P, Yu L-J (2009) An endophytic taxol-producing fungus from taxus media, *Cladosporium cladosporioides* MD2. Curr Microbiol 59(3):227–232

Zhang G, Sun S, Zhu T, Lin Z, Gu J, Li D et al (2011) Antiviral isoindolone derivatives from an endophytic fungus *Emericella* sp. associated with *Aegiceras corniculatum*. Phytochemistry 72(11–12):1436–1442

Zhang Y, Han T, Ming Q, Wu L, Rahman K, Qin L (2012) Alkaloids produced by endophytic fungi: a review. Nat Prod Commun 7(7):963–968

Zhang H, Liu R, Yang J, Li H, Zhou F (2017a) Bioactive alkaloids of *Aspergillus fumigatus*, an endophytic fungus from *Astragalus membranaceus*. Chem Nat Compd 53(4):802–805

Zhang S, Wang XN, Zhang XL, Liu XZ, Zhang YJ (2017b) Complete mitochondrial genome of the endophytic fungus *Pestalotiopsis fici*: features and evolution. Appl Microbiol Biotechnol 101(4):1593–1604

Zhao J, Mou Y, Shan T, Li Y, Zhou L, Wang M et al (2010) Antimicrobial metabolites from the endophytic fungus *Pichia guilliermondii* isolated from *Paris polyphylla* var. yunnanensis. Molecules 15(11):7961–7970

Zhao J, Shan T, Mou Y, Zhou L (2011) Plant-derived bioactive compounds produced by endophytic fungi. Mini Rev Med Chem 11(2):159–168

Zuccaro A, Lahrmann U, Güldener U, Langen G, Pfiffi S, Biedenkopf D et al (2011) Endophytic life strategies decoded by genome and transcriptome analyses of the mutualistic root symbiont *Piriformospora indica*. PLoS Pathog 7(10):e1002290

Chapter 5
Natural Products from Endophytic Fungi: Synthesis and Applications

Parasuraman Paramanantham, Subhaswaraj Pattnaik, and Busi Siddhardha

5.1 Introduction

Endophytes are organisms that infect living plant tissues without affecting the physiology of the host plant by establishing a symbiotic relationship that provides advantages in interacting with the host and associated environment (Casella et al. 2013). The term endophyte was coined initially by De Bary to refer to any organism found within living plant tissues without creating disease symptoms in the host plant (Nisa et al. 2015). All plants harbor a diverse range of microbial communities such as bacteria, archaebacteria, fungi, and protista that possess endophytic attributes and play a substantial role in the physiological and metabolic activity of the host plants (Hardoim et al. 2015). In the symbiotic association between plant host and their associated endophytes, plants provide organic nutrients to the colonizing endophytes, whereas the endophytes mimic biosynthetic pathways for the synthesis of bioactive metabolites that have widespread application in agricultural sectors such as the management of growth and development of host plants, advanced resistance mechanism against phytopathogens, and maintenance of systemic resistance. Among the endophytes living in close association with host plants, endophytic fungi represent a highly diverse and polyphyletic group of microorganisms that colonize internal plant tissues without causing any disease symptoms in the host (Kusari et al. 2013). The ubiquity and diversity of endophytes colonizing plant tissues are well established from the fact of the occurrence of more than 300,000 plant species, and each individual plant harbors one or more groups of endophytes (Yu et al. 2010). Endophytic fungi have received considerable attention in the last few decades owing to the untapped reservoir of novel bioactive metabolites they harbor, which long went unnoticed. Though endophytes are recognized as a source of novel metabolites, enzymes for widespread industrial applications, and agents that

P. Paramanantham · S. Pattnaik · B. Siddhardha (✉)
Department of Microbiology, School of Life Sciences, Pondicherry University, Puducherry, India

B. P. Singh (ed.), *Advances in Endophytic Fungal Research*, Fungal Biology,
https://doi.org/10.1007/978-3-030-03589-1_5

enhance the abiotic or biotic stress tolerance of their plant hosts, information regarding other perspectives in the field of biology remains unexploited (Suryanarayanan 2013).

5.1.1 An Insight into the World of Endophytes

Though there are abundant reports on the diversity of epiphytic microorganisms associated with host plants, endophytes have attracted significant interest in the last few years due to their closer biological associations with their host plants compared to their epiphytic counterparts (Strobel 2003). In the environment, the majority of plant species, from nonvascular plants to angiosperms, are closely associated with a diverse range of endophytic fungi (Venugopalan and Srivastava 2015).

According to recent findings of the National Center for Biotechnology Information (NCBI) and internal transcribed spacer sequence (ITS) analysis, among all the eukaryotic endophytes, the majority of endophytes (two-thirds) belong to the phyla Glomeromycota and Ascomycota, followed by the phyla Basidiomycota and Zygomycota. Arbuscular mycorrhizal fungi (AMFs), important members of the phylum Glomeromycota, form ubiquitous endosymbiotic relationships with host plants and play a crucial role in plant physiological processes. Among the Ascomycota and Basidiomycota, the class Dothideomycetes and Agaricomycetes represent the majority of endophytes, respectively (Hardoim et al. 2015).

5.1.2 Different Types of Endophytes and Their Diversity

Fungal endophytes are found in the majority of land plants in both natural and anthropogenic communities and can survive in different climatic conditions ranging from tropical to arctic climates, suggesting their ubiquity, cryptic lifestyle, and widespread diversity. Their diversity and ecological function within individual host plants have gained considerable interest in connection with exploring fungal ecology in association with host plant, fungal classification in terms of evolutionary relationship with host plants, and applied biomedical and pharmaceutical applications (Arnold 2007). Endophytic microbes can be divided into several classes based upon their interaction with the host plants. The major classes of endophytes include the family of endophytic Clavicipitaceae; fungal endophytes of dicots, lichens, bryophytes and ferns, tree bark, xylem, and roots; and prokaryotic endophytes of plants (Zhang et al. 2006). In addition, based on the colonization characteristics of fungal endophytes, they are categorized into different classes. Clavicipitaceous endophytes represent class I endophytes, which colonize rhizospheres, endorhizae, and aerial tissues of host plants, whereas fungal endophytes, which have horizontal/vertical transmittance, represent class II endophytes. The members of Ascomycota and Basidiomycota represent class III endophytes, which colonize the aerial tissues of host plants, whereas class IV endophytes are restricted to roots represented by mycorrhizal fungi (Hardoim et al. 2015). The study of endophytes has two important objectives: bioprospecting for the exploitation of fungal endophytes leading to the production of hosts, as well as novel microbial bioactive metabolites for widespread pharmaceutical and biotechnological applications; and understanding plant–microbe interactions that account for stress tolerance, adaptation to harsh environmental conditions, improved productivity, and modulation of host as well as microbial metabolic activities (Wani et al. 2015).

5.1.3 Ecosystem Functioning of Plant Endophytes

Though the current understanding of the life history and classification of fungal endophytes has increased, their evolutionary origins and specific ecological and ecosystem services remain undefined and need to be investigated to acquire sound knowledge of their ecological functions. Microbial endophytes are capable of mimicking the physiological and metabolic pathways of host plants and produce phytohormones and vitamins and supply other nutrients to host plants. Microbial endophytes are also important for the proper management of ecosystem functioning by modulating the ecological understanding of plant communities and mediating intense ecological interactions (Zhang et al. 2006).

5.1.4 Importance of Endophytic Fungi in Natural Product Research

Fungi are among the prolific resources for exploitation of bioactive metabolites with widespread biomedical and pharmaceutical applications. Among fungi, endophytes remain the most important organisms for the screening of biologically active metabolites in relation with their host plants. A diverse class of chemical moieties such as phenolic compounds and their derivatives, terpenes and their derivatives, steroids, isocoumarins, and cytochalasins have been isolated from fungal endophytes inhabiting different plant hosts by different biosynthetic pathways (Nisa et al. 2015). In the last few decades, fungal endophytes have attracted attention owing to their potential to produce various bioactive natural products of immense therapeutic and biomedical value in the treatment of numerous diseases (Kusari et al. 2014). In recent years, fungal endophytes have been actively characterized for their inherent ability to control phytopathogen-mediated disease severity and the maintenance of systemic resistance. In addition, endophytic fungi also imitate the production of bioactive metabolites that are also biosynthesized by their respective host plants (Kusari et al. 2013). The plant-endophyte-mediated production of bioactive metabolites not only confers resistance and protection to host plants but also provides a path for treating dreadful human diseases. In recent years, the increased emergence of drug-resistant microbes and constant reoccurrence of diseases have prompted the scientific community to exploit the untapped potential of plant-associated endophytes for next-generation drugs to target microbial infections and other serious diseases (Alvin et al. 2014). Endophytic fungi are ubiquitous in nature and constitute a rich lineage of varied novel bioactive compounds with promising industrial and pharmaceutical applications in, for example, agrochemicals, biopesticides, antimicrobials, antioxidants, and anticancer agents (Uzma et al. 2016, 2018; Mishra et al. 2017a). In recent years, the production of extracellular enzymes such as amylases, cellulases, pectinases, and asparaginases from fungal endophytes has also been explored for their widespread industrial and biotechnological applications. Owing to the desirable functional traits shown by endophytic fungi in association with host plants, more feasible and sustainable strategies are being developed for enhanced biotechnological and pharmaceutical applications. Plant-associated endophytic fungi have a rich heritage of producing certain novel and very rare phytochemicals originally produced by host plants. In this context, fungal endophytes could be explored for the production of rare bioactive compounds as their hosts, thereby minimizing selective pressures on rare plant species, suggesting simultaneous preservation strategies along with widespread biological applications (Fig. 5.1) (Strobel and Daisy 2003).

Medicinal plants have served as a reservoir of immense therapeutically active metabolites and have been used in folk medicines since ancient times. However, extensive and indiscriminate use of plant resources for bioactive compound produc-

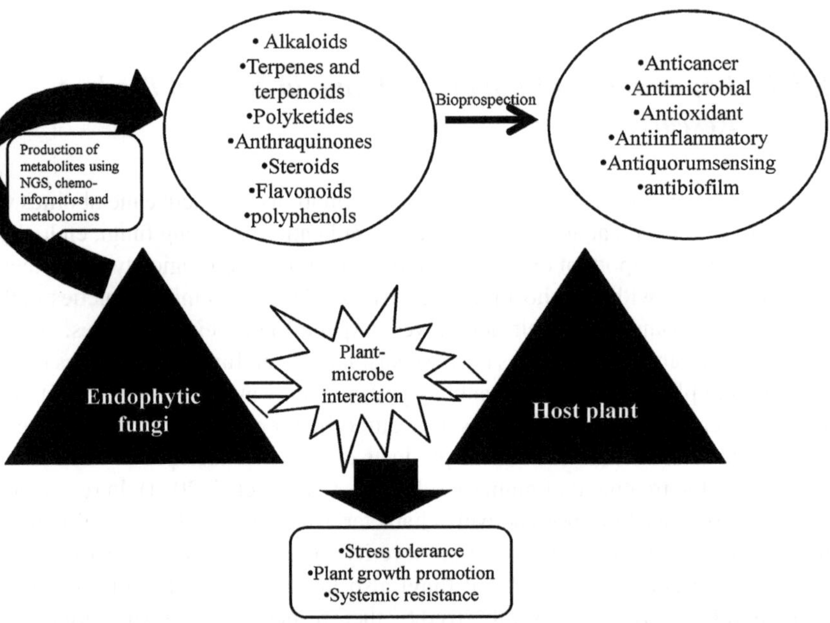

Fig. 5.1 Schematic representation of symbiotic association between endophytic fungi and host plants and its role in determining plant–microbe interactions and bioprospecting the production of bioactive secondary metabolites for therapeutic applications

tion has resulted in certain issues of global concern. In this context, to counteract the associated problems, the scientific community has focused on alternative strategies involving endophytic fungi. Medicinal plants harbor a diverse range of endophytic fungi that have the inherent potential to mimic their host plant in producing secondary metabolites of immense therapeutic efficacy (Venieraki et al. 2017).

5.2 Metabolic Pathways in the Biosynthesis of Natural Products

Recently, various studies have been conducted to elucidate the biosynthetic pathways involved in the production of plant secondary metabolites. Additionally, the behavior of endophytic fungi that mimic the biosynthetic pathways of host plants in synthesizing the secondary metabolites of the plants has been well characterized. Partial knowledge has allowed an understanding that endophytic fungi living symbiotically in association with plants are developing the ability to mimic host chemical diversity and are also believed to be the reason for the production of biologically active metabolites once they establish a relationship with plants (Le Cocq et al. 2017). Secondary plant metabolites are synthesized by three important metabolic pathways such as polyketide, shikimate, and mevalonate pathways. In most cases, polyketide pathways are involved in the synthesis of certain groups of secondary metabolites such as polyphenols, quinine, and prostaglandins. The second pathway, shikimate, contributes to the synthesis of amino acids that possess an aromatic ring in their structure. The aforementioned pathways are widely observed in plants, bacteria, and fungi but not observed in animals (Kumara et al. 2014). This section covers steps and genes involved in the aforementioned three pathways.

5.2.1 Polyketide Synthase Pathway

Polyketides are a large class of natural products that are structurally diverse and are used in a vast array of biological and pharmacological activities, including anticancer, antifungal, antibacterial, anti-parasitic, and immunosuppressive applications. All structurally and chemically versatile polyketides are synthesized by continuous condensation reactions between a thioesterified malonic acid derivative and an acyl thioester. This condensation reaction catalyzes by an enzyme called polyketide synthases (PKSs), which are classified into three different groups: types I, II, and III (Refaei et al. 2014; Feng et al. 2015; Rao and Satish 2015).

Type I PKSs are further categorized into modular and iterative types. In microorganisms, the modular class of type I PKSs plays a significant role in the assembly of a 6-deoxyerythronolide B scaffold of erythromycin A. Usually these PKSs are involved in the assembly of seven precursors, including propionyl-coenzyme A (CoA) starter unit along with six (2*S*)-methylmalonyl-coenzyme A polyketide structural extender units. The primary precursor unit initiates polyketide synthesis; in the meantime, the extender units stretch for the successful completion of polyketide backbone development. A module is a group of catalytic domains that have control over the insertion of precursors into their appropriate polyketide backbone. In type I modular PKSs, modules exhibit enormously as same as the amount of precursors placed into the polyketide backbone. Once the modules are bound with their key starter unit, they activate various catalytic domains and the modules that are integrated with extender units generally have three important domains that are involved in the polyketide extension process, and three other auxiliary domains actively participate in β-keto processing. Ketosynthase, acyltransferase, and acyl carrier proteins are the three core domains of PKSs. The domain acyltransferase acts as a gatekeeper for modules and controls the growth of polyketide chains. This also facilitates the malonyl derivative of extender units to form covalent bonds with the sulfhydryl group of the acyl carrier protein prosthetic group that participates as a 4′-phosphopantetheinyl (4′-Ppant) component of CoA. The domain ketosynthase is involved in catalytic activity against the decarboxylative Claisen and promotes condensation reactions with a neighboring acyl carrier-protein-linked molonate derivative and an acyl carrier protein that combine with acyl thioester to enhance the polyketide structural chain. The other optional domains such as ketoreductase, dehydratase, and enoylreductase are involved in the alteration of the oxidative state of the β-keto group developed after ketosynthase-mediated condensation. The repeated type I PKSs use the modular type I PKSs to produce the complete polyketide backbone structures (Refaei et al. 2014; Zhang et al. 2015; Santiago et al. 2014).

The type II PKSs constitute the core catalytic domain as observed in type I PKSs, with the exclusion of two ketosynthase domains such as α-ketosynthase and β-ketosynthase. The type II PKSs generally govern polyketide length and enzymatic activities. Herein, no reductive assembling of the β-keto groups exhibits until after complete synthesis of the polyketide. As similar type I and type II PKSs, even type III PKSs have a starter unit with a sequence of extender units to synthesize poly-β-keto chains. The major difference between type III PKSs and types I and II is generally the lack of a complex catalytic domain and the use of an acyl carrier-protein-independent mechanism. Rather, a single enzyme supports acyl-CoA thioester and catalyzes acyl radicals change between CoA and an active site cysteine (Chan et al. 2009; Miller et al. 2012; Yao et al. 2016).

5.2.2 *Shikimate Pathway*

The shikimate pathway is recognized as being an essential pathway in plants and microbial systems; it coordinates carbohydrate metabolism and facilitates the production of aromatic compounds. It is an array of seven-step enzymatic processes where phosphoenolpyruvate and erythrose 4-phosphate are changed to chorismate. This chorismate acts as a precursor for aromatic amino acids and several secondary metabolites holding aromatic rings in their structure such as tryptophan, phenylalanine, and tyrosine, which frames the major parts of metabolism found in plants and microorganisms. Initially seven enzymatic steps of the shikimate pathway were found in bacteria, namely *Escherichia coli* and *Salmonella typhimurium*. However, the substrate, product of these enzymes, and intermediate of the pathway are identical for prokaryotic and eukaryotic organisms. In rare cases, a greater dissimilarity is observed in the primary structure and behaviors of the enzymes in the prokaryotic and eukaryotic system (Herrmann and Weaver 1999; Lunardelli Negreiros de Carvalho et al. 2016).

The shikimate pathway is also widely known as a biosynthetic pathway of chorismate and is composed of a cascade of seven reactions catalyzed by six different enzymes that coordinate the central carbon metabolism process and the aromatic amino acid interconnected by changing the phosphoenol pyruvate produced in the glycolysis cycle and erythrose 4-phosphate originating from the pentose phosphate pathway to form chorismate. This chorismate acts as a universal precursor for all amino acids holding aromatic rings in their structures and numerous secondary metabolites of plants and microorganisms. The shikimate pathway starts with the condensation of phosphoenol pyruvate and erythrose-4-phosphate under the influence of 3-deoxy-D-arabino-heptulosonate-7-phosphate synthase (DAHPS) to form a 7-carbon six-group heterocyclic compound, 3-deoxy-D-arabiono-heptulosonate7-phosphate (DAHP), which is also considered a 2-deoxy-d-glucose-6-phosphate derivative. The second step of this pathway involves the exchange of ring oxygen in the exocyclic seventh carbon of DAHP to develop an intensely replaced cyclohexane derivative, 3-dehydroquinate, by the action of a 3-dehydroquinate synthase enzyme. The remaining five enzymatic steps are involved in actions like introducing a side chain and three double bonds that modify the aforementioned cyclohexane into a benzene structure (Herrmann 1995). The next two enzymatic processes involved in the shikimate pathway are the dehydration of 3-dehydroquinate to produce 3-dehydroshikimate and its reduction to form shikimate by employing nicotinamide adenine dinucleotide phosphate (NADPH). 3-dehydroquinate dehydratase and shikimate dehydrogenase catalyze the respective reaction and facilitate the difference in enzyme organization in the three kingdoms (Maeda and Dudareva 2012). The fifth enzymatic process of the shikimate biosynthesis pathway is promoted by shikimate kinase that converts shikimate to shikimate 3-phosphate. The activity of shikimate kinase depends upon the presence of divalent cations such as Mg^{2+} or Mn^{2+} and possess shikimate bonding and nucleotide bonding domains that experience significant conformational conversion upon bonding of shikimate and ATP. The sixth enzyme, namely 5-enolpyruvylshikimate 3-phosphate synthase, is involved in the catalytic action, the penultimate step of the shikimate pathway, and the development of 5-enolpyruvylshikimate 3-phosphate by transporting the enolpyruvyl moiety of phosphoenolpyruvate to the 5-hydroxyl position of shikimate 3-phosphate. The enolpyruvyl group gradually becomes a side chain of phenylalanine and tyrosine and is cleared in the time of the biosynthesis of tryptophan. The final enzymatic process of the shikimate pathway is the elimination of the 3-phosphate from the 5-enolpyruvylshikimate 3-phosphate, to develop chorismate (Fig. 5.2). Chorismate needs flavin mononucleotide as a cofactor that contributes one electron to the 5-enolpyruvylshikimate 3-phos-

Fig. 5.2 Schematic representation of steps involved in shikimate pathway

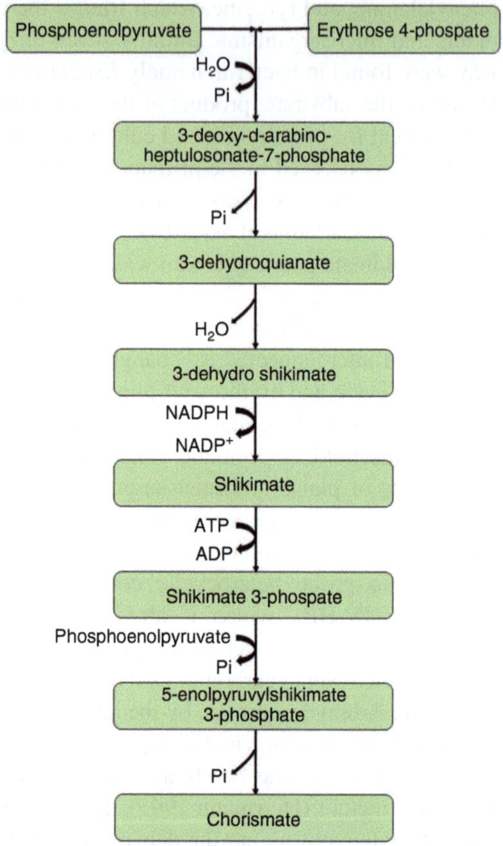

phate basic compound to yield phosphate reduction and it participates in C6 hydrogen bonding abstraction (Tzin and Galili 2010; Maeda and Dudareva 2012).

5.2.3 *Mevalonate Pathway*

The mevalonate pathway is an essential biometabolic pathway that helps the cells to produce important bioactive compounds and helps in multiple cellular processes. In this pathway, mevalonate is converted into sterol isoprenoids like cholesterol, steroid hormones, indispensable precursors of bile acids, and lipoproteins. The intermediates from this pathway play a significant role in the posttranslational modification of protein involved in intracellular communication and cell growth and differentiation, molecular expression, protein glycosylation, and cytoskeletal differentiation (Buhaescu and Izzedine 2007). The first step in the mevalonate pathway is the biosynthesis of 3-hydroxy-3-methylglutaryl-CoA to produce three molecules of acetyl-CoA (Fig. 5.3). Actoacetyl-CoA is formed through acetoacetyl-CoA thiolase in a first

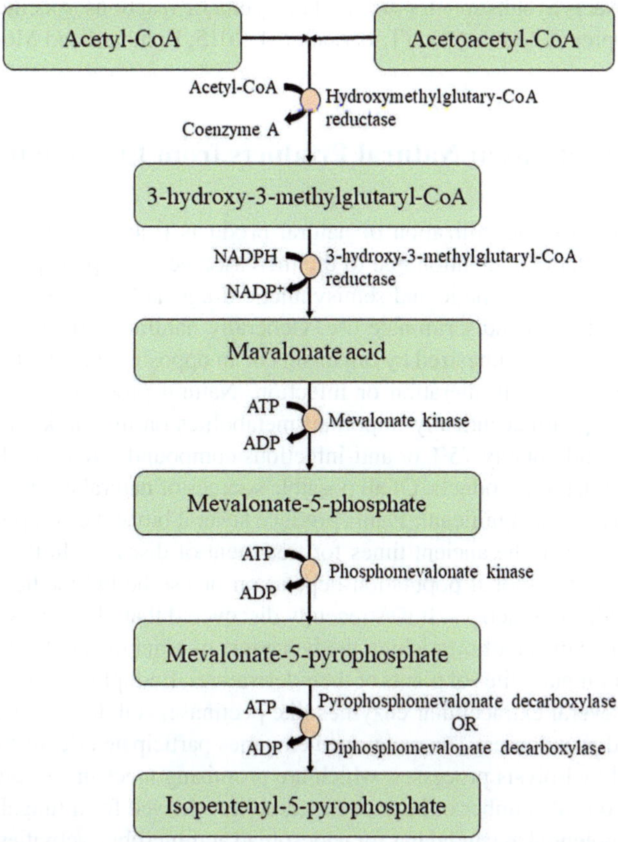

Fig. 5.3 Simplified sequential cascade of reactions involved in mevalonate pathway

condensation reaction, whereas a second condensation reaction occurs between acetoacetyl-CoA and a third acetyl-CoA molecule is catalyzed by 3-hydroxy-3-methylglutaryl-CoA synthase. The second step begins with the breakdown of 3-hydroxy-3-methylglutaryl-CoA to mavalonate acid by NADPH where the reaction is catalyzed by a 3-hydroxy-3-methylglutaryl-CoA reductase enzyme. This enzyme is also involved in the regulation of the posttranslational level by phosphorylation arbitrated through AMP-activated kinase. The third enzymatic reaction of this pathway promotes the conversion of mevalonic acid into mevalonate-5-phosphate by the catalytic action of the enzyme mevalonate kinase. The fourth key enzyme of the mevalonate pathway is phosphomevalonate kinase, which is primarily responsible for the conversion of mevalonate-5-phosphate into mevalonate-5-pyrophosphate using ATP as the phosphate and energy source. The enzyme pyrophosphomevalonate decarboxylase or diphosphomevalonate decarboxylase is the fifth enzyme of the mevalonate pathway; it is involved in the conversion of mevalonate-5-pyrophosphate into isopentenyl-5-pyrophosphate, the end metabolic outcome of the mevalonate pathway that acts as substrate for different biosynthetic reactions, specifically cholesterol and isoprenoid production (Tricarico et al. 2015; Lombard and Moreira 2011).

5.3 Applications of Natural Products from Endophytic Fungi

In the recent past, the utilization of natural products from biological sources has attracted considerable attention due to the increased negative perceptions about the long-term safety of synthetic and semisynthetic drugs and environmental concerns associated with their indiscriminate use. Generally natural products are secondary metabolites that are synthesized by organisms in an opposing reaction to the external enforce like nutritional alteration or infection. Natural products constitute nearly 50% of novel pharmaceutically important metabolites on the market introduced in recent years, and notably 75% of anti-infectious compounds are natural products or derivatives of natural products. Of all possible sources of natural products, plants are considered the most significant. Plants produce several bioactive compounds, which has been used from the ancient times for treatment of diseases. In the modern era, almost 80% of the global population depend on or use herbal medicines for their primary healthcare practices. It was recently discovered that plant-associated microorganisms are capable of mimicking the behaviors of plant metabolism and producing secondary metabolites of plants or their derivatives. Endophytes have the capacity to produce several extracellular enzymes like pectinase, cellulase, lipase, amylases, laccases, and proteinases. The mentioned enzymes participate effectively in biodegradation and hydrolysis processes, which are promising mechanisms against pathogenic infections. A number of bioactive metabolites derived from fungal endophytes have shown potential as candidates for widespread antimicrobial activities (Table 5.1).

5.3.1 Plant Disease Management and Sustainable Agriculture

Endophytes are a group of microorganisms residing in inter- and intracellular regions of plant tissues. In most cases, they develop a symbiotic relationship with plants and function as a defense mechanism against phytopathogens. They protect plants directly by synthesizing and releasing metabolites to counteract a variety of pathogenic microorganisms and induce systemic host defense mechanisms by promoting plant growth and development. Fungal endophytes have a rich lineage in synthesizing novel bioactive compounds with widespread applications such as antagonistic activities toward plant pathogenic microorganisms and herbivores. These bioactive compounds mainly belong to different chemical moieties such as alkaloids, steroids, terpenoids, peptides, polyketones, flavonoids, quinols, and phenols (Hardoim et al. 2015). Additionally, they are also capable of releasing different kinds of antibiotics and hydrolytic enzymes to protect from the colonization of plant pathogenic microorganisms, insects, and nematodes from infected plants. The enhancement of plant

Table 5.1 List of high-value secondary metabolites from plant-associated endophytic fungi and their biomedical applications

Sl no.	Compound name	Chemical group	Endophyte strain	Host plant	Biological activity	Reference
1.	Azadirachtin A and B	Tetranotriterpenoid	*Eupenicillium parvum*	*Azadirachta indica*	Insecticidal	Kusari et al. (2012a)
2.	Ginkgolide B	Diterpenoids	*Fusarium oxysporum* SYP0056	*Ginkgo biloba*	Cardiovascular treatment	Cui et al. (2012)
3.	Kaemferol	Flavonol	*Fusarium chlamydosporum*	*Tylophora indica*	Antioxidant and anticancer	Chaturvedi et al. (2014)
4.	Emodin	Anthraquinone	*Thielavia subthermophila* INFU/Hp/KF/34B	*Hypericum perforatum*	Antiviral, anticancer, antidiabetic	Venugopalan and Srivastava (2015)
5.	Gymnemagenin	Triterpenoid aglycone	*Penicillium oxalicum*	*Gymnema sylvestre* R.Br.	Antidiabetic	Parthasarathy and Sathiyabama (2014)
6.	Lapachol	Phenolic compound	*Aspergillus niger, Alternaria alternata*	*Tabebuia argentea*	Anticancer	Nirupama et al. (2011)
7.	Piperine	Alkaloid	*Periconia* sp.	*Piper longum*	Antimicrobial, antioxidant, and anticancer	Verma et al. (2011)
8.	Rhein	Anthraquinone	*Fusarim solani* R13	*Rheum palmatum*	Antitumor, antimicrobial, and anti-inflammatory	You et al. (2013)
9.	Sanguinarine	Polycyclic ammonium ion	*Fusarium proliferatum* BLH51	*Macleaya cordata*	Anticancer	Wang et al (2014)
10.	Cytochalasins	Alkaloid	*Rhinocladiella* sp.	*Tripterygium wilfordii*	Anticancer	Strobel and Daisy (2003)
11.	Diosgenin	Phytosteroid	*Paecilomyces* sp.	*Paris polyphylla* var. *yunnanensis*	Cholesterol lowering activity	Cao et al. (2007)
12.	Rohitukine	Alkaloid	*Fusarium proliferatum*	*Dysoxylum binectariferum*	Anti-inflammatory and anticancer	Kumara et al. (2012)

growth and development by endophytes could be achieved in different ways including the synthesis of phytohormones, the synthesis of siderophores, the action of nitrogen fixation, and the solubilization of minerals like phosphorus by enzymatic activity. Additionally, they also benefit plants by their natural resistance to soil contaminants like xenobiotics using degrading enzymes, which promotes sustainable phytoremediation (Alvin et al. 2014; Le Cocq et al. 2017; Pimentel et al. 2011; Fouda et al. 2015; Lugtenberg et al. 2016). Hence, in recent years numerous studies have been conducted to investigate the potential of endophytes for plant growth promotion and developments that could facilitate sustainable agricultural practices (Rai et al. 2014; Nair and Padmavathy 2014). Waqas and coworkers investigated the potential of endophytic fungi on the stimulation of plant growth and their role in the promotion of disease resistance in plants. Class I and II endophytes have an inherent property of enhancing the ecophysiology of host plants, thereby enabling host plants to counteract different versions of abiotic stress conditions (Rodriguez et al. 2009). The endophytic fungi *Penicillium cireinum* LWL4 and *Aspergillus terreus* LWL5 were evaluated for their efficacy in the promotion of plant growth and protection from infection in sunflower plants. The result of one study concluded that the plant inoculated with fungal endophytes altered plant growth during disease condition by modulating responses and coordinating with plant defense mechanisms. Additionally, the study postulated that management techniques involved in endophytic symbiosis could facilitate sustainability in agriculture by decreasing excessive agriculture-related chemicals (Waqas et al. 2015). Similar work has been performed to examine the participation of endophytic fungi in *Verticillum* wilt, a soilborne disease, and in suppressing the development of the cotton plant. Herein *Penicillium simplicissimum*, *Leptosphaeria* spp., *Talaromyces flavus*, and *Acremonium* spp. were isolated from cotton roots and investigated their effect against cotton wilt infection caused by a defoliating agent, *V. dahlia*. The test result indicated that the fungal isolates are capable of delaying infection and reducing wilt symptoms in cotton. *P. simplicissimum* and *Leptosphaeria* spp. treatment increased the transcript level of essential plant growth promoters that facilitate the increase of the cotton defense reaction. Additionally, they support the biocontrol of *V. dahlia* and enhance cotton seed production in cotton fields. This study also provided a better understanding of plant (cotton) endophyte interactions that promote the development of growth of the plant and protect from *Verticillium* wilt of cotton (Yuan et al. 2017).

Two endophytes, *Penicillium chrysogenum* Pc_25 and *Alternaria alternata* Aa_27, isolated from the medicinal plant of *Asclepias sinaica* not only produce bioactive metabolites for biomedical application but also have the tendency to produce ammonia and a phytohormone, indole acetic acid (IAA), thereby having a profound impact on plant growth parameters such as root elongation and plant growth promotion (Fouda et al. 2015). Antibiotics or hydrolytic enzymes produced by plant-associated endophytes prevent the colonization of pathogenic microorganisms and protect from the attack of insects and nematodes. In addition, endophytes provide systemic resistance to host plants by releasing metabolites that promote the host defense mechanism against other plant pathogenic microorganisms. Endophytes also characteristically promote plant growth by several mechanisms, such as the production of phytohormones, synthesis of siderophores, nitrogen fixation, solubilization of minerals such as phosphorus, or via enzymatic activities, such as ethylene suppression by 1-aminocyclopropane-1-carboxylate deaminase (Alvin et al. 2014). Plant-associated endophytic fungi thrive in various environmental conditions and produce some novel bioactive constituents that enable them to survive in biotic and abiotic stress conditions and protect host plants from phytopathogens (Jalgaonwala et al. 2011). The close interactions between endophytic fungi with their host plants play a crucial role in the health management of host plants. Endophytic fungi play several beneficial roles such as competing with phytopathogens for colonization sites and nutrient availability and thereby inducing the resistance mechanism in host plants against pathogens, providing protection to hosts from attacks by herbivorous insects by producing a wide variety of toxins, and greatly inducing plant growth promotion by synthesizing plant-growth-promoting hormones such as IAA (Souza and dos Santos 2017).

5.4 Biomedical and Therapeutic Applications

5.4.1 Anticancer Compounds

In recent times, cancer has emerged as one of the most important and devastating human diseases causing severe loss of human life and economic damage. Though conventional chemotherapeutic strategies have been found to be effective, their relative toxicity and associated side effects spurred the scientific community to search for alternative drugs with greater therapeutic efficacy. The rich repository of deoxypodophyllotoxin and podophyllotoxin in *Aspergillus fumigatus* Fresenius isolated from *Juniperus communis* L. Horstmann and *Trametes hirsuta* suggests its potential as an anticancer prodrug (Demain and Sanchez 2009; Kusari et al. 2009, 2013). Akone et al. (2016) have shown that the genus *Chaetomium* isolated from healthy leaves of *Sapium ellipticum* produces chaetocochins A-C, which showed significant anticancer activities (Akone et al. 2016). Paclitaxel isolated from the bark of *Taxus brevifolia* demonstrated a high level of efficacy in combating the progression of different types of metastatic cancers. In this context, the endophytic fungus *Taxomyces andreanae*, isolated from *T. brevifolia*, also mimics the synthesis of paclitaxel, suggesting a novel and promising approach to produce this valuable compound. Since those studies, current drug discovery programs on natural product research have focused on exploring the potential of endophytic fungi in producing paclitaxel and other anticancer drugs (Zhao et al. 2010).

5.4.2 Antimicrobial Compounds

The ability of microbial endophytes to synthesize an array of bioactive metabolites with potent antimicrobial activities has been attributed since ancient times to their coevolution in diverse ecological niches and natural habitats. Endophytic fungi that inhabit medicinal plants have a rich heritage of producing diverse bioactive compounds with widespread applications. They also imitate host plant metabolic pathways and synthesize a new range of bioactive compounds. The fungal endophyte *Ampelomyces* sp. isolated from *Urospermum picroides* produces an array of bioactive compounds that exhibit significant cytotoxic activity in vitro against L5178Y cells. In addition, these bioactive fractions also contribute to enhanced antimicrobial activity against bacterial pathogens such as *Staphylococcus aureus*, *S. epidermidis*, and *Enterococcus faecalis* (Aly et al. 2008). The genus *Xylaria* is an important endophytic fungus inhabiting diverse medicinal plants and produces a wide range of bioactive compounds with diverse biomedical applications. For example, multiplolides, glucoside derivatives, xylarosides A, xylarosides B, and sordaricin isolated from *Xylaria* sp. exhibited biocidal activity against *Candida albicans*. In addition, *Xylaria* also produce cytotoxic cytochalasins and acetyl choline esterase inhibitors known as xyloketals (Pongcharoen et al. 2008).

The endophytic fungus *Cryptosporiosis* sp. isolated from *Clidemia hirta* produces an array of bioactive metabolites and exhibits antimicrobial properties against *Bacillus cereus*, *S. aureus*, *Escherichia coli*, and *Pseudomonas fluorescens* (Zilla et al. 2013). Pyrrocidine C, a novel bioactive compound isolated from *Lewia infectoria* SNB-GTC2402, exhibited potential antifungal activity against *C. albicans* (ATCC10213) (Casella et al. 2013). Recently, furano-polyene 3-epi-aureonitol isolated from *Chaetomium* sp. colonizing healthy leaves of *Sapium ellipticum* collected in Cameroon showed tremendous antibacterial activity (Akone et al. 2016). Two endophytes, *Penicillium chrysogenum* Pc_25 and *Alternaria alternata* Aa_27, isolated from the medicinal plant of *Asclepias sinaica* produce extracellular enzymes such as cellulase, gelatinase, and xylanase and exhibited significant antimicrobial potential against *S. aureus*, *B. subtilis*, *E. coli*, *P. aeruginosa*, *Salmonella typhimurium*, and *C. albicans* (Fouda et al. 2015). In the early twenty-first century, two compounds, pestacin and isopestacin, obtained from *Pestalotiopsis microspora* colonizing *Terminalia morobensis* also displayed potent antimicrobial activity (Strobel and Daisy 2003).

Recently, *Phaeoacremonium* sp. isolated from the leaves of *Senna spectabilis* produced a wide range of lactone derivatives such as isoaigialones A, B, and C and aigialone. These lactone derivatives showed potent antifungal activity against *Cladosporium cladosporioides* and *C. sphaerospermum* (Silva et al. 2017). The presence of bioactive compounds such as viridicatol, tenuazonic acid, alternariol, and alternariol monomethyl ether in *Eleusine coracana* showed antimicrobial activity against *Fusarium graminearum* (Mousa et al. 2015).

5.4.3 Antimalarial Compounds

Azadirachtin A and its structural analogs are a class of natural insecticides found exclusively in *Azadirachta indica*. In 2012, for the first time, *Eupenicillium parvum* colonizing *A. indica* reported for the production of azadirachtin A and B similar to that of the host plant under optimized conditions. These compounds showed significant insecticidal properties, suggesting the close association of endophytes with their hosts (Kusari et al. 2012b).

5.4.4 Anti-Quorum Sensing and Anti-Biofilm Compounds

The increase in the complexity of the treatment process of infectious diseases due to the continuous development of resistance mechanisms by microorganisms to currently available treatments and pharmaceutical drugs has lent urgency to the search for new therapeutic agents. The manifestations of microbial pathogenic factors like biofilm formation is directly regulated by a quorum sensing (QS) mechanism in microorganisms that causes complications in the field of healthcare, pharmaceuticals, food processing, and packing industries (Mookherjee et al. 2018). Studies have shown that endophytic fungal secondary metabolites could be potential agents for inhibiting the QS system in pathogenic microbes and inhibiting virulence factors including biofilm formation. Targeting the QS system of pathogenic microorganisms using natural products including secondary metabolites of endophytes has attracted significant attention because of the low selective pressure to develop resistance, to target bacterial virulence by QS inhibitory agents compared to traditional antibiotics, and the comparative safety of using natural drug molecules versus synthetic drugs for human health and the natural environmental. However, the therapeutic effects of QS-antagonistic in vivo conditions are not universally accepted, despite several successful studies on anti-QS agents in therapeutic practice in controlling infectious diseases (Christensen et al. 2012; Brackman et al. 2011). As discussed in the previous section, endophytic fungi are potent inhibitors of plant and animal pathogens by means of their secondary metabolites. With that background knowledge on secondary metabolites of various endophytic fungi and their secondary metabolites were isolated and investigated for their efficacy in blocking the QS circuit of pathogenic microorganisms and inhibiting their virulence factors. The endophytic fungus *Penicillium restrictum* was isolated from the stems of a milk thistle plant. In cultured broth, the isolate produces distinct red colored guttates, identified as new polyhydroxyanthraquinones, which have not been investigated, particularly from the standpoint of chemistry. Polyhydroxyanthraquinones were found to be effective QS inhibitors and potential antivirulence molecules against the clinical isolates of methicilillin-resistant *S. aureus* (Figueroa et al. 2014). Another study was conducted by Zhou and coworkers using crude metabolites from *Plectosphaerella cucumerina* (endophytic fungus from *Orychophragmus violaceus*); the researchers investigated their potential in blocking the QS, biofilm formation, and virulence factors of *P. aeruginosa*. The result revealed that at sub-MIC concentrations, the extract inhibited biofilm formation and distorted the biofilm of *P. aeruginosa.* Nevertheless, the crude metabolites are also involved in the inhibition of several QS-dependent virulence determines of the test organisms. This could be a fascinating example of the identification and screening of potential bioactive compounds from endophytic fungi (Zhou et al. 2017).

In recent times, preserving bioactive compounds from marine environments has attracted greater attention because of their versatility, richness, and obscurity. The endophytic fungus *Coniothyrium cereal* was isolated from the green alga *Enteromorpha* sp. and found to be capable of producing new phenalenone derivatives, well-studied compounds with anti-biofilm activity (Elsebai et al. 2011). Similar work was conducted aimed at the extraction of metabolites from aquatic fungi to investigate the effective anti-QS potentiality. Herein, large numbers of samples were isolated from reef organisms and their secondary metabolites were extracted and screened for QS inhibition. The experimental result revealed that endophytic filamentous fungi exhibited QS disruption activity without showing any antibacterial activity against the reporter strain *Chromobacterium violaceum* (Martín-Rodríguez et al. 2014).

5.4.5 *Other Biomedical Applications*

Besides the applications of plant-associated endophytic fungi in the generation of novel and alternative antimicrobial and anticancer agents, their other biomedical applications also account for greater understanding (Mishra et al. 2016b). The fungal metabolite (L-783, 281), isolated from an endophytic fungus, *Pseudomassaria* sp., acts as an insulin mimetic agent and has the ability to lower blood glucose levels significantly, suggesting its role as potential antidiabetic agent in the near future. In addition, the endophytic fungus *Fusarium subglutinans* produces diterpene pyrones such as subglutinol A and B, which showed potential immunomodulatory properties (Strobel and Daisy 2003).

5.5 Future Prospects

In recent times, the increasing trend toward resistance to available drugs has intensified the quest for novel and highly effective drugs in combating microbial infections and other serious diseases. Though bioactive compounds obtained from natural sources are effective at minimizing disease severity, the discovery of next-generation drugs remains an uphill task. In this context, the intervention of synthetic chemistry in developing novel drug molecules is attracting considerable attention (Deepika et al. 2016). Plants and plant products are the most exploited resources for the discovery of potent bioactive drugs. Microbial endophytes living in close association with host plants chemically bridge the gap between plants and microbes and has caught the attention of the scientific community due to their relatively high metabolic versatility and ability to produce a wide range of bioactive constituents similar to those of host plants (Mishra et al. 2016a; Deepika et al. 2016). It is imperative that the interaction between endophytic fungi with their host plants remains unique and even slight variations in culture conditions may result in the production of different kinds of secondary metabolites. From an intense literature survey, it has been elucidated that the optimization of culture conditions such as media composition, aeration, temperature, and other environmental parameters has a profound impact on the production of novel and desirable natural products by endophytes. Coculture systems can be further assisted by emerging and innovative biotechnological platforms such as genomics, transcriptomics, proteomics, metabolomics, high-throughput and next-generation sequencing (NGS) technologies, and bioinformatics tools to isolate and improve the metabolite production. These next-generation strategies will enhance our current understanding of endophytic molecular interactions with plant hosts with special reference to the regulation of the expression of specific gene clusters for enhanced production of desired bioactive compounds (Kusari et al. 2012a).

In the era of frontiers in drug discovery, endophytic fungi remain the most untapped source of bioactive constituents for therapeutics and industrial applications. As endophytic fungi are a rich source of natural products displaying a broad spectrum of biological activities, they remain the most bioprospected reservoir in current drug programs aimed at discovering novel compounds. Recently, advances in high-throughput genome-sequencing technology and other biotechnological interfaces such as biotransformation processes have emerged as the most efficient tools for the production of desired bioactive compounds (Pimentel et al. 2011; Rajamanikyam et al. 2017). Hence, it is imperative to understand the close association between fungal endophytes and host plants for the production of novel and highly proactive drugs with widespread applications (Mishra et al. 2017b; Jia et al. 2016).

5.6 Conclusion

Research on endophytic fungi facilitates the discovery of bioactive natural compounds and provides better knowledge about secondary metabolites producing microorganisms with their biosynthesis pathways. Advances in biotechnology resulting in identification at the molecular level have led to the development of simple techniques and methods for screening of potential fungal isolates from the environment for the successive extraction of novel bioactive compounds. Additionally, new technologies facilitates the overexpression of particular gene clusters to promote the utilization of microorganisms in the industrial sector for the bulk synthesis of biologically important molecules. Moreover, the production of small molecules by fungi could encourage the isolation of novel therapeutics and enzymes to boost national economies. Hence, endophytes are an interesting area for the exploration of novel natural metabolites as therapeutic agents, food additives, economically important enzymes, agriculturally important molecules, and other applications in environmental sustainability.

References

Akone SH, Mandi A, Kurtan T, Hartmann R, Lin W, Daletos G, Proksch P (2016) Inducing secondary metabolite production by the endophytic fungus *Chaetomium* sp. through fungalebacterial co-culture and epigenetic modification. Tetrahedron 72:6340–6347

Alvin A, Miller KI, Neilan BA (2014) Exploring the potential of endophytes from medicinal plants as sources of antimycobacterial compounds. Microbiol Res 169:483–495

Aly AH, Edrada-Ebel RA, Wray V, Muller WEG, Kozytska S, Hentschel U, Proksch P, Ebel R (2008) Bioactive metabolites from the endophytic fungus *Ampelomyces* sp. isolated from the medicinal plant *Urospermum picroides*. Phytochemistry 69:1716–1725

Arnold AE (2007) Understanding the diversity of foliar endophytic fungi: progress, challenges, and frontiers. Fungal Biol Rev 21:51–66

Brackman G, Cos P, Maes L, Nelis HJ, Coenye T (2011) Quorum sensing inhibitors increase the susceptibility of bacterial biofilms to antibiotics in vitro and in vivo. Antimicrob Agents Chemother 55(6):2655–2661

Buhaescu I, Izzedine H (2007) Mevalonate pathway: a review of clinical and therapeutical implications. Clin Biochem 40(9–10):575–584

Cao X, Li J, Zhou L, Xu L, Li J, Zhao J (2007) Determination of diosgenin content of the endophytic fungi from *Paris polyphylla* var. *yunnanensis* by using an optimum ELISA. Nat Prod Res Dev 19:1020–1023

Casella TM, Eparvier V, Mandavid H, Bendelac A, Odonne G, Dayan L, Duplais C, Espindola LS, Stien D (2013) Antimicrobial and cytotoxic secondary metabolites from tropical leaf endophytes: isolation of antibacterial agent pyrrocidine C from *Lewia infectoria* SNB-GTC2402. Phytochemistry 96:370–377

Chan YA, Podevels AM, Kevany BM, Thomas MG (2009) Biosynthesis of polyketide synthase extender units. Nat Prod Rep 26(1):90–114

Chaturvedi P, Gajbhiye S, Roy S, Dudhale R, Chowdhary A (2014) Determination of kaemferol in extracts of *Fusarium chlamydosporum*, an endophytic fungi of *Tylophora indica* (Asclepeadaceae) and its anti-microbial activity. IOSR J Pharm Biol Sci 9:51–55

Christensen LD, van Gennip M, Jakobsen TH, Alhede M, Hougen HP, Hoiby N, Bjarnsholt T, Givskov M (2012) Synergistic antibacterial efficacy of early combination treatment with tobramycin and quorum-sensing inhibitors against Pseudomonas aeruginosa in an intraperitoneal foreign-body infection mouse model. J Antimicrob Chemother 67(5):1198–1206

Cui Y, Yi D, Bai X, Sun B, Zhao Y, Zhang Y (2012) Ginkgolide B produced endophytic fungus (*Fusarium oxysporum*) isolated from *Ginkgo biloba*. Fitoterapia 83:913–920

Deepika VB, Murali TS, Satyamoorthy K (2016) Modulation of genetic clusters for synthesis of bioactive molecules in fungal endophytes: a review. Microbiol Res 182:125–140

Demain AL, Sanchez S (2009) Microbial drug discovery: 80 years of progress. J Antibiot 62:5–16

Elsebai MF, Kehraus S, Lindequist U, Sasse F, Shaaban S, Gütschow M, Josten M, Sahl HG, König GM (2011) Antimicrobial phenalenone derivatives from the marine-derived fungus *Coniothyrium cereale*. Org Biomol Chem 9(3):802–808

Feng P, Shang Y, Cen K, Wang C (2015) Fungal biosynthesis of the bibenzoquinone oosporein to evade insect immunity. Proc Natl Acad Sci 112(36):11365–11370

Figueroa M, Jarmusch AK, Raja HA, El-Elimat T, Kavanaugh JS, Horswill AR, Cooks RG, Cech NB, Oberlies NH (2014) Polyhydroxyanthraquinones as quorum sensing inhibitors from the guttates of *Penicillium restrictum* and their analysis by desorption electrospray ionization mass spectrometry. J Nat Prod 77(6):1351–1358

Fouda AH, Hassan SE, Eid AM, Ewais EE (2015) Biotechnological applications of fungal endophytes associated with medicinal plant *Asclepias sinaica* (Bioss.). Ann Agric Sci 60(1):95–104

Hardoim PR, van Overbeek LS, Berg G, Pirttila AM, Compant S, Campisano A, Doring M, Sessitsch A (2015) The hidden world within plants: ecological and evolutionary considerations for defining functioning of microbial endophytes. Microbiol Mol Biol Rev 79(3):293–320

Herrmann KM (1995) The shikimate pathway: early steps in the biosynthesis of aromatic compounds. Plant Cell 7(7):907–919

Herrmann KM, Weaver LM (1999) The shikimate pathway. Annu Rev Plant Physiol Plant Mol Biol 50(1):473–503

Jalgaonwala RE, Mohite BV, Mahajan RT (2011) A review: natural products from plant associated endophytic fungi. J Microbiol Biotechnol Res 1(2):21–32

Jia M, Chen L, Xin HL, Zheng CJ, Rahman K, Han T, Qin LP (2016) A friendly relationship between endophytic fungi and medicinal plants: a systematic review. Front Microbiol 7:906

Kumara PM, Zuehlke S, Priti V, Ramesha BT, Shweta S, Ravikanth G, Vasudeva R, Santhoshkumar TR, Spiteller M, Uma Shaanker R (2012) *Fusarium proliferatum*, an endophytic fungus from *Dysoxylum binectariferum* Hook.f, produces rohitukine, a chromane alkaloid possessing anti-cancer activity. Antonie Van Leeuwenhoek 101(2):323–329

Kumara PM, Shweta S, Vasanthakumari MM, Sachin N, Manjunatha BL, Jadhav SS, Ravikanth G, Ganeshaiah KN, Shaanker RU (2014) Endophytes and plant secondary metabolite synthesis: molecular and evolutionary perspective. In: Verma VC, Gange AC (eds) Advances in endophytic research. Springer India, New Delhi, pp 177–190

Kusari S, Lamshoft M, Spiteller M (2009) *Aspergillus fumigatus* Fresenius, an endophytic fungus from *Juniperus communis* L. Horstmann as a novel source of the anticancer pro-drug deoxypodophyllotoxin. J Appl Microbiol 107:1019–1030

Kusari S, Hertweck C, Spiteller M (2012a) Chemical ecology of endophytic fungi: origins of secondary metabolites. Chem Biol 19:792–798

Kusari S, Verma VC, Lamshoeft M, Spiteller M (2012b) An endophytic fungus from *Azadirachta indica* A. Juss. that produces azadirachtin. World J Microbiol Biotechnol 28:1287–1294

Kusari S, Pandey SP, Spiteller M (2013) Untapped mutualistic paradigms linking host plant and endophytic fungal production of similar bioactive secondary metabolites. Phytochemistry 91:81–87

Kusari S, Singh S, Jayabaskaran C (2014) Biotechnological potential of plant-associated endophytic fungi: hope versus hype. Trends Biotechnol 32(6):297–303

Le Cocq K, Gurr SJ, Hirsch PR, Mauchline TH (2017) Exploitation of endophytes for sustainable agricultural intensification. Mol Plant Pathol 18(3):469–473

Lombard J, Moreira D (2011) Origins and early evolution of the mevalonate pathway of isoprenoid biosynthesis in the three domains of life. Mol Biol Evol 28(1):87–99

Lugtenberg BJJ, Caradus JR, Johnson LJ (2016) Fungal endophytes for sustainable crop production. FEMS Microbiol Ecol 92(12). Edited by G. Muyzer; pii: fiw194

Lunardelli Negreiros de Carvalho P, de Oliveira Silva E, Aparecida Chagas-Paula D, Honorata Hortolan Luiz J, Ikegaki M (2016) Importance and implications of the production of phenolic secondary metabolites by endophytic fungi: a mini-review. Mini-Rev Med Chem 16(4):259–271

Maeda H, Dudareva N (2012) The shikimate pathway and aromatic amino acid biosynthesis in plants. Annu Rev Plant Biol 63(1):73–105

Martín-Rodríguez A, Reyes F, Martín J, Pérez-Yépez J, León-Barrios M, Couttolenc A, Espinoza C, Trigos Á, Martín V, Norte M, Fernández J (2014) Inhibition of bacterial quorum sensing by extracts from aquatic fungi: first report from marine endophytes. Mar Drugs 12(11):5503–5526

Miller KI, Qing C, Sze DMY, Neilan BA (2012) Investigation of the biosynthetic potential of endophytes in traditional Chinese anticancer herbs. PLoS One 7(5):35953

Mishra VK, Singh G, Passari AK, Yadav MK, Gupta VK, Singh BP (2016a) Distribution and antimicrobial potential of endophytic fungi associated with ethnomedicinal plant Melastoma malabathricum L. J Environ Biol 37(2):229–237

Mishra VK, Passari AK, Singh BP (2016b) In vitro antimycotic and biosynthetic potential of fungal endophytes associated with Schima Wallichii. In: Kumar P et al (eds) Current trends in disease diagnostics. Springer, Cham, pp 367–381

Mishra VK, Passari AK, Chandra P, Leo VV, Kumar B, Gupta VK, Singh BP (2017a) Determination and production of antimicrobial compounds by Aspergillus clavatonanicus strain MJ31, an endophytic fungus from Mirabilis jalapa L. using UPLC-ESI-MS/MS and TD GC-MS. PLoS One 12(10):1–24. https://doi.org/10.1371/journal.pone.0186234

Mishra VK, Passari AK, Leo VV, Singh BP (2017b) Molecular diversity and detection of endophytic fungi based on their antimicrobial biosynthetic genes. In: Singh BP, Gupta VK (eds)

Molecular markers in mycology, fungal biology. Springer, Cham, pp 1–35. https://doi.org/10.1007/978-3-319-34106-4_1

Mookherjee A, Singh S, Maiti MK (2018) Quorum sensing inhibitors: can endophytes be prospective sources? Arch Microbiol 200(2):355–369

Mousa WK, Schwan A, Davidson J, Strange P, Liu H, Zhou T, Auzanneau FI, Raizada MN (2015) An endophytic fungus isolated from finger millet (*Eleusine coracana*) produces anti-fungal natural products. Front Microbiol 6:1157

Nair DN, Padmavathy S (2014) Impact of endophytic microorganisms on plants, environment and humans. Sci World J 2014:1–11

Nirupama R, Chaithra K, Govindappa M, Chandrappa CP, Vinay RB (2011) Antimicrobial and antioxidant activities of endophytes from *Tabebuia argentea* and identification of anticancer agent (lapachol). J Med Plant Res 5:3643–3652

Nisa H, Kamili AN, Nawchoo IA, Shafi S, Shameem N, Bandh SA (2015) Fungal endophytes as prolific source of phytochemicals and other bioactive natural products: a review. Microb Pathog 82:50–59

Parthasarathy R, Sathiyabama M (2014) Gymnemagenin-producing endophytic fungus isolated from a medicinal plant *Gymnema sylvestre* R.Br. Appl Biochem Biotechnol 172(6):3141–3152

Pimentel MR, Molina G, Dionísio AP, Maróstica Junior MR, Pastore GM (2011) The use of endophytes to obtain bioactive compounds and their application in biotransformation process. Biotechnol Res Int 2011:576286

Pongcharoen W, Rukachaisirikul V, Phongpaichit S, Kuhn T, Pelzing M, Sakayaroj J, Taylor WC (2008) Metabolites from the endophytic fungus *Xylaria* sp. PSU-D14. Phytochemistry 69:1900–1902

Rai M, Rathod D, Agarkar G, Dar M, Brestic M, Pastore GM, Junior MRM (2014) Fungal growth promotor endophytes: a pragmatic approach towards sustainable food and agriculture. Symbiosis 62(2):63–79

Rajamanikyam M, Vadlapudi V, Amanchy R, Upadhyayula SM (2017) Endophytic fungi as novel resources of natural therapeutics. Braz Arch Biol Technol 60:e17160542

Rao HCY, Satish S (2015) Genomic and chromatographic approach for the discovery of polyketide antimicrobial metabolites from an endophytic *Phomopsis liquidambaris* CBR-18. Front Life Sci 8(2):200–207

Refaei J, Hussein ZAM, Santhanam J (2014) Diversity of type i polyketide synthase genes in a bioactive Phoma sp. endophytic fungus isolated from *Cinnamomum molissimum* in Malaysia. J Pure Appl Microbiol 8(2):1173–1182

Rodriguez RJ, White JF Jr, Arnold AE, Redman RS (2009) Fungal endophytes: diversity and functional roles. New Phytol 182:314–330

Santiago C, Sun L, Munro MHG, Santhanam J (2014) Polyketide and benzopyran compounds of an endophytic fungus isolated from *Cinnamomum mollissimum*: biological activity and structure. Asian Pac J Trop Biomed 4(8):627–632

Silva GH, Zeraik ML, de Oliveira CM, Teles HL, Trevisan HC, Pfenning LH, Nicolli CP, Young MCM, Mascarenhas YP, Abreu LM, Saraiva AC, Medeiros AI, Bolzani VS, Araujo AR (2017) Lactone derivatives produced by a *Phaeoacremonium* sp., an endophytic fungus from *Senna spectabilis*. J Nat Prod 80:1674–1678

Souza BS, dos Santos TT (2017) Endophytic fungi in economically important plants: ecological aspects, diversity and potential biotechnological applications. J Bioenergy Food Sci 4(2):113–126

Strobel GA (2003) Endophytes as sources of bioactive products. Microb Infect 5:535–544

Strobel G, Daisy B (2003) Bioprospecting for microbial endophytes and their natural products. Microbiol Mol Biol Rev 67(4):491–502

Suryanarayanan TS (2013) Endophyte research: going beyond isolation and metabolite documentation. Fungal Ecol 6:561–568

Tricarico P, Crovella S, Celsi F (2015) Mevalonate pathway blockade, mitochondrial dysfunction and autophagy: a possible link. Int J Mol Sci 16(7):16067–16084

Tzin V, Galili G (2010) The biosynthetic pathways for shikimate and aromatic amino acids in *Arabidopsis thaliana*. Arab Book 8(i):0132

Uzma F, Konappa NM, Chowdappa S (2016) Diversity and extracellular enzyme activities of fungal endophytes isolated from plants of Western Ghats, Karnataka. Egypt J Basic Appl Sci 3:335–342

Uzma F, Hashem A, Murthy N, Mohan HD, Kamath PV, Singh BP, Venkataramana M, Gupta VK, Siddaiah CN, Chowdappa S, Alqaeawi AA, Abd Allah EF (2018) Endophytic fungi—alternative sources of cytotoxic compounds: a review. Front Pharmacol 9(309):1–37. https://doi.org/10.3389/fphar.2018.00309

Venieraki A, Dimou M, Katinakis P (2017) Endophytic fungi residing in medicinal plants have the ability to produce the same or similar pharmacologically active secondary metabolites as their hosts. Hellenic Plant Protect J 10:51–66

Venugopalan A, Srivastava S (2015) Endophytes as in vitro production platforms of high value plant secondary metabolites. Biotechnol Adv 33:873–887

Verma VC, Lobkovsky E, Gange AC, Singh SK, Prakash S (2011) Piperine production by endophytic fungus *Periconia* sp. isolated from *Piper longum* L. J Antibiot 64:427–431

Wang XJ, Min CL, Ge M, Zuo RH (2014) An endophytic sanguinarine-producing fungus from *Macleaya cordata, Fusarium proliferatum* BLH51. Curr Microbiol 68:336–341

Wani ZA, Ashraf N, Mohiuddin T, Riyaz-Ul-Hassan S (2015) Plant-endophyte symbiosis, an ecological perspective. Appl Microbiol Biotechnol 99:2955–2965

Waqas M, Khan AL, Hamayun M, Shahzad R, Kang SM, Kim JG, Lee IJ (2015) Endophytic fungi promote plant growth and mitigate the adverse effects of stem rot: an example of *Penicillium citrinum* and *Aspergillus terreus*. J Plant Interact 10(1):280–287

Yao L, Tan C, Song J, Yang Q, Yu L, Li X (2016) Isolation and expression of two polyketide synthase genes from *Trichoderma harzianum* 88 during mycoparasitism. Braz J Microbiol 47(2):468–479

You X, Feng S, Luo S, Cong D, Yu Z, Yang Z, Zhang J (2013) Studies on a rhein-producing endophytic fungus isolated from *Rheum palmatum* L. Fitoterapia 85:161–168

Yu H, Zhang L, Li L, Zheng C, Guo L, Li W, Sun P, Qin L (2010) Recent developments and future prospects of antimicrobial metabolites produced by endophytes. Microbiol Res 165:437–449

Yuan Y, Feng H, Wang L, Li Z, Shi Y, Zhao L, Feng Z, Zhu H (2017) Potential of endophytic fungi isolated from cotton roots for biological control against Verticillium wilt disease. PLoS One 12(1):e0170557

Zhang HW, Song YC, Tan RX (2006) Biology and chemistry of endophytes. Nat Prod Rep 23:753–771

Zhang D, Tao X, Chen R, Liu J, Li L, Fang X, Yu L, Dai J (2015) Pericoannosin A, a polyketide synthase–nonribosomal peptide synthetase hybrid metabolite with new carbon skeleton from the endophytic fungus Periconia sp. Org Lett 17(17):4304–4307

Zhao J, Zhou L, Wang J, Shan T, Zhong L, Liu X, Gao X (2010) Endophytic fungi for producing bioactive compounds originally from their host plants. In: Mendez-Vilas A (ed) Current research, technology and education topics in applied microbiology and microbial biotechnology. Formatex Research Center, Badajoz, pp 567–576

Zhou J, Bi S, Chen H, Chen T, Yang R, Li M, Fu Y, Jia AQ (2017) Anti-biofilm and antivirulence activities of metabolites from *Plectosphaerella cucumerina* against *Pseudomonas aeruginosa*. Front Microbiol 8(MAY):1–17

Zilla MK, Qadri M, Pathania AS, Strobel GA, Nalli Y, Kumar S, Guru SK, Bhushan S, Singh SK, Vishwakarma RA, Riyaz-Ul-Hassan S, Ali A (2013) Bioactive metabolites from an endophytic *Cryptosporiopsis* sp. inhabiting *Clidemia hirta*. Phytochemistry 95:291–297

[illegible]

Chapter 6
Biodiversity of Endophytic Fungi from Diverse Niches and Their Biotechnological Applications

Kusam Lata Rana, Divjot Kour, Imran Sheikh, Neelam Yadav, Ajar Nath Yadav, Vinod Kumar, Bhim Pratap Singh, Harcharan Singh Dhaliwal, and Anil Kumar Saxena

6.1 Introduction

Endophytes are defined as microbes living asymptomatically in tissues of plants, causing no harm to the host plants and isolated from surface sterilized explants. The word endophytes means "in the plant" (from Greek *endon* = within, *phyton* = plant), and endophytes originated mostly from the rhizosphere and phyllosphere (Ryan et al. 2008). The use of this term is as wide as its actual definition and scope of potential hosts and inhabitants, for example, bacteria (Kobayashi and Palumbo 2000), fungi (Stone et al. 2000), plants (Marler et al. 1999), and insects in plants (Feller 1995). Different authors have defined endophytes in comparatively distinct ways (Fig. 6.1) (Bary 1866; Petrini 1991; Rollinger and Langwnheim 1993; Mostert et al. 2000; Wilson 1995; Hirsch and Braun 1992; Carroll 1977; Mercado-Blanco and Lugtenberg 2014; Rosenblueth and Martínez-Romero 2006; Schulz and Boyle 2005). Microbial endophytes can be segregated from exterior-sterilized plant tissue or obtained from the interior tissue of plants (Hallmann et al. 1997). Signaling molecules such as flavonoids, isoflavonoids, and phenolics are ejected from the plant roots, attracting fungi from the rhizosphere to colonize inside the plant as

K. L. Rana · D. Kour · I. Sheikh · A. N. Yadav (✉) · V. Kumar · H. S. Dhaliwal
Department of Biotechnology, Akal College of Agriculture, Eternal University, Baru Sahib, Himachal Pradesh, India
e-mail: ajar@eternaluniversity.edu.in

N. Yadav
Gopi Nath P.G. College, VBSP University, Ghazipur, Uttar Pradesh, India

B. P. Singh
Molecular Microbiology and Systematics Laboratory, Department of Biotechnology, Mizoram University, Aizawl, Mizoram, India

A. K. Saxena
ICAR-National Bureau of Agriculturally Important Microorganisms, Mau, Uttar Pradesh, India

B. P. Singh (ed.), *Advances in Endophytic Fungal Research*, Fungal Biology,
https://doi.org/10.1007/978-3-030-03589-1_6

Fig. 6.1 Endophytic microbial world and its biotechnological applications in agriculture

endophytes. Endophytes are ubiquitous and reported in most plant studies and also considered niche specific. Plant-associated fungi may be sorted in different groups on the basis of their survival and role such as mycorrhizal, pathogenic, epiphytic, endophytic, and saprotrophic fungi (Porras-Alfaro and Bayman 2011). A relatively small number of organisms, such as mycorrhizal fungi, endophytic fungi, and pathogenic fungi, cross the boundary at the rhizosphere and reach deeper into the roots of plants (Bais et al. 2008).

Fungal endophytes colonize the tissues of hosts; transmission of endophytes follows two routes, vertical and horizontal (Carroll 1988). Vertical transmission of endophytic fungi takes place from mother plant to offspring, via seeds, often referred to as "true endophytes." *Undifilum oxytropisi* fungal endophytes isolated from locoweed plant species provide good evidence of vertical transmission; (Cook et al. 2009). A study done by Hodgson et al. (2014) in six forb species, cornflower, lesser knapweed, common poppy, ribwort plantain, sheep's sorrel, and groundsel, and two endophyte species, *Alternaria alternata* and *Cladosporium sphaerospermum*, reported on the vertical transmission of species. Horizontal transmission occurs via soil- or airborne spores. To date, endophytes in forbs affect the leaves of their hosts via horizontal transmission.

Earlier fossil records and the evolutionary process revealed an association between endophytic fungi and diverse groups of plants. Plant–microbe interactions resulted in adaptation, plant growth promotion, uptake of micronutrients, and the production of various groups of secondary metabolites and bioactive compounds with promising applications in agriculture, therapeutics, and industry. Because they protect plants against biotic and abiotic stresses, beneficial endophytic fungi are considered eco-friendly bioresources. Endophytic fungi may take part in plant growth promotion solubilization of phosphorus, potassium, and zinc; the production of phytohormones (indole acetic acids, gibberrellic acids, and cytokinin), Fe-chelating compounds, hydrolytic enzymes, hydrogen cyanide, and ammonia (Rai et al. 2014). Different groups of fungi such as Ascomycota, Basidiomycota, Mucoromycota, and Oomycota have been reported as plant growth promoters and protectants under normal natural and abiotic stress conditions. Arbuscular mycorrhizae (AMs) are fungi that are well known to promote activities that can enhance agricultural improvements. Thus, these microorganisms have emerged as a main research subject in connection with sustainability targets. Over 90% of the roots of all plant species are reported to associate with mycorrhizal fungi. Between host plants and the soil, endomycorrhizae assist in the interchange of nutrients. Various inorganic phosphorus, mineral, or organic nitrogen and amino acids are captured from the soil by mycorrhizae (Bonfante and Genre 2010; Finlay 2008; Mathur et al. 2011; Suman et al. 2016b; Verma et al. 2017b; Mishra et al. 2015). Fungal endophytes producing natural products have been discovered to hinder the extensive types of diseases caused in humans, animals, and plants by phytopathogens, bacteria, fungi, viruses, and protozoans, and the metabolites and bioactive compounds produced by fungal endophyte possess antimicrobial, anticancer, and insecticidal properties significant in pharmaceutical science and in biotechnological applications (Uzma et al. 2018; Mishra et al. 2017a; Aly et al. 2010; Kusari and Spiteller 2011; Marinho et al. 2005; Wang et al. 2007; Zhang et al. 2006).

Endophytic microbes (archaea, bacteria, and eukarya) remain in plant tissues without inducing significant damage to the host. They subsist largely inside the living tissues of plant species in a pattern of symbiotic association to somewhat pathogenic microbes. These microbes have been segregated from a distinct variety of plants, including *Triticum* (Yadav et al. 2018a; Verma et al. 2015c, 2016a, b; Yadav 2009), *Oryza sativa* (Mano and Morisaki 2007; Naik et al. 2009a; Piromyou et al. 2015), *Capsicum annuum L.* (Kang et al. 2007; Yang et al. 2009), *Saccharum officinarum* (Mendes et al. 2007), *Zea mays* (Araújo et al. 2000; Montanez et al. 2012; Thanh and Diep 2014), mustard (Lee et al. 2008; Poonguzhali et al. 2006; Sheng et al. 2008), citrus (Andreote et al. 2008; Kasotia and Choudhary 2014), chilli (Kang et al. 2007), *Solanum tuberosum* (Manter et al. 2010; Rado et al. 2015), *Solanum lycopersicum* (Hallmann et al. 1997), *Glycine max* (Hung and Annapurna 2004; Mingma et al. 2014), *Pisum sativum* (Narula et al. 2013; Tariq et al. 2014), *Phaseolus vulgaris* (Suyal et al. 2015), *Helianthus* (Ambrosini et al. 2012; Forchetti et al. 2010), *Gossypium* (Quadt-Hallmann et al. 1997), *Cicer arietinum* (Saini et al. 2015), *Pennisetum glaucum* (Beatriz Sánchez et al. 2014), and *Fragaria × ananassa* (Hardoim et al. 2012). Fungal endophytes pertaining to diverse genera, including *Acremonium, Alternaria, Aspergillus, Berkleasmium, Chaetomium, Cladosporium, Claviceps, Colletotrichum, Cryptococcus, Curvularia, Fusarium, Geomyces, Glomus, Leptospora, Metarhizium, Microdochium, Neotyphodium, Ophiognomonia,*

Paecilomyces, Penicillium, Phaeomoniella, Phyllosticta, Piriformospora, Rhizoctonia, Rhizopus, Rhodotorula, Talaromyces, Trichoderma, Wallemia, and *Xylaria*, have been isolated from various host plants (Suman et al. 2016b; Verma et al. 2017b; Yadav et al. 2018a, c).

Biotechnology has opened up new opportunities for biotechnological uses of beneficial endophytic microbes for agriculture, industry, and medicine. An understanding of endophytic fungal diversity from diverse niches and their various uses in agriculture is essential and functional for the growth of plants, protection, and yield (Yadav 2018; Singh et al. 2017). Because of their capability to assist the growth of plants and adapt under diverse, extreme abiotic stresses, endophytic microbes caputured the attention of the scientific community (Soni et al. 2018). Endophytic fungi, with the capacity to yield novel secondary metabolites and bioactive compounds, will find application in a wide area of industrial, horticultural, and pharmaceutical processes (Joseph and Priya 2011). Endophytic fungal consortiums with beneficial endophytic bacterial strains will probably be useful in agriculture as bio-inoculants and biocontrol agents and for the biofortification of micronutrients. This chapter defines the process of isolation of endophytic fungi from diverse niches and their characterization, identification, biodiversity, and biotechnological applications in agriculture, pharmaceuticals, and industry.

6.2 Isolation and Characterization of Fungal Endophytes

Fungi that colonize plant tissues are mostly affected by the environmental conditions surrounding the host plants such as the type of soil and its pH, rain, alkalizing property of the soil, and climate. Endophytic fungi may develop in small amounts and occasionally in confined locations inside plants, so that it is difficult to determine their particular relation to their host plant. For the separation of fungal endophytes, care must be taken to prevent contamination by undesired epiphytic fungi (Singh et al. 2016a, b). Endophytic fungi may be isolated from different parts of plants such as roots, shoots, stems, leaves, flowers, bark, and meristems. To isolate the fungi, plant samples should be collected in a sterile polythene bag under aseptic conditions and transferred to the laboratory (Lu et al. 2012). The plant parts should be washed with tap water to eliminate any adhering soil particles. Surface disinfection is an essential step for the elimination of epiphytic fungi (Nefzi et al. 2018), which may be isolated using various methods of surface sterilization or growth media.

Samples of plants are generally sterilized by subsequent submergence in 70% ethanol for 1–3 min and 1–3% sodium hypochlorite for 3–5 min, followed by repetitive washing in sterile water to eliminate any remaining sodium hypochlorite (Miché and Balandreau 2001). Double or triple surface sterilization with a mixture of ethanol and other disinfectants is also recommended to eliminate epiphytic microbes (Suman et al. 2016a, b). All the desired samples should be drenched separately with 10 mL sterile 0.85% NaCl using a pestle and mortar and then blended by vortices running at full speed for 60 s. The products should then be used to isolate the endophytic fungi, which may be isolated using an enrichment method and a standard serial dilution plating technique. To isolate different groups of fungal strains, selective, differential, and complex media may be used with original constituents or by diluting into 10, 20, 50, and 100 times in their constituents. Along with natural endophytic fungi, extremophilic fungi may also be isolated through enrichments, other techniques, and conditions, for example, acidophilic fungi (pH 3–5), alkaliphilic fungi (pH 8–11), halophilic fungi (with 5–20% NaCl concentration), psychrophilic fungi (incubation at >5 °C), thermophilic fungi (incubation at >45 °C), and xerophilic fungi (7–10% polyethylene glycol) (Yadav et al. 2015c, d). The plates should be incubated for up to 5, 15, and 30 days for the isolation of fast-, medium-, and slow-growing endophytic fungi.

For the identification and characterization of fungal endophytes, both morphological and molecular techniques should be used. The morphological characterization of fungi may be carried out on the basis of color, size of spores, colony diameter, texture, shape, growth rate, morphology of hyphae, and conidia. For long-term preservation, mycelia and spores are collected in 20% glycerol in water (v/v) and stored at −80 °C. For the identification of fungal endophytes, modern techniques of molecular biology using genomic DNA isolation, amplifications of ITS genes, and sequencing of desired genes may be used. For phylogeny and taxonomical affiliation, polymerase chain reaction (PCR)-amplified Internal transcribed spacer (ITS) genes are sequenced and compared with sequences accessible in the NCBI database and with help of MEGA software.

To identify plant growth promoting capability and other agricultural and biotechnological uses of endophytic fungi (Yadav et al. 2018b), for plant-growth-promoting (PGP) attributes endophytic fungi should be screened, for instance, iron-chelating agents (Schwyn and Neilands 1987), production of phytohormones indole-3-acetic acid (Bric et al. 1991), gibberellic acid (Brown and Burlingham 1968), production of ammonia (Cappucino and Sherman 1992), hydrogen cyanide (HCN) (Bakker and Schippers 1987), 1-aminocyclopropane-1-carboxylate (ACC) deaminase (Jacobson et al. 1994), solubilization of potassium (Hu et al. 2006), phosphorus solubilization, (Pikovskaya 1948), and zinc solubilization (Fasim et al. 2002), hydrolytic enzyme production (Yadav et al. 2016b), and biocontrol against different microbial pathogens (Sijam and Dikin 2005).

6.3 Biodiversity of Fungal Endophytes

The microbes belonging to various groups have been described as endophytes such as archaea, bacteria, and fungi (Yadav et al. 2018a). Among fungi, the members of different phyla such as Ascomycota, Basidiomycota, Mucoromycota, and Oomycota have been identified as endophytic from various crops (Fig. 6.2). Overall, the distribution of endophytic fungi has been reported to be mostly from the phylum Ascomycota followed by Basidiomycota. The least number of fungal strains have

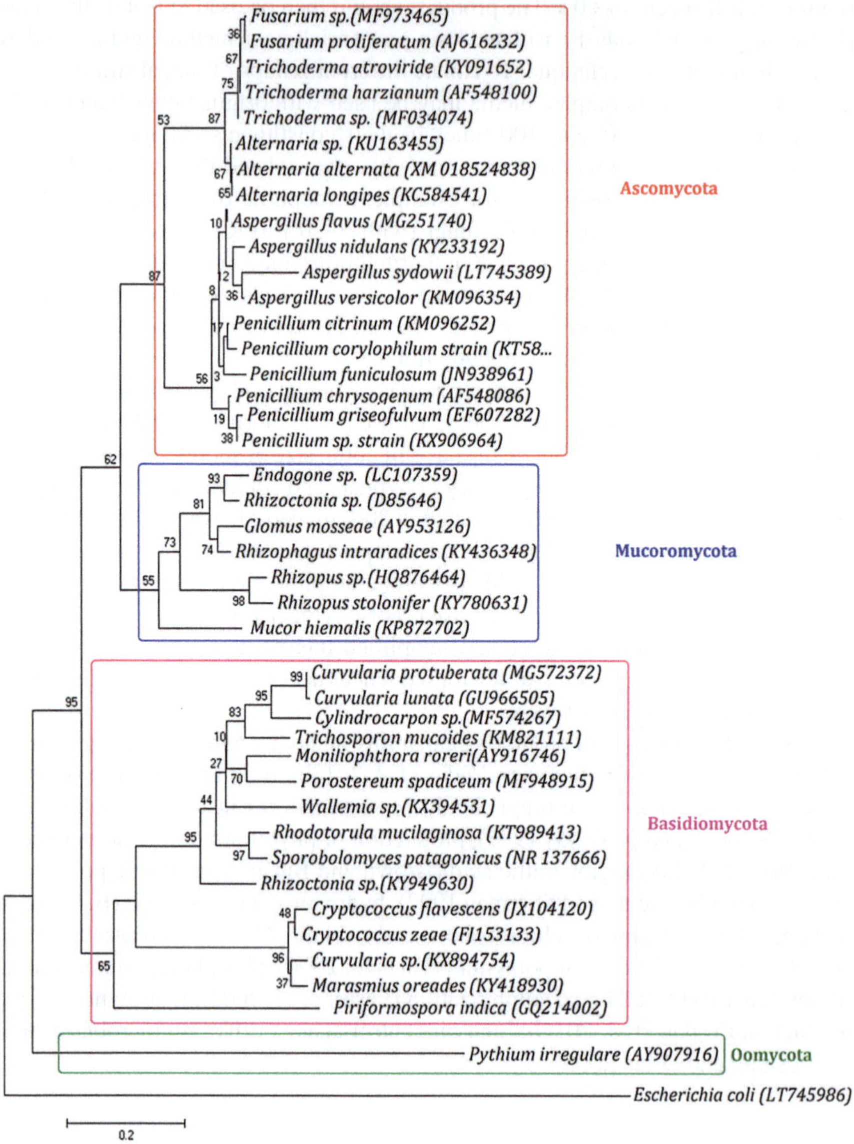

Fig. 6.2 Phylogenetic tree showing relationship among endophytic fungi associated with various crops

been reported from the phylum Oomycota (Fig. 6.3). A review of various studies on the biodiversity of endophytic fungi from different crops (Nisa et al. 2015) revealed that the dominant genera include *Aspergillus, Fusarium, Penicillium*, and *Piriformospora*. Along with predominant genera some niche-specific fungal strains have been reported, for example, *Penicillium brevicompactum* and *Penicillium*

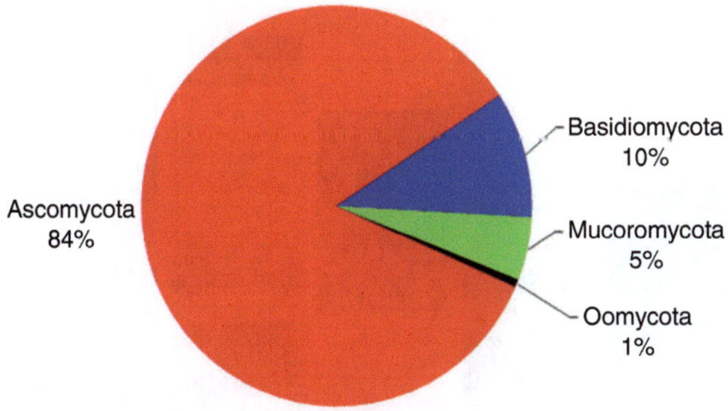

Fig. 6.3 Abundance of endophytic fungi belonging to diverse phyla isolated from various plants

glabrum isolated from barley (*Hordeum vulgare*); *Berkleasmium, Chaetomium, Cryptococcus, Exophiala, Gibberella zeae, Helminthosporium, Kernia pachypleura, Marasmius oreades, Paecilomyces, Periconia macrospinosa, Phaeomoniella capensis,* and *Wallemia* isolated from maize (*Zea mays*); *Colletotrichum boninense, Colletotrichum capsici, Colletotrichum fragariae, Colletotrichum gloeosporioides, Colletotrichum truncatum, Diaporthe helianthi, Diaporthe phaseolorum, Didymella bryoniae, Fusarium proliferatum, Gibberella moniliformis, Guignardia mangiferae, Guignardia vaccinii, Leptospora rubella, Magnaporthegrisea, Nectriamauritiicola, Paraconiothyrium variabile, Penicillium funiculosum,* and *Phoma herbarum* isolated from soybean (*Glycine max*); *Diaporthe endophytica, Epicoccum huancayense, Fusarium bactridioides, Fusarium verticillioides,* and *Metarhizium brunneum* isolated from sugarcane (*Saccharum officinarum*); and *Pleospora herbarum, Rhodotorula rubra, Septoriatritici, Setosphaeria rostrata, Stemphylium, Stemphylium botryosum, Talaromyces flavus, Trichoderma atroviride,* and *Trichoderma hamatum* isolated from wheat (*Triticum aestivum*) (Fig. 6.4). Very few studies have been published on niche-specific microbes but there are several on the niche specificity of microbes from various extreme habitats (Saxena et al. 2016; Yadav et al. 2015a, 2016b, 2017b). Upon review of six different crops, it was found that of those endophytic fungi that have been studied, most belong to the phylum Ascomycota. Among different reported genera from 6 cereal crops, 14 different genera, *Acremonium, Alternaria, Aspergillus, Cladosporium, Cryptococcus, Curvularia, Fusarium, Metarhizium, Mucor, Ophiognomonia, Penicillium, Phomopsis, Trichoderma*, and *Xylaria*, with more than 3 distinct species from one or more crops have been reported (Fig. 6.5).

Since the nineteenth century, there have been significant reports on the biodiversity of endophytic microbes from diverse crops. In plants, the highly diverse group of endophytic fungi provide tolerance against biotic and abiotic stresses. Fisher et al. (1992) isolated fungal endophytes from healthy maize plants from the core and epidermis of the stem and identified them as *Aureobasidium pullulans* and *Alternaria alternata*. Larran et al. (2002) investigated the most efficient sterilization technique

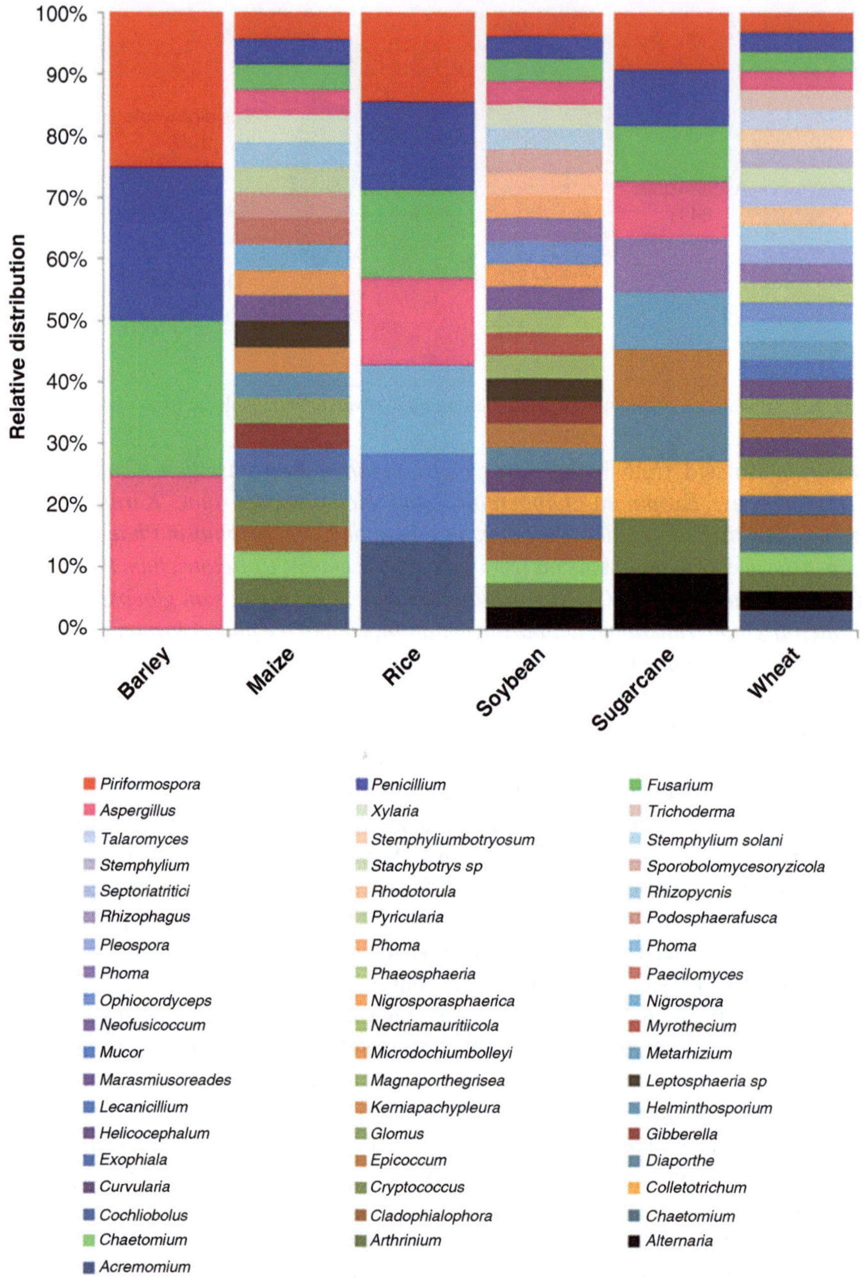

Fig. 6.4 Diversity and distribution of endophytic fungi reported from various host plants

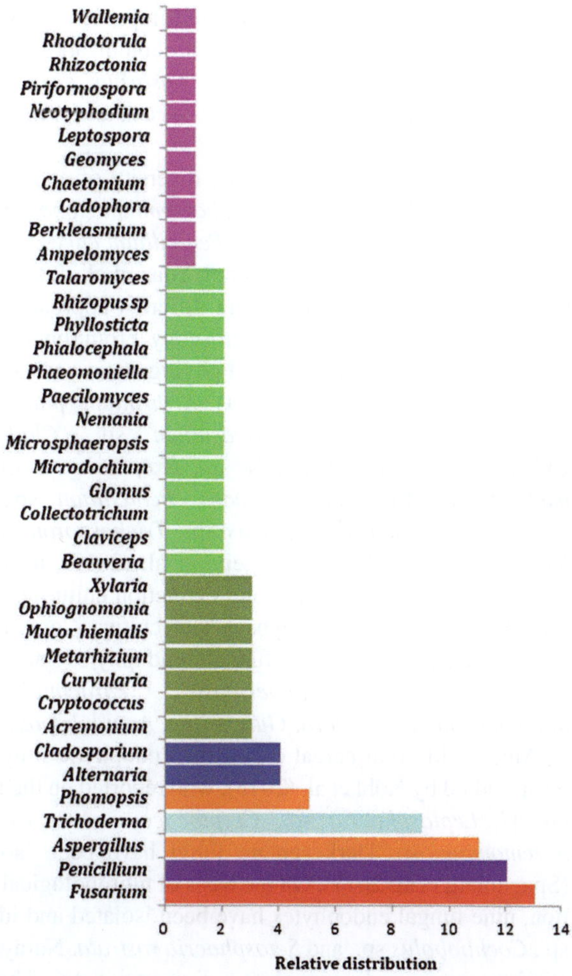

Fig. 6.5 Distribution of predominant genera of endophytic fungi isolated from various host plants

for the elimination of epiphytes using a sodium hypochlorite solution for the separation of true endophytic microbes from wheat crops. About 130 fungal endophytes have been isolated and identified as *Rhodotorula rubra, Alternaria alternata, Cladosporium herbarum*, and *Epicoccum nigrum* with the highest frequency. Tian et al. (2004) reported the fungal endophytes *Aspergillus, Fusarium, Helminthosporium, Paecilomyces, Penicillium*, and *Pyricularia* as endophytes from rice in South China. Wakelin et al. (2004) isolated 223 fungal endophytes from wheat plants growing in southern Australia and identified them as *Penicillium bilaiae, Penicillium simplicissimum, Penicillium griseofulvum, Talaromyces flavus, Penicillium radicum,* and *Penicillium.* Deshmukh et al. (2006) reported that *Piriformospora indica*, from the roots of barley, proliferates in dead host cells, increases with tissue maturation, and establishes mutualistic symbiosis with barley plants. After establishment, mutual symbiosis confers resistance to the host against disease and abiotic stress factors and improved growth of the host. Larran et al. (2007) reported that the highest colonization frequency was shown by *Rhodotorularubra, Penicillium* sp., *Fusarium graminearum, Epicoccum nigrum, Cryptococcus* sp., *Cladosporium herbarum, and Alternaria alternata* in cultivars of wheat.

In another study looking at the diversity of endophytic fungi in *Oryza sativa*, Naik et al. (2009b) reported that *Chaetomium globosum, Cladosporium cladosporioides, Fusarium oxysporum,* and *Penicillium chrysogenum* were the dominant fungal strains among 19 isolated fungi. Yuan et al. (2010) studied the root endophytic microbiota of rice in China using different approaches such as cultivation, direct PCR, ecological indices, and microscopy. From root tissue of *Oryza* about 58 fungal isolates have been retrieved, and phylogenetic analysis revealed that fungal endophytes have been identified as *Arthrinium phaeospermum, Aspergillus* sp., *Aspergillus vitricola, Berkleasmium* sp., *Cladophialophora chaetospira, Cylindrocarpon* sp., *Diaporthe* sp., *Exophiala pisciphila, Fusarium* sp., *Kernia pachypleura, Marasmius oreades, Penicillium* sp., *Phaeomoniella capensis, Podosphaera fusca, Rhizopycnis* sp., *Trichosporon mucoides, Wallemia* sp., and *Xylaria venosula.* De Souza Leite et al. (2013), utilizing two different isolation techniques, fragment plating and extinction culturing, described diverse genera of fungal endophytes from soybean leaf cultivars grown in Brazil. In soybean, the niche-specific genera of fungal endophytes are *Xylaria, Sporobolomyces, Rhodotorula, Phaeosphaeriopsis, Paraconiothyrium, Ophiognomonia, Magnaporthe, Leptospora, Guignardia, Annulohypoxylon,* and *Ampelomyces.*

Among different cereal crops, the endophytic fungal diversity from maize has been studied by Köhl et al. (2015), who reported on the fungal strains *Microdochium bolleyi*, *Leptosphaeria* sp., *Cryptococcus zeae*, *Cryptococcus flavescens*, and *Acremomium* sp. Dark septate fungi have been isolated from roots of wheat (Spagnoletti et al. 2017). On the basis of morphological and molecular characterization, nine fungal endophytes have been isolated and identified as *Ophiosphaerella* sp., *Cochliobolus* sp., and *Setosphaeria rostrata.* Narayan et al. (2017) reported that *Piriformospora indica* inhabits a diverse variety of hosts and, due to its valuable nature, is used as a biocontrol agent and in biofertilizers. Mohd et al. (2017) reported that the association of terrestrial plants with arbuscular mycorrhizal fungi resulted in beneficial plant–microbes associations. The colonization of rice plants with *Piriformospora indica* reestablished the growth of the plants following contamination by toxic arsenic. Bhagyasree et al. (2018) isolated and characterized *Trichoderma* sp., *Rhizopus* sp., *Penicillium* sp., *Mucor* sp., *Metarhizium* sp., *Lecanicillium* sp., *Fusarium* sp., *Beauveria* sp., and *Aspergillus* sp. as endophytes from maize.

6.4 Biotechnological Applications of Endophytic Fungi

Plants play a significant role in choosing and elevating the types of microbes by the components of exudates secreted from their roots. Thus, relying on the exudates' nature and concentrations, which are organic, and the analogous capability of the microbes to consume these as energy sources, the microbes establish an endophytic relationship. The association of endophytic fungi with cereal crops is of agronomic significance because, by using various mechanisms of plant growth promotion, fungal endophytes enrich plant growth (Yadav and Yadav 2018a, b; Yadav et al. 2018a). Endophytic fungi may increase crop yields, remove impurities, and constrain pathogens and novel substances. Endophytes stimulate the growth of plants by synthesizing phytohormones, for instance, indole acetic acid (IAA) and cytokinins (UmaMaheswari et al. 2013), and through the biocontrol of phytopathogens by means of the production of antifungal or antibacterial agents (Xu et al. 2015), siderophores (Verma et al. 2011), competition for nutrients or improvements in the bioavailability of minerals, and the introduction of acquired host resistance (Vallad and Goodman 2004). The use of microbial plant growth promoters is an alternative to conventional agronomic technologies (Timmusk et al. 2017). Endophytic fungi can influence the development of plants directly or indirectly by mediating the uptake of specific nutrients from the atmosphere and indirectly by decreasing or inhibiting the harmful consequences of one or more phytopathogenic organisms.

Secondary metabolites are organic compounds that often play a significant role in plant defense against herbivory and other interspecies defenses and also provide benefits to plants, such as protection, antagonism, and species correlations, but they are not essential for existence (Tiwari and Rana 2015). Humans use secondary metabolites as medications, seasonings, and recreational drugs. Endophytes were described to supply wide-ranging varieties of bioactive secondary metabolites, including xanthones, tetralones, terpenoids, steroids, quinones, phenolic acids, flavonoids, benzopyranones, alkaloids, and others (Tan and Zou 2001; Rana et al. 2016a, b; Yadav 2017; Yadav et al. 2017a). Such bioactive metabolites are extensively used in agrochemicals, antitoxins, immunosuppressive drugs, antioxidants, anticancer agents, treatment of parasitic diseases, for example those caused by helminths, amoebas, ectoparasites, parasitic fungi, and protozoa (Strobel 2003). Over the past 20 years, secondary metabolite production from fungal endophytes has expanded dramatically. In inter- or intracellular healthy tissues of plants, fungal endophytes spend all or part of their life cycle in symbiotic association. Association of endophytic fungus with plants stimulates secondary metabolite synthesis. Secondary metabolites are low-molecular-weight compounds that are produced as an adaptation to a specific function in nature, for example, interaction between fungi and host, protection, and management of symbiosis. The synthesis of secondary metabolites depends on the biotope in which the fungal endophyte grows and adapts. The secondary metabolites secreted by endophytic fungi have potential therapeutic use because they possess antimicrobial, anticancerous, antioxidant, and antirheumatic properties. Plant–fungal interaction is one of the most efficient and economical processes.

6.4.1 Potential Role of Endophytic Fungi in Agriculture

Endophytic fungi are agronomically important because they can enhance the growth and nutrition of plants by using various direct and indirect PGP attributes. Growth-stimulating endophytic fungi reside inside apoplastic spaces in host plants. Sustainable agriculture involves the use of approaches to proliferating or retaining the present rate of production of food while eliminating harm to the environment and human well-being (Hamm 2008). The practice of microbial plant growth supporters is an alternative to conventional farming techniques. Endophytes started as rhizospheric microbes or soil microbes. Fungal endophytes directly or indirectly promote the growth and yield of plants. Plants must cope with a series of environmental and biotic pressures like aridity, wintriness, saltiness, or pathogenesis. By some indirect mechanisms, for example, through induced systemic resistance (ISR), bioremediation, or biocontrol, fungal endophytes support plants in overcoming such stresses. Fungal endophytes possess a collection of biocontrol mechanisms to counter phytopathogens and insects (Biswas et al. 2018; Mishra et al. 2016a, b; Gao et al. 2010).

6.4.1.1 Role of Fungal Endophytes in Plant Growth Promotion

The growth of plants can be actively or passively promoted by endophytic microbes, through various and multiple mechanisms, providing fitness to host, resistance among both biotic and abiotic stresses in plants. Fungal endophytes stimulate the growth of plants directly or indirectly by triggering the solubilization of phosphorus, potassium, and zinc, triggering host plant defense responses against phytopathogens by various mechanisms, such as niche exclusion and competition, direct antagonism of pathogens by antibiosis, parasitism, or predation, altering plant hormonal levels such as gibberellins, auxin, ethylene, and IAA, siderophore production, ACC deaminase production, and the supply of vital vitamins to plants (Ji et al. 2014; Khan et al. 2010).

Phosphorus is an essential nutrient for the growth and development of plants. It is generally classified as a major nutrient. It maintains the general health and vigor of plants, for example, it stimulates the development of roots, increases stalk and stem strength, promotes the formation of flowers, enhancements in crop excellence, and improved immunity of plants to diseases. Phosphorus is an essential constituent of adenosine triphosphate (ATP), the “energy unit” of plants, and a constituent of genetic material such as genes and chromosomes. It is the vital part of transferring genetic information from one generation to another, delivering the “blueprint” for all conditions of plant progress and improvements in DNA and RNA (Nyoki and

Ndakidemi 2014). Phosphorus is found in large quantities in seeds and fruits because it is important for the development of seeds. The major storage form of phosphorus in seeds is phytin. A deficiency of phosphorus in plants results in delayed maturation, lowered property of fodder, fruits, vegetables, and grains, reduced resistance to disease, reduced growth, number, and surface area of leaf; in addition, the growth of shoots is affected more than the growth of roots, and in some plants leaves that are deficient in phosphorus develop a purple color.

Soil contains about 0.5% phosphorus, and for plant absorption only a small amount of phosphorus is available; the rest remains as insoluble salts (Rodrıguez and Fraga 1999). Phosphorus in the soil is grouped into two classes, organic form and inorganic form. In plant debris, compost, and microbial tissues, phosphorus is present in organic form (Costa et al. 2017). Apatite, iron complexes, aluminium phosphates, and phosphorus absorbed onto clay particles are sources of inorganic phosphorus (Burke et al. 1989). Aluminium (Al^{3+}), iron (Fe^{3+}), and calcium (Ca^{2+}) ions react with inorganic phosphate, which leads to the development of insoluble complexes. Phosphorus occurs as tricalcium phosphate [Ca_3 $(PO_4)_2$] in alkaline soils and $FePO_4$ and $AlPO_4$ in acidic soil. Phosphorus in insoluble form is simply cannot be used by plants. The replacement of soil P reserves through chemical fertilization is a common but long-term practice (Priyadharsini and Muthukumar 2017). Many studies have reported that fungi are most proficient in solubilizing essential nutrients from soil. *Penicillium* and *Aspergillus* were reported as two important genera identified with high mineral-solubilizing activity, related to a reduction in pH of Pikovskaya's medium. *Penicillium* sp. showed better efficiency for the solubilization of $Ca_3(PO_4)_2$ and rock phosphate (Gupta et al. 2007). The solubilization of inorganic insoluble phosphate salts by microbes results in the release of organic acid, which subsequently decreases the pH of surrounding (Wakelin et al. 2004; Yadav et al. 2015b, 2016a).

The different genera of fungal strains in endophytic association described to boost plant development and improvement are advantageous for sustainable agriculture. An investigation by Nath et al. (2012) explained the phosphate-solubilizing activity of two diverse species of endophytic *Penicillium* isolated from leaves of tea. At up to 8 days, there was a rise in the acidity of the cultural medium represented by both isolates that showed the remarkable activity of phosphate solubilization. The value of phosphate-solubilizing activity of species 1 ranged from 39.22 ± 1.17 to 86.10 ± 1.20 μg/mL, while that of species 2 varied between 32.57 ± 1.41 and 84.25 ± 1.5 μg/mL; at the same time, the pH of the medium declined from 6.25 ± 0.26 to 3.22 ± 0.22 and 6.22 ± 0.45 to 3.54 ± 0. Isolates have the potential to be used as biofertilizer agents.

Fungal endophytes have been isolated from the lateral roots of *Lens esculenta* Moench and microscopic features, morphology, and 18S rDNA sequence analysis revealed a fungus identified as *Trichoderma gamsii*. In culture medium, the fungus solubilized up to 17% of TCP with a decrease in pH. During qualitative analysis, fungus was absolute for the production of ammonia, chitinase (0.1 U/mL), and salicylic acid (20.05 mg/mL). Under greenhouse bioassay using four test crops, *Trichoderma gamsii* showed potential for plant growth (Rinu et al. 2014). Fungal endophytes stimulate the growth of plants through the solubilization of phosphorus and other PGP attributes. Tomato (*Solanum lycopersicum*) plants were used for the isolation of endophytic fungi from a field station in the Pithoragarh district of Uttrakhand, India, and identified as *Chaetomium globosum, Fusarium oxysporum, Fusarium semitectum, Fusarium solani, Fusarium fusarioides, Fusarium moniliforme, Mucor* sp., *Aspergillus niger, Aspergillus* sp., *Aspergillus versicolor, Mucor hiemalis,* and *Trichoderma peudokoningii rifai.* On Pikovskaya's solid agar and liquid medium fungal endophytes solubilized phosphate entirely. The solubilization of phosphate increased from day 2 to day 8; the highest range was observed in *Trichoderma peudokoningii* (37.45 ± 2.78 to 64.32 ± 2.87) μg/mL, seven were positive for siderophore, four for HCN, and three showed ammonia production. IAA production was found to reach a maximum in *Fusarium fusarioides*. All isolates showed maximum siderophore production at day 21 after inoculation. These isolated root endophytic fungi possess PGP properties and thus have the potential to be used as biofertilizers (Chadha et al. 2015).

Embryophyta (land plants) mostly depend on AM fungi and scavenge a macronutrient like phosphorus (P) from the soil. Almario et al. (2017) studied the root-associated fungal microbiome of *Arabis alpina* (Brassicaceae). They isolated and then reinstated a sample in its natural P-poor soil, which resulted in better growth of plant and P uptake. The fungus displayed traits like mycorrhiza; analysis of the genome revealed a connection in its endophytic existence and the enlargement of its collection of carbohydrate-active enzymes.

Potassium is assessed as a "quality nutrient" after only nitrogen when it comes to nutrients required by plants. Potassium is a vital macronutrient necessary for the growth and development of plants. Potassium regulates the opening and closing of stomata, prompts the initiation of enzymes and the formation of ATP, advances immunity against drought; in addition, the production of protein and starch also requires potassium, and it yields grain higher in starch content, decreases respiration and preventing loss of energy, boosts the transfer of sugars and starch, helps build cellulose, and delays the onset of crop diseases (Wang et al. 2013a, b).

Potassium exists in three different forms in soil: (a) unavailable form, (b) slowly accessible or stable form, and (c) exchangeable form. Approximately 90–98% of total soil K is found in unavailable form. Potassium present in soil is mostly in fixed form in mica, orthoclase, biotite, illite, and feldspar. Potassium-solubilizing microbes use certain mechanisms for the solubilization of potassium acidolysis, chelation, complexolysis, and the production of organic acid (Suman et al. 2016b; Verma et al. 2015b). The study by Nath et al. (2015) deals with the isolation of fungal endophytes from roots, stems, and leaves of tea (*Camellia sinensis*) gathered from tea gardens of Assam, India, and under in vitro conditions evaluated for PGP activities. *Aspergillus niger* (36.49 ± 1.17 μg/mL) was detected with the maximum IAA activity followed by *Penicillium sclerotiorum* (36.35 ± 2.07 μg/mL). *Fusarium oxysporum* (12.46 ± 0.84 μg/mL) displayed maximum GA3 activity followed by *Penicillium chrysogenum* F1 (10.95 ± 0.37 μg/mL). The most effective phosphate solubilizer is *Penicillium sclerotiorum* (215.98 ± 0.2 μg/mL), the best potassium solubilizer is *Aspergillus niger* (1.7 4 ± 0.2), and the best zinc solubilizer is *Penicillium sclerotiorum*.

Spagnoletti et al. (2017) evaluated the capability of dark septate endophytes (DSEs) to solubilize calcium, aluminium, and iron phosphates. DSEs were isolated from the roots of wheat (*Triticum aestivum*) and the forages *Panicum coloratum* and *Chloris gayana*. Two in vitro methodologies, solid and liquid media, were used to determine the ability of DSEs to solubilize phosphate. *Ophiosphaerella* sp. and *Cochliobolus* sp. were most efficient in solubilizing calcium phosphate. The strains *Drechslera* sp. (P6), *Ophiosphaerella herpotricha*, and *Drechslera* sp. (12–15) were capable of solubilizing aluminium phosphate. In liquid medium, the most capable strain in solubilizing calcium phosphate was *Drechslera* sp. (12–15), *Curvularia* spp. solubilize aluminium phosphate, and *Ophiosphaerella* spp. solubilize iron phosphate. However, the result showed to manage sustainable agroecosytems DSEs have the potential applications to be used as biofertilizers.

Phytohormones are defined as natural substances secreted in certain organs of plants that can be translocated to other sites, where they activate exact biological, physical, and morphological responses. Growth regulators such as gibberellic acid, IAA, and cytokinin are also produced by plant microbiomes (Verma et al. 2015a, 2016a, b; Yadav 2009, 2017; Yadav et al. 2017a). Plant microbiomes specifically endophytic fungi producing phytohormones participate in the development process of various plant parts such as flowers, stems, leaves, and fruit growth and maturation, and they also affect gene expression and transcription levels. At small concentrations, plant hormones stimulate and effect the growth, development, and differentiation of cells and tissues. There are five main classes of phytohormones: (a) abscisic acid is one of the essential plant growth regulators, (b) formation of roots and buds and the development of cells is influenced by auxins, (c) the development of shoots and division of cells are influenced by cytokinins, (d) ethylene is the hormone responsible for adaptions under extreme conditions, (e) the growth of pollen tubes, expansion of stems, and the mobilization of depository substances in seeds are initiated by gibberellins.

Studies by Hamayun et al. (2009) demonstrated that certain fungal endophytes encourage the development of hosts through the synthesis of phytohormones such as gibberellins (GAs), IAA, and cytokinins. *Cladosporium sphaerospermum* endophytic fungi isolated from root samples of *Glycine max* (L) have been reported to produce gibberellins. The culture filtrate displayed the existence of greater amounts of GA3, GA4, and GA7 (6.62, 2.1, and 1.26 ng/mL) and promoted the growth of rice. A fungal endophyte, *Cadophora malorum*, producing GAs was isolated from the roots of *Calystegia soldanella* (You et al. 2013). Fungal endophyte species RSF-4 L and RSF6L, identified as *Fusarium* sp. and *Alternaria* sp., were isolated from *Solanum nigrum* leaves. The Salkowski assays specify both *Fusarium* sp. and *Alternaria* sp. produce 54 and 30 μg/mL IAA respectively. Culture Filtrates (CFs) from each isolate were evaluated for their ability to promote the development of Dongjin rice plants. Upon treatment with fungal CFs, plant growth attributes such as plant biomass, chlorophyll content, and extent of root-shoot, for example, were enhanced. Both fungal endophytes play a vital role in promoting the growth of plants (Khan et al. 2015).

Fungal endophytes enhance the fitness of plants and enable the adaptation of plant hosts to biotic and abiotic stresses. These fungi also produce secondary metabolites, some of which are bioactive compounds that play a rote in interactions with plant hosts and protect plants from certain pests and diseases. There is sufficient evidence to prove the association of fungal endophytes with hosts that results in the enhancement of host fitness, protection against insects, pests, herbivores, and root nematodes, improves drought tolerance, and improves root growth, for example; in turn, fungal endophytes acquire shelter and nutrition and are vertically transmitted via host seeds (Schardl et al. 2004). Endophytic microbes directly or indirectly influence the progress of plants by the production of secondary metabolites. Bioactive substances produced by endophytes provide plants with energy and defense capabilities. Endophytic microbes protect host plants and produce mycotoxin against pathogenic microbes from natural enemies (Carroll 1988). The root of the common dune plant has been used for the isolation of fungal endophytes and screening for the secretion of secondary metabolites. Through investigation of the 18S rRNA gene sequence, the strain IR-3-3 has been identified as *Penicillium citrinum* and produces GA_5 gibberellins (Khan et al. 2008). Enzymes that are extracellular in nature secreted by fungi are gaining importance in biotechnology, pharmaceuticals, the food industry, beverages, leather, agriculture, bioremediation, and biotransformation of organic compounds (Benjamin and Pandey 1998; Pimentel et al. 2011; Verma et al. 2017b; Yadav et al. 2018c).

6.4.1.2 Biofertilizers: Microbes with Multifarious PGP Attributes

Biofertilizers are microorganisms that improve the nutrient quality of the soil by enhancing the accessibility of the nutrients to plants. To meet the demands of an expanding population, there is a greater need to produce healthy crops as most agricultural production is dependent on chemical fertilizers, which are damaging the environment while simultaneously affecting the health of humans. Therefore, biofertilizers are the best substitute for chemical fertilizers for improving the fertility as well as enhancement in the yield of crops. The use of these microbes as biofertilizers represent a novel tool, and they have greater potential to provide benefits to sustainable agriculture. The microbes colonize roots, where they stimulate the growth of plants by diverse direct and indirect mechanisms, including through the production of various phytohormones, biological nitrogen fixation, solubilization of phosphorus, potassium, and zinc, production of iron-chelating agent, HCN, and various lytic enzymes. Microbes that provide benefits to plants are referred to as PGP microbes. Extensive work on biofertilizers clearly reveals that these microbes have the potential to provide the required nutrients to crops in sufficient amounts to enhance their productivity (Table 6.1) (Srivastava et al. 2013; Suman et al. 2016a; Kour et al. 2017a, b;Yadav et al. 2017c).

Table 6.1 Multifunctional PGP attributes of endophytic fungi

Fungal endophyte	Host	P	K	Zn	IAA	Reference
Aspergillus Niger	Tomato	+	–	–	+	Chadha et al. (2015)
Aspergillus sp.	Tomato	+	–	–	+	Chadha et al. (2015)
Aspergillus sp.	Sugarcane	+	–	–	–	Sane and Mehta (2015)
Aspergillus versicolor	Tomato	+	–	–	+	Chadha et al. (2015)
Chaetomium globosum	Tomato	+	–	–	+	Chadha et al. (2015)
Cochliobolus sp.	Wheat	+	–	–	–	Spagnoletti et al. (2017)
Curvularia sp.	Wheat	+	–	–	–	Spagnoletti et al. (2017)
Drechslera sp.	Wheat	+	–	–	–	Spagnoletti et al. (2017)
Fusarium equiseti	Pea	+	–	–	–	Šišić et al. (2017)
Fusarium fusarioides	Tomato	+	–	–	+	Chadha et al. (2015)
Fusarium moniliforme	Tomato	+	–	–	+	Chadha et al. (2015)
Fusarium oxysporum	Tomato	+	–	–	+	Chadha et al. (2015)
Fusarium semitectum	Tomato	+	–	–	+	Chadha et al. (2015)
Fusarium solani	Tomato	+	–	–	+	Chadha et al. (2015)
Glomus intraradices, BEG72	Wheat	+	+	+	–	Colla et al. (2015)
Glomus mosseae	Wheat	+	+	+	–	Colla et al. (2015)
Mucor hiemalis	Tomato	+	–	–	+	Chadha et al. (2015)
Mucor sp.	Tomato	+	–	–	+	Chadha et al. (2015)
Ophiosphaerella herpotricha	Wheat	+	–	–	–	Spagnoletti et al. (2017)
Ophiosphaerella sp.	Wheat	+	–	–	–	Spagnoletti et al. (2017)
Penicillium bilaiae, RS7B-SD1	Wheat	+	–	–	–	Wakelin et al. (2004)
Penicillium funiculosum	Soybean	+	–	–	+	Khan et al. (2011)
Penicillium griseofulvum	Wheat	+	–	–	–	Wakelin et al. (2004)
Penicillium radicum, FRR4718	Wheat	+	–	–	–	Wakelin et al. (2004)
Penicillium simplicissimum	Wheat	+	–	–	–	Wakelin et al. (2004)
Penicillium sp.	Sugarcane	+	–	–	–	Sane and Mehta (2015)
Piriformospora indica	Rice	+	–	–	–	Mohd et al. (2017)
Porostereum spadiceum, AGH786	Soybean	–	–	–	+	Hamayun et al. (2017)
Setosphaeria rostrata	Wheat	+	–	–	–	Spagnoletti et al. (2017)
Talaromyces flavus	Wheat	+	–	–	–	Wakelin et al. (2004)
Trichoderma atroviride	Wheat	+	+	+	–	Colla et al. (2015)
Trichoderma peudokoningii rifa	Tomato	+	–	–	+	Chadha et al. (2015)

A microbial consortium is two or more microbial groups living symbiotically. The consortium can be endosymbiotic or ectosymbiotic. The combination of multifarious PGP endophytic microbes for sustainable agriculture may be a replacement of chemical fertilizers. There are many reports on developments of microbial consortiums and microbes having multifarious PGP attributes, including archaea, bacteria, or fungi. These consortiums may be epiphytic, endophytic, or rhizospheric (Suman et al. 2016b; Verma et al. 2017b; Yadav et al. 2018b). Endophytic fungi with multifunctional PGP attributes have been reported from diverse host plants worldwide. Khalmuratova et al. (2015) identified endophytic fungi by sequencing internal transcribed spacer (ITS) regions in halophilic plants from South Korea. All 160 strains belonged to Ascomycota and Basidiomycota, with *Fusarium* being the dominant genus followed by *Penicillium* and *Alternaria*. Further, rice seedlings have been treated with culture filtrate for PGP effects. Maximum plant length has been observed in plants treated with fungal strain *Talaromyces pinophilus* (Su-3-4-3). Further, the chromatographic analysis of the culture filtrate of Su-3-4-3 showed the presence of physiologically active gibberellins, including GA_1 and GA_3, along with other physiologically inactive one, GA_9 and GA_{24}. A microbial consortium based on the arbuscular mycorrhizal fungi *Glomus intraradices* BEG72, *Glomus mosseae*, and *Trichoderma atroviride* MUCL 45632 could improve seedling establishment, yield, and the grain quality of wheat (*Triticum durum* Desf.) (Colla et al. 2015).

6.5 Role of Fungal Endophytes in Biofortification of Micronutrients

The production of various wild, traditional, or ancient food crops, which are genetically very diverse and rich in nutritional compounds, has greatly decreased. In the present scenario, only 30 plant species account for approximately 95% of the world's food energy supply. To meet the daily requirements of humans, wheat cultivars have high-yielding capacity but are poor sources of micronutrients, particularly Fe and Zn. In commercial wheat cultivars, 20–35 mg/k is not adequate to accomplish human energy and nutrient requirements. Over a period of time, such wheat-based diets can result in malnutrition and health-related issues such as anemia. At present, biofortification and the use of PGP microbes is an attractive method. PGP microbes carry almost the mobilization of nutrients by diverse mechanisms including acidification, chelation, exchange reactions, and release of organic acids (Verma et al. 2017a). More than half of the global human population is affected by dietary deficiencies of essential micronutrients such as Fe and Zn (White and Broadley 2009). The micronutrient deficiency also referred to as 'Hidden Hunger' effects in impaired physical growth and cognitive development, there is compact psychomotor development of children, weakness, fatigue, immunity is reduced, infertility, sickness and even demise in severe circumstances (Pfeiffer and McClafferty 2007; Stein 2010). Hence, there is a need to improve mineral concentrations in important cereal crops such as rice, wheat, and maize, as well as common beans and other legumes. There are many possible strategies to alleviate micronutrient deficiencies, including dietary diversification, mineral supplementation, post-harvest food fortification, biofortification, and microbial interventions.

Biofortification is the most sustainable, targeted, and cost-effective approach to overcoming these difficulties by informative the micronutrient content and nutritional quality of cereal crops themselves by increasing mineral levels and bioavailability in edible parts, more specifically in the endosperm (Borrill et al. 2014; Tiwari et al. 2010; Verma et al. 2017a). The exploitation of microorganisms to improve nutrient use productivity and biofortify the grains of cereal crops is the least applied strategy. An endophyte is defined as an endosymbiont, which could either be a bacterium or fungus, colonizing inside a plant for a minimum part of its life cycle. These are universal and are known in all species of plants. Endophytes afford numerous benefits to plants, for example, they may enhance the growth of the host plant, help in the acquisition of nutrients, improve the ability of plants to endure abiotic stresses, including drought, and reduce biotic stresses by augmenting plant resistance to insects, pathogens, and herbivores. Even though there is abundant information on the use of rhizospheric microbes to boost the acquisition of micronutrients (Gosal et al. 2010; Sharma et al. 2012), work on the use of endophytic microorganisms is limited (Ren et al. 2012; Wang et al. 2014).

A number of mechanisms have been reported on the inoculation of endophytes for enhancing the uptake of Fe and Zn such as chelation of Fe by siderophore-producing microorganisms, the production of organic acids in root exudates, proton extrusion by rhizospheric microorganisms, and the production of phytohormones (Chen et al. 2014; Desai and Archana 2011; Fasim et al. 2002; Kobayashi et al. 2014; Kobayashi and Nishizawa 2012; Li et al. 2010). The root exudates of plants significantly affects the solubilization and mobilization of soil nutrients by modulating the biological and chemical processes operating in the rhizosphere (Zhang et al. 2010). Microbial inoculation results in the increased production of organic acids that lowers the pH of soil near roots, which directly contributes to the solubilization of diverse insoluble nutrient sources like P, K, and Zn (Goswami et al. 2014; Sirohi et al. 2015).

It has been mentioned in several published reports that inoculating plants with PGP endophytes notably increases the surface area and volume of roots, the number of root tips, and root biomass (Delaplace et al. 2015; Vacheron et al. 2013). Plants with fertile root systems are reported to have enhanced nutrient use proficiency (Dong et al. 1995; Genc et al. 2007). Recent reports show that in addition to root construction, structural features of roots are also affected by soil microflora such as *Burkholderia pyrrocinia* (R-46) + *Pseudomonas fluorescens* (R-55) and *Trichoderma asperellum* (Rêgo et al. 2014). Significant variations in the uptake of micronutrients as well as their accumulation in root, shoot and grains have been reported (Chatzistathis et al. 2009); the most significant difference has been suggested to be attributed to the vital part of metal transporter families including zinc-regulated transporter (ZRT), iron-regulated transporter (IRT), cation diffusion facilitator (CDF), and zinc–iron transporter-like protein (ZIP) that are involved in the uptake, translocation, and sequestration of Fe and Zn from roots to grain (Colangelo and Guerinot 2006). In rice, wheat, maize, and *Arabidopsis thaliana*, metal transporter families in plants including *ZIP* family proteins have been studied (Xu et al. 2010), and the overexpression of ZIP proteins led to the sequestration of additional amounts of Zn in the seeds of bare emmer wheat (Durmaz et al. 2011).

Further, mycorrhizal fungi also improve the uptake of all essential mineral nutrients from soils for most plants, comprising all the major grain crops and nearly all vegetables and fruits that are linked, and, consequently, improve the growth and yield of plants. There are six categories of mycorrhizae, including arbuscular, arbutoid, ecto, ericoid, monotropoid, and orchid (Wang and Qiu 2006). Among these, AM is the most widespread and major. It is characterized by arbuscules, which are little tree-shaped fungal arrangements inside the cortical cells of roots that function as the key locations of nutrient interchange among plants and fungi. AM consists of external hyphae that provide a wide surface area or network for nutrient from soils. These symbiotic fungi thus alter, directly or indirectly, the mineral nutrition of plant products that are also vital for humans and also play a role in agriculture.

6.6 Potential Applications of Endophytic Fungi in Industry

Beneficial microbes are used for the production of industrial products in mass quantities, for example, the production of antimicrobial drugs, antibiotics, riboflavin, and enzymes. Microbes are largely used in the industrial sector, including in the making of a variety of metabolites, such as ethanol, butanol, lactic acid, and riboflavin, as well as in the biotransformation of numerous chemicals to enable the lessening of environmental pollution by various methods such as wastewater management, bioremediation, mycoremediation, and composting. Microbes can also be used to make certain nonmicrobial products, such as insulin. Griseofulvin is a useful antifungal agent isolated from *Penicillium griseofulvin*, which is cheaply produced industrially. A number of industrial applications use the biological activity of fungi involved in modifications of plant cell walls. Fungi are able to disrupt the cell walls of plants by producing a wide variety of lytic enzymes. In textile processing, enzymes are used to treat and modify fibers such as cellulases from *Trichoderma*. The pulp and paper industry also benefits from fungi that produce enzymes such as xylanase and peroxidase. Enzymes are both economically and environmentally beneficial as they create little or no waste.

6.6.1 Industrially Important Bioactive Compounds

Over the years, natural products from microorganisms, plants, or animals are playing a chief role in the search for new drugs. The naturally derived products are nontoxic and inexpensive and have been exploited for human use. Fungal endophytes are the largest storehouse of bioactive compounds. Bioactive compounds, including carotenoids and flavonoids such as anthocyanins, phenolic acids, and polyphenols, provide a lot of benefits such as lowering the risk of heart disease, cancer, and various other diseases.

Alexander Fleming in 1928 discovered the first bioactive compound from *Penicillium notatum*, or penicillin. During the 1990s, one of the most useful anticancer drugs was paclitaxel. An endophyte of *T. brevifolia*, *Taxomyces andreanae*, has been reported to produce paclitaxel. Later research suggested the transfer of lateral genes from host to fungus (Stierle et al. 1993). The fungal endophyte *Fusarium* has been reported to produce subglutinol A and diterpene pyrones showing immunosuppressive activity. The endophyte was isolated from stem of *Tripterygium wilfordii* (Strobel and Pliam 1997). Isobenzofuranone, as Isopestacin, obtained from the fungal endophyte *Pestalotiopsis microspore*, possesses antifungal and antioxidant activity (Strobel et al. 2002). The antimicrobial activity of fungal endophytes was screened against the pathogenic organisms *Candida albicans, Cryptococcus neoformans,* and *Staphylococcus aureus.* Fungal endophytes have been isolated from leaves and branches of five different species of *Garcinia* plants. The fungal endophyte *Phomopsis* sp. and *Botryosphaeria* sp. showed antibacterial activity against *Staphylococcus aureus* (Phongpaichit et al. 2006). Developments in screening technologies have received much attention, so that fungal endophytes are an excellent source of biologically active compounds with applications in medicine and agriculture (Aly et al. 2011; Rana et al. 2017, 2018) (Table 6.2).

Antimicrobial compounds produced by fungal endophytes have attracted much attention from researchers for the screening of fungal endophytes producing novel metabolites. Fungal endophytes have been reported to produce bioactive metabolites similar to their host plant (Mishra et al. 2017b; Stierle et al. 1993; Xu et al. 2009). Bioactive compounds isolated from the genus *Xylaria* showed antifungal activity against *Candida albicans* (Pongcharoen et al. 2008). Fifty-three fungal endophytes isolated from *Dendrobium devonianum* and *D. thyrsiflorum* in which the potential fungal strain *Fusarium tricinctum,* and *Phoma* displayed strong inhibitory activity against pathogens, including *Aspergillus fumigates, Bacillus subtilis, Candida albicans, Cryptococcus neoformans, Escherichia coli,* and *Staphylococcus aureus. Epicoccum nigrum* exhibited strong antibacterial activity against three pathogens, *S. aureus, E. coli,* and *B. subtilis* (Xing et al. 2011).

A large number of bioactive compounds have been known to be produced by fungal endophytes, including alkaloids, benzopyranones, chinones, cytochalasines, depsipeptides, enniatines, furandiones, flavonoids, isocumarines, peptides, perylene derivatives, phenols, polyketones, quinols, steroids, terpenoids, tetralones, and xanthones (Elfita et al. 2011; Tenguria et al. 2011; Kusari and Spiteller 2012). Su et al. (2014) reported 193 endophytic microbes from Chinese medicinal plants *Camptotheca cuminata* Decne, *Gastrodia elata* Blume, and *Pinellia ternata.* On the basis of morphological and rDNA sequence analysis, fungal isolates have been found to belong to Ascomycota, Basidiomycota, and Mucoromycotina. Using submerged culture, fungal strains have been screened for the production of bioactive compounds and have been found to produce camptothecin, 10-hydroxycamptothecin, gastrodin, and ephedrine hydrochloride as bioactive compounds.

Intracellular and extracellular extracts of the fungal endophyte *Aspergillus flavus* exhibited a broad spectrum of antibacterial activity against human pathogenic bacteria (*Lactococcus lactis* NCTC 497, *Bacillus subtilis* 168, *Staphylococcus aureus* SG511, *Staphylococcus carnosus* TM300, *Bacillus pseudomycoides* DSM 12442, and *Escherichia coli* BL21 (DE3)). Gas chromatography-mass spectrometry (GC-MS) showed the presence of a variety of bioactive components such as 4-nitrobenzoic acid, 3-chlorophenyl ester (27.23%), and (+)-salsolidine (21.82%). The fungal endophyte *Aspergillus flavus* isolated from jute can be used as a commercial and

Table 6.2 Secondary metabolites producing fungal strains and their biotechnological applications

Microorganism	Secondary metabolite	Application	Reference
Acremonium sp.	Cordyheptapeptides C-E	Cytotoxic	Chen et al. (2012)
Acremonium sp.	Leucinostatin A	Antifungal	Strobel et al. (1997)
Alternaria alternata	Capsaicin	Anticancer	Clark and Lee (2016)
Alternaria sp.	Altenusin	Antimicrobial	Kjer et al. (2009)
Alternaria sp.	Alternariol	Antibacterial	Tian et al. (2017)
Ampelomyces sp.	Altersolanol A	Antimicrobial	Aly et al. (2008)
Aspergillus fumigatus	Deoxypodophyllotoxin	Antimicrobial	Kusari et al. (2009)
Cephalosporium sp.	Cephalosol	Antimicrobial	Zhang et al. (2008)
Chaetomium elatum	Xanthoquinodin	Cytotoxic	Chen et al. (2013)
Chloridium sp.	Javanicin	Antibacterial	Kharwar et al. (2009)
Coniochaeta sp.	Coniothiepinol A	Cytotoxic	Wang et al. (2010)
Cryptosporiopsis sp.	Cryptocandin	Antifungal	Strobel et al. (1999)
Cytonaema sp.	Cytonic acid A	Antiviral	Guo et al. (2000)
Cytonaema sp.	Cytonic acid B	Antimicrobial	Li et al. (2007b)
Entrophospora sp.	Camptothecin	Anticancer drug	Puri et al. (2005)
Eupenicillium parvum	Azadirachtin A	Antifeedant	Kusari et al. (2012)
Fusarium oxysporum	4-hydroxybenzoic acid	Nematicidal	Bogner et al. (2017)
Fusarium solani	Camptothecin	Anticancer drugs	Shweta et al. (2010)
Fusarium subglutinans	Subglutinol A and B	Immunosuppressive	Lee et al. (1995)
Geotrichum sp. AL4	1,3-oxazinane	Nematicidal	Li et al. (2007a)
Microsphaeropsis sp.	Microsphaerophtalide	Antifungal	Sommart et al. (2012)
Microsphaeropsis sp.	Botralin	Antimicrobial	Li et al. (2007b)
Muscodor sutura	Isocaryophyllene	Antimicrobial	Kudalkar et al. (2012)
Muscodor tigerii	Octadecylmorpholine	Antimicrobial	Saxena et al. (2015)
Myxotrichum sp.	Myxodiol A	Antifungal	Yuan et al. (2013)
Myxotrichum sp.	Myxotrichin A	Cytotoxic	Yuan et al. (2013)
Neurospora terricola	Terricollene A	Cytotoxic	Zhang et al. (2009)
Nigrospora sp.	Solanapyrone C	Antifungal	Wu et al. (2009)
Penicillium janthinellum	Polyketide citrinin	Antibacterial	Marinho et al. (2005)
Penicillium melinii	Ginsenocin	Cytotoxic	Zheng et al. (2013)
Penicillium sp.	Penicillenols A_1	Cytotoxic	Lin et al. (2008)
Pestalotiopsis jester	Jesterone	Antifungal	Li and Strobel (2001)
Pestalotiopsis sp.	Torreyanic acid	Anticancer	Lee et al. (1996)
Pestalotiopsis sp.	Pestacin	Antioxidant	Strobel et al. (2002)
Pestalotiopsis neglecta	Phenols, flavonoids	Antibacterial	Sharma et al. (2016)
Pestalotiopsis sp.	Ambuic acid derivative	Antibacterial	Ding et al. (2008)
Pestalotiopsis sp.	Pestalol 37	Antiviral	Sun et al. (2014)
Phaeosphaeria sp.	Phaeosphaerin A	Cytotoxic	Li et al. (2012a)
Phialocephala fortinii	Podophyllotoxin	Anticancer	Eyberger et al. (2006)
Phoma pinodella	Phomodione	Antibacterial	Hoffman et al. (2008)
Phomopsis longicolla	Dicerandrol C	Antibacterial	Erbert et al. (2012)
Preussia africana	Preussochrome C	Cytotoxic	Zhang et al. (2012)

(continued)

Table 6.2 (continued)

Microorganism	Secondary metabolite	Application	Reference
Talaromyces pinophilus	Siderophores ferrirubin	Aphicide	Vinale et al. (2017)
Ulocladium sp.	Ophiobolin P	Antibacterial	Wang et al. (2013a, b)

natural resource having biological and therapeutic activity (Wulandari and Suryantini 2018).

6.6.2 Industrially Important Hydrolytic Enzymes

For the last two decades, fungal endophytes isolated from various plant sources have proven to be an invaluable source of organic products for agriculture, industry, biomedical development, and so forth and also produce extracellular hydrolase enzymes, such as pectinases, cellulases, lipases, amylases, laccases, xylanase, and proteases, as one of the mechanisms to resist pathogenic organisms and to acquire nutrients from their host. Enzymes are an eco-friendly substitute for harsh chemicals. Enzymes work at a neutral pH and moderate temperature. In recent years, enzyme research using microbes has been very dynamic and hopeful. Further, there are numerous records on the diversity of microbes from extreme environments and their biotechnological use in agronomy and medical industry (Yadav 2015;Verma et al. 2015c; Yadav et al. 2016b, 2017b, d; Sahay et al. 2017; Singh et al. 2016a, b; Suman et al. 2015). Fungal endophytes that create extracellular enzymes have been described in diverse cereal, woody, and medicinal plants. Extracellular enzymes target various macromolecules, break down lignin, proteins, carbohydrates, and sugar-based polymers, to break down into simpler ones.

Rhizopus oryzae isolated from Mediterranean plants has been found to produce lipases that are membrane bound (Torres et al. 2003). The seeds of *Brucea javanica* are used as medicine for the treatment of dysentery, malaria, and cancer. Fungal endophytes have been isolated from leaves, stems, and branches of *B. javanica* and have been screened for the production of extracellular enzymes including amylase, cellulase, ligninase, xylanase, and pectinase using different media including starch agar, cellulose azure agar, poly R agar, xylan agar, and pectin agar prepared and autoclaved. All fungal endophytes produce amylase and cellulase; only one sterile mycelium produces ligninase, and no isolate produces pectinase (Choi et al. 2005).

Periconia sp. produces a thermotolerant β-glucosidase having high activity toward cellobiose and carboxymethylcellulose. The β-glucosidase hydrolyzes rice straw into simple sugars. The enzymes have the potential to convert lignocellulosic biomass into biofuels and chemicals (Harnpicharnchai et al. 2009). The fungal endophytes *Curvularia brachyspora, C. verruciformis, Phyllosticta* sp., *Colletotrichum crassipes, Lasiodiplodia theobromae, Cladosporium cladosporioides, Colletotrichum falcatum, Colletotrichum gloeosporioides, Drechslera hawaiiensis, Nigrospora sphaerica,* and *Xylariales* have been isolated from medicinal plants, viz., *Adhatoda vasica, Costus igneus, Coleus aromaticus,* and *Lawsonia inermis*, and screened for their ability to synthesize extracellular enzymes, i.e., amylase, cellulase, laccase, lipase, and protease. *Cladosporium cladosporioides, Curvularia brachyspora, Colletotrichum crassipes, Colletotrichum gloeosporioides, Drechslera hawaiiensis, Nigrospora sphaerica, Phyllosticta* sp., and *Xylariales* have been shown to produce amylase.

Laccases, manganese peroxidase and lignin peroxidases, hemicelluloses, and oxidoreductases are needed for the complete degradation of lignocelluloses (Correa et al. 2014). *Pereskia bleo*, *Oldenlandia diffusa, Murraya koenigii*, *Cymbopogon citratus*, and plants possess anticancer properties. These plants host fungal endophytes that produce the anticancer enzyme L-asparaginase. L-Asparaginase-producing endophytes form pink zones on agar as a result of the hydrolysis of asparagine into aspartic acid and ammonia. Asparaginase-producing isolates were identified as *Colletotrichum*, *Fusarium*, *Phoma,* and *Penicillium* (Chow and Ting 2015). *Asclepias sinaica*, a medicinal plant, is used for the isolation of fungal endophytes. These fungal strains were identified as *Penicillium chrysogenum, Sterile hyphae*, and *Alternaria alternata.* These fungal endophytes generate numerous extracellular enzymes comprising amylase, pectinase, cellulase, gelatinase, xylanase, and tyrosinase and had a considerable impact on the development of plants by the synthesis of ammonia and IAA (Fouda et al. 2015).

Fungal endophytes have been acknowledged as potential sources of bioactive secondary metabolites, and 93.33% of fungal endophytes, including *Aspergillus japonicas, Curvularia lunata, Nigrospora* sp., *C. gloeosporioides, Trichoderma* sp., *Xylaria* sp., *Rhizoctonia* sp., *Fusarium chlamydosporum, Penicillium citrinum, Helminthosporium* sp., *Aspergillus sydowii, Cladosporium* spp., *Aspergillus terreus, Alternaria alternata, Fusarium oxysporum, Penicillium chrysogenum, Aspergillus sydowii, Colletotrichum truncatum, Bipolaris* spp., *Talaromyces rotundus, Penicillium purpurogenum,* and *Cylindrocephalum* spp., obtained from the roots, leaves, and flowers of Cymbidium aloifolium L. have been reported to produce the phosphatase enzyme.

6.7 Potential Role of Endophytic Fungi in Environmental Sustainability

Endophytic microbes colonize the healthy tissues of plants inter- or intracellularly. Fungi utilize a broad range of organic compounds as sources of nutrition. The organic compounds comprise cellulose, pectin, lignin, chitin, starch, xenobiotic, hydrocarbons, and pesticides. The research role of endophytic fungi has been reported worldwide, and it has been concluded that fungal endophytes play an important role in the degradation of plant debris. Endophytes produce certain enzymes, for example, cellulose, lipoidase, proteinase, phenoloxidase, and pectinase, and when plants die and fall to the ground, fungal endophytes use glucose, oligosaccharides, cellulose, lignin, keratin, pectin, lipids, and components of plant residues and decompose rapidly (Lumyong et al. 2002; Tomita 2003). Because of industrialization or anthropogenic activity, a huge variety of pollutants, for instance, petroleum hydrocarbons, polycyclic aromatic hydrocarbons (PAHs), halogenated hydrocarbons, pesticides, solvents, salt, and heavy metals, have been introduced into the environment that cause environmental pollution (Rajkumar et al. 2010). PAHs are carcinogenic or toxic to living organisms. Newly emerging technologies, such as phytoremediation and bioremediation, are gaining a considerable amount of research attention (Li et al. 2012b).

Bioremediation relies on the breakdown of a variety of pollutants with biological activities and is divided into two types: intrinsic/natural attenuation/passive bioremediation and engineered bioremediation. Microorganisms having a high capacity to degrade PAHs are important for the adequate remediation of PAH contamination. White rot fungus *Phanerochaete chrysosporium* has been reported to extensively degrade PAHs (Bhatt et al. 2002; Šašek et al. 2003; Zheng and Obbard 2000). The achievement of bioremediation depends on certain factors, such as the physicochemical and biological properties of the soil (Alexander 1994), contaminant types and their concentrations (Providenti et al. 1993), the ability of particular microorganisms to grow in a given soil and to degrade the target contaminants (Pointing 2001), and the bioavailability of contaminants (Semple et al. 2003). An endophytic fungus, *Ceratobasidium stevensii*, isolated from *Bischofia polycarpa* has been reported to effectively degrade phenanthrene. Phenanthrene has been used as a model PAH compound (Dai et al. 2010).

Tian et al. (2007) studied the disintegration of phenanthrene by the endophytic fungus *Phomopsis* sp. in *Oryza sativa*. Russell et al. (2011) demonstrated the degradation of synthetic polymer polyester polyurethane (PUR) by *Pestalotiopsis microspora*. Fungal endophytes isolated from *Pterocarpus macrocarpus* were reported to degrade benzo(a)pyrene (Juhasz and Naidu 2000). Plants have generated numerous procedures to overpower microbial diseases by producing secondary compounds such as phytoanticipins and phytoalexins. Endophytes that live inside plants must subsist with certain toxic compounds generated by plants, and such endophytes have developed a variety of tolerant mechanisms, for instance, the synthesis of exo-enzymes and mycotoxins (Pinto et al. 2000). Endophytic fungi such as *Phomopsis* sp., *Glomerella cingulate*, *Diaporthe phaseolorum*, and *Aspergillus fumigatus* have been reported to make alterations to the three-dimensional structures of compounds and shown stereoselective biotransformation potential (Borges et al. 2008).

6.8 Conclusion and Future Prospects

With the passage of the past few decades, the human population has also increased, and to feed the population, the production of food has increased. The use of chemical fertilizers, fungicides, bactericides, and pesticides to increase crop yields has also increased, which has had certain negative impacts on Earth's atmosphere, causing air and groundwater pollution. In agriculture, new alternative methods were developed to improve the supply of nutrients. Organic farming is one of the strategies that aid in the longer shelf life of plants; it generally depends on the natural microflora of soil, and PGP microbes including endophytes, epiphytes, and rhizospheric microbes. Endophytes establish long-term or lifelong associations with their host. Fungal endophytes have attracted considerable attention for their ability to promote the growth of plants by direct or indirect mechanisms. They have been known to colonize all parts of plant hosts. After several decades of research, the scientific community has become familiar with microbial endophytes and their vibrant and potent role in sustainable agriculture, pharmaceuticals, ecology, and biotechnology. The involvement of fungal endophytes in element cycling has significant consequences for living organisms and human health. To promote sustainable agriculture, endophytic microbes can be used as biocontrol agents and biofertilizers. The significance of fungal endophytes that produce enzymes when remediating environmental pollutants such as polychlorinated hydrocarbons and polyaromatic hydrocarbons is understood. Metabolites produced by endophytes can also be of importance.

Endophytes that produce bioactive compounds that originate from their host plants have attracted increased interest from many investigators. Endophytic microbes, despite enormous biotechnological applications, have revealed to a lesser extent that they are a rich source of bioactive metabolites. Future research will need to take into account the development of genomic tools and metabolomics tools, both allowing for further studies on the life of endophytes inside plants and plant–microbe interactions. While there are fungal endophytes to be explored, genetic transformation should be considered a mode to group foremost attributes found in divergent strains. Future research in this field will have significant environmental and economic implications.

Acknowledgements The authors acknowledge the Department of Biotechnology, Government of India, for the financial support provided (Grant BT/AGR/BIOFORTI/PHII/NIN/2011), Ministry of Food Processing Industries (MoFPI), Government of India, grant for infrastructural facility development (F. No. 5-11/2010-HRD), and Prof. H.S. Dhaliwal, Vice-Chancellor, Eternal University, Baru Sahib, for providing necessary infrastructural facilities and constant encouragement.

References

Alexander M (1994) Biodegradation and bioremediation. Academic Press Inc, San Diego, CA, pp 267–269

Almario J, Jeena G, Wunder J, Langen G, Zuccaro A, Coupland G, Bucher M (2017) Root-associated fungal microbiota of nonmycorrhizal *Arabis alpina* and its contribution to plant phosphorus nutrition. Proc Natl Acad Sci 114:E9403–E9412

Aly AH, Edrada-Ebel R, Wray V, Müller WE, Kozytska S, Hentschel U, Proksch P, Ebel R (2008) Bioactive metabolites from the endophytic fungus *Ampelomyces* sp. isolated from the medicinal plant *Urospermum picroides*. Phytochemistry 69:1716–1725

Aly AH, Debbab A, Kjer J, Proksch P (2010) Fungal endophytes from higher plants: a prolific source of phytochemicals and other bioactive natural products. Fungal Divers 41:1–16

Aly AH, Debbab A, Proksch P (2011) Fungal endophytes: unique plant inhabitants with great promises. Appl Microbiol Biotechnol 90:1829–1845

Ambrosini A, Beneduzi A, Stefanski T, Pinheiro F, Vargas L, Passaglia LP (2012) Screening of plant growth promoting rhizobacteria isolated from sunflower (*Helianthus annuus* L.). Plant Soil 356:245–264. https://doi.org/10.1007/s11104-011-1079-1

Andreote FD, Rossetto PB, Souza LC, Marcon J, Maccheroni W, Azevedo JL, Araujo WL (2008) Endophytic population of *Pantoea agglomerans* in citrus plants and development of a cloning vector for endophytes. J Basic Microbiol 48:338–346

Araújo JMD, Silva ACD, Azevedo JL (2000) Isolation of endophytic actinomycetes from roots and leaves of maize (*Zea mays* L.). Braz Arch Biol Technol 43:447. https://doi.org/10.1590/S1516-89132000000400016

Bais HP, Broeckling CD, Vivanco JM (2008) Root exudates modulate plant—microbe interactions in the rhizosphere. In: Karlovsky P (ed) Secondary metabolites in soil ecology. Springer, Berlin, pp 241–252

Bakker AW, Schippers B (1987) Microbial cyanide production in the rhizosphere in relation to potato yield reduction and *Pseudomonas* SPP-mediated plant growth-stimulation. Soil Biol Biochem 19:451–457

Beatriz Sánchez D, Gómez RM, García AM, Bonilla RR (2014) Phosphate solubilizing bacteria isolated from Pennisetum clandestinum associate to livestock systems in the andean area. Rev UDCA Actual Divul Cient 17:423–431

Benjamin S, Pandey A (1998) *Candida rugosa* lipases: molecular biology and versatility in biotechnology. Yeast 14:1069–1087

Bhagyasree S, Ghosh S, Thippaiah M, Rajgopal N (2018) Survey on natural occurrence of endophytes in maize (*Zea mays* L.) ecosystem. Int J Curr Microbiol App Sci 7:2526–2533

Bhatt M, Cajthaml T, Šašek V (2002) Mycoremediation of PAH-contaminated soil. Folia Microbiol 47:255–258

Biswas S, Kundu DK, Mazumdar SP, Saha AR, Majumdar B, Ghorai AK, Ghosh D, Yadav AN, Saxena AK (2018) Study on the activity and diversity of bacteria in a New Gangetic alluvial soil (Eutrocrept) under rice-wheatjute cropping system. Journal of Environmental Biology 39 (3):379–386

Bogner CW, Kamdem RS, Sichtermann G, Matthäus C, Hölscher D, Popp J, Proksch P, Grundler FM, Schouten A (2017) Bioactive secondary metabolites with multiple activities from a fungal endophyte. Microb Biotechnol 10:175–188

Bonfante P, Genre A (2010) Mechanisms underlying beneficial plant–fungus interactions in mycorrhizal symbiosis. Nat Commun 1:48

Borges KB, Borges WDS, Pupo MT, Bonato PS (2008) Stereoselective analysis of thioridazine-2-sulfoxide and thioridazine-5-sulfoxide: an investigation of rac-thioridazine biotransformation by some endophytic fungi. J Pharm Biomed Anal 46:945–952

Borrill P, Connorton JM, Balk J, Miller AJ, Sanders D, Uauy C (2014) Biofortification of wheat grain with iron and zinc: integrating novel genomic resources and knowledge from model crops. Front Plant Sci 5:53

Bric JM, Bostock RM, Silverstone SE (1991) Rapid in situ assay for indoleacetic acid production by bacteria immobilized on a nitrocellulose membrane. Appl Environ Microbiol 57:535–538

Brown ME, Burlingham SK (1968) Production of plant growth substances by *Azotobacter chroococcum*. J Gen Microbiol 53:135–144

Burke AJ, Waller PA, Pickering WF (1989) The evaluation of inorganic phosphate species in salt water lake sediments. Chem Spec Bioavail 1:47–57

Bary A (1866) Morphologie und Physiologie der Pilze, Flechten, und Myxomyceten, (Vol. II), Hofmeister's Handbook of Physiological Botany, Leipzig, Germany

Cappucino JC, Sherman N (1992) Nitrogen cycle. In: Microbiology: a laboratory manual, 4th edn. Benjamin/Cumming Pub. Co., New York, pp 311–312

Carroll G (1988) Fungal endophytes in stems and leaves: from latent pathogen to mutualistic symbiont. Ecology 69:2–9

Chadha N, Prasad R, Varma A (2015) Plant promoting activities of fungal endophytes associated with tomato roots from central Himalaya, India and their interaction with *Piriformospora indica*. Int J Pharm Bio Sci 6:333–343

Chatzistathis T, Therios I, Alifragis D (2009) Differential uptake, distribution within tissues, and use efficiency of manganese, iron, and zinc by olive cultivars kothreiki and koroneiki. Hort Sci 44:1994–1999

Chen Z, Song Y, Chen Y, Huang H, Zhang W, Ju J (2012) Cyclic heptapeptides, cordyheptapeptides C–E, from the marine-derived fungus *Acremonium persicinum* SCSIO 115 and their cytotoxic activities. J Nat Prod 75:1215–1219

Chen GD, Chen Y, Gao H, Shen LQ, Wu Y, Li XX, Li Y, Guo LD, Cen YZ, Yao XS (2013) Xanthoquinodins from the endolichenic fungal strain *Chaetomium elatum*. J Nat Prod 76:702–709

Chen B, Shen J, Zhang X, Pan F, Yang X, Feng Y (2014) The endophytic bacterium, *Sphingomonas* SaMR12, improves the potential for zinc phytoremediation by its host, Sedum alfredii. PLoS One 9:e106826

Choi Y, Hodgkiss I, Hyde K (2005) Enzyme production by endophytes of *Brucea javanica*. J Agric Technol 1:55–66

Chow Y, Ting AS (2015) Endophytic L-asparaginase-producing fungi from plants associated with anticancer properties. J Adv Res 6:869–876

Clark R, Lee S-H (2016) Anticancer properties of capsaicin against human cancer. Anticancer Res 36:837–843

Colangelo EP, Guerinot ML (2006) Put the metal to the petal: metal uptake and transport throughout plants. Curr Opin Plant Biol 9:322–330

Colla G, Rouphael Y, Bonini P, Cardarelli M (2015) Coating seeds with endophytic fungi enhances growth, nutrient uptake, yield and grain quality of winter wheat. Int J Plant Prod 9:171–189

Cook D, Gardner DR, Ralphs MH, Pfister JA, Welch KD, Green BT (2009) Swainsoninine concentrations and endophyte amounts of Undifilum oxytropis in different plant parts of *Oxytropis sericea*. J Chem Ecol 35:1272. https://doi.org/10.1007/s10886-009-9710-9

Correa A, Pacheco S, Mechaly AE, Obal G, Béhar G, Mouratou B, Oppezzo P, Alzari PM, Pecorari F (2014) Potent and specific inhibition of glycosidases by small artificial binding proteins (Affitins). PLoS One 9:e97438

Costa ADR, Silva Júnior ML, Kern DC, Ruivo MDLP, Marichal R (2017) Forms of soil organic phosphorus at black earth sites in the Eastern Amazon. Rev Ciênc Agron 48:1–12

Carroll F (1977) Preliminary studies on the incidence of needle endophytes in some European conifers. Sydowia 29:87-103.

de Souza Leite T, Cnossen-Fassoni A, Pereira OL, Mizubuti ESG, de Araújo EF, de Queiroz MV (2013) Novel and highly diverse fungal endophytes in soybean revealed by the consortium of two different techniques. J Microbiol 51:56–69

Dai CC, Tian LS, Zhao YT, Chen Y, Xie H (2010) Degradation of phenanthrene by the endophytic fungus *Ceratobasidum stevensii* found in *Bischofia polycarpa*. Biodegradation 21:245–255

Delaplace P, Delory BM, Baudson C, de Cazenave MMS, Spaepen S, Varin S, Brostaux Y, du Jardin P (2015) Influence of rhizobacterial volatiles on the root system architecture and the production and allocation of biomass in the model grass *Brachypodium distachyon* (L.) P. Beauv. BMC Plant Biol 15:195

Desai A, Archana G (2011) Role of siderophores in crop improvement. In: Maheshwari DK (ed) Bacteria in agrobiology: plant nutrient management. Springer, Heidelberg, pp 109–139

Deshmukh S, Hückelhoven R, Schäfer P, Imani J, Sharma M, Weiss M, Waller F, Kogel K-H (2006) The root endophytic fungus *Piriformospora indica* requires host cell death for proliferation during mutualistic symbiosis with barley. Proc Natl Acad Sci 103:18450–18457

Ding G, Li Y, Fu S, Liu S, Wei J, Che Y (2008) Ambuic acid and torreyanic acid derivatives from the endolichenic fungus *Pestalotiopsis* sp. J Nat Prod 72:182–186

Dong B, Rengel Z, Graham RD (1995) Root morphology of wheat genotypes differing in zinc efficiency. J Plant Nutr 18:2761–2773

Durmaz E, Coruh C, Dinler G, Grusak MA, Peleg Z, Saranga Y, Fahima T, Yazici A, Ozturk L, Cakmak I (2011) Expression and cellular localization of ZIP1 transporter under zinc deficiency in wild emmer wheat. Plant Mol Biol Rep 29:582–596

Elfita E, Muharni M, Munawar M, Legasari L, Darwati D (2011) Antimalarial compounds from endophytic fungi of Brotowali (*Tinaspora crispa* L). Indo J Chem 11:53–58

Erbert C, Lopes AA, Yokoya NS, Furtado NA, Conti R, Pupo MT, Lopes JLC, Debonsi HM (2012) Antibacterial compound from the endophytic fungus *Phomopsis longicolla* isolated from the tropical red seaweed *Bostrychia radicans*. Bot Mar 55:435–440

Eyberger AL, Dondapati R, Porter JR (2006) Endophyte fungal isolates from *Podophyllum peltatum* produce podophyllotoxin. J Nat Prod 69:1121–1124

Fasim F, Ahmed N, Parsons R, Gadd GM (2002) Solubilization of zinc salts by a bacterium isolated from the air environment of a tannery. FEMS Microbiol Lett 213:1–6

Feller IC (1995) Effects of nutrient enrichment on growth and herbivory of dwarf red mangrove (*Rhizophora mangle*). Ecol Monogr 64:477–505

Finlay RD (2008) Ecological aspects of mycorrhizal symbiosis: with special emphasis on the functional diversity of interactions involving the extraradical mycelium. J Exp Bot 59:1115–1126

Fisher P, Petrini O, Scott HL (1992) The distribution of some fungal and bacterial endophytes in maize (*Zea mays* L.). New Phytol 122:299–305

Forchetti G, Masciarelli O, Izaguirre MJ, Alemano S, Alvarez D, Abdala G (2010) Endophytic bacteria improve seedling growth of sunflower under water stress, produce salicylic acid, and inhibit growth of pathogenic fungi. Curr Microbiol 61:485–493

Fouda AH, Hassan SED, Eid AM, Ewais EED (2015) Biotechnological applications of fungal endophytes associated with medicinal plant *Asclepias sinaica* (Bioss.). Ann Agric Sci 60:95–104

Gao F-K, Dai C-C, Liu X-Z (2010) Mechanisms of fungal endophytes in plant protection against pathogens. Afr J Microbiol Res 4:1346–1351

Genc Y, Huang CY, Langridge P (2007) A study of the role of root morphological traits in growth of barley in zinc-deficient soil. J Exp Bot 58:2775–2784

Gosal S, Karlupia A, Gosal S, Chhibba I, Varma A (2010) Biotization with *Piriformospora indica* and *Pseudomonas fluorescens* improves survival rate, nutrient acquisition, field performance and saponin content of micropropagated *Chlorophytum* sp. Indian J Biotechnol 09:289–297

Goswami D, Dhandhukia P, Patel P, Thakker JN (2014) Screening of PGPR from saline desert of Kutch: growth promotion in *Arachis hypogea* by *Bacillus licheniformis* A2. Microbiol Res 169:66–75

Guo B, Dai JR, Ng S, Huang Y, Leong C, Ong W, Carté BK (2000) Cytonic acids A and B: novel tridepside inhibitors of hCMV protease from the endophytic fungus *Cytonaema* species. J Nat Prod 63:602–604

Gupta N, Sabat J, Parida R, Kerkatta D (2007) Solubilization of tricalcium phosphate and rock phosphate by microbes isolated from chromite, iron and manganese mines. Acta Bot Croat 66:197–204

Hallmann J, Quadt-Hallmann A, Mahaffee W, Kloepper J (1997) Bacterial endophytes in agricultural crops. Can J Microbiol 43:895–914

Hamayun M, Khan SA, Ahmad N, Tang D-S, Kang S-M, Na C-I, Sohn E-Y, Hwang Y-H, Shin D-H, Lee B-H (2009) *Cladosporium sphaerospermum* as a new plant growth-promoting endophyte from the roots of *Glycine max* (L.) *Merr*. World J Microbiol Biotechnol 25:627–632

Hamayun M, Hussain A, Khan SA, Kim HY, Khan AL, Waqas M, Irshad M, Iqbal A, Rehman G, Jan S (2017) Gibberellins producing endophytic fungus *Porostereum spadiceum* AGH786 rescues growth of salt affected soybean. Front Microbiol 8:686. https://doi.org/10.3389/fmicb.2017.00686

Hamm MW (2008) Linking sustainable agriculture and public health: opportunities for realizing multiple goals. J Hun Environ Nutr 3:169–185

Hardoim P, Nissinen R, van Elsas JD (2012) Ecology of bacterial endophytes in sustainable agriculture. In: Maheshwari DK (ed) Bacteria in agrobiology: plant probiotics. Springer, Berlin, pp 97–126
Harnpicharnchai P, Champreda V, Sornlake W, Eurwilaichitr L (2009) A thermotolerant β-glucosidase isolated from an endophytic fungi, *Periconia* sp., with a possible use for biomass conversion to sugars. Protein Expr Purif 67:61–69
Hodgson S, Cates C, Hodgson J, Morley NJ, Sutton BC, Gange AC (2014) Vertical transmission of fungal endophytes is widespread in forbs. Ecol Evol 4:1199–1208
Hoffman AM, Mayer SG, Strobel GA, Hess WM, Sovocool GW, Grange AH, Harper JK, Arif AM, Grant DM, Kelley-Swift EG (2008) Purification, identification and activity of phomodione, a furandione from an endophytic *Phoma* species. Phytochemistry 69:1049–1056
Hu X, Chen J, Guo J (2006) Two phosphate- and potassium-solubilizing bacteria isolated from Tianmu Mountain, Zhejiang, China. World J Microbiol Biotechnol 22:983–990. https://doi.org/10.1007/s11274-006-9144-2
Hung PQ, Annapurna K (2004) Isolation and characterization of endophytic bacteria in soybean (*Glycine* sp.). Omonrice 12:92–101
Hirsch G, Braun U (1992) Communities of parasitic microfungi. In: Fungi in vegetation science. Springer, pp 225–250
Jacobson CB, Pasternak J, Glick BR (1994) Partial purification and characterization of 1-aminocyclopropane-1-carboxylate deaminase from the plant growth promoting rhizobacterium *Pseudomonas putida* GR12-2. Can J Microbiol 40:1019–1025
Ji SH, Gururani MA, Chun S-C (2014) Isolation and characterization of plant growth promoting endophytic diazotrophic bacteria from Korean rice cultivars. Microbiol Res 169:83–98
Joseph B, Priya RM (2011) Bioactive compounds from endophytes and their potential in American. J Biochem Mol Biol 1:291–309
Juhasz AL, Naidu R (2000) Bioremediation of high molecular weight polycyclic aromatic hydrocarbons: a review of the microbial degradation of benzo [a] pyrene. Int Biodeter Biodegr 45:57–88
Kang SH, Cho H, Cheong H, Ryu C, Kim JF, Park S (2007) Two bacterial entophytes eliciting both plant growth promotion and plant defense on pepper (*Capsicum annuum* L.). J Microbiol Biotechnol 17:96–103
Kasotia A, Choudhary DK (2014) Role of endophytic microbes in mitigation of abiotic stress in plants. In: Ahmad P, Rasool S (eds) Emerging technologies and management of crop stress tolerance. Elsevier, New York, pp 97–108
Khalmuratova I, Kim H, Nam YJ, Oh Y, Jeong MJ, Choi HR, You YH, Choo YS, Lee IJ, Shin JH (2015) Diversity and plant growth promoting capacity of endophytic fungi associated with halophytic plants from the west coast of Korea. Mycobiology 43:373–383
Khan SA, Hamayun M, Yoon H, Kim HY, Suh SJ, Hwang SK, Kim JM, Lee IJ, Choo YS, Yoon UH (2008) Plant growth promotion and *Penicillium citrinum*. BMC Microbiol 8:231
Khan MS, Zaidi A, Ahemad M, Oves M, Wani PA (2010) Plant growth promotion by phosphate solubilizing fungi–current perspective. Arch Agron Soil Sci 56:73–98
Khan AL, Hamayun M, Kim YH, Kang SM, Lee IJ (2011) Ameliorative symbiosis of endophyte (*Penicillium funiculosum* LHL06) under salt stress elevated plant growth of *Glycine max* L. Plant Physiol Biochem 49:852–861
Khan AR, Ullah I, Waqas M, Shahzad R, Hong SJ, Park GS, Jung BK, Lee IJ, Shin JH (2015) Plant growth-promoting potential of endophytic fungi isolated from *Solanum nigrum* leaves. World J Microbiol Biotechnol 31:1461–1466
Kharwar RN, Verma VC, Kumar A, Gond SK, Harper JK, Hess WM, Lobkovosky E, Ma C, Ren Y, Strobel GA (2009) Javanicin, an antibacterial naphthaquinone from an endophytic fungus of neem, *Chloridium* sp. Curr Microbiol 58:233–238
Kjer J, Wray V, Edrada-Ebel R, Ebel R, Pretsch A, Lin W, Proksch P (2009) Xanalteric acids I and II and related phenolic compounds from an endophytic *Alternaria* sp. isolated from the mangrove plant *Sonneratia alba*. J Nat Prod 72:2053–2057

Kobayashi T, Nishizawa NK (2012) Iron uptake, translocation, and regulation in higher plants. Annu Rev Plant Biol 63:131–152

Kobayashi DY, Palumbo JD (2000) Bacterial endophytes and their effects on plants and uses in agriculture. In: Bacon CW, White JF (eds) Microbial endophytes. Marcel Dekker, Inc., New York, NY, pp 199–233

Kobayashi T, Itai RN, Nishizawa NK (2014) Iron deficiency responses in rice roots. Rice 7:27

Köhl J, Lombaers C, Moretti A, Bandyopadhyay R, Somma S, Kastelein P (2015) Analysis of microbial taxonomical groups present in maize stalks suppressive to colonization by toxigenic *Fusarium* spp.: a strategy for the identification of potential antagonists. Biol Control 83:20–28

Kour D, Rana KL, Verma P, Yadav AN, Kumar V, Dhaliwal HS (2017a) Biofertilizers: Eco-friendly technologies and bioresources for sustainable agriculture. In: Proceeding of International Conference on Innovative Research in Engineering Science and Technology. p 14.

Kour D, Rana KL, Verma P, Yadav AN, Kumar V, Dhaliwal HS (2017b) Drought tolerant phosphorus solubilizing microbes: Diversity and biotechnological applications for crops growing under rainfed conditions. In: Proceeding of National Conference on Advances in Food Science and Technology, p 166

Kudalkar P, Strobel G, Riyaz-Ul-Hassan S, Geary B, Sears J (2012) *Muscodor sutura*, a novel endophytic fungus with volatile antibiotic activities. Mycoscience 53:319–325

Kusari S, Spiteller M (2011) Are we ready for industrial production of bioactive plant secondary metabolites utilizing endophytes? Nat Prod Rep 28:1203–1207

Kusari S, Spiteller M (2012) Metabolomics of endophytic fungi producing associated plant secondary metabolites: progress, challenges and opportunities. In: Roessner U (ed) Metabolomics. InTech, Rijeka. https://doi.org/10.5772/31596

Kusari S, Lamshöft M, Spiteller M (2009) *Aspergillus fumigatus Fresenius*, an endophytic fungus from *Juniperus communis* L. Horstmann as a novel source of the anticancer prodrug deoxypodophyllotoxin. J Appl Microbiol 107:1019–1030

Kusari S, Verma VC, Lamshoeft M, Spiteller M (2012) An endophytic fungus from *Azadirachta indica* A. Juss. that produces azadirachtin. World J Microbiol Biotechnol 28:1287–1294

Larran S, Perello A, Simon M, Moreno V (2002) Isolation and analysis of endophytic microorganisms in wheat (*Triticum aestivum* L.) leaves. World J Microbiol Biotechnol 18:683–686

Larran S, Perelló A, Simón MR, Moreno V (2007) The endophytic fungi from wheat (*Triticum aestivum* L.). World J Microbiol Biotechnol 23:565–572

Lee JC, Lobkovsky E, Pliam NB, Strobel G, Clardy J (1995) Subglutinols A and B: immunosuppressive compounds from the endophytic fungus *Fusarium subglutinans*. J Org Chem 60:7076–7077

Lee JC, Strobel GA, Lobkovsky E, Clardy J (1996) Torreyanic acid: a selectively cytotoxic quinone dimer from the endophytic fungus *Pestalotiopsis microspora*. J Org Chem 61:3232–3233

Lee SO, Choi GJ, Choi YH, Jang KS, Park DJ, Kim CJ, Kim JC (2008) Isolation and characterization of endophytic actinomycetes from Chinese cabbage roots as antagonists to *Plasmodiophora brassicae*. J Microbiol Biotechnol 18:1741–1746

Li JY, Strobel GA (2001) Jesterone and hydroxy-jesterone antioomycete cyclohexenone epoxides from the endophytic fungus *Pestalotiopsis jesteri*. Phytochemistry 57:261–265

Li GH, Yu ZF, Li X, Wang XB, Zheng LJ, Zhang KQ (2007a) Nematicidal metabolites produced by the endophytic fungus *Geotrichum* sp. AL4. Chem Biodivers 4:1520–1524

Li W-C, Zhou J, Guo S-Y, Guo L-D (2007b) Endophytic fungi associated with lichens in Baihua mountain of Beijing, China. Fungal Divers 25:69–80

Li W, Ye Z, Wong M (2010) Metal mobilization and production of short-chain organic acids by rhizosphere bacteria associated with a Cd/Zn hyperaccumulating plant, *Sedum alfredii*. Plant Soil 326:453–467

Li G, Wang H, Zhu R, Sun L, Wang L, Li M, Li Y, Liu Y, Zhao Z, Lou H (2012a) Phaeosphaerins A–F, cytotoxic perylenequinones from an endolichenic fungus, *Phaeosphaeria* sp. J Nat Prod 75:142–147

Li HY, Wei DQ, Shen M, Zhou ZP (2012b) Endophytes and their role in phytoremediation. Fungal Divers 54:11–18

Lin Z-J, Lu Z-Y, Zhu T-J, Fang Y-C, Gu Q-Q, Zhu W-M (2008) Penicillenols from *Penicillium* sp. GQ-7, an endophytic fungus associated with *Aegiceras corniculatum*. Chem Pharm Bull 56:217–221

Lu Y, Chen C, Chen H, Zhang J, Chen W (2012) Isolation and identification of endophytic fungi from *Actinidia macrosperma* and investigation of their bioactivities. Evid Based Complement Alternat Med 2012:382742. https://doi.org/10.1155/2012/382742

Lumyong S, Lumyong P, McKenzie EH, Hyde KD (2002) Enzymatic activity of endophytic fungi of six native seedling species from Doi Suthep-Pui National Park, Thailand. Can J Microbiol 48:1109–1112

Mano H, Morisaki H (2007) Endophytic bacteria in the rice plant. Microbes Environ 23:109–117

Manter DK, Delgado JA, Holm DG, Stong RA (2010) Pyrosequencing reveals a highly diverse and cultivar-specific bacterial endophyte community in potato roots. Microb Ecol 60:157–166

Marinho AM, Rodrigues-Filho E, Moitinho MDLR, Santos LS (2005) Biologically active polyketides produced by *Penicillium janthinellum* isolated as an endophytic fungus from fruits of *Melia azedarach*. J Braz Chem Soc 16:280–283

Marler M, Pedersen D, Mitchell-Olds T, Callaway R (1999) A polymerase chain reaction method for detecting dwarf mistletoe infection in Douglas-fir and western larch. Can J For Res 29:1317–1321

Mathur N, Vyas P, Joshi N, Choudhary K, Purohit DK (2011) Mycorrhiza: a potent bioinoculant for sustainable agriculture. In: Pathak H, Sharma A (eds) Microbial technology "the emerging era" lap lambert. Academic Publishing Ag & Co. Kg, Dudweiller Landstr, pp 230–245

Mendes R, Pizzirani-Kleiner AA, Araujo WL, Raaijmakers JM (2007) Diversity of cultivated endophytic bacteria from sugarcane: genetic and biochemical characterization of *Burkholderia cepacia* complex isolates. Appl Environ Microbiol 73:7259–7267

Mercado-Blanco J, Lugtenberg B (2014) Biotechnological applications of bacterial endophytes. Curr Biotechnol 3:60–75

Miché L, Balandreau J (2001) Effects of rice seed surface sterilization with hypochlorite on inoculated *Burkholderia vietnamiensis*. Appl Environ Microbiol 67:3046–3052

Mingma R, Pathom-aree W, Trakulnaleamsai S, Thamchaipenet A, Duangmal K (2014) Isolation of rhizospheric and roots endophytic actinomycetes from Leguminosae plant and their activities to inhibit soybean pathogen, *Xanthomonas campestris* pv. glycine. World J Microbiol Biotechnol 30:271–280

Mishra S, Singh A, Keswani C, Saxena A, Sarma B, Singh H (2015) Harnessing plant-microbe interactions for enhanced protection against phytopathogens. In: Arora N (ed) Plant microbes symbiosis: applied facets. Springer, New Delhi, pp 111–125

Mishra VK, Singh G, Passari AK, Yadav MK, Gupta VK, Singh BP (2016a) Distribution and antimicrobial potential of endophytic fungi associated with ethnomedicinal plant *Melastoma malabathricum* L. J Environ Biol 37(2):229–237

Mishra VK, Passari AK, Singh BP (2016b) In vitro antimycotic and biosynthetic potential of fungal endophytes associated with *Schima Wallichii*. In: Kumar P et al (eds) Current trends in disease diagnostics. Springer, Cham, pp 367–381

Mishra VK, Passari AK, Chandra P, Leo VV, Kumar B, Gupta VK, Singh BP (2017a) Determination and production of antimicrobial compounds by *Aspergillus clavatonanicus* strain MJ31, an endophytic fungus from *Mirabilis jalapa* L. using UPLC-ESI-MS/MS and TD GC-MS. PLoS One 12(10):1–24. https://doi.org/10.1371/journal.pone.0186234

Mishra VK, Passari AK, Leo VV, Singh BP (2017b) Molecular diversity and detection of endophytic fungi based on their antimicrobial biosynthetic genes. In: Singh BP, Gupta VK (eds) Molecular markers in mycology, fungal biology. Springer, Cham, pp 1–35. https://doi.org/10.1007/978-3-319-34106-4_1

Mohd S, Shukla J, Kushwaha AS, Mandrah K, Shankar J, Arjaria N, Saxena PN, Narayan R, Roy SK, Kumar M (2017) Endophytic fungi *Piriformospora indica* mediated protection of host from arsenic toxicity. Front Microbiol 8:754

Montanez A, Blanco AR, Barlocco C, Beracochea M, Sicardi M (2012) Characterization of cultivable putative endophytic plant growth promoting bacteria associated with maize cultivars (*Zea mays* L.) and their inoculation effects *in vitro*. Appl Soil Ecol 58:21–28

Mostert L, Crous P, Petrini O (2000) Endophytic fungi associated with shoots and leaves of Vitis vinifera, with specific reference to the Phomopsis viticola complex. Sydowia 52:46–58

Naik BS, Shashikala J, Krishnamurthy Y (2009a) Study on the diversity of endophytic communities from rice (*Oryza sativa* L.) and their antagonistic activities in vitro. Microbiol Res 164:290–296

Naik BS, Shashikala J, Krishnamurthy Y (2009b) Study on the diversity of endophytic communities from rice (*Oryza sativa* L.) and their antagonistic activities in vitro. Microbiol Res 164:290–296

Narayan OP, Verma N, Singh AK, Oelmüller R, Kumar M, Prasad D, Kapoor R, Dua M, Johri AK (2017) Antioxidant enzymes in chickpea colonized by *Piriformospora indica* participate in defense against the pathogen *Botrytis cinerea*. Sci Rep 7:13553. https://doi.org/10.1038/s41598-017-12944-w

Narula S, Anand R, Dudeja S, Pathak D (2013) Molecular diversity of root and nodule endophytic bacteria from field pea (*Pisum sativum* L.). Legum Res 36:344–350

Nath R, Sharma G, Barooah M (2012) Efficiency of tricalcium phosphate solubilization by two different endophytic *Penicillium* sp. isolated from tea (*Camellia sinensis* L.). Eur J Exp Biol 2:1354–1358

Nath R, Sharma G, Barooah M (2015) Plant growth promoting endophytic fungi isolated from tea (Camellia sinensis) shrubs of Assam, India. Appl Ecol Environ Res 13:877–891

Nefzi A, Aydi Ben Abdallah R, Jabnoun-Khiareddine H, Ammar N, Somai L, Hamada W, Haouala R, Daami-Remadi M (2018) Investigation on biosuppression of *Fusarium* crown and root rot of tomato (*Solanum lycopersicum* L.) and growth promotion using fungi naturally associated to *Solanum linnaeanum* L. Afr J Microbiol Res 12:152–170

Nisa H, Kamili AN, Nawchoo IA, Shafi S, Shameem N, Bandh SA (2015) Fungal endophytes as prolific source of phytochemicals and other bioactive natural products: a review. Microb Pathog 82:50–59

Nyoki D, Ndakidemi PA (2014) Effects of *Bradyrhizobium japonicum* and phosphorus supplementation on the productivity of legumes. Int J Plant Soil Sci 3:894–910

Petrini O (1991) Fungal endophytes of tree leaves. In: Andrews JH, Hirano SS (eds) Microbial ecology of leaves. Springer, New York, pp 179–197

Pfeiffer WH, McClafferty B (2007) HarvestPlus: breeding crops for better nutrition. Crop Sci 47:S-88–S-105

Phongpaichit S, Rungjindamai N, Rukachaisirikul V, Sakayaroj J (2006) Antimicrobial activity in cultures of endophytic fungi isolated from *Garcinia* species. FEMS Immunol Med Microbiol 48:367–372

Pikovskaya R (1948) Mobilization of phosphorus in soil in connection with vital activity of some microbial species. Mikrobiologiya 17:362–370

Pimentel MR, Molina G, Dionísio AP, Maróstica Junior MR, Pastore GM (2011) The use of endophytes to obtain bioactive compounds and their application in biotransformation process. Biotechnol Res Int 2011:1. https://doi.org/10.4061/2011/576286

Pinto LSRC, Azevedo JL, Pereira JO, Vieira MLC, Labate CA (2000) Symptomless infection of banana and maize by endophytic fungi impairs photosynthetic efficiency. New Phytol 147:609–615

Piromyou P, Greetatorn T, Teamtisong K, Okubo T, Shinoda R, Nuntakij A, Tittabutr P, Boonkerd N, Minamisawa K, Teaumroong N (2015) Preferential association of endophytic *Bradyrhizobia* with different rice cultivars and its implications for rice endophyte evolution. Appl Environ Microbiol 81:3049–3061

Pointing S (2001) Feasibility of bioremediation by white-rot fungi. Appl Microbiol Biotechnol 57:20–33

Pongcharoen W, Rukachaisirikul V, Phongpaichit S, Kühn T, Pelzing M, Sakayaroj J, Taylor WC (2008) Metabolites from the endophytic fungus *Xylaria* sp. PSU-D14. Phytochemistry 69:1900–1902

Poonguzhali S, Madhaiyan M, Sa T (2006) Cultivation-dependent characterization of rhizobacterial communities from field grown Chinese cabbage *Brassica campestris* ssp pekinensis and screening of traits for potential plant growth promotion. Plant Soil 286:167–180

Porras-Alfaro A, Bayman P (2011) Hidden fungi, emergent properties: endophytes and microbiomes. Annu Rev Phytopathol 49:291–315

Priyadharsini P, Muthukumar T (2017) The root endophytic fungus *Curvularia geniculata* from *Parthenium hysterophorus* roots improves plant growth through phosphate solubilization and phytohormone production. Fungal Ecol 27:69–77

Providenti MA, Lee H, Trevors JT (1993) Selected factors limiting the microbial degradation of recalcitrant compounds. J Ind Microbiol 12:379–395

Puri SC, Verma V, Amna T, Qazi GN, Spiteller M (2005) An endophytic fungus from *Nothapodytes foetida* that produces camptothecin. J Nat Prod 68:1717–1719

Quadt-Hallmann A, Kloepper J, Benhamou N (1997) Bacterial endophytes in cotton: mechanisms of entering the plant. Can J Microbiol 43:577–582

Rado R, Andrianarisoa B, Ravelomanantsoa S, Rakotoarimanga N, Rahetlah V, Fienena F, Andriambeloson O (2015) Biocontrol of potato wilt by selective rhizospheric and endophytic bacteria associated with potato plant. Afr J Food Agric Nutr Dev 15:9762–9776

Rai M, Rathod D, Agarkar G, Dar M, Brestic M, Pastore GM, Junior MRM (2014) Fungal growth promotor endophytes: a pragmatic approach towards sustainable food and agriculture. Symbiosis 62:63–79

Rajkumar M, Ae N, Prasad MNV, Freitas H (2010) Potential of siderophore-producing bacteria for improving heavy metal phytoextraction. Trends Biotechnol 28:142–149

Rana KL, Kour D, Yadav AN, Kumar V, Dhaliwal HS (2016a) Biotechnological applications of endophytic microbes associated with barley (Hordeum vulgare L.) growing in Indian Himalayan regions. In: In: Proceeding of 86th Annual Session of NASI & Symposium on "Science, Technology and Entrepreneurship for Human Welfare in The Himalayan Region", p 80

Rana KL, Kour D, Yadav AN, Kumar V, Dhaliwal HS (2016b) Endophytic microbes from wheat: Diversity and biotechnological applications for sustainable agriculture. In: In: Proceeding of 57th Association of Microbiologist of India & International symposium on "Microbes and Biosphere: What's New What's Next", p 453

Rana KL, Kour D, Verma P, Yadav A, Kumar V, Singh D (2017) Diversity and biotechnological applications of endophytic microbes associated with maize (Zea mays L.) growing in Indian Himalayan regions In: Proceeding of national conference on advances in food science and technology, p 41

Rana KL, Kour D, Yadav AN (2018) Endophytic microbiomes: Biodiversity, ecological significance and biotechnological applications. Res J Biotechnol 14:1–30

Rêgo MCF, Ilkiu-Borges F, Filippi MCCD, Gonçalves LA, Silva GBD (2014) Morphoanatomical and biochemical changes in the roots of rice plants induced by plant growth-promoting microorganisms. J Bot 2014:818797. https://doi.org/10.1155/2014/818797

Ren X, Zhang N, Cao M, Wu K, Shen Q, Huang Q (2012) Biological control of tobacco black shank and colonization of tobacco roots by a *Paenibacillus polymyxa* strain C5. Biol Fertil Soils 48:613–620

Rinu K, Sati P, Pandey A (2014) *Trichoderma gamsii* (NFCCI 2177): a newly isolated endophytic, psychrotolerant, plant growth promoting, and antagonistic fungal strain. J Basic Microbiol 54:408–417

Rodrıguez H, Fraga R (1999) Phosphate solubilizing bacteria and their role in plant growth promotion. Biotechnol Adv 17:319–339

Rollinger JL, Langenheim JH (2018) Geographic Survey of Fungal Endophyte Community Composition in Leaves of Coastal Redwood. Mycologia 85 (2):149–156

Rosenblueth M, Martínez-Romero E (2006) Bacterial endophytes and their interactions with hosts. Mol Plant-Microbe Interact 19:827–837

Russell JR, Huang J, Anand P, Kucera K, Sandoval AG, Dantzler KW, Hickman D, Jee J, Kimovec FM, Koppstein D (2011) Biodegradation of polyester polyurethane by endophytic fungi. Appl Environ Microbiol 77:6076–6084

Ryan RP, Germaine K, Franks A, Ryan DJ, Dowling DN (2008) Bacterial endophytes: recent developments and applications. FEMS Microbiol Lett 278:1–9

Sahay H, Yadav AN, Singh AK, Singh S, Kaushik R, Saxena AK (2017) Hot springs of Indian Himalayas: potential sources of microbial diversity and thermostable hydrolytic enzymes. 3. Biotech 7:1–11

Saini R, Dudeja SS, Giri R, Kumar V (2015) Isolation, characterization, and evaluation of bacterial root and nodule endophytes from chickpea cultivated in Northern India. J Basic Microbiol 55:74–81

Sane S, Mehta S (2015) Isolation and evaluation of rock phosphate solubilizing fungi as potential biofertilizer. J Fertil Pestic 6:156–160

Saxena S, Meshram V, Kapoor N (2015) *Muscodor tigerii* sp. nov.-volatile antibiotic producing endophytic fungus from the northeastern Himalayas. Ann Microbiol 65:47–57

Saxena AK, Yadav AN, Rajawat M, Kaushik R, Kumar R, Kumar M, Prasanna R, Shukla L (2016) Microbial diversity of extreme regions: an unseen heritage and wealth. Indian J Plant Gen Res 29:246–248

Schardl CL, Leuchtmann A, Spiering MJ (2004) Symbioses of grasses with seedborne fungal endophytes. Annu Rev Plant Biol 55:315–340

Schulz B, Boyle C (2005) The endophytic continuum. Mycol Res 109:661–686

Schwyn B, Neilands J (1987) Universal chemical assay for the detection and determination of siderophores. Anal Biochem 160:47–56

Semple KT, Morriss A, Paton GI (2003) Bioavailability of hydrophobic organic contaminants in soils: fundamental concepts and techniques for analysis. Eur J Soil Sci 54:809–818

Sharma SK, Sharma MP, Ramesh A, Joshi OP (2012) Characterization of zinc-solubilizing *Bacillus* isolates and their potential to influence zinc assimilation in soybean seeds. J Microbiol Biotechnol 22:352–359

Sharma D, Pramanik A, Agrawal PK (2016) Evaluation of bioactive secondary metabolites from endophytic fungus *Pestalotiopsis neglecta* BAB-5510 isolated from leaves of *Cupressus torulosa* D.Don. 3 Biotech 6:210. https://doi.org/10.1007/s13205-016-0518-3

Sheng X-F, Xia J-J, Jiang C-Y, He L-Y, Qian M (2008) Characterization of heavy metal-resistant endophytic bacteria from rape (*Brassica napus*) roots and their potential in promoting the growth and lead accumulation of rape. Environ Pollut 156:1164–1170

Shweta S, Zuehlke S, Ramesha B, Priti V, Kumar PM, Ravikanth G, Spiteller M, Vasudeva R, Shaanker RU (2010) Endophytic fungal strains of *Fusarium solani*, from *Apodytes dimidiata* E. Mey. ex Arn (Icacinaceae) produce camptothecin, 10-hydroxycamptothecin and 9-methoxycamptothecin. Phytochemistry 71:117–122

Sijam K, Dikin A (2005) Biochemical and physiological characterization of *Burkholderia cepacia* as biological control agent. Int J Agric Biol 7:385–388

Singh DP, Singh HB, Prabha R (2016a) Microbial inoculants in sustainable agricultural productivity. Springer, New Delhi

Singh RN, Gaba S, Yadav AN, Gaur P, Gulati S, Kaushik R, Saxena AK (2016b) First, high quality draft genome sequence of a plant growth promoting and cold active enzymes producing psychrotrophic *Arthrobacter agilis* strain L77. Stand Genomic Sci 11:54. https://doi.org/10.1186/s40793-016-0176-4

Singh DP, Singh HB, Prabha R (2017) Plant-microbe interactions in agro-ecological perspectives. Springer, New Delhi

Sirohi G, Upadhyay A, Srivastava PS, Srivastava S (2015) PGPR mediated zinc biofertilization of soil and its impact on growth and productivity of wheat. J Soil Sci Plant Nutr 15:202–216

Sommart U, Rukachaisirikul V, Tadpetch K, Sukpondma Y, Phongpaichit S, Hutadilok-Towatana N, Sakayaroj J (2012) Modiolin and phthalide derivatives from the endophytic fungus *Microsphaeropsis arundinis* PSU-G18. Tetrahedron 68:10005–10010

Soni R, Yadav SK, Rajput AS (2018) ACC-deaminase producing rhizobacteria: prospects and application as stress busters for stressed agriculture. In: Panpatte DG, Jhala YK, Shelat HN, Vyas RV (eds) Microorganisms for green revolution. Springer, New Delhi, pp 161–175

Spagnoletti F, Tobar N, Di Pardo AF, Chiocchio V, Lavado R (2017) Dark septate endophytes present different potential to solubilize calcium, iron and aluminium phosphates. Appl Soil Ecol 111:25–32

Srivastava AK, Kumar S, Kaushik R, Saxena AK, Padaria JC, Gupta A, Pal KK, Gujar GT, Sharma A, Singh P (2013) Diversity analysis of Bacillus and other predominant genera in extreme environments and its utilization in Agriculture. https://doi.org/10.13140/2.1.1357.3927

Stein AJ (2010) Global impacts of human mineral malnutrition. Plant Soil 335:133–154

Stierle A, Strobel G, Stierle D (1993) Taxol and taxane production by *Taxomyces andreanae,* an endophytic fungus of Pacific yew. Science 260:214–216

Stone JK, Bacon CW, White J (2000) An overview of endophytic microbes: endophytism defined. In: Bacon CW, White JF (eds) Microbial endophytes. Marcel Dekker, New York, pp 3–29

Strobel GA (2003) Endophytes as sources of bioactive products. Microbes Infect 5:535–544

Strobel GA, Pliam NB (1997) Immunosuppressant diterpene compound. Google Patents

Strobel GA, Torczynski R, Bollon A (1997) *Acremonium* sp.—a leucinostatin A producing endophyte of European yew (*Taxus baccata*). Plant Sci 128:97–108

Strobel GA, Miller RV, Martinez-Miller C, Condron MM, Teplow DB, Hess W (1999) Cryptocandin, a potent antimycotic from the endophytic fungus *Cryptosporiopsis* cf. quercina. Microbiology 145:1919–1926

Strobel G, Ford E, Worapong J, Harper JK, Arif AM, Grant DM, Fung PC, Chau RMW (2002) Isopestacin, an isobenzofuranone from *Pestalotiopsis microspora,* possessing antifungal and antioxidant activities. Phytochemistry 60:179–183

Su H, Kang J, Cao J, Mo L, Hyde KD (2014) Medicinal plant endophytes produce analogous bioactive compounds. Chiang Mai J Sci 41:1–13

Suman A, Verma P, Yadav AN, Saxena AK (2015) Bioprospecting for extracellular hydrolytic enzymes from culturable thermotolerant bacteria isolated from Manikaran thermal springs. Res J Biotechnol 10:33–42

Suman A, Verma P, Yadav AN, Srinivasamurthy R, Singh A, Prasanna R (2016a) Development of hydrogel based bio-inoculant formulations and their impact on plant biometric parameters of wheat (*Triticum aestivum* L.). Int J Curr Microbiol Appl Sci 5:890–901

Suman A, Yadav AN, Verma P (2016b) Endophytic microbes in crops: diversity and beneficial impact for sustainable agriculture. In: Singh DP, Abhilash P, Prabha R (eds) Microbial inoculants in sustainable agricultural productivity, research perspectives. Springer-Verlag, New Delhi, pp 117–143. https://doi.org/10.1007/978-81-322-2647-5_7

Sun JF, Lin X, Zhou XF, Wan J, Zhang T, Yang B, Yang XW, Tu Z, Liu Y (2014) Pestalols A–E, new alkenyl phenol and benzaldehyde derivatives from endophytic fungus *Pestalotiopsis* sp. AcBC2 isolated from the Chinese mangrove plant *Aegiceras corniculatum.* J Antibiot 67:451–457

Suyal DC, Yadav A, Shouche Y, Goel R (2015) Bacterial diversity and community structure of Western Indian Himalayan red kidney bean (*Phaseolus vulgaris*) rhizosphere as revealed by 16S rRNA gene sequences. Biologia 70:305–313

Šašek V, Cajthaml T, Bhatt M (2003) Use of fungal technology in soil remediation: a case study. Water Air Soil Pollut Focus 3:5–14

Šišić A, Baćanović J, Finckh MR (2017) Endophytic *Fusarium equiseti* stimulates plant growth and reduces root rot disease of pea (*Pisum sativum* L.) caused by *Fusarium avenaceum* and *Peyronellaea pinodella.* Eur J Plant Pathol 148:271–282

Tan RX, Zou WX (2001) Endophytes: a rich source of functional metabolites. Nat Prod Rep 18:448–459

Tariq M, Hameed S, Yasmeen T, Zahid M, Zafar M (2014) Molecular characterization and identification of plant growth promoting endophytic bacteria isolated from the root nodules of pea (*Pisum sativum* L.). World J Microbiol Biotechnol 30:719–725

Tenguria RK, Khan FN, Quereshi S (2011) Endophytes-mines of pharmacological therapeutics. World J Sci Technol 1:127–149

Thanh DTN, Diep CN (2014) Isolation, characterization and identification of endophytic bacteria in maize (*Zea mays* L.) cultivated on Acrisols of the Southeast of Vietnam. Am J Life Sci 2:224–233

Tian X, Cao L, Tan H, Zeng Q, Jia Y, Han W, Zhou S (2004) Study on the communities of endophytic fungi and endophytic actinomycetes from rice and their antipathogenic activities in vitro. World J Microbiol Biotechnol 20:303–309

Tian LS, Dai C, Zhao Y, Zhao M, Yong Y, Wang X (2007) The degradation of phenanthrene by endophytic fungi *Phomopsis* sp. single and co-cultured with rice. China Environ Sci 27:757–762

Tian J, Fu L, Zhang Z, Dong X, Xu D, Mao Z, Liu Y, Lai D, Zhou L (2017) Dibenzo-α-pyrones from the endophytic fungus *Alternaria* sp. Samif01: isolation, structure elucidation, and their antibacterial and antioxidant activities. Nat Prod Res 31:387–396

Timmusk S, Behers L, Muthoni J, Muraya A, Aronsson A-C (2017) Perspectives and challenges of microbial application for crop improvement. Front Plant Sci 8:49

Tiwari R, Rana C (2015) Plant secondary metabolites: a review. Int J Eng Res Gen Sci 3:661–670

Tiwari VK, Rawat N, Neelam K, Kumar S, Randhawa GS, Dhaliwal HS (2010) Substitutions of 2S and 7U chromosomes of *Aegilops kotschyi* in wheat enhance grain iron and zinc concentration. Theor Appl Genet 121:259–269

Tomita F (2003) Endophytes in Southeast Asia and Japan: their taxonomic diversity and potential applications. Fungal Divers 14:187–204

Torres M, Dolcet MM, Sala N, Canela R (2003) Endophytic fungi associated with Mediterranean plants as a source of mycelium-bound lipases. J Agric Food Chem 51:3328–3333

UmaMaheswari T, Anbukkarasi K, Hemalatha T, Chendrayan K (2013) Studies on phytohormone producing ability of indigenous endophytic bacteria isolated from tropical legume crops. Int J Curr Microbiol Appl Sci 2:127–136

Uzma F, Hashem A, Murthy N, Mohan HD, Kamath PV, Singh BP, Venkataramana M, Gupta VK, Siddaiah CN, Chowdappa S, Alqaeawi AA, Abd Allah EF (2018) Endophytic fungi—alternative sources of cytotoxic compounds: a review. Front Pharmacol 9(309):1–37. https://doi.org/10.3389/fphar.2018.00309

Vacheron J, Desbrosses G, Bouffaud M-L, Touraine B, Moënne-Loccoz Y, Muller D, Legendre L, Wisniewski-Dyé F, Prigent-Combaret C (2013) Plant growth-promoting rhizobacteria and root system functioning. Front Plant Sci 4:356

Vallad GE, Goodman RM (2004) Systemic acquired resistance and induced systemic resistance in conventional agriculture. Crop Sci 44:1920–1934

Verma V, Singh S, Prakash S (2011) Bio-control and plant growth promotion potential of siderophore producing endophytic Streptomyces from *Azadirachta indica* A. Juss. J Basic Microbiol 51:550–556

Verma P, Yadav AN, Khannam KS, Panjiar N, Kumar S, Saxena AK, Suman A (2015a) Assessment of genetic diversity and plant growth promoting attributes of psychrotolerant bacteria allied with wheat (*Triticum aestivum*) from the northern hills zone of India. Ann Microbiol 65:1885–1899

Verma P, Yadav AN, Shukla L, Saxena AK, Suman A (2015b) Alleviation of cold stress in wheat seedlings by *Bacillus amyloliquefaciens* IARI-HHS2-30, an endophytic psychrotolerant K-solubilizing bacterium from NW Indian Himalayas. Natl J Life Sci 12:105–110

Verma P, Yadav AN, Shukla L, Saxena AK, Suman A (2015c) Hydrolytic enzymes production by thermotolerant *Bacillus altitudinis* IARI-MB-9 and *Gulbenkiania mobilis* IARI-MB-18 isolated from Manikaran hot springs. Int J Adv Res 3:1241–1250

Verma P, Yadav AN, Khannam KS, Kumar S, Saxena AK, Suman A (2016a) Molecular diversity and multifarious plant growth promoting attributes of bacilli associated with wheat (*Triticum aestivum* L.) rhizosphere from six diverse agro-ecological zones of India. J Basic Microbiol 56:44–58

Verma P, Yadav AN, Khannam KS, Mishra S, Kumar S, Saxena AK, Suman A (2016b) Appraisal of diversity and functional attributes of thermotolerant wheat associated bacteria from the peninsular zone of India. Saudi J Biol Sci. https://doi.org/10.1016/j.sjbs.2016.01.042

Verma P, Yadav AN, Kumar V, Kumar K, Dhaliwal HS (2017a) Microbes mediated biofortification of wheat (*Triticum aestivum* L.) for micronutrients by Fe-chelating and Zn-solubilizing bacteria. In: Proceeding of national conference on advances in food science and technology, pp 199–200

Verma P, Yadav AN, Kumar V, Singh DP, Saxena AK (2017b) Beneficial plant-microbes interactions: biodiversity of microbes from diverse extreme environments and its impact for crops improvement. In: Singh DP, Singh HB, Prabha R (eds) Plant-microbe interactions in agro-ecological perspectives. Springer Nature, Singapore, pp 543–580. https://doi.org/10.1007/978-981-10-6593-4_22

Vinale F, Nicoletti R, Lacatena F, Marra R, Sacco A, Lombardi N, d'Errico G, Digilio M, Lorito M, Woo S (2017) Secondary metabolites from the endophytic fungus *Talaromyces pinophilus*. Nat Prod Res 31:1778. https://doi.org/10.1080/14786419.2017.1290624

Wakelin SA, Warren RA, Harvey PR, Ryder MH (2004) Phosphate solubilization by *Penicillium* spp. closely associated with wheat roots. Biol Fertil Soils 40:36–43

Wang B, Qiu Y-L (2006) Phylogenetic distribution and evolution of mycorrhizas in land plants. Mycorrhiza 16:299–363

Wang F, Jiao R, Cheng A, Tan S, Song Y (2007) Antimicrobial potentials of endophytic fungi residing in *Quercus variabilis* and brefeldin A obtained from *Cladosporium* sp. World J Microbiol Biotechnol 23:79–83

Wang Y, Niu S, Liu S, Guo L, Che Y (2010) The first naturally occurring thiepinols and thienol from an endolichenic fungus *Coniochaeta* sp. Org Lett 12:5081–5083

Wang M, Zheng Q, Shen Q, Guo S (2013a) The critical role of potassium in plant stress response. Int J Mol Sci 14:7370–7390

Wang QX, Bao L, Yang XL, Liu DL, Guo H, Dai HQ, Song FH, Zhang LX, Guo LD, Li SJ (2013b) Ophiobolins P–T, five new cytotoxic and antibacterial sesterterpenes from the endolichenic fungus *Ulocladium* sp. Fitoterapia 90:220–227

Wang Y, Yang X, Zhang X, Dong L, Zhang J, Wei Y, Feng Y, Lu L (2014) Improved plant growth and Zn accumulation in grains of rice (*Oryza sativa* L.) by inoculation of endophytic microbes isolated from a Zn Hyperaccumulator, *Sedum alfredii* H. J Agric Food Chem 62:1783–1791

White PJ, Broadley MR (2009) Biofortification of crops with seven mineral elements often lacking in human diets–iron, zinc, copper, calcium, magnesium, selenium and iodine. New Phytol 182:49–84

Wilson D (1995) Endophyte: The Evolution of a Term, and Clarification of Its Use and Definition. Oikos 73 (2):274

Wu SH, Chen YW, Shao SC, Wang LD, Yu Y, Li ZY, Yang LY, Li SL, Huang R (2009) Two new solanapyrone analogues from the endophytic fungus *Nigrospora* sp. YB-141 of *Azadirachta indica*. Chem Biodivers 6:79–85

Wulandari RS, Suryantini R (2018) Growth of Albizia in Vitro: Endophytic Fungi as Plant Growth Promote of Albizia. Int Sch Sci Res Inn 12(1)8

Xing Y-M, Chen J, Cui J-L, Chen X-M, Guo S-X (2011) Antimicrobial activity and biodiversity of endophytic fungi in *Dendrobium devonianum* and *Dendrobium thyrsiflorum* from Vietman. Curr Microbiol 62:1218–1224

Xu L-L, Han T, Wu J-Z, Zhang Q-Y, Zhang H, Huang B-K, Rahman K, Qin L-P (2009) Comparative research of chemical constituents, antifungal and antitumor properties of ether extracts of *Panax ginseng* and its endophytic fungus. Phytomedicine 16:609–616

Xu Y-G, Wang B-S, Yu J-J, Ao G-M, Zhao Q (2010) Cloning and characterisation of ZmZLP1, a gene encoding an endoplasmic reticulum-localised zinc transporter in *Zea mays*. Funct Plant Biol 37:194–205

Xu L, Meng W, Cao C, Wang J, Shan W, Wang Q (2015) Antibacterial and antifungal compounds from marine fungi. Mar Drug 13:3479–3513

Yadav AN (2009) Studies of Methylotrophic Community from the Phyllosphere and Rhizosphere of Tropical Crop Plants. M.Sc. thesis, Bundelkhand University, p 66. https://doi.org/10.13140/2.1.5099.0888

Yadav AN (2015) Bacterial diversity of cold deserts and mining of genes for low temperature tolerance. Ph.D. thesis, Indian Agricultural Research Institute, New Delhi and Birla Institute of Technology, Ranchi. p 234. https://doi.org/10.13140/RG.2.1.2948.1283/2

Yadav AN (2017) Agriculturally important microbiomes: biodiversity and multifarious PGP attributes for amelioration of diverse abiotic stresses in crops for sustainable agriculture. Biomed J Sci Tech Res 1:1–4

Yadav AN, Sachan SG, Verma P, Saxena AK (2015a) Prospecting cold deserts of north western Himalayas for microbial diversity and plant growth promoting attributes. J Biosci Bioeng 119:683–693

Yadav AN, Sharma D, Gulati S, Singh S, Kaushik R, Dey R, Pal KK, Saxena AK (2015b) Haloarchaea endowed with phosphorus solubilization attribute implicated in phosphorus cycle. Sci Rep 5:12293. https://doi.org/10.1038/srep12293

Yadav AN, Sachan SG, Verma P, Tyagi SP, Kaushik R, Saxena AK (2015c) Culturable diversity and functional annotation of psychrotrophic bacteria from cold desert of Leh Ladakh (India). World Journal of Microbiology and Biotechnology 31(1):95–108

Yadav AN, Verma P, Kumar M, Pal KK, Dey R, Gupta A, Padaria JC, Gujar GT, Kumar S, Suman A, Prasanna R, Saxena AK (2015d) Diversity and phylogenetic profiling of niche-specific Bacilli from extreme environments of India. Annals of Microbiology 65(2):611–629

Yadav AN, Rana KL, Kumar V, Dhaliwal HS (2016a) Phosphorus solubilizing endophytic microbes: potential application for sustainable agriculture. EU Voice 2:21–22

Yadav AN, Sachan SG, Verma P, Kaushik R, Saxena AK (2016b) Cold active hydrolytic enzymes production by psychrotrophic bacilli isolated from three sub-glacial lakes of NW Indian Himalayas. J Basic Microbiol 56:294–307

Yadav AN, Verma P, Kour D, Rana KL, Kumar V, Singh B, Chauahan VS, Sugitha T, Saxena AK, Dhaliwal HS (2017a) Plant microbiomes and its beneficial multifunctional plant growth promoting attributes. Int J Environ Sci Nat Resour 3:1–8. https://doi.org/10.19080/IJESNR.2017.03.555601

Yadav AN, Verma P, Kumar V, Sachan SG, Saxena AK (2017b) Extreme cold environments: a suitable niche for selection of novel psychrotrophic microbes for biotechnological applications. Adv Biotechnol Microbiol 2:1–4

Yadav AN, Verma P, Sachan SG, Kaushik R, Saxena AK (2017c) Psychrotrophic microbiomes: molecular diversity and beneficial role in plant growth promotion and soil health. In: Panpatte DG, Jhala YK, Shelat HN, Vyas RV (eds) Microorganisms for green revolution, Microbes for sustainable agro-ecosystem, vol 2. Springer Singapore, Singapore, pp 197–240. https://doi.org/10.1007/978-981-10-7146-1_11

Yadav AN, Verma P, Sachan SG, Saxena AK (2017d) Biodiversity and biotechnological applications of psychrotrophic microbes isolated from Indian Himalayan regions. EC Microbiol ECO 01:48–54

Yadav AN, Kumar V, Prasad R, Saxena AK, Dhaliwal HS (2018a) Microbiome in crops: diversity, distribution and potential role in crops improvements. In: Prasad R, Gill SS, Tuteja N (eds) Crop improvement through microbial biotechnology. Elsevier, New York, pp 305–332

Yadav AN, Verma P, Kumar S, Kumar V, Kumar M, Singh BP, Saxena AK, Dhaliwal HS (2018b) Actinobacteria from rhizosphere: molecular diversity, distributions and potential biotechnological applications. In: Singh BP, Gupta VK, Passari AK (eds) New and future developments in microbial biotechnology and bioengineering. Elsevier, New York, pp 13–41. https://doi.org/10.1016/B978-0-444-63994-3.00002-3

Yadav AN, Verma P, Kumar V, Sangwan P, Mishra S, Panjiar N, Gupta VK, Saxena AK (2018c) Biodiversity of the genus *Penicillium* in different habitats. In: Gupta VK, Rodriguez-Couto S (eds) New and future developments in microbial biotechnology and bioengineering, *Penicillium* system properties and applications. Elsevier, Amsterdam, pp 3–18. https://doi.org/10.1016/B978-0-444-63501-3.00001-6

Yang JW, Yu SH, Ryu C-M (2009) Priming of defense-related genes confers root-colonizing bacilli-elicited induced systemic resistance in pepper. Plant Pathol J 25:389–399

You YH, Yoon H, Kang SM, Woo JR, Choo YS, Lee IJ, Shin JH, Kim JG (2013) *Cadophora malorum* Cs-8-1 as a new fungal strain producing gibberellins isolated from *Calystegia soldanella*. J Basic Microbiol 53:630–634

Yuan ZL, Zhang CL, Lin F, Kubicek CP (2010) Identity, diversity, and molecular phylogeny of the endophytic mycobiota in the roots of rare wild rice (*Oryza granulate*) from a nature reserve in Yunnan, China. Appl Environ Microbiol 76:1642–1652

Yuan C, Wang HY, Wu CS, Jiao Y, Li M, Wang YY, Wang SQ, Zhao ZT, Lou HX (2013) Austdiol, fulvic acid and citromycetin derivatives from an endolichenic fungus, *Myxotrichum* sp. Phytochem Lett 6:662–666

Yadav AN (2018) Biodiversity and biotechnological applications of host-specific endophytic fungi for sustainable agriculture and allied sectors. Acta Scientific Microbiology 1:01–05.

Yadav AN, Yadav N (2018a) Stress-adaptive microbes for plant growth promotion and alleviation of drought stress in plants. Acta Scientific Agriculture 2:85–88.

Yadav N, Yadav A (2018b) Biodiversity and biotechnological applications of novel plant growth promoting methylotrophs. J Appl Biotechnol Bioeng 5:342–344.

Zhang HW, Song YC, Tan RX (2006) Biology and chemistry of endophytes. Nat Prod Rep 23:753–771

Zhang HW, Huang WY, Chen JR, Yan WZ, Xie DQ, Tan RX (2008) Cephalosol: an antimicrobial metabolite with an unprecedented skeleton from endophytic *Cephalosporium acremonium* IFB-E007. Chem Eur J 14:10670–10674

Zhang F, Liu S, Lu X, Guo L, Zhang H, Che Y (2009) Allenyl and alkynyl phenyl ethers from the endolichenic fungus *Neurospora terricola*. J Nat Prod 72:1782–1785

Zhang F, Shen J, Zhang J, Zuo Y, Li L, Chen X (2010) Rhizosphere processes and management for improving nutrient use efficiency and crop productivity: implications for China. In: Advances in agronomy, vol 107. Elsevier, pp 1–32

Zhang F, Li L, Niu S, Si Y, Guo L, Jiang X, Che Y (2012) A thiopyranchromenone and other chromone derivatives from an endolichenic fungus, *Preussia africana*. J Nat Prod 75:230–237

Zheng Z, Obbard JP (2000) Removal of polycyclic aromatic hydrocarbons from soil using surfactant and the white rot fungus *Phanerochaete chrysosporium*. J Chem Technol Biotechnol 75:1183–1189

Zheng CJ, Xu LL, Li YY, Han T, Zhang QY, Ming QL, Rahman K, Qin LP (2013) Cytotoxic metabolites from the cultures of endophytic fungi from *Panax ginseng*. Appl Microbiol Biotechnol 97:7617–7625

Chapter 7
Secondary Metabolites from Endophytic Fungi: Chemical Diversity and Application

Himani Meena, Sairengpuii Hnamte, and Busi Siddhardha

7.1 Introduction

7.1.1 Antibiotic Resistance

Worldwide antibiotic resistance development in pathogenic microbial communities is an immense challenge for scientists. Microbial resistance to existing therapeutic agents and the evolution of multidrug-resistant microbes require a dedicated scientific approach to fight microorganisms. Antibiotic resistance developed by pathogenic microorganisms toward conventional antibiotics led to the discovery of new drug candidates. The development of microbial resistance is associated with some significant crucial steps such as the transformation of drug molecules inside cells, the modification of target receptors or ligand binding sites for drugs, remodeling microbial metabolic pathways, the impairment of drug entry into cells and the release of the accumulated drug through an efflux pump, mutations in gene content, and biofilm formation (Gupta and Birdi 2017). Harmful microbial communities confer pathogenicity through different routes, specifically the production of toxic substances, bacteriocins, lytic enzyme secretion, quorum sensing (QS) systems, and biofilm formation. The production of virulence factors in pathogenic microorganisms cause a magnification of the disease from acute form to a chronic condition. Resistance to antibiotics depends on some key elements, such as the time required to develop resistance, pressure in the selection of resistant microbes and the rate of mutation of particular genes. Horizontal gene transfer mediates the movement of mobile resistance genetic elements among microorganisms, whereas mutations, i.e., insertions, deletions, duplications, and inversions, speed up the rate of development of resistance of microbes to therapeutic agents (Durao et al. 2018). Self-medication

H. Meena · S. Hnamte · B. Siddhardha (✉)
Department of Microbiology, School of Life Sciences, Pondicherry University, Kalapet, Puducherry, India
e-mail: siddhardha.mib@pondiuni.edu.in

B. P. Singh (ed.), *Advances in Endophytic Fungal Research*, Fungal Biology,
https://doi.org/10.1007/978-3-030-03589-1_7

in developing countries has also increased the consumption of drugs without prescriptions and led to antibiotic resistance. *Pseudomonas aeruginosa* is a nosocomial pathogen that causes infection in hospitalized patients suffering from severe burns, lung diseases (cystic fibrosis) and immunodeficient diseases. *P. aeruginosa* possesses antibiotic resistance against amikacin, meropenem and levofloxacin according to a genome-based study. Uptakes of environmental DNA have increased the chances of bacterial resistance through the incorporation of antibiotic-resistant mobile elements in the *P. aeruginosa* genome. A genomic approach reveals the presence of resistome in cells that regulate horizontal gene transfer among bacteria. Genomic elements and intrinsic genes were identified on antibiotic resistance-specific loci in a resistome-wide association study that found high levels of integrons. The integron dynamic confirms the role of an outer membrane protein, OprM, in the meropenem resistance of *P. aeruginosa* (Jaillard et al. 2017). A QS system contributes to the antibiotic resistance mechanism in bacteria by regulating the production of virulence factors, cell lytic enzymes and biofilm formation. The QS system in *P. aeruginosa* is composed of two regulatory systems, LasI/R and RhlI/R, where an inducer gene produces a signal molecule that further binds to its transcriptional regulator. When the signaling molecule reaches its threshold level, it begins the process of bacterial communication and forms multiple layers of bacterial communities. Biofilm is made up of exopolysaccharides (EPSs), eDNA, alginate, amino acids, and proteins that protect bacterial cells from external harsh conditions, increase the survival rate, hinder antimicrobial entry and hence, increase bacterial persistence and decrease susceptibility to antimicrobial agents (Ganesh and Rai 2018). Alginate is one of the main components of a biofilm matrix that supply nutrients and water to the multiple layered bacterial biofilm. The linear polymer is made up of 1-4 linked saccharides, β-D mannuronic acid (M), and α-L-guluronic acid epimers whose production is regulated by three main genes, algU, algL, and algD. *P. aeruginosa* possesses AmpC β-lactamase enzymes that hydrolyze β-lactams, i.e., cephalosporins, monobactams, carbapenems and penicillin, with increased susceptibility to antibiotics. Limited uptake of drug molecules, efflux pump activation and modification of target sites or receptors contribute to antimicrobial resistance in *P. aeruginosa* (Gholami et al. 2017). *Salmonella,* a foodborne microbes cause gastroenteritis and exhibit antibiotic resistance to first-line drugs such as chloramphenicol, tetracycline, and trimethoprim-sulfamethoxazole (Luo et al. 2018). Pathogenicity islands (PAIs) are mobile genomic elements with pathogenic properties that transfer among bacterial species through horizontal gene transfer. The genetic transfer of PAIs depends on the guanine-cytosine content of both donor and recipient microorganisms. PAIs are short genome regions that carry virulence genes in the form of plasmid, which can be adhesins, toxins, or apoptotic genes and superantigens in microorganisms (Nieto et al. 2016). The different types of PAIs in microorganisms vary from microbial strain to strain, for example, *Salmonella* has five types of PAI, whereas *E. coli* has seven (Nieto et al. 2016; Phillips-Houlbracq et al. 2018). Phillips-Houlbracq et al. (2018) assessed the presence of a particular virulence gene that is encoded by PAI sequences responsible for *E. coli* pathogenicity in ventilated hospital patients. The authors reported the predominant specific gene markers susceptible to microbial pathogenicity as Antigen 43, HmuR-like hemin receptor precursor, F17 fimbrial protein precursor, and extracellular serine protease SepA. Plasmids transmit virulent genes from a pathogenic microbe to the environment or other bacterial cells and magnify the infectious disease condition in the host body. For example, the CTX M-type plasmid, mostly found in *E. coli*, *Klebsiella pneumoniae* and *Shigella*, encodes for the β-lactamase enzyme, and the CTX-M-type cefotaximase hydrolyzes β-lactam drug molecules. Hernandez-Flores et al. (2018) studied pMEX01 in *E. coli*, which confers drug resistance against β-lactams due to the presence of the *blaCTX-M-14* gene. *E. coli* maintains a high abundance of efflux pump system *TetA/B* that mediates the extrusion of tetracycline from inside cells to the surrounding environment. Due to the presence of tetracycline, *E. coli* causes significant alterations in bacterial cell metabolism (catabolic) and enzymatic activity (ATPase activity, phosphorylation, and anaerobic conditions), alterations in peptidoglycan anchored cell wall proteins and ribosomal translation at the proteomic level (Jones-Dias et al. 2017). One of the most studied microbes in connection with antibiotic resistance is methicillin-resistant *S. aureus* (MRSA), which mainly affects surgical sites and causes diseases related to the blood circulatory system and respiratory system at a high risk level. MRSA can affect persons who have undergone invasive surgery, those in long-term-care units, and those who experience unhygienic medical practices. MRSA infects patients via two different virulence production systems, primarily surface proteins (e.g., protein A, coagulase, fibronectin binding protein, collagen binding protein) and a secondary secretory system (enterotoxin A, type-II secretory system, and α-toxin) (Paling et al. 2017; Loftus et al. 2018). Antibiotic resistance developed by microbes can be natural or acquired, depending on the environmental conditions encountered by the microbial species. The use of conventional antibiotics in different drug combinations may provide some relief, but the microbial world should be explored for novel drug molecules.

7.1.2 Microbial Sources of Bioactive Metabolite Production

The emergence of antibiotic resistance to already reported drugs has elevated up the search for antimicrobial agents from microbial sources. Microorganisms are well known for the secretion of beneficial bioactive compounds such as antibiotics (e.g., penicillin, streptomycin, and erythromycin), antifungals (e.g., nystatin, amphotericin, and cycloheximide), antiprotozoans (e.g., monensin, salinomycin, and trichostatins), herbicides (e.g. herbicidin), alkaloids, pyrrolizidines, terpenoids, quinines, steroids, polyphenol, peptides, indole and isocoumarins. Bioactive compounds derived from microbial origin have the capability to attach to target protein domains and interrupt protein–protein interactions in pathogenic microbial communication mechanisms. A microbial metabolite may act as a potential compound as an antimicrobial candidate or limit antibiotic resistance. Bacteria from soil ecological background are promising agents for the production of bioactive compounds. Soil bacterial species produce various natural compounds with chemically diversified structures. Based on different biological and physiological properties, bioactive compounds are divided into two different classes, volatile compounds and soluble compounds. Volatile compounds are terpenes, nitrogen-containing molecules, indole acetic acid, and pyrazines that can evaporate in the environment. Bacteriocins and peptides (siderophore and lipopeptides) are able to solubilize in soil and are easily diffusable in groundwater (Tyc et al. 2017). EPSs are among the most studied extracellular compounds released by microorganisms that survive under extreme conditions and display numerous biological activities such as antioxidant, antitumor, and anti-inflammatory activities. Homopolysaccharides and heteropolysaccharide are further categorized into three kinds based on their physiochemical properties: intracellular, structural, and extracellular polysaccharides. Researchers have identified an EPS variant, capsular polysaccharide, that covalently binds with cell membranes and provides protection to cells from harsh environmental conditions. Capsular exopolysaccharides, is loosely bound with cell membranes, whereas lipopeptide (protein A molecule) is a protein anchored in the cell membranes of *S. aureus* (Liu et al. 2017c). *Bacillus thuringiensis* is a rod-shaped, soil bacterium that can survive under anaerobic conditions and produce bacteriocins and endospores in rough conditions. *B. thuringiensis* harbors *Cry* genes, which produce crystal proteins that act as pesticides against Lepidoptera, Diptera and Coleoptera pests. Chandrasekaran et al. (2018) examined the presence of secondary metabolites in *B. thuringiensis* that were responsible for insecticide and pesticide activity. Gas chromatography-mass spectrometry (GC-MS) analysis identified the presence of 32 secondary metabolite compounds, a mixture of linear [hexane, 3-ethyl-, hexadecane, 1,1-bis (dodecyloxy)] and cyclic compounds (dibutyl phthalate, butanoic acid, 3-methyl-, 2-phenyl, histamine-2-carboxylic acid) with a wide range of lipopeptide derivatives. A *Bacillus amyloliquefaciens* strain, CEIZ-11, was identified as a potential producer of fengycin, iturin, and surfactin lipopeptide controlled by nonribosomal genes such as *fenD*, *ituC*/*ituD*, or *sfp*, respectively. The bioactive compounds produced by *B. amyloliquefaciens* have antifungal effects on *Pythium aphanidermatum* (Zouari et al. 2016). Siahmoshteh et al. (2017) found that *Bacillus subtilis* and *B. amyloliquefaciens* are quiescent candidates for the biocontrol of *A. parasiticus* by producing antifungal compounds. *Rhodobacter sphaeroides* is a well-known photosynthetic bacterium that synthesizes a divergent group of chemical molecules, i.e., carotenoids, 5-aminolevulinic acid, biopolymers, and coenzymes, that exhibit antioxidant and anti-inflammatory activity. *A. parasiticus* is the causative agent of plant disease, especially in nut and crop plants. *A. parasiticus* produces alpha toxin group A/B, which is highly toxic for food crop plants and severely affects animals after consumption through plants. Wang et al. (2017b) studied a genetically modified photosynthetic bacterium, *R. sphaeroides* WL-APD911, that generates the carotenoid Lycogen as a result of the condensation of double molecules of geranylgeranyl pyrophosphate converted into phytoene, followed by desaturation using the phytoene desaturase (CrtI) enzyme. Pyocyanin pigment produced by *P. aeruginosa* is a green-blue colored phenazine compound that produces free reaction oxygen species. Pyocyanin production is controlled by two genes in which the *MvfR* gene generates transcription factors that act as activators of *phnAB* genes (Depke et al.

2017). Narenkumar et al. (2017) observed the effect of pyocyanin produced by *P. aeruginosa* TBH2 as an anticorrosive agent against microbe-influenced corrosion caused by biofilm-forming *Bacillus* spp Various fungal metabolites have performed diverse biological activities (Table 7.1).

7.2 Endophytic Fungi and Secondary Metabolite Production with Their Application

7.2.1 Host–Endophytic Fungus Interaction

Endophytic fungi are associated with plant microbiomes that inhabit plant endospheric regions and produce plenty of chemically diversified bioactive compounds in response to environmental stress as well as plant defense mechanisms. Plant–microbe interactions are the key elements for the determination of microbial diversity that depends on physiological and geological conditions. Endophytes belong to angiosperm and gymnosperm groups that produce a variety of antimicrobial, antitumor, and anti-insecticidal metabolites. Secondary metabolite production can be enhanced by environmental factors regulated by host plants. Endophytic fungi are classified based on their transmission and biological activities, which affect plant systems. Group I endophytic fungi are transmitted through vertical gene transfer, which carries a genetic element within plant systems, whereas Group II endophytes provide resistance to environmental stresses. Group III endophytes require horizontal gene transfer in which fungi acquire genes from other fungal species that produce bioactive molecules. Endophytes present in plant systems produce secondary metabolites based on plant health and age. Fungi present in different regions of plants have distinct fungal species with their own diversified chemical production. Liu et al. (2017a) studied the microbial diversity of *Paris polyphylla var. yunnanensis* plants and discovered that, among rhizospheric endophytes, the dominant species *Fusarium oxysporum* accompanied by *Leptodontidium* sp. and *Trichoderma viride,* was present. Besides these three dominant fungal species, *Alternaria* sp., *Cylindrocarpon* sp., *Chaetomium* sp., *Pyrenochaeta* sp., *Penicillium swiecickii, T. viride,* and *Truncatella angustata* presence were also reported.

7.2.2 Chemical Diversification of Endophytic Fungal Secondary Metabolites

Fungal secondary metabolites originate from a few clearly defined compound classes via their corresponding biosynthetic pathways, but for each compound class a variation of building blocks, enzymatic mechanisms, and tailoring steps can lead to an extremely diverse array of chemical structures. Each bioactive metabolite has

Table 7.1 Bioactive compounds derived from various microbial sources and their biological activities

S. No.	Microbial sources	Bioactive compound	Biological activity	Reference
1	*Bacillus thuringiensis*	Dibutyl phthalate, butanoic acid, 3-methyl-, 2-phenyl, histamine-2-carboxylic acid	Insecticide Pesticide	Chandrasekaran et al. (2018)
2	*Bacillus amyloliquefaciens*	Fengycin, Iturin Surfactin lipopeptide	Antifungal	Zouari et al. (2016)
3	*Rhodobacter sphaeroides*	Carotenoids, 5-aminolevulinic acid	Antioxidant Anti-inflammatory	Wang et al. (2017a, b)
4	*Pseudomonas aeruginosa*	Phenazine compound	Pigmentation	Depke et al. (2017) Narenkumar et al. (2017)
5	*Pseudocercospora* sp. ESL 02	Terreic acid 6-methylsalicylic acid	Antioxidant	Prihantini and Tachibana (2017)
6	*Achaetomium* sp.	Phenolic compound Tennins	Antioxidant	Anitha and Mythili (2017)
7	*Streptomyces samsunensis* M1463T	Efomycins M, G, Oxohygrolidin, Abierixin and 29-O-methylabierixin	Antimicrobial	Supong et al. (2016)
8	*Fusarium* sp.	Equisetin	Antimicrobial	Ratnaweera et al. (2015)
9	*Penicillium canescens Alternaria alternata*	3-methyl-1-butanol and phenylethylalcohol	Antimicrobial	Malhadas et al. (2017)
10	*Exserohilum rostratum*	Polyketide monocerin	Antimicrobial	Pinheiro et al. (2017)
11	*Plectranthus amboinicus (Lour.) Spreng Plectranthus Nicoletta*	Carvacol	Antimicrobial	Zhang et al. (2017)
12	*Xylaria* sp.	Cytochalasin D	Antimicrobial	Khan et al. (2016)
13	*Pestalotiopsis* sp.	Pestallic acid E (þ)-ambuic acid	Antitumor/ anticancer/ antiproliferative	Li et al. (2017)
14	*Fusarium equseti* EF-32	(E)-3-(2,3-dihydroxy-phenyl) acrylic acid	Antitumor/ anticancer/ antiproliferative	Venkateswarulu et al. (2017)
15	*Zygomycota* and *Ascomycota*		Biocontrol agent	Zheng et al. (2017)
16	*Phoma, Phomopsis, Altwenaria, Acremonium*		Biocontrol agent	Zhang et al. (2014)
17	*Lasiodiplodia Pseudotheobromae*		Biocontrol agent	Xiang et al. (2016)
18	*Muscodor heveae*		Biocontrol agents	Siri-Udom et al. (2017)

its own chemical structure attached to different side chain groups, which have a significant effect on the plant immune system and in the pharmaceutical field. Due to the production of secondary metabolites in large amounts, *Trichoderma* spp. can be used as biocontrol agents for phytopathogens. *Trichoderma* spp. stimulate the vegetative growth of host plants and provide protection against phytopathogens. The fungi provide protection against phytopathogen via three main mechanisms such as competition within the biological niche, production of antibiotics and mycoparasitism that stimulates plant immune systems and enhances resistance against virulence factors secreted by pathogenic microbes. *Trichoderma* sp. secretes harzianic acid (HA) and 6-pentyl-a-pyrone (6PP), which are abundant and constituted of secondary metabolites. HA and 6PP have been found to active compounds and showed antimicrobial activity against powdery mildew caused by *Uncinula necator* and increased crop production (Pascale et al. 2017).

Indrianingsih and Tachibana (2017) assessed the production of α-Glucosidase inhibitor in an endophytic fungus, *Xylariaceae* sp. QGS 01, isolated from *Quercusgilva blume*. They compared the enzymatic activity with α-glucosidase inhibitor produced by *Saccharomyces cerevisiae* against α-glucosidase, 8-hydroxy-6,7-dimethoxy-3-methylisocoumarine. An endophytic fungus, *Aspergillus* sp. EJC04, is able to produce six bioactive compounds, i.e., ergosterol, ergosterol peroxide, mevalonolactone, cytochalasin B, and cytochalasin H, isolated from the *Bauhinia guianensis* plant. Feitosa et al. (2016) studied cytochalasin B and its diacetate derivative, which showed lethal activity against *Artemia salina* in brine shrimp assay. Song et al. (2017) isolated a mixture of a chemically diversified structure based on high-performance liquid chromatography and nuclear magnetic resonance (NMR) analysis from the endophyte fungus *Phomopsis asparagi* SWUKJ5 of the *Kadsura angustifolia* plant. The most abundant bioactive compound was found to be phomaspyrone followed by macommelin-8,9-diol, macommelin-9-o, and macommelin in mycelium broth culture. A recent study by Bai et al. (2017) showed the production of asperterzine, a member of the diketopiperazine (DKP) group, which exhibits significant biological activity, along with two more bioactive fractions, bis-dethiobis(methylthio)—acetylaranotin and bis-dethiobis (methylthio)-acetylapoaranotin compounds. The bioactive compounds were secreted by the endophytic fungus *Aspergillus terreus* PR-P-2 isolated from *Camellia sinensis var. assamica*.

Tang et al. (2017) evaluated the chemical diversity of secondary metabolites in endophytic fungi, *Phomopsis* sp. sh917, isolated from *Isodoneriocalyx var. laxiflora*, by adopting a one strain/many compounds (OSMAC) strategy. Polyketide molecules, such as phomopsis ketones, (10S)-10-*O*-b-D-40-methoxymannopyranosyldiaporthin, and clearanol H compounds, were more abundant in the fungal crude extract. *Alternaria* sp. Samif 01, harboring the roots of *Salvia miltiorrhiza Bunge*, is a potent producer of Alternariol 9-methyl ether, which may be a promising candidate for biological activity against nematodes *Bursaphelenchus xylophilus* and *Caenorhabditis elegans*. Alternariol 9-methyl ether also exerts an inhibitory effect against the spore germination of the microbial strain *Magnaporthe oryzae* (Lou et al. 2016). Hu et al. (2017) isolated an endophytic fungus, *P. citrinum* 46, derived from *Salicornia herbacea* Torr and is able to produce novel antimicrobial compounds. The isolated novel compounds pencitrin and pencitrinol, with two known dimeric molecules, penicitrinone A and penicitrinone E, exhibited a promising cytotoxicity in the cell lines A549 human lung cancer cells and HepG2 human liver cancer cells. 3-nitropropionic acid (3-NPA) is a toxic compound produced by fungal species as a defense mechanism against insects and animals. *Diaporthe citri* is able to produce 3-nitropropionic acid (3-NPA), which is beneficial for plant defense systems and in plant–microbe interactions (Polonio et al. 2016). Mangrove fungi are a rich source of bioactive compounds with numerous biological activities such as antitumor, antinematode and antioxidant activity. Two tremulane sesquiterpene lactones, coriolopsin A and coriolopsin B, along with two known sesquiterpene lactone analogues, conocenol C and ceriponol E were found in the endophytic fungus *Coriolopsis* sp. J5 isolated from *Ceriops tagal* (Chen et al. 2017). *Pseudolagarobasidium acaciicola*, isolated from the mangrove tree *Bruguiera gymnorrhiza*, produce a large amount of sesquiterpene, novel tricyclic, and spirobicyclic compounds with (nor) sesquiterpene endoperoxides (merulin or steperoxide). These isolated sesquiterpenes showed a higher efficacy against the HL-60 cell line for cytotoxicity activity. The endoperoxide moiety of bioactive compounds supports a cytotoxic nature and provides a prominent structural element for biological activities (Wibowo et al. 2016). Cai et al. (2017) isolated depsidones talaromyones A and B from an endophytic fungus, *Talaromyces stipitatus* SK-4, associated with the mangrove plant *Acanthus ilicifolius*. *T. stipitatus* was studied as a potent producer of cytotoxic nor-sesquiterpene peroxides, insecticidal meroterpenoids and anti-Aβ42 aggregation coumarins. Bioactive depsidones showed potent antimicrobial activity against *Bacillus subtilis*. Saetang et al. (2017) identified two bioactive compounds from an endophytic fungus, *Simplicillium* sp. PSU-H41, residing within *Hevea brasiliensis* leaf. The bioactive compounds were found to be depsidones, simplicildone A, α-pyrone, and botryorhodine C, which showed higher antimicrobial activity against MRSA and *Cryptococcus neoformans*. Monoterpenoids play a significant role in polyketide production by initiating biosynthesis as a precursor molecule. Duana et al. (2016) isolated seven monoterpenoids based on structural studies and biological activities from the endophytic fungus *Penicillium* sp. T2-8 harbored on *Gastrodia elata*. They identified the bioactive compounds preaustinoid A1, dehydroaustinol, Austin, (S)-18,19-dihydroxyneogrifolin, and neogrifolin and their derivatives (Fig. 7.1 and Table 7.2).

7.3 Metabolic Pathways: Enzymatic Reactions for Secondary Metabolite Production

This chapter reviews the biosynthetic pathways involved in the synthesis of secondary metabolites via a well-organized series of chemical reactions. The production of any secondary metabolite of microbial origin requires a precursor for the initiation of a systematic reaction. The enzymatic reaction chain involves a precursor, substrate and end product with the release of a byproduct. *Trichoderma* spp. produce a

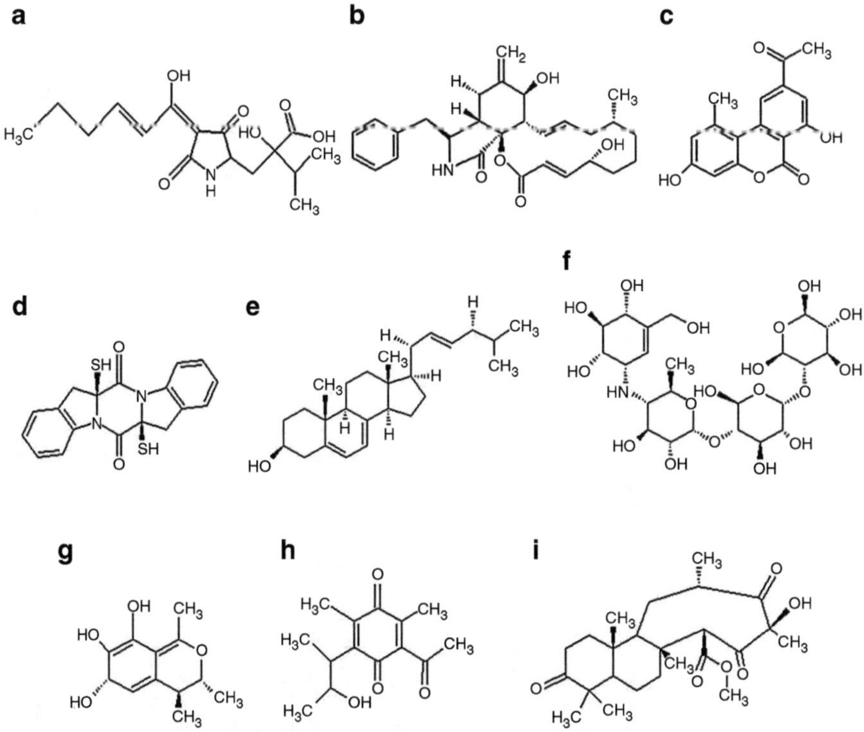

Fig. 7.1 Secondary metabolite production in endophytic fungi bioactive metabolites with diversified chemical structures. (**a**) Harzianic acid (**b**) Cytochalasin B (**c**) Alternariol 9-methyl ether (**d**) Asperterzine (**e**) Ergosterol (**f**) α-Glucosidase inhibitor (**g**) Penicitrin (**h**) Penicitron (**i**) Preaustinoid A1

tremendous amount of secondary metabolites with antimicrobial activity and also promote plant growth and increase crop yields. Zeilinger et al. (2016) explored the biosynthetic pathway for the synthesis of nonribosomal peptides such as peptaibiotics, siderophores and diketopiperazines, polyketides, terpenes, pyrones, and isocyane metabolites and determined the role of acetyl-CoA, mevalonate and amino acids as reaction precursor. Acetyl-Coenzyme A and malonyl-Coenzyme A act as a precursor that is further catalyzed by a group of polyketide synthases, viz. ketoacyl synthase, an acyl transferase and a phosphopantetheine attachment site domain.

Upadhyay et al. (2016) investigated the role of polyketide synthase pathway in melanin synthesis and its deposition in extracellular vesicles as secondary metabolite in *Aspergillus fumigatus*. They explained the involvement of the specific gene *mvp1* in the synthesis process of melanin metabolite via melanosome biogenesis and trafficking through cell membranes. Melanin compounds (g-glutaminyl-3,4-dihydroxybenzene) (GDHB), pyomelanin, p-aminophenol (PAP)-melanin, heterogeneous melanins and aspulvinone E are important in morphogenesis, stress resistance, virulence, and energy transduction during cell cycle and other biological activities (Toledo

Table 7.2 Secondary metabolite production in potential endophytic fungi with significant chemical structural diversity and properties

Serial no.	Endophytic fungi	Metabolite nature	Fungal secondary metabolite	Harboring source of endophytes	Reference
1	*Trichoderma* sp.	Alkaloid	Harzianic acid (HA) 6-pentyl-a-pyrone (6PP)	–	Pascale et al. (2017)
2	*Xylariaceae* sp. QGS 01	Monosaccharide	α-Glucosidase inhibitor	*Quercusgilva Blume*	Indrianingsih and Tachibana (2017)
3	*Aspergillus* sp. EJC04	Sterol Alkaloid	Ergosterol Ergosterol peroxide Mevalonolactone Cytochalasin B Cytochalasin H	*Bauhinia guianensis*	Feitosa et al. (2016)
4	*Phomopsis asparagi* SWUKJ5	Pyrone	Phomaspyrones Macommelin-8, 9-diol Macommelin-9-o Macommelin	*Kadsura angustifolia*	Song et al. (2017)
5	*Phomopsis* sp. sh917	Polyketide	Phomopsiketones Clearanol H 10S)-10-O-b-D-40-methoxymannopyranosyl Diaporthin	*Isodoneriocalyx var. laxiflora*	Tang et al. (2017)
6	*Aspergillu sterreus* PR-P-2	Diketopiperazines	Asperterzine	*Camellia sinensis var. assamica*	Bai et al. (2017)
7	*Alternaria* sp. Samif01	Mycoestrogen	Alternariol 9-methyl ether	*Salvia miltiorrhiza Bunge*	Lou et al. (2016)

8	*P. citrinum*46	Mycotoxin	Pencitrin Pencitrinol Penicitrinone A Penicitrinone E	*Salicornia herbacea* Torr	Hu et al. (2017)
9	*Diaporthe citri*	Mycotoxin	3-nitropropionic acid (3-NPA)	–	Polonio et al. (2016)
10	*Coriolopsis* sp. J5	Tremulane sesquiterpenes	Coriolopsin A Coriolopsin B Conocenol C Ceriponol E	*Ceriopstagal*	Chen et al. (2017)
11	*Pseudolagarobasidium acaciicola*	Sesquiterpene	Tricyclic Spirobicyclic (nor) sesquiterpene endoperoxides (merulin or steperoxide	*Bruguiera gymnorrhiza*also	Wibowo et al. (2016)
12	*Talaromyces stipitatus* SK-4	Polyketide	Talaromyones A Talaromyones B	*Acanthus ilicifolius*	Cai et al. (2017)
13	*Simplicillium* sp. PSU-H41	Depsidones	Simplicildone A Botryorhodine C	*Hevea brasiliensis*	Saetang et al. (2017)
14	*Penicillium* sp. T2-8	Polyketide-terpenoid	Preaustinoid A1, Dehydroaustinol, Austin (S)-18,19-dihydroxyneogrifolin neogrifolin	*Gastrodia elata*	Duana et al. (2016)

et al. 2017). Shikimate pathways, which is the most studied metabolic pathways, is composed of seven unique enzymes that catalyze typical chemical reactions. These enzymes are 3-deoxy-D-arabinoheptulosonate 7-phosphate (DAHP) synthase, 3-dehydroquinate synthase, 3-dehydroquinate dehydratase, shikimate dehydrogenase, shikimate kinase, 5-enolpyruvylshikimate-3-phosphate (EPSP) synthase and chorismate synthase. The shikimate pathway begins with a condensation process of two specific substances, phosphoenolpyruvate and erythrose-4-phosphate, followed by the production of chorismate with a byproduct of DAHP and Pi, an enzymatic reaction catalyzed by DAHP synthase. Chorismate acts as substrate for the synthesis of three aromatic amino acids, phenylalanine, tyrosine and tryptophan. Shikimate kinase and EPSP synthase play a significant role via an ATP phosphorylation process and the coupling of phosphoenolpyruvate to synthesize 5-enolpyruvylshikimate-3-phosphate (Peek and Christendat 2015; Tohge et al. 2013).

The isoprenoid biosynthesis pathway is among the most studied pathways for the production of carotenoids, taxadiene, isoprene, monoterpenoids and tanshinone. Isoprenoid biosynthesis is regulated by the condensation of two five-carbon precursors, isopentenyl diphosphate (IPP) and dimethylallyl diphosphate (DMAPP), catalyzed by prenyl diphosphate synthases (Dannert 2015). Taxol is an anticancer drug isolated from the tree *Taxus brevifolia*, but due to environmental issues, scientists are searching for an alternative to taxol production. An endophytic fungus, *Taxomyces andreanae*, isolated from the *Taxus* tree, was found to be a potent producer of a particular diterpenoid as an alternative approach. Taxol biosynthesis is initiated by the condensation of two diterpene prenyl precursors, followed by cyclization, depending on an enzyme, that is, taxane-5α-hydroxylase terpene synthase (Heinig et al. 2013).

Recently Uzma et al. (2018) reviewed the anticancer potential of endophytic fungi. *Penicillium* species produce a vast variety of alkaloids, cyclic diketopiperazines, quinolines and polyketides with various biological activities. Cyclic diketopiperazines are highly active compounds that work as precursors for pharmaceutical agents, for example, Roquefortine (Kozlovsky et al. 2013). Yang et al. (2014) examined the production of hazel *Penicillium aurantiogriseum* NRRL 62431and sequenced the gene responsible for the production of bioactive compounds with anticancer activity. They reported that genes available for the biosynthesis of Taxol showed a similarity with genes encoding for phenylalanine aminomutase (PAM), geranylgeranyl diphosphate synthase (GGPPS), taxane 5α-hydroxylase (T5OH), taxane 13α-hydroxylase (T13OH), taxane 7β-hydroxylase (T7OH), taxane 2α-hydroxylase (T2OH), and taxane 10β-hydroxylase (T10OH) present in *Taxus* trees. An additional gene was also identified in *P. aurantiogriseum* NRRL 62431 that encodes for an enzyme and plays an important role in pathways. The formation of precursor acetyl CoA is regulated by the mevalonic acid pathway. Researchers reported gibberellic acid (GA) as a phytotoxic compound secreted by the fungus *Gibberella fujikuroi* associated with rice cultivation. It causes bakanae (foolish seedling) disease in rice plants, that affects the height of a plant by causing diseased condition like etiolated and chlorotic in which plants are unable to support their own weight and die due to topple over. Hydroxy methyl glutaryl (HMG) coenzyme A via mevalonic acid, isopentenyl diphosphate, geranyl diphosphate (GDP), farnesyl diphosphate (FDP), and geranylgeranyl diphosphate (GGDP) act as precursor and substrate for GA synthesis as well as for carotenoid production (Cerezo et al. 2018). GA biosynthesis in fungi is regulated by the mevalonic acid pathway via the condensation of acetyl-CoA and an isopentenyl diphosphate unit. The actual process of GA synthesis is regulated by the production of an intermediate unit, farnesyl diphosphate (FDP). One of the most valuable intermediates, GGDP, is synthesized by the catalytic activity of GGDP synthase (GGS1/2). Double-step cyclization is applied to the formation of a first intermediate ent-kaurene molecule via ent-copalyl diphosphate (CPP), which later oxidizes into ent-kaurenoic acid (KA). KA further converts to ent-7-α-hydroxy KA via 7β-hydroxylation followed by oxidation at C-6 β for the contraction of ring B and generates a GA_{12}-aldehyde molecule. Cerezo et al. (2018) described GA synthesis in *Fusarium fujikoroi* regulated by a collection of genes localized on a gene cluster. The GGDP synthase gene (ggs2), the bifunctional ent-copalyl diphosphate synthase/ent-kaurene synthase (cps/ks), 2-oxoglutarate-dependent dioxygenase, and cytochrome P450 monooxygenases genes are required for the conversion of GA_4 to GA_7. Metabolic pathways possess specific mechanisms for the synthesis of bioactive compounds, such as meroterpenoid sterretonin, andrastin A, austinoland anditomin in *Fusarium* sp. In enzymatic reactions, an aromatic polyketide, 3, 5-dimethylorsellinic acid (DMOA), serves as a precursor molecule, and any further reaction is catalyzed by membrane-bound terpene cyclases (TPCs).

The systematic arrangement of a bioactive compound is composed of polyketide starter moieties, prenyl chain lengths, and cyclization reactions. The folding of a synthesized juvenile protein is based on the stereochemistry of TPCs that recognize only preexisting motifs, for example, an aspartate-rich domain (DDXXD) for biosynthesis. Oxygenase enzymes such as flavin adenine dinucleotide-dependent monooxygenases (FMOs), non-hemeiron-dependent dioxygenases, and cytochrome P450 monooxygenases are essential for the postcyclization process of newly synthesized compounds. The postcyclization process is necessary for assuring structural complexity through oxidative rearrangements followed by a carbonate-forming reaction and, finally, the cyclization of a terpenoid moiety (Matsuda et al. 2016). Aflatoxin production in an endophytes fungi is necessary to understand the mechanism behind its risky behavior towards pathogenic microbial strains. Liu et al. (2017b) revealed the presence of gene clusters associated with the biosynthesis of aflatoxin in *Aspergillus flavus*. They documented a profile for the expression of biosynthetic genes in a pathway cluster, viz. the two key regulatory genes *aflR* and *aflS* with a regulatory gene *laeA* for secondary metabolite production. Wang et al. (2017a) reported on a new biosynthetic pathway for the production of novel dimeric (epi) polythio dioxopiperazines, verticillin compounds, by an endophytic fungus, *Clonostachys rogersoniana*. They documented the important gene *verP* for the biosynthetic process of a secondary metabolite. Gao et al. (2018) disclosed the gene cluster required for the biosynthesis of functionalized bioactive molecules, active sesterterpenes, through the genome mining process of the bifunctional terpene synthase (BTS) enzyme region against *Phomabetae* and *Colletotrichum orbiculare*. Indole acetic acid (IAA) bioactive compounds are highly important for plant growth promotion secreted by endophytic fungi and for their own physiological patterns. Luo et al. (2016) explored a new route called the L-tryptophan-dependent pathway for IAA production in endophytic fungi, *Fusarium graminearum,* along with existing metabolic pathways such as the indole-3-pyruvic acid (IPA), the tryptamine (TAM) and the indole-3-acetonitrile (IAN) pathway. The L-TRP IAA pathway was found to be a potent track for IAA synthesis in endophytic fungi. Endophytic fungi can be used as cell biofactories for the synthesis of numerous chemical molecules with diversified chemical structures and catalytic moieties (Table 7.3 and Fig. 7.2).

7.4 Modification of Biosynthetic Pathways: Epigenetic Modification and Activation of Silent Gene Cluster

Natural products originating from endophytic fungi are highly valuable for the pharmaceutical industry for the generation of lead drug molecules as well as in agriculture fields for crop protection. The microbial biosynthesis of bioactive compounds occurs much less compared to utilization, so increased production of secondary metabolites is required to meet demand on a large scale. Scientists have discovered the presence of biosynthetic gene clusters (BGCs) that maintain secondary metabolite synthesis, the regulatory domain of particular genomic elements and resistance to unwanted compounds during growth periods. Some BGCs are un-tapped due to their non-expressive nature, which results in a variety of a protein which are unique in nature that remains un-interrupted by the scientist known as cryptic or orphan molecule (Li et al. 2016). Under a bioinformatics approach, bioinformatic tools are necessary to recognize the actual process of microbial biosynthesis, which leads to the discovery of novel biosynthetic pathways for a particular natural compound, study the genomes of endophytic fungi, and manipulate existing pathways to achieve the enhanced production of a diverse group of molecules. Modification of biosynthetic gene clusters is required for understanding the mechanism behind biosynthesis, the construction of expression hosts, the activation of silent genes, and improved yield of specific molecules (Lee and Medema 2016). Biosynthetic pathways involve a precursor, a group of enzymes to regulate enzymatic reactions to produce natural end products and extra byproducts. Polyketide synthases and nonribosomal peptide synthetases are exclusive enzymes available for the synthesis of specifically designed biomolecules, i.e., alkaloids, terpenes, peptides, monoterpenoids, polyketides and polysaccharides. Bioinformatics tools for the identification of BGCs are highly sensitive and able to identify the presence of tremendous numbers of genes related to the biosynthesis of natural products such bioinformatics tools are ClusterFinder, antiSMASH, PRISM, NP.searcher, EvoMining, and GRAPE (Loureiro et al. 2018). To increase the expression of unique identified gene sequences, some strategies must be applied to solve the problem of lower bioactive compound production such as by inducing the excessive expression of specific transcription factor regions, a global regulator gene, and the interchange of biosynthetic genes. An *in silico* approach can be utilized to predict metabolite products using information about gene clusters known as a forward approach, and a

Table 7.3 A biosynthetic pathway is a chain of chemical reactions regulated by different enzymes and genes for the production of natural bioactive molecules

S. no.	Biosynthetic pathway	Precursor	Enzyme involved	Bioactive metabolite	Reference
1	Shikimate pathway	Phosphoenolpyruvate Erythrose-4-phosphate	3-Deoxy-D-arabinoheptulosonate 7-phosphate (DAHP) synthase, 3-dehydroquinate synthase, 3-dehydroquinate dehydratase, Shikimate dehydrogenase, shikimate kinase, 5-enolpyruvylshikimate-3-phosphate (EPSP) synthase Chorismate synthase	Polyketide Aromatic compound Alkaloid	Peek and Christendat (2015) Tohge et al. (2013)
2	Isoprenoid pathway	Acetyl CoA Malonyl CoA	Acetoacetyl-CoA thiolase HMG-CoA synthase HMG-CoA reductase Mevalonate kinase Phosphomevalonate kinase Mevalonate phosphate decarboxylase	Sesquiterpenoid Diterpenoid Carotenoid	Dannert (2015) Upadhyay et al. (2016) Toledo et al. (2017) Kozlovsky et al. (2013) Yang et al. (2014)
3	GA pathway	Hydroxy methyl glutaryl (HMG) coenzyme A via mevalonic acid, Isopentenyl diphosphate, geranyl diphosphate (GDP), Farnesyl diphosphate (FDP) Geranylgeranyl diphosphate (GGDP)	Geranylgeranyldiphosphate synthase, Copalyldiphosphate synthase/ent-kaurene synthase 2-oxoglutarate-dependent dioxygenase, Cytochrome P450 monooxygenases	Gibberellic acid	Cerezo et al. (2018)

The table lists metabolic pathways with the enzymes involved and end products

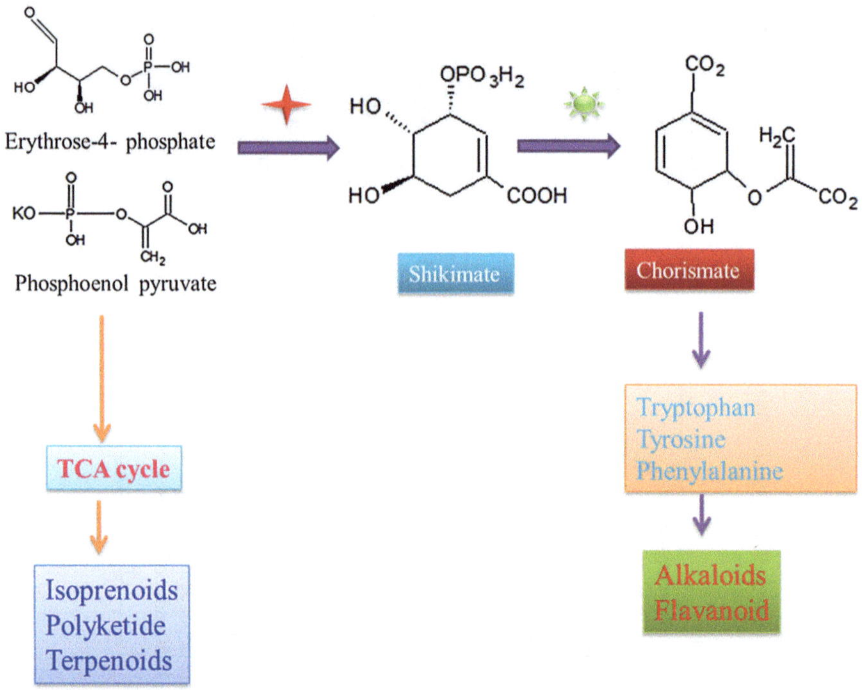

Fig. 7.2 Graphical representation of metabolic pathways involved in fungal secondary metabolite production: **Shikimate pathway**—condensation of E4P with PEP → Shikimate → Chorismate → aromatic amino acid (phenylalanine, tryptophan, tyrosine); **TCA** mediated **MVA + MEP pathway**—isoprenoids, polyketides, and terpenoids

retro-biosynthetic method reveals the respective gene clusters based on information about the metabolites (Khater et al. 2016). Rearrangement of gene clusters provides the opportunity to researchers to determine the specific location of genetic material in close proximity to particular loci.

Epigenetic remodeling of a fungal genome is mediated by treating it with DNA methyl transferase and histone deacetylase inhibitors. These epigenetic modifiers enhance the production of natural compounds by increasing the complexity of the chemical structure of molecules. They also increase the expression level of genes responsible for secondary metabolite production and produce new bioactive compounds. This method reduces the cost of using molecule- or culture-dependent techniques and maximizes the chances of studying the route of cryptic gene products. Activation of such gene clusters composed of genes encoding for transcription regulatory proteins in particular BGCs. The inactive form of a silent gene cluster can be activated via the identification of LysR-type transcriptional regulator (LTTR) genes required for the induction of cryptic genes. Manipulating pathway-specific regulators, reporter-guided mutant selection, refactoring, and heterologous expression are specific methods for the manipulation of pathways and their enzymes for the activation of silent gene clusters (Valayil 2016). Sun et al. (2018) added the chemical modifier 10 μM trichostatin A (TSA) to the culture of *Aspergillus terreus* OUCMDZ-2739, a marine algicolous fungus, and identified the chemical diversity in secondary metabolites. The bioactive chemicals were characterized using NMR and identified as novel meroterpenoids as a (4S)-4-decarboxylflavipesolide C, 1-(2,2-dimethylchroman-6-yl)-3-(4-hydroxyphenyl)propan-2-one, (R,E)-3-(2,2-dimethylchroman-6-yl)-4-hydroxy-5-((2-(2-hydroxypropan-2-yl)-2,3-dihydrobenzofuran-5-yl)methylene) furan-2(5H)-one, methyl (R)-2-(2-(2-hydroxypropan-2-yl)-2,3-dihydrobenzofuran-5-yl) acetate compound. The addition of chemical modifier to the culture broth enhanced the chemically diversified molecules that possessed cytotoxicity against MCF-12 cells and showed an inhibitory effect for α-glucosidase.

Akone et al. (2016) performed two individual experiments for the enhanced production of secondary metabolites in endophytic fungi, *Chaetomium* sp., using an epigenetic method. In the first experiment, they introduced *B. subtilis* in fungal culture medium and identified the secondary metabolites produced due to the coculture strategy. In the second experiment, they added epigenetic modifier, suberoylanilide hydroxamic acid to the fungal culture. Identification of active metabolite produced in the presence of epigenetic modifier was performed using a NMR analytical tool. Among 12 active metabolites, isosulochrin metabolite showed elevated production after treatment with suberoylanilide hydroxamic acid. Bai et al. (2018) reported the production of new chemical molecules from the mutant *Calcarisporium arbuscula* that lacked expression of histone deacetylase after the addition of epigenetic modifier. Following the deletion of 3-histone deacetylase, the mutant started producing novel bioactive compounds. These compounds were further investigated using the NMR tool, and the presence of diterpenoid in the fungal crude extract, i.e. three cassanes, one cleistanthane, six pimaranes, and two isopimaranes with cleistanthane analogs was reported. These compounds showed a greater higher inhibitory effect on the expression of the matrix metallo-proteinase gene family (MMP1 and MMP2) in human breast cancer (MCF-7) cells. Zhang et al. (2016) studied the role of putative methyltransferase LaeA and transcription factor CreA in asexual development and controlling secondary metabolic gene cluster expression in fungal species. They explained the importance of both CreA and LaeA in the production of pigment molecules and spore formation. LaeA played an effective role in the gene silencing of the *CreA* gene in a subtelomeric region.

7.5 Applications

In this chapter we reviewed the potential of endophytic fungi for the production of bioactive compounds and their chemical diversity. Endophytic fungi play an important role in miscellaneous fungal communities and have found wide use in pharmaceuticals, neutraceuticals, crop protection and industry.

7.5.1 Pharmaceutical Approach

Antibiotic resistance is a life-threatening problem that has reached a level where using a conventional drug is useless. To enhance drug susceptibility in pathogenic cells and treat infections in a minimal amount of time, microbe-derived compounds have evolved as potential drug molecules. Endophytic fungi have the ability to produce a large number of natural compounds with various activities, i.e., antimicrobial, antioxidant and antitumor (Uzma et al. 2018; Mishra et al. 2017a, b).

7.5.1.1 Antioxidant Compounds

Antioxidant properties possessed by bioactive compounds regulate their activity by scavenging the free radicals through an electron localization process. Secondary metabolites were isolated from endophytic *Pseudocercospora* sp. ESL 02 harboured in *Elaeocarpus sylvestris* and were evaluated for their antioxidant activity. Bioactive fractions were separated by column chromatography and subjected to NMR for the determination of structural conformation and active moieties of the compounds. The data showed the presence of two bioactive terreic acid and 6-methylsalicylic acid in abundant amounts, which were further examined for antioxidant properties. The ability of bioactive compounds to carry out antioxidant activities was confirmed by DPPH radical scavenging activity and β-carotene bleaching assay at 1 mg/mL concentration. The latter compound's structure showed a similarity with salicylic acid, and was reported for the first time as being produced by a microbial source (Prihantini and Tachibana 2017). Anitha and Mythili (2017) investigated the antioxidant properties of secondary metabolites isolated from *Achaetomium* sp., an endophytic fungus associated with *Euphorbia hirta*. A variety of metabolic compounds were found and estimated using a standard protocol. The presence of total phenolic compounds was determined as 44.02 ± 1.57 μg, total flavonoid content was 54.540 ± 1.820 μg, and total tannin content was observed to be 18.790 ± 1.018 μg. Based on the phytochemical profile of the endophytic fungus, ethyl acetate crude extract was assayed for antioxidant, antimicrobial, and antitumor activity. Antioxidant activity was measured and found to be 66.890 ± 1.385% to 87.340 ± 0.289% in the presence of total phenolic, total flavonoid, and total tannin acid. Antimicrobial activity against gram-positive and gram-negative bacteria was measured and shown to be significant for the inhibition of microorganisms.

7.5.1.2 Antimicrobial Compounds

Supong et al. (2016) studied the biosynthesis of diverse chemical structures produced by *Streptomyces samsunensis* M1463T such as efomycins M, G, oxohygrolidin, abierixin, and 29-*O*-methylabierixin bioactive molecules, in association with the plant leaves of *Oryza sativa*. They examined the antimicrobial capacity of bioactive compounds against *Mycobacterium tuberculosis*, *B. cereus*, *Colletotrichum gloeosporioides* and *Colletotrichum capsici* with IC_{50} values ranging from 1.40 to 5.23 μg/mL. A tetramic acid derivative, Equisetin, from endophytic *Fusarium* sp., showed excellent antimicrobial activity against *B. subtilis, Staphylococcus aureus* and MRSA at 8 and 16 μg/mL minimum inhibitory concentration, respectively (Ratnaweera et al. 2015). Antimicrobial activity varied from species to species in endophytic fungi isolated from the same sources such as *Penicillium commune*, *Penicillium canescens*, and *Alternaria alternata* associated with *Olea europaea* L. tree. *P. commune*, *P. canescens* and *A. alternata* exhibited antimicrobial activity against gram-positive/negative bacteria and yeast at 2.7-fold and 8.0-fold, respectively. NMR and GC-MS studies supported and confirmed the presence of 3-methyl-1-butanol and phenylethyl alcohol compound as potential antimicrobial agents (Malhadas et al. 2017). Pinheiro et al. (2017) isolated the polyketide monocerin from the endophytic fungus *Exserohilum rostratum* found in a typical Amazonian plant, *Bauhinia guianensis.* The potential of the bioactive polyketide monocerin was investigated against *E. coli* (ATCC 25922), *P. aeruginosa* (ATCC 27853), *S. aureus* (ATCC 25923), *B. subtilis* (ATCC 6633) and *S. typhimurium* (ATCC14028). *Plectranthus amboinicus (Lour.) Spreng* and *Plectranthus Nicoletta* displayed efficient antistreptococcal activity and fractions containing different bioactive compounds, i.e., carvacol, total carotenoid, and phenolic content (Zhang et al. 2017). Marine seaweed endophytic fungus *Xylaria* sp. assisted with Brazilian marine red algae *Bostrychia tenella* (Ceramiales) secretes cytochalasin D, a potential candidate that exhibits antifungal, antibacterial, and antitumor activity (Felicio et al. 2015). The bioactive compounds naphtha-quinones, anhydro-fusarubin, and methyl ether of fusarubin from *Cladosporium sp.* linked with *Rauwolfia serpentina* (L.) Benth. ex Kurz. showed higher antimicrobial activity against *S. aureus*, *E. coli*, *P. aeruginosa*, and *Bacillus megaterium* (Khan et al. 2016).

7.5.1.3 Antitumor/Anticancer/Antiproliferative Compounds

Li et al. (2017) studied bioactive compounds, pestallic acid E and (þ)-ambuic acid, biosynthesized by the Hawaiian endophytic fungus *Pestalotiopsis* sp. The chemical structure and active site of the bioactive compounds were confirmed by NMR and mass spectrometric studies and examined for their antiproliferative activity against A2780 and cisplatin-resistant A2780 (A2780CisR) cell lines. The IC_{50}values ranged from 3.3 to 17.0 μM, and they showed great potential as antiproliferative agents. Venkateswarulu et al. (2017) reported on the anticancer and anticandida activity of (E)-3-(2, 3-dihydroxyphenyl) acrylic acid, a bioactive molecule from the endophytic fungus *Fusarium equseti* EF-32 isolated from *Terminalia pallida*. The bioactive molecules showed potential anticandida activity against *Candida tropicalis* and *Candida albicans*. The anticancer activity of the compound was determined using MTT (3-(4, 5-dimethylthiazol-2-yl)-2,5-diphenyltetrazolium bromide) assay in breast cancer cell lines. Thus, the secondary metabolite was found to be a potent antiproliferative agent through cytotoxicity studies using MCF7 breast cancer cell lines. Minarni et al. (2017) isolated an endophytic fungus from *Annona muricata* L. and evaluated it for its anticancer activity against MCF-7 (Michigan Cancer Foundation-7). A total of four fungal isolates were found based on their biological activity; the highest-ranked fungal strain was studied further for molecular characterization. Phylogenetic studies confirmed that the endophytic fungi belong to *Phomopsis* sp. and showed higher anticancer activity at IC_{50} = 19.20 μg/mL concentration.

7.5.2 Endophytic fungi as Biocontrol Agents

Endophytic fungi play a significant role in ecosystem balancing and boosting host growth. Zheng et al. (2017) studied the endophytic fungi of *Panax notoginseng*. About 89 fungi were acquired from the roots, stems, seeds, and leaves of *P. notoginseng* and 41 isolates were selected for taxonomic characterization, which represented different morphotypes. The fungal isolates belongs to Zygomycota and Ascomycota. Their results revealed that *P. notoginseng* was capable of diversifying endophytic fungi that might be effective in the biocontrol of notoginseng root rot and a new compound was identified. Zhang et al. (2014) identified an endophytic fungus from *Brassica napus*, i.e., oilseed rape. They studied its ability to suppress the plant pathogenic fungi *Botrytis cinerea* and *Sclerotinia sclerotiorum.* Some of the selected isolates of the fungal endophytes were evaluated in plotting experiments to promote the growth of oilseed rape. The researchers' final findings suggest that the isolated fungal endophytes are promising biocontrol agents against *Sclerotinia sclerotiorum.* Endophytic fungi isolated from Chinese medicinal plants were examined for their potential in biocontrol and were further tested for their antimicrobial activities. A total of 208 isolates of fungal endophytes were obtained from the leaves, flowers and stems of the medicinal plants. Most of them belong to *Phoma*, *Phomopsis*, *Altwenaria* and *Acremonium.* The researchers concluded that of the 208 endophytic fungal isolates, 15 showed antifungal activity. The strain *Lasiodiplodia pseudotheobromae* is a vigorous biocontrol agent against the powdery mildew of wheat (Xiang et al. 2016). Active and volatile secondary metabolites produced by endophytic fungi, i.e., *Muscodor heveae* were studied by Siri-Udom et al. (2017). The volatile compounds of *M. heveae* exhibited antimicrobial activities. The volatile compounds were tested for inhibiting the seed germination and growth of seedling tested plants. Biofumigation of *M. heveae* also revealed the suppression of the disease. The researchers concluded that the bioactive volatile organic compounds of *M. heveae* can be used as an alternative method in controlling the white rot disease of rubber trees.

7.6 Conclusion and Future Perspective

Endophytic fungi are ubiquitous microorganisms that survive in the epidermal and aerial parts of plants that harbor vast chemical diversity. Endophytic fungi produce secondary metabolites, i.e., terpenes, alkaloids, monoterpenoids, peptides, and polyketides, with significant biological activities. Secondary metabolites are chemical substances with diverse structures and active moieties that possess antibacterial, antifungal, anticandida, anticancer and antiproliferative activities. The production of bioactive compounds is mediated by systematic chemical reaction chains controlled by significant precursors and their important enzymes, known as biosynthetic or metabolic pathways. Modification of these pathways via epigenetic remodeling and silent gene cluster activation may enhance secondary metabolite production. Secondary metabolites can be utilized as pharmaceutical agents (antioxidant, antimicrobial, and anticancer agents), in the enzyme industry and as biocontrol agents for phytopathogens. Endophytic fungi have metabolic diversity and possess unique secondary metabolite pathways, which will pave the way to novel metabolite isolation and applications in medicine.

References

Akone SH, Mandi A, Kurtan T, Hartmann R, Lin W, Daletos G, Proksch P (2016) Inducing secondary metabolite production by the endophytic fungus *Chaetomium sp.* through fungale bacterial co-culture and epigenetic modification. Tetrahedron 72:6340–6347

Anitha KPGU, Mythili S (2017) Antioxidant and hepatoprotective potentials of novel endophytic fungus *Achaetomium* sp., from *Euphorbia hirta*. Asian Pac J Trop Med 10(6):588–593

Bai J, Mu R, Dou M, Yan D, Liu B, Wei Q, Wan J, Tang Y, Hu Y (2018) Epigenetic modification in histone deacetylase deletion strain of *Calcarisporiumarbuscula* leads to diverse diterpenoids. Acta Pharm Sin B 8(4):687–697. https://doi.org/10.1016/j.apsb.2017.12.012

Bai J, Guo F, Wang R, Chen G, Li Z, Shao M, Xue C, Hua H (2017) Asperterzine, a symmetric aromatized derivative of epipolythiodioxopiperazine, from the endophytic fungus Aspergillusterreus PR-P-2. Chin J Chem 29:535–537

Cai R, Chen S, Long Y, Li C, Huang X, She Z (2017) Depsidones from *Talaromycesstipitatus* SK-4, an endophytic fungus of themangrove plant *Acanthus ilicifolius*. Phytochem Lett 20:196–199

Cerezo SS, Montiel NM, Sanchez JG, Perez-y-Terron R, Contreras RDM (2018) Gibberellin biosynthesis and metabolism: a convergent route for plants, fungi and bacteria. Microbial Res 208:85–98

Chandrasekaran R, Revathi K, Senthil-Nathan S, Kalaivani K, Hunter WB, Duraipandiyan V, Al-Dhabi NA, Esmail GA (2018) Eco-friendly formulation of wild *Bacillus thuringiensis* secondary metabolites through molecular characterization against the lepidopteranpest. Physiol Mol Plant Path 101:93–104

Chen LL, Kong FD, Wang P, Yuan JZ, Guo ZK, Wang H, Dai HI, Mei WL (2017) Two new tremulanesesquiterpenes from a mangrove endophytic fungus, *Coriolopsis* sp. J5. Chin J Chem 28:222–225

Dannert CS (2015) Biosynthesis of Terpenoid natural products in fungi. Adv Biochem Eng Biotechnol 148:19–61

Depke T, Franke R, Bronstrup B (2017) Clustering of MS2 spectra using unsupervised methods to aid the identification of secondary metabolites from *Pseudomonas aeruginosa*. J Chromatogr B 1071:19–28

Duana R, Zhoua H, Yang Y, Li H, Dong J, Li X, Chen G, Zhao L, Ding Z (2016) Antimicrobial meroterpenoids from the endophytic fungus *Penicillium* sp. T2-8 associated with *Gastrodiaelata*. Phytochem Lett 18:197–201

Durao P, Balbontin R, Gordo I (2018) Evolutionary mechanisms shaping the maintenance of antibiotic resistance. Trends Microbiol 26(8):677–691. https://doi.org/10.1016/j.tim.2018.01.005

Feitosa ADO, Dias ACS, Ramos GDC, Bitencourt HR, Siqueira JES, Marinho PSB, Barison A, Ocampos FMM, Marinho AMR (2016) Lethality of cytochalasin B and other compounds isolated from fungus *Aspergillus* sp. (Trichocomaceae) endophyte of *Bauhinia guianensis* (Fabaceae). Rev Argent Microbiol 48(3):259–263

Felicio RD, Pavao GB, Oliveira ALL, Erbert C, Conti R, Pupo MT, Furtado NAJC, Ferreira EG, Lotufo LVC, Young MCM, Yokoya NS, Debonsi HM (2015) Antibacterial, antifungal and cytotoxic activities exhibited by endophytic fungi from the Brazilian marine red alga *Bostrychiatenella* (Ceramiales). Rev Bras Farmacogn 25:641–650

Ganesh PS, Rai VR (2018) Attenuation of quorum-sensing-dependent virulence factors and biofilm formation by medicinal plants against antibiotic resistant *Pseudomonas aeruginosa.* J Tradit Complement Med 8:170–177

Gao L, Narita K, Ozaki T, Kumakura N, Gan P, Minami A, Liu C, Lei X, Shirasu K, Oikawa H (2018) Identification of novel sesterterpenes by genome mining of phytopathogenic fungi *Phoma* and *Colletotrichumsp*. Tetrahedron Lett 59:1136–1139

Gholami S, Mohammad T, Sohrabi N (2017) Comparison of biofilm formation and antibiotic resistance pattern of *Pseudomonas aeruginosa* in human and environmental isolates. Microb Pathog 109:94–98

Gupta PD, Birdi TJ (2017) Development of botanicals to combat antibiotic resistance. J Ayurveda Integr Med 8:266–275

Heinig U, Scholz S, Jennewein S (2013) Getting to the bottom of Taxol biosynthesis by fungi. Fungal Divers 60:161–170

Hernandez-Flores JL, Perez JC, Gutierrez CS, Hernandez AC, Alonsoc GS, Hernandez SP, Gomez SR, Fernandez F, Loske AM, Guillen JC (2018) pMEX01, a 70 kb plasmid isolated from *Escherichia coli* that confers resistance to multiple β-lactam antibiotics. Braz J Microbiol 49(3):569–574. https://doi.org/10.1016/j.bjm.2017.11.002

Hu Y, Zhang J, Liu D, Guo J, Liu T, Xin Z (2017) Pencitrin and pencitrinol, two new citrinin derivatives from an endophytic fungus *Penicilliumcitrinumsalicorn* 46. Phytochem Lett 22:229–234

Indrianingsih AW, Tachibana S (2017) α-Glucosidase inhibitor produced by an endophytic fungus, *Xylariaceae* sp. QGS 01 from *Quercusgilva Blume*. Food Sci Hum Wellness 6:88–95

Jaillard M, Belkum AV, Cady KC, Creely D, Shortridge D, Blanc B, Barbu EM, Dunne WM Jr, Zambardi G, Enright M, Mugnier N, Le Priol C, Schicklin S, Guigon G, Veyrieras JB (2017) Correlation between phenotypic antibiotic susceptibility and the resistome in *Pseudomonas aeruginosa.* Int J Antimicrob Agents 50:210–218

Jones-Dias D, Carvalho AS, Moura IB, Manageiro V, Igrejas G, Caniça M, Matthiesen R (2017) Quantitative proteome analysis of an antibiotic resistant *Escherichia coli* exposed to tetracycline reveals multiple affected metabolic and peptidoglycan processes. J Proteome 156:20–28

Khan MIH, Sohrab MH, Rony SR, Tareq FS, Hasan CM, Mazid MA (2016) Cytotoxic and antibacterial naphthoquinones from an endophytic fungus, *Cladosporium* sp. Toxicol Rep 3:861–865

Khater S, Anand S, Mohanty D (2016) *Insilico* methods for linking genes and secondary metabolites: the way forward. Synth Syst Biotechnol 1:80–88

Kozlovsky AG, Zhelifonova VP, Antipova TV (2013) Biologically active metabolites of Penicillium fungi. J Org Biomol Chem 1:11–21

Lee TAJ, Medema MH (2016) Computational strategies for genome-based natural product discovery and engineering in fungi. Fungal Genet Biol 89:29–36

Li YF, Tsai KJS, Harvey CGB, Li JJ, Ary BE, Berlew EE, Boehman BL, Findley DM, Friant AG, Gardner CA, Gould MP, Ha JH, Lilley BK, McKinstry EL, Nawal S, Parry RC, Rothchild KW, Silbert SD, Tentilucci MD, Thurston AM, Wai RB, Yoon Y, Aiyar RS, Medema MH, Hillenmeyer ME, Charkoudian LK (2016) Comprehensive curation and analysis of fungal biosynthetic gene clusters of published natural products. Fungal Genet Biol 89:18–28

Li CS, Yang BJ, Turkson J, Cao S (2017) Anti-proliferative ambuic acid derivatives from Hawaiian endophytic fungus *Pestalotiopsis sp*. FT172. Phytochemistry 140:77–82

Liu T, Greenslade A, Yang S (2017a) Levels of rhizome endophytic fungi fluctuate in *Paris polyphylla var. yunnanensis* as plants age. Plant Divers 39:60–64

Liu X, Guan X, Xing F, Lv C, Dai X, Liu Y (2017b) Effect of water activity and temperature on the growth of *Aspergillusflavus*, the expression of aflatoxin biosynthetic genes and aflatoxin-production in shelled peanuts. Food Control 82:325–332

Liu J, Wang X, Pu H, Liu S, Kan J, Jin C (2017c) Recent advances in endophytic exopolysaccharides: production, structural characterization, physiological role and biological activity. Carbohydr Polym 157:1113–1124

Loftus RW, Dexter F, Robinson ADM (2018) Methicillin-resistant *Staphylococcus aureus* has greater risk of transmission in the operating room than methicillin-sensitive *S. aureus*. Am J Infect Control 46(5):520–525. https://doi.org/10.1016/j.ajic.2017.11.002

Lou J, Yu R, Wang X, Mao Z, Fu L, Liu Y, Zhou L (2016) Alternariol 9-methyl ether from the endophytic fungus *Alternaria* sp. Samif01 and its bioactivities. Braz J Microbiol 47:96–101

Loureiro C, Medema MH, Oost JV, Sipkema D (2018) Exploration and exploitation of the environment for novel specialized metabolites. Curr Opin Biotechnol 5:206–213

Luo K, Rocheleau H, Qi PF, Zheng YL, Zhao HY, Ouellet T (2016) Indole-3-acetic acid in *Fusarium graminearum*: identification of biosynthetic pathways and characterization of physiological effects. Fungal Biol 120(9):1135–1145

Luo Y, Yi W, Yao Y, Zhu N, Qin P (2018) Characteristic diversity and antimicrobial resistance of *Salmonella* from gastroenteritis. J Infect Chemother 24:251–255

Malhadas C, Malheiro R, Pereira JP, Pinho PG, Baptista P (2017) Antimicrobial activity of endophytic fungi from olive tree leaves. World J Microbiol Biotechnol 33:46

Matsuda Y, Awakawa T, Mori T, Abe I (2016) Unusual chemistries in fungal meroterpenoid biosynthesis. Curr Opin Chem Biol 31:1–7

Minarni, Artika IM, Julistiono H, Bermawie N, Riyanti EI, Hasim, Hasan AEZ (2017) Anticancer activity test of ethyl acetate extract of endophytic fungi isolated from soursop leaf (*Annonamuricata L.*). Asian Pac J Trop Med 10(6):566–571

Mishra VK, Passari AK, Chandra P, Leo VV, Kumar B, Gupta VK, Singh BP (2017a) Determination and production of antimicrobial compounds by *Aspergillus clavatonanicus* strain MJ31, an endophytic fungus from Mirabilis jalapa L. using UPLC-ESI-MS/MS and TD GC-MS. PLoS One 12(10):1–24. https://doi.org/10.1371/journal.pone.0186234

Mishra VK, Passari AK, Leo VV, Singh BP (2017b) Molecular diversity and detection of endophytic fungi based on their antimicrobial biosynthetic genes. In: Singh BP, Gupta VK (eds) Molecular markers in mycology, fungal biology. Springer International Publishing, Switzerland, pp 1–35. https://doi.org/10.1007/978-3-319-34106-4_1

Narenkumar J, Sathishkumar K, Sarankumar RK, Murugan K, Rajasekar A (2017) An anticorrosive study on potential bioactive compound produced by *Pseudomonas aeruginosa* TBH2 against the biocorrosive bacterial biofilm on copper metal. J Mol Liq 243:706–713

Nieto PA, Pardo-Roa C, Salazar-Echegarai FJ, Tobar HE, Coronado-Arrazola I, Riedel CA, Kalergis MA, Bueno SM (2016) New insights about excisable pathogenicity islands in *Salmonella* and their contribution to virulence. Microbes Infect 18:302–309

Paling FP, Wolkewitz M, Bode LGM, Klein Klouwenberg PMC, Ong DSY, Depuydt P, Bus LD, Sifakis F, Bonten MJM, Kluytmans JAJW (2017) *Staphylococcus aureus* colonization at ICU admission as a risk factor for developing *S. aureus* ICU pneumonia. Clin Microbiol Infect 23:49.e9–49.e14

Pascale A, Vinale F, Manganiello G, Nigro M, Lanzuise S, Ruocco M, Marra R, Lombardi N, Woo SL, Lorito M (2017) *Trichoderma* and its secondary metabolites improve yield and quality of grapes. Crop Prot 92:176–181

Peek J, Christendat D (2015) The shikimate dehydrogenase family: functional diversity within a conserved structural and mechanistic framework. Arch Biochem Biophys 566:85–99

Phillips-Houlbracq M, Ricarda JD, Foucriera A, Yoder-Himes D, Gaudrya S, Bex J, Messika J, Margetis D, Chatel J, Dobrindt U, Denamur E, Roux D (2018) Pathophysiology of *Escherichia coli* pneumonia: respective contribution of pathogenicity islands to virulence. Int J Med Microbiol 308:290–296

Pinheiro EAA, Pina JRA, Feitosa AO, Carvalho JM, Borges FC, Marinho PSB, Marinho AMR (2017) Bioprospecting of antimicrobial activity of extracts of endophytic fungi from *Bauhinia guianensis*. Rev Argent Microbiol 49(1):3–6

Polonio JC, Ribeiro MAS, Rhoden SA, Sarragiotto MH, Azevedo JL, Pamphille JA (2016) 3-Nitropropionic acid production by the endophytic *Diaporthecitri*: molecular taxonomy, chemical characterization, and quantification under pH variation. Fungal Biol 120:1600–1608

Prihantini AI, Tachibana S (2017) Antioxidant compounds produced by *Pseudocercospora* sp. ESL 02, an endophytic fungus isolated from *Elaeocarpussylvestris*. Asian Pac J Trop Biomed 7(2):110–115

Ratnaweera PB, Silva ED, Williams DE, Andersen RJ (2015) Antimicrobial activities of endophytic fungi obtained from the arid zone invasive plant *Opuntiadillenii* and the isolation of equisetin, from endophytic *Fusarium sp*. BMC Complement Altern Med 15:220

Saetang P, Rukachaisirikul V, Phongpaichit S, Preedanon S, Sakayaroj J, Borwornpinyo S, Seemakhan S, Muanprasat C (2017) Depsidones and an a-pyrone derivative from *Simpilcillium* sp. PSU-H41, an endophytic fungus from *Heveabrasiliensis* leaf. Phytochemistry 143:115–123

Siri-Udom S, Suwannarach N, Lumyong S (2017) Applications of volatile compounds acquired from Muscodor heveae against white root rot disease in rubber trees (Hevea brasiliensis Müll. Arg.) and relevant allelopathy effects. Fungal Biol 121(6-7):573–581

Siahmoshteh F, Esfahani ZH, Spadaro D, Ghahfarokhi MS, Abyaneh MR (2017) Unraveling the mode of antifungal action of Bacillus subtilis and Bacillus amyloliquefaciens as potential biocontrol agents against aflatoxigenic *Aspergillus parasiticus*. Food Control 89:300–307

Song HC, Qin D, Han MJ, Wang L, Zhang K, Dong JY (2017) Bioactive 2-pyrone metabolites from an endophytic *Phomopsisasparagi* SWUKJ5. 2020 of *Kadsuraangustifolia*. Phytochem Lett 22:235–240

Sun K, Zhu G, Hao J, Wang Y, Zhu W (2018) Chemical-epigenetic method to enhance the chemodiversity of the marine algicolous fungus, *Aspergillusterreus* OUCMDZ-2739. Tetrahedron 74:83–87

Supong K, Thawai C, Choowong W, Kittiwongwattana C, Thanaboripat D, Laosinwattana C, Koohakan P, Parinthawong N, Pittayakhajonwut P (2016) Antimicrobial compounds from endophytic *Streptomyces* sp. BCC72023 isolated from rice (*Oryza sativa* L.). Res Microbiol 167:290–298

Tang JW, Wang WG, Li A, Yan BC, Chen R, Li ZN, Du X, Sun HD, Pu JX (2017) Polyketides from the endophytic fungus *Phomopsis* sp. sh917 by using the one strain/many compounds strategy. Tetrahedron 73:3577–3584

Tohge T, Watanabe M, Hoefgen R, Fernie AR (2013) Shikimate and phenylalanine biosynthesis in the green lineage. Front Plant Sci 4:62. https://doi.org/10.3389/fpls.2013.00062

Toledo AV, Franco MEE, Lopez SMY, Troncozo MI, Saparrat MCN, Balatti PA (2017) Melanins in fungi: types, localization and putative biological roles. Physiol Mol Plant Path 99:2–6

Tyc O, Song C, Dickschat JS, Vos M, Garbeva P (2017) The ecological role of volatile and soluble secondary metabolites produced by soil bacteria. Trends Microbiol 25(4):280–292

Upadhyay S, Xu X, Lowry D, Jackson JC, Roberson RW, Lin X (2016) Subcellular compartmentalization and trafficking of the biosynthetic machinery for fungal melanin. Cell Rep 14:2511–2518

Uzma F, Hashem A, Murthy N, Mohan HD, Kamath PV, Singh BP, Venkataramana M, Gupta VK, Siddaiah CN, Chowdappa S, Alqaeawi AA, Abd Allah EF (2018) Endophytic fungi—alternative sources of cytotoxic compounds: a review. Front Pharmacol 9(309):1–37. https://doi.org/10.3389/fphar.2018.00309

Valayil JM (2016) Activation of microbial silent gene clusters: genomics driven drug discovery approaches. Biochem Anal Biochem 5:276

Venkateswarulu N, Chari PVB, Basha SKT, Nagaraju C, Vijaya T (2017) Isolation and purification of (E)-3- (2, 3- dihydroxyphenyl) acrylic acid fromendophytic fungi *Fusariumequseti* EF-32 and its anti-candidal and anticancer activities. Biocatal Agric Biotechnol 11:294–301

Wang Y, Hu P, Pan Y, Zhu Y, Liu X, Che Y, Liu G (2017a) Identification and characterization of the verticillin biosynthetic genecluster in *Clonostachys rogersoniana*. Fungal Genet Biol 103:25–33

Wang CC, Huang SY, Huang SH, Wen ZH, Huang ZY, Liu WS, Wang HMD (2017b) A synthetic biological secondary metabolite, Lycogen TM, produced and extracted from *Rhodobactersphaeroides* WL-APD911 in an optimizatioal scale-up strategy. Food Sci Hum Wellness 6:195–201

Wibowo M, Prachyawarakorn V, Aree T, Mahidol C, Ruchirawat S, Kittakoop P (2016) Cytotoxic sesquiterpenes from the endophytic fungus *Pseudolagarobasidiumacaciicola*. Phytochemistry 122:126–138

Xiang L, Gong S, Yang L, Hao J, Xue MF, Zeng FS, Zhang XJ, Shi WQ, Wang H, Yu D (2016) Biocontrol potential of endophytic fungi in medicinal plants from Wuhan botanical garden in China. Biol Control 94:47–55

Yang Y, Zhao H, Barrero RA, Zhang B, Sun G, Wilson IW, Xie F, Walker KD, Parks JW, Bruce R, Guo G, Chen L, Zhang Y, Huang X, Tang X, Liu H, Bellgard MI, Qiu D, Lai J, Hoffman A (2014) Genome sequencing and analysis of thepaclitaxel-producing endophytic fungus *Penicilliumaurantiogriseum* NRRL 62431. BMC Genomics 15:69

Zeilinger S, Gruber S, Bansal R, Mukherjee PK (2016) Secondary metabolism in *Trichoderma*-chemistry meets genomics. Fungal Biol Rev 30:74–90

Zhang Q, Zhang J, Yang L, Zhang L, Jiang D, Chen W, Li G (2014) Diversity and biocontrol potential of endophytic fungi in *Brassica napus*. Biol Control 72:98–108

Zhang X, Zhu Y, Bao L, Gao L, Yao G, Li Y, Yang Z, Li Z, Zhong Y, Li F, Yin H, Qu Y, Qin Y (2016) Putative methyltransferase LaeA and transcription factor CreA are necessary for proper asexual development and controlling secondary metabolic gene cluster expression. Fungal Genet Biol 94:32–46

Zhang B, Wijesundara NM, Abbey L, Rupasinghe HSV (2017) Growing medium amendments effect on growth, secondary metabolites and anti-streptococcal activity of two species of *Plectranthus*. J Appl Res Med Aromat Plants 5:53–59

Zheng YK, Miao CP, Hua-Hong Chen HH, Huang FF, Xia YM, Chen YW, Zhao LX (2017) Endophytic fungi harbored in *Panaxnotoginseng*: diversity and potential as biological control agents against host plant pathogens of root-rot disease. J Ginseng Res 41:353–360

Zouari I, Jlaiel L, Tounsi S, Trigui M (2016) Biocontrol activity of the endophytic *Bacillus amyloliquefaciens*strain CEIZ-11 against *Pythiumaphanidermatum* and purification of its bioactive compounds. Biol Control 100:54–62

Chapter 8
The Bull Effect of Endophytic Fungi: An Approach with Quorum Sensing

Subhoshmita Mondal and Sarangam Majumdar

8.1 Introduction

Microbes and eukaryotic organisms can communicate with each other via chemical signaling molecules. Such signaling-molecule-mediated "conversation" is known as quorum sensing (QS). Organisms, i.e., bacteria (Miller and Bassler 2001; Majumdar and Mondal 2016; Shapiro 1998, Williams et al. 2007), yeast (Avbelj et al. 2015; Sprague and Winans 2006), slime molds (Golé et al. 2011; Brock and Gomer 1999), and social insects (Seeley and Visscher 2004), are emitted and sensed biochemical signaling molecules, which are considered autoinducers (AIs) or quorum sensing molecules (QSMs). When cell density decreases, the production of QSMs is also lowered. Organisms also sense high concentrations of QSMs at high cell population densities in their vicinity and switch on QSM production (Fig. 8.1). When the cell number density reach a certain level, a coordinated change in organism behaviour is initiated.

This AI-mediated communication process is well studied in the bacterial world, and QS regulates other series of actions like extracellular polymeric substance production, biofilm formation, virulence factors, swarming, antibiotic production, sporulation, and bioluminescence (Majumdar and Pal 2016, 2017a, b, 2018; Majumdar and Roy 2017, 2018; Majumdar et al. 2012, 2017). In a recent advancement in this field, researchers also developed synthetic bacterial communication systems that reveal different unseen episodes in biochemical communication that can be significant in biotechnology, Biocomputing, and medical sciences (Gardner et al. 2000; Elowitz and Leibler 2000; You et al. 2004; Basu et al. 2005; Brenner

S. Mondal
Department of Chemical Engineering, Jadavpur University, Kolkata, India

S. Majumdar (✉)
Dipartimento di Ingegneria Scienze Informatiche e Matematica, Università degli Studi di L' Aquila, L' Aquila, Italy

B. P. Singh (ed.), *Advances in Endophytic Fungal Research*, Fungal Biology,
https://doi.org/10.1007/978-3-030-03589-1_8

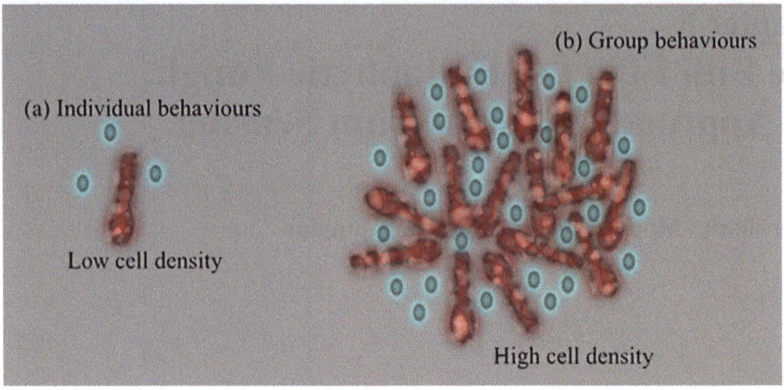

Fig. 8.1 Illustration of fungal communication process through chemical signaling molecules (at low density and high density)

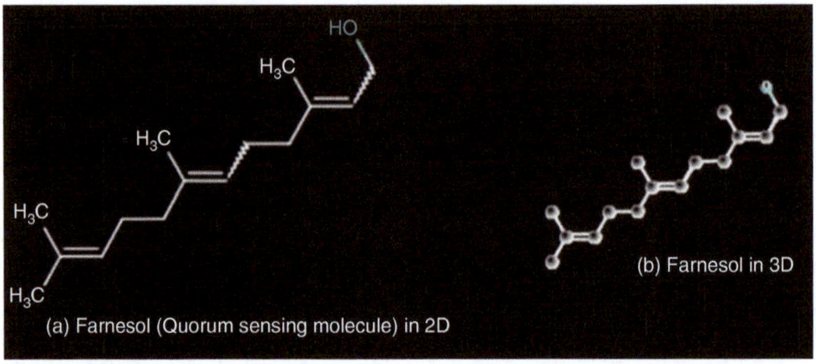

Fig. 8.2 Chemical structure of farnesol (quorum sensing molecule)

et al. 2007; Shou et al. 2007; Stricker et al. 2008; Danino et al. 2010; Chen et al. 2015; Scott and Hasty 2016; Balagaddé et al. 2008; Song et al. 2009; Baym et al. 2016; Datla et al. 2017).

QS mechanisms are not only confined to bacteria. They have also been observed in the pathogenic fungus *Candida albicans*. The communication process of *C. albicans* is transmitted by farnesol (QSM) (Fig. 8.2) (Hornby et al. 2001). Other than farnesol, there are several other QSMs such as tyrosol (second QSM in *C. albicans*) (Fig. 8.3) (Chen et al. 2004), phenylethanol, and tryptophol (QSMs of *Saccharomyces cerevisiae*) (Chen and Fink 2006).

Fungal communication activities have also been explored in fungi such as *Histoplasma capsulatum* (thermo-dimorphic pathogenic fungus) (Kügler et al. 2000), *Ceratocystis ulmi* (phytopathogenic fungus) (Hornby et al. 2004), *Neurospora crassa* (Roca et al. 2005), *Saccharomyces cerevisiae* (Severin et al. 2008), and *Cryptococcus neoformans* (Lee et al. 2007). Fernesol effectuates the regulation of *C. albicans* filamentation, biofilm formation, oxidative stress, modulation of drug efflux, and various microbes (*Aspergillus nidulans*, *Saccharomyces cerevisiae*, *Aspergillus niger*, *Aspergillus fumigatus*, *Fusarium graminearum*, *Candida*

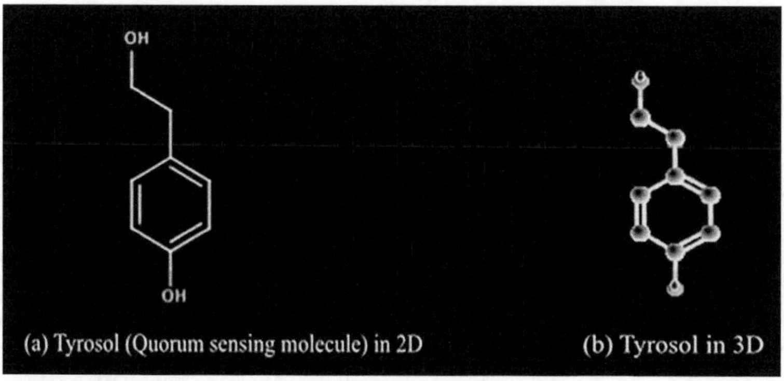

Fig. 8.3 Chemical structure of tyrosol (quorum sensing molecule)

Table 8.1 Secreted bacterial molecules and their effects on *C. albicans*

Bacterium	Secreted molecules	*C. albicans* response
A. baumannii	Unknown, found in cell-free supernatant	Filament inhibition, biofilm inhibition
B. cenocepacia	Cis-2-dodecenoic acid	Growth inhibition and filament inhibition
P. aeruginosa	Phenazines	Cell death
P. aeruginosa	3-oxo-c12-homoserine lactone, dodecanol, and C12-acyl homoserine lactone	Filament inhibition and reversion to yeast morphology
Salmonella enterica serovar Typhimurium	Unknown, found in cell-free supernatant	Filament and biofilm inhibition, reduced viability
S. gordonii	Unknown, found in cell-free supernatant	Induction of filamentous growth, suppression of farnesol-mediated filament inhibition
X. campestris	Cis-11-methyl-2-dodecenoic acid	Filament inhibition

dubliniensis, *Candida parapsilosis*, *Paracoccidioides brasiliensis*, *Mycobacterium smegmatis*, *Pseudomonas aeruginosa*) (Albuquerque and Casadevall 2012). *C. albicans* also interacts with other bacteria (Table 8.1).

8.2 Quorum Sensing Mechanism of *Candida albicans*

Candida albicans is a pathogenic fungus that is commonly found in human microbiota, and its virulence depends on the polymorphism between yeast, hyphae, and pseudohyphae forms. *C. albicans* cells develop into filamentous forms at densities lower than 10^6 cells/ml and the fungus grows as budding yeast in highly dense cells. It has been noticed that the cells of *C. albicans* are able to communicate with each other after they reach a certain threshold of QSMs (in high cell density), and Hornby and his collaborators identified the first QSMs in a fungal communication system (isoprenoid farnesol) (Hornby et al. 2001). They emphasized farnesol as a chemical signaling molecule that regulates *C. albicans* filamentation. Moreover, they investigated the conditions in which media-purified farnesol inhibited the yeast-to-mycelium conversion activated by three distinct germ tube formation inducers such as L-proline, serum, and *N*-acetylglucosamine. But the isoprenoid had no effects on the fungal growth rate (Hornby et al. 2001). This QSM (farnesol) also modulated *C. albicans* drug efflux, which is governed by ABC multidrug transporters, and potentiated the effects of azoles and polyenes (Sharma and Prasad 2011). The condition medium farnesol protects from oxidative stress, which is actually induced by plumbagin and the superoxide anion-generating agents menadione and hydrogen peroxide (Westwater et al. 2005).

QS controls gene expression, but the signaling pathways of the QS mechanism in fungi is unclear (Langford et al. 2009). Previously it was thought that Ras-cAMP-PKA was a possible pathway with a general repressor TUP1 (in farnesol-mediated QS) (Hall et al. 2009; Kebaara et al. 2008; Sato et al. 2004). Later, Ras-cAMP-Efg1 signaling pathways (Fig. 8.4) were discovered that are involved in hyphal growth

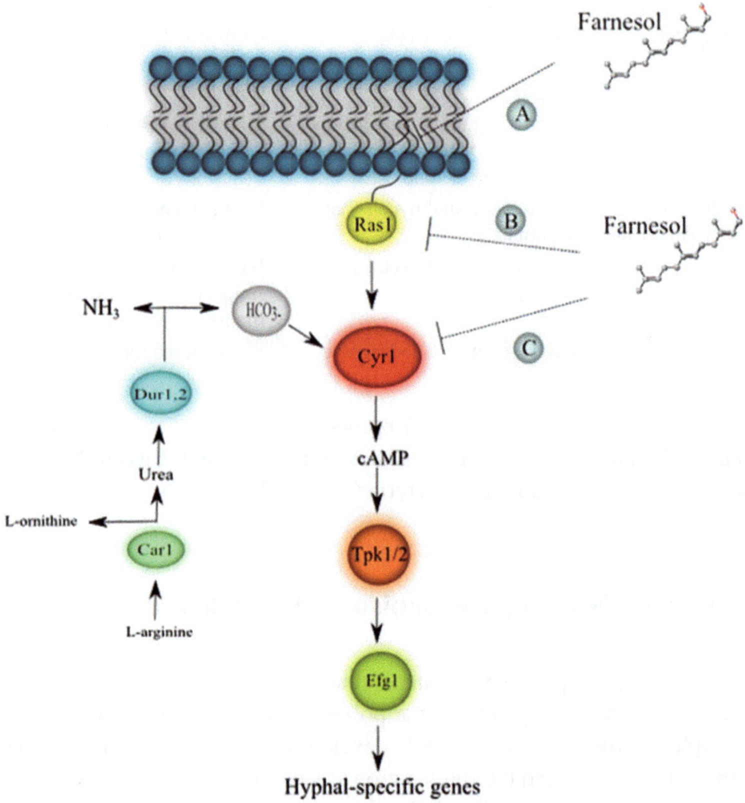

Fig. 8.4 Schematic diagram of cAMP pathway in *C. albicans*, where A, B, and C indicate possible points of inhibition by farnesol

and hyphal formation (Davis-Hanna et al. 2008). Ras1-adenylate cyclase signaling pathways are involved in case of oxidative stress, which is induced by farnesol (Deveau et al. 2010). Other signaling pathways are involved in the QS mechanism, such as a cyclic adenosine monophosphate (cAMP) signaling pathway and a mitogen-activated protein kinase pathway (Kebaara et al. 2008; Davis-Hanna et al. 2008).

Other QSMs (tyrosol) were discovered later. Tyrosol is a QSM in *C. albicans* that minimizes the length of the lag phase of growth and regulates biofilm formation and filamentation. This QSM also has inhibitory activity against neutrophils. It has been observed that the addition of condition medium shortens the lag phase and active molecules (aromatic alcohol tyrosol) in condition media. This chemical signaling molecule is secreted by *C. albicans* during growth. Tyrosol excites filamentation in opposition to farnesol and promotes germ tube formation as well (Kügler et al. 2000).

8.3 Biofilm Formation

In general, biofilms are defined as clumps of microbes and/or fungi in which cells are often fixed firmly in a self-produced matrix of extracellular polymeric substances that adhere to each other and/or to surface. The fungal cell-to-cell communication mechanism regulates biofilm formation, biofilm growth, and biofilm dispersion. The properties of farnesol have been observed at the moment of biofilm development (Ramage et al. 2002). Farnesol inhibits biofilm formation, and the inhibition rate is dependent on the time taken by the cells to adhere before farnesol is added. But the biofilm structure is not affected by QSMs. Farnesol regulates genes associated with drug resistance, cell surface hydrophilicity, cell wall maintenance, iron transport, and heat shock proteins (Cao et al. 2005). On the other hand, tyrosol (QSM) concentration corresponds to an increase in the biomass of *C. albicans* biofilms and stimulates hyphal growth at the early stage of biofilm formation (Alem et al. 2006).

C. albicans biofilm is composed of yeast from cells, pseudohyphal cells and hyphal cells, all of which are surrounded by an extracellular matrix (Fig. 8.5). *C. albicans* biofilm usually forms on host surfaces, mucosal surfaces, parenchymal organs, and epithelial cell linings. This fungal biofilm causes various diseases and can damage the kidneys and liver. Fungal biofilm also functions as a repository of drug-resistant cells, which can actually detach, divide, and implant bloodstream infections (Lohse et al. 2018).

8.4 Quorum Sensing and Endophytic Fungi

Endophytic fungi colonize the inside part of plants (e.g., the intercellular spaces of aerial plant parts, leaf sheaths, bark and root systems) and exchange nutrients with plants. This type of fungus decreases the damage caused by pathogens. Endophytes are secondary metabolites and a rich form of enzyme. Thus, they are very important

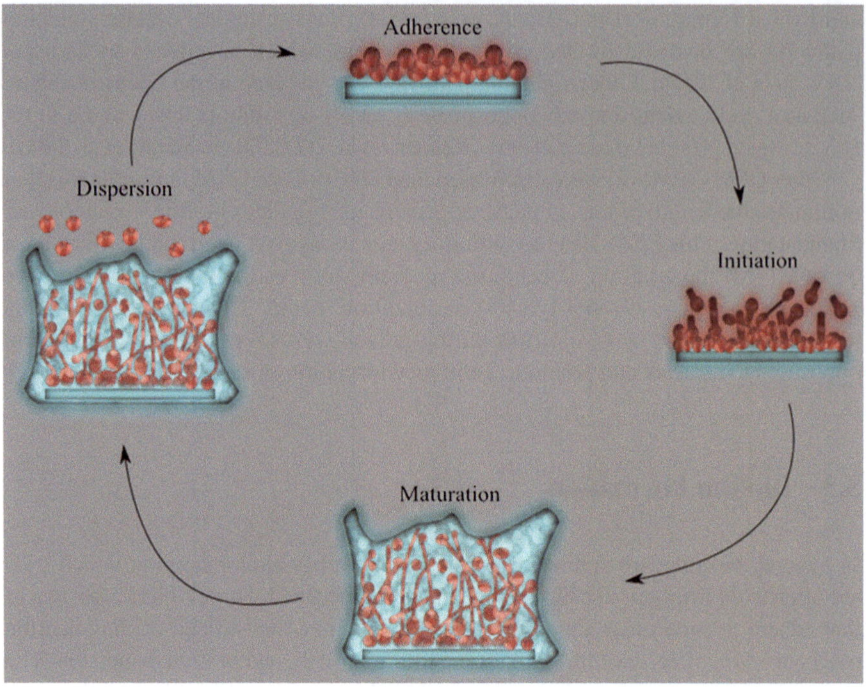

Fig. 8.5 Fungal biofilm formation process. Biofilm formation is a multistage process consisting of adherence, initiation, maturation, and dispersion

for industry for higher enzyme production as well as higher production of anti-QSMs. Researchers use secondary metabolites such as anti-QSMs, antimicrobials, and antioxidants. Moreover, they have discovered a new method of significant molecule production for industrial purposes. An example is the Australian red alga *Delisea pulchra*, which contributes secondary metabolites called halogenated furanones (with a structure similar to that of acyl homoserine lactone molecules) as a QSM analogs and used as anti-QSMs.

Endophytic fungi (*Fusarium graminearum* and *Lasiodiplodia* sp.) can be isolated from *Ventilago madraspatana*, and QS inhibitors are also screened, which suppress the expression of violacein production in *Chromobacterium violaceum*. This experimental observation can be applied in future therapy models and for anti-QS drugs (Rajesh and Rai 2013).

Polyhydroxyanthraquinones are QS inhibitors of *Penicillium restrictum* (isolated from *Silybum marianum*) (Figueroa et al. 2014), which has antivirulence action. Endophytic fungi isolated from marine habitats also produce QS-disrupting metabolites and are a source of antifouling agents (Martin-Rodriguez et al. 2014). This anti-QS is formally known as quorum quenching. In a recent review, Kusari and other researchers (Kusari et al. 2015) mention some important new challenges in the field, which include microbial population density and multiple signaling systems, the exact role of biochemical signaling mechanism, host plant gene expression and regulation, rehearsed subculturing of endophytes outside host plants, host plant environment, climate conditions, and difference in nutrition.

8.5 Mathematical Therapy Models

Substitutes for antibiotics constitute one of the major current areas of research. Since pathogenic QS bacteria are closely connected with human health, several mathematical models have been proposed recently whose major focus is to use a QS mechanism as a target for the treatment of quorum quenching. Likewise, Anguige and collaborators considered *P. aeruginosa*, one of the significant human pathogenic bacteria, and developed a new model to study an alternative form of antibiotic therapy. They revealed that AI concentration can be decreased by the application of anti-QS agents, in particular dosage, even if the remaining parameters are equally important for successful treatment (Anguige et al. 2004). Later, these researchers elaborated their work to incorporate well-mixed bacterial populations (Anguige et al. 2005). In contrast, Viretta and Fussenegger proposed a deterministic model (three QS systems, *las*, *rhl*, and mvfR-PQS) to understand virulence and the QS response to pharmacological interference. They determined that the interference of a pseudomonas quinolone signal (PQS) signaling system can act as an effective antivirulence agent (Viretta and Fussenegger 2004). Fagerlind and collaborators developed a model (quorum quenching) for the autoinducer antagonist(quorumsensing blocker) which predicted that quorum sensing blockers induced degradation of LasR is crucial for developing successful quorum quenching (Fagerlind et al. 2005). Additionally, Ward studied an anti-QS model in batch culture and biofilms and demonstrated that the efficacy of the treatment perhaps depended on the time of application of QSI and putative anti-*LasI* treatment is very potent (Ward 2008). Apart from these examples, various other therapy-related models (Beckmann et al. 2012; Anand et al. 2013) will supply an alternative path to antibiotic treatment in the near future.

8.6 Future Direction of Research

Our current understanding about cell-to-cell communication in fungi is the tip of the iceberg. We must investigate the QS system of different fungal communication processes and QS signaling networks. Recent knowledge does not give us a much clearer picture of this biochemically complex phenomenon. It will require more experimental evidence and theoretical investigation. Mathematical modeling of fungal communication systems can be very useful for understanding the overall phenomenon and be validated by experimentation. One can also think about synthetic fungal communication processes, which may be an important idea to unlock the biological design principle of QS in fungi. Regarding synthetic biology, the basic science is merging with engineering, which will help us to design artificial biological systems for a precise understanding of different types of complex biological facts and their applications in the medical sciences and industry to reveal the underlying mechanisms of different microscopic and macroscopic biological systems. The challenge is to recognize general, measurable strategies that combine the fabrication of increasingly complex gene circuits with reliable performance, as well as to construct an original and significant technological foundation for quantitative circuit characterization. On the other hand, fungal biofilms are also regulated by QS, which is clinically important. Endophytic fungi and their communication systems can be useful in research on anti-QS systems. Fungal communication (including endophytic fungi) can reveal different unseen phenomena in biochemical communication, which can be relevant in areas of biotechnology, biocomputing, medical sciences, and industry.

References

Albuquerque P, Casadevall A (2012) Quorum sensing in fungi—a review. Med Mycol 50(4):337–345

Alem MA, Oteef MD, Flowers TH, Douglas LJ (2006) Production of tyrosol by Candida albicans biofilms and its role in quorum sensing and biofilm development. Eukaryot Cell 5:1770–1779

Anand R, Rai N, Thattai M (2013) Interactions among quorum sensing inhibitors. PLoS One 8(4):e62254

Anguige K, King JR, Ward JP, Williams P (2004) Mathematical modelling of therapies targeted at bacterial quorum sensing. Math Biosci 192(1):39–83. https://doi.org/10.1016/j.mbs.2004.06.008

Anguige K, King JR, Ward JP (2005) Modelling antibiotic- and anti-quorum sensing treatment of a spatially-structured Pseudomonas aeruginosa population. J Math Biol 51(5):557–594. https://doi.org/10.1007/s00285-005-0316-8

Avbelj M, Zupan J, Kranjc L, Raspor P (2015) Quorum-sensing kinetics in Saccharomyces cerevisiae: a symphony of ARO genes and aromatic alcohols. J Agric Food Chem 63:8544–8550

Balagaddé FK, Song H, Ozaki J, Collins CH, Barnet M, Arnold FH, Quake SR, You L (2008) A synthetic Escherichia coli predator–prey ecosystem. Mol Syst Biol 4(1):187

Basu S, Gerchman Y, Collins CH, Arnold FH, Weiss R (2005) A synthetic multicellular system for programmed pattern formation. Nature 434(7037):1130–1134

Baym M, Lieberman TD, Kelsic ED, Chait R, Gross R, Yelin I, Kishony R (2016) Spatiotemporal microbial evolution on antibiotic landscapes. Science 353(6304):1147–1151

Beckmann BE, Knoester DB, Connelly BD, Waters CM, McKinley PK (2012) Evolution of resistance to quorum quenching in digital organisms. Artif Life 18(3):291–310

Brenner K, Karig DK, Weiss R, Arnold FH (2007) Engineered bidirectional communication mediates a consensus in a microbial biofilm consortium. Proc Natl Acad Sci U S A 104(44):17300–17304

Brock DA, Gomer RH (1999) A cell-counting factor regulating structure size in dictyostelium. Genes Dev 13:1960–1969

Cao YY, Cao YB, Xu Z et al (2005) cDNA microarray analysis of differential gene expression in Candida albicans biofilm exposed to farnesol. Antimicrob Agents Chemother 49:584–589

Chen H, Fink GR (2006) Feedback control of morphogenesis in fungi by aromatic alcohols. Genes Dev 20(9):1150–1161

Chen H, Fujita M, Feng Q, Clardy J, Fink GR (2004) Tyrosol is a quorum-sensing molecule in *Candida albicans*. Proc Natl Acad Sci U S A 101(14):5048–5052

Chen Y, Kim JK, Hirning AJ, Josić K, Bennett MR (2015) Emergent genetic oscillations in a synthetic microbial consortium. Science 349(6251):986–989

Danino T, Mondragón Palomino O, Tsimring L, Hasty J (2010) A synchronized quorum of genetic clocks. Nature 463(7279):326–330

Datla US, Mather WH, Chen S, Shoultz IW, Täuber UC, Jones CN, Butzin NC (2017) The spatiotemporal system dynamics of acquired resistance in an engineered microecology. Sci Rep 7(1):16071

Davis-Hanna A, Piispanen AE, Stateva LI, Hogan DA (2008) Farnesol and dodecanol effects on the Candida albicans Ras1-cAMP signalling pathway and the regulation of morphogenesis. Mol Microbiol 67:47–62

Deveau A, Piispanen AE, Jackson AA, Hogan DA (2010) Farnesol induces hydrogen peroxide resistance in Candida albicans yeast by inhibiting the Ras-cyclic AMP signaling pathway. Eukaryot Cell 9:569–577

Elowitz MB, Leibler S (2000) A synthetic oscillatory network of transcriptional regulators. Nature 403(6767):335–338

Fagerlind MG, Nilsson P, Harlén M, Karlsson S, Rice SA, Kjelleberg S (2005) Modeling the effect of acylated homoserine lactone antagonists in Pseudomonas aeruginosa. Biosystems 80(2):201–213

Figueroa M, Jarmusch AK, Raja HA, El-Elimat T, Kavanaugh JS, Horswill AR, Cooks RG, Cech NB, Oberlies NH (2014) Polyhydroxyanthraquinones as quorum sensing inhibitors from the guttates of Penicillium restrictum and their analysis by desorption electrospray ionization mass spectrometry. J Nat Prod 77(6):1351–1358

Gardner TS, Cantor CR, Collins JJ (2000) Construction of a genetic toggle switch in Escherichia coli. Nature 403(6767):339–342

Golé L, Rivière C, Hayakawa Y, Rieu JP (2011) A quorum-sensing factor in vegetative Dictyostelium discoideum cells revealed by quantitative migration analysis. PLoS One 6(11):e26901

Hall RA, Cottier F, Muhlschlegel FA (2009) Molecular networks in the fungal pathogen Candida albicans. Adv Appl Microbiol 67:191–212

Hornby JM, Jensen EC, Lisec AD, Tasto JJ, Jahnke B, Shoemaker R, Dussault P, Nickerson KW (2001) Quorum sensing in the dimorphic fungus Candida albicans is mediated by farnesol. Appl Environ Microbiol 67(7):2982–2992

Hornby JM, Jacobitz-Kizzier SM, McNeel DJ, Jensen EC, Treves DS, Nickerson KW (2004) Inoculum size effect in dimorphic fungi: extracellular control of yeast-mycelium dimorphism in Ceratocystis ulmi. Appl Environ Microbiol 70(3):1356–1359

Kebaara BW, Langford ML, Navarathna DH et al (2008) Candida albicans Tup1 is involved in farnesol-mediated inhibition of filamentous-growth induction. Eukaryot Cell 7:980–987

Kügler S, Sebghati TS, Eissenberg LG, Goldman WE (2000) Phenotypic variation and intracellular parasitism by Histoplasma capsulatum. Proc Natl Acad Sci U S A 97(16):8794–8798

Kusari P, Kusari S, Spiteller M, Kayser O (2015) Implications of endophyte-plant crosstalk in light of quorum responses for plant biotechnology. Appl Microbiol Biotechnol 99(13):5383–5390

Langford ML, Atkin AL, Nickerson KW (2009) Cellular interactions of farnesol, a quorum-sensing molecule produced by Candida albicans. Future Microbiol 4:1353–1362

Lee H, Chang YC, Nardone G, Kwon-Chung KJ (2007) TUP1 disruption in Cryptococcus neoformans uncovers a peptide-mediated density-dependent growth phenomenon that mimics quorum sensing. Mol Microbiol 64:591–601

Lohse MB, Gulati M, Johnson AD, Nobile CJ (2018) Development and regulation of single-and multi-species Candida albicans biofilms. Nat Rev Microbiol 16(1):19–31

Majumdar S, Mondal S (2016) Conversation game: talking bacteria. J Cell Commun Signal 10(4):331–335

Majumdar S, Pal S (2016) Quorum sensing: a quantum perspective. J Cell Commun Signal 10(3):173–175

Majumdar S, Pal S (2017a) Cross-species communication in bacterial world. J Cell Commun Signal 11(2):187–190
Majumdar S, Pal S (2017b) Bacterial intelligence: imitation games, time-sharing, and long-range quantum coherence. J Cell Commun Signal 11(3):281–284
Majumdar S, Pal S (2018) Information transmission in microbial and fungal communication: from classical to quantum. J Cell Commun Signal 12(2):491–502. https://doi.org/10.1007/s12079-018-0462-6
Majumdar S, Roy S (2017) Spatiotemporal patterns and chaos in non-equilibrium bacterial communication. In: 17th BIOMAT international symposium on mathematical and computational biology, Moscow, Russia
Majumdar S, Roy S (2018) Relevance of quantum mechanics in bacterial communication. Neuroquantology 16(3):1–6
Majumdar S, Datta S, Roy S (2012) Mathematical modelling of quorum sensing and bioluminescence in bacteria. Int J Adv Appl Sci 1(3):139–146
Majumdar S, Roy S, Llinas R (2017) Bacterial conversations and pattern formation. bioRxiv. https://doi.org/10.1101/098053
Martin-Rodriguez AJ, Reyes F, Martin J, Perez-Yepez J, Leon-Barrios M, Couttolenc A, Espinoza C, Trigos A, Martin VS, Norte M, Fernandez JJ (2014) Inhibition of bacterial quorum sensing by extracts from aquatic fungi: first report from marine endophytes. Mar Drugs 12:5503–5526. https://doi.org/10.3390/md12115503
Miller MB, Bassler BL (2001) Quorum sensing in bacteria. Annu Rev Microbiol 55:165–199
Rajesh PS, Rai VR (2013) Hydrolytic enzymes and quorum sensing inhibitors from endophytic fungi of Ventilago madraspatana Gaertn. Biocatal Agric Biotechnol 2(2):120–124
Ramage G, Saville SP, Wickes BL, Lopez-Ribot JL (2002) Inhibition of Candida albicans biofilm formation by farnesol, a quorum-sensing molecule. Appl Environ Microbiol 68:5459–5463
Roca MG, Arlt J, Jeffree CE, Read ND (2005) Cell biology of conidial anastomosis tubes in Neurospora crassa. Eukaryot Cell 4(5):911–919
Sato T, Watanabe T, Mikami T, Matsumoto T (2004) Farnesol, a morphogenetic autoregulatory substance in the dimorphic fungus Candida albicans, inhibits hyphae growth through suppression of a mitogen-activated protein kinase cascade. Biol Pharm Bull 27:751–752
Scott SR, Hasty J (2016) Quorum sensing communication modules for microbial consortia. ACS Synth Biol 5(9):969–977
Seeley TD, Visscher PK (2004) Group decision making in nest-site selection by honey bees. Apidologie (Celle) 35:101–116
Severin FF, Meer MV, Smirnova EA, Knorre DA, Skulachev VP (2008) Natural causes of programmed death of yeast Saccharomyces cerevisiae. Biochim Biophys Acta 1783(7):1350–1353
Shapiro JA (1998) Thinking about bacterial populations as multicellular organisms. Annu Rev Microbiol 52:81–104
Sharma M, Prasad R (2011) The quorum-sensing molecule farnesol is a modulator of drug efflux mediated by ABC multidrug transporters and synergizes with drugs in Candida albicans. Antimicrob Agents Chemother 55:4834–4843
Shou W, Ram S, Vilar JM (2007) Synthetic cooperation in engineered yeast populations. Proc Natl Acad Sci U S A 104(6):1877–1882
Song H, Payne S, Gray M, You L (2009) Spatiotemporal modulation of biodiversity in a synthetic chemical-mediated ecosystem. Nat Chem Biol 5(12):929–935
Sprague GF Jr, Winans SC (2006) Eukaryotes learn how to count: quorum sensing by yeast. Genes Dev 20:1045–1049
Stricker J, Cookson S, Bennett MR, Mather WH, Tsimring LS, Hasty J (2008) A fast, robust and tunable synthetic gene oscillator. Nature 456(7221):516–519
Viretta AU, Fussenegger M (2004) Modeling the quorum sensing regulatory network of human-pathogenic Pseudomonas aeruginosa. Biotechnol Prog 20:670–678. https://doi.org/10.1021/bp0343231

Ward J (2008) Mathematical modeling of quorum-sensing control in biofilms. In: Balaban N (ed) Control of biofilm infections by signal manipulation, vol. 2 of Springer Series on Biofilms. Springer, Berlin, pp 79–108

Westwater C, Balish E, Schofield DA (2005) Candida albicans-conditioned medium protects yeast cells from oxidative stress: a possible link between quorum sensing and oxidative stress resistance. Eukaryot Cell 4:1654–1661

Williams P, Winzer K, Chan WC, Camara M (2007) Look who's talking: communication and quorum sensing in the bacterial world. Philos Trans R Soc Lond Ser B Biol Sci 362(1483):1119–1134

You L, Cox RS III, Weiss R, Arnold FH (2004) Programmed population control by cell–cell communication and regulated killing. Nature 428(6985):868–871

Chapter 9
Endophytic Fungi: Role in Phosphate Solubilization

Preeti Mehta, Rashmi Sharma, Chayanika Putatunda, and Abhishek Walia

9.1 Introduction

Phosphorus (P) is the vital nutrient element for overall plant growth and productivity after nitrogen. Its structural and chemical properties decrease its free availability and make a limiting nutrient for plant growth (Mehta et al. 2013a). It is present in soils as organic and inorganic forms. Despite the large reserve of phosphorus in soil, its availability is greatly reduced by various fixation reactions occurring during biogeochemical cycling of this element (Martinazzo et al. 2007; Kumar et al. 2015). Only 75% of the externally added P is adsorbed by the plants, and this adsorption-precipitation phenomenon is strongly influenced by soil type and pH (Lin et al. 2006). In acidic soils, some metal oxides like iron and aluminium adsorb the free phosphorus and form iron and aluminium phosphates. Whereas in alkaline soils, calcium causes precipitation and results in poor solubilization of phosphatic fertilizers (Igual et al. 2001). Since phosphorus is the major nutrient element required by plants, this has led to the profound use of chemical-based phosphatic fertilizers by agriculturalists for optimum crop production (Sharma et al. 2015). However, employing chemical-based fertilizers have concerns associated with environment degradation and human health. An alternative approach to lessen the overall use and demand of chemical fertilizers is to exploit indigenous microflora, particularly those that are able to solubilize insoluble mineral elements in soil.

Pikovskaya (1948) reported insoluble phosphate solubilization by microorganisms. Various phosphate-solubilizing microbes have been reported in past three decades and numerous bacteria and fungi have been evaluated for their

P. Mehta
Centre of Advance Bioenergy Research, Research & Development Centre, Indian Oil Corporation Limited, Faridabad, India

R. Sharma · C. Putatunda · A. Walia (✉)
Department of Microbiology, DAV University, Jalandhar, India

B. P. Singh (ed.), *Advances in Endophytic Fungal Research*, Fungal Biology,
https://doi.org/10.1007/978-3-030-03589-1_9

P-solubilization potential (Sharma et al. 2017). Among these P-solubilizing microorganisms isolated and identified, the species of *Pseudomonas* and *Bacillus* in bacteria (Mehta et al. 2015) and *Aspergillus* and *Penicillium* among fungi were most prevalent (Wakelin et al. 2004). Mineral phosphate-solubilizing microorganisms are ubiquitous and have variable cell numbers in different soil that differ in their mineral phosphate-solubilizing (mps) ability from one medium to another. Qualitative estimation of P-solubilization in both bacterial and fungal strains can be done by measuring zone of solubilization on Pikovskaya's agar medium containing an insoluble P-source (Chauhan et al. 2014). However, detection of halo zone on agar medium is not an authentic test for P-solubilization; quantitative assays in liquid medium are required to be done. The viable microbial cells exhibiting P-solubilization potential are generally termed as microphos (Zaidi et al. 2009). The microbes showing significant P-solubilization potential under in vitro conditions are further examined under in planta before their mass multiplication as bioinoculants.

There is a sharp increase in the use of endophytic microorganisms as agricultural inoculants over a past few years. They promote plant growth; antagonize phytopathogens along with production of many industrial metabolites (Azevedo et al. 2000; Chauhan et al. 2016). In soil, although phosphate-solubilizing fungi constitute only 0.1–0.5% of total fungal populations but impart great benefits towards overall plant nutrition (Kucey 1983). Unlike bacteria the P-solubilization potential of fungal strains is not lost even upon repeated sub-culturing. Fungal hyphae can easily go over long distances in soil (Kucey 1983) and release more organic acids than bacteria (Venkateswarlu et al. 1984). Hence, they may be more efficient phosphate solubilizers in soils compared to indigenous rhizobacterial population. Among rhizospheric fungi the most common P-solubilizing strains are *Aspergillus*, *Penicillium*, *Trichoderma*, and *Rhizoctonia solani* (Wakelin et al. 2004; Sharma et al. 2013). A nematofungus *Arthrobotrys oligospora* was identified with a good ability to solubilize mineral phosphorus under in vitro as well as under *in planta* conditions (Duponnois et al. 2006).

In recent years, huge range of P-solubilizing endophytic fungi have been identified including the genera *Penicillium*, *Aspergillus*, *Piriformospora*, *Curvularia*, and other class of endophytic symbionts arbuscular mycorrhizal (AM) fungi (Khan and Khan 2002). Symbiotic root colonizers, AM fungi support plant mineral nutrition in exchange for photosynthetic carbon (Smith and Read 2008). The major AM fungi belong to genera *Glomus, Gigaspora, Acaulospora, Entrophospora, Sclerocystis*, and *Scutellospora*. Gallaud (1905) has described two types of morphologically variant symbionts of AM fungi named Arum and Paris. Arum is characterized by hyphal distribution throughout the cortex, then spreading into apoplast and formation of specialized structures called arbuscules. Whereas the Paris type include extensive network of intracellular mycelium with formation of coils that include extensive intracellular mycelial growth, forming coils that may or may not be interconnected with arbuscules. Plant growth promotion by AM fungi is mainly associated with their ability to solubilizing and mobilizing phosphorus from soil. Besides AM fungi, a lot other endophytic fungi are able to colonize and penetrate the plant roots and impact plant performances (Harman et al. 2004). Their impact on plants may be positive or negative depending on the fungal partner and host interaction and soil physico-chemical properties (Mayerhofer et al. 2012). Root endophytic fungi have been well characterized for their association and growth promotion of broad varieties of plants (Verma et al. 1998; Varma et al. 2001; Deshmukh et al. 2006). Growth promotion in such endophytic fungal colonized plants has been primarily linked to their contribution towards the plant P nutrition (Neubert et al. 2006; Franken 2012; Vitorino et al. 2016).

Given the beneficial effect of endophytic fungi on plant P nutrition, the question that whether these non-pathogenic endophytic fungi flourish due to similar or some other P-solubilization strategies has led to our interest in the area. Here, we have integrated this concept and incorporated the role of root associated fungi and AM fungi in plant P nutrition. This chapter gives detailed insight into endophytic phosphate-solubilizing fungi, their application, rhizosphere colonization, and mechanism of P-solubilization.

9.2 Endophytic Fungi vs. Mycorrhizal Fungi in Plant P Nutrition

Rhizosphere is a warfield where microbes are under the influence of plant roots. Roots are the key sites for uptake of mineral elements and exudation of organic compounds, those act as carbon and energy sources for the indigenous microflora. There is a higher microbial cell count in rhizosphere than in the bulk soil that form "rhizosphere effect" (Hinsinger et al. 2009). Plant–microbe interactions in the rhizosphere are critical for regulating biogeochemical recycling of mineral elements and maintaining the microbial community structure in the rhizosphere (Singh et al. 2007). In several cases, plant–microbe interactions are evolved in such a way that some fungi appear to live non-pathologically inside plant roots as endophytes and some form symbiotic relationship with roots called mycorrhizae (Gehring et al. 2006). These microbes have been considered to originate in the rhizosphere and then enter into roots through natural openings, wounds, and at the site of seedling emergence (Gaiero et al. 2013). Colonization of plant roots by these endophytic microbes may be local or systemic or both inter and intracellular. The endophytic microenvironment provides a certain degree of protection from environmental stress factors to the inhabiting fungi and to be biologically efficient inoculants, the endophytes must be able to colonize root environment more vigorously and efficiently.

Endophytic microbes grow in and throughout the host plant tissues. There they release certain chemicals that impart disease resistance and aids in the mineral nutrition of host plant. Earlier, phosphorus mobilization was considered a characteristic of mycorrhizal fungi only. However, a number of fungi that grow in plant tissues are not mycorrhizal. The term "endophyte" represents all the microbes that grow inside the plant tissues without posing any damage to the host (Chanway 1996). Certain endophytic fungi other than mycorrhizae are able to solubilize P and provide it to their non-mycorrhizal partner, promoting its growth under low-P conditions (Harman et al. 2004; Mandyam et al. 2010; Radhakrishnan et al. 2015). Among these non-mycorrhizal endophytic fungi are the forms that release conidia and microsclerotia inside host tissues and possess melainized, dark septate hyphae. These endophytic fungal forms are called dark septate endophytes (DSE). They improve overall plant growth and health under environmental stress (Khan et al. 2015), mineralize organic compounds in soil (Mandyam et al. 2010) besides their important role in solubilizing insoluble nutrients (particularly phosphorus). A dark septate root endophytic fungus *Curvularia geniculata* isolated from *Parthenium hysterophorus* roots was known to improve plant growth through P-solubilization and phytohormone production (Priyadharsini and Muthukumar 2017).

Mycorrhizae, on the other hand, are much more specific. This interaction implies a true symbiotic relationship between smaller fungal partner in soil and the larger partner the roots of plant. Mycorrhizae are categorized into different types depending upon the plant and fungus relationship and their structural features. Distinctions are made on the basis if the hyphae colonizing the plant penetrate root cells (endomycorrhizae) or grow on or between root cells (ectomycorrhizae) or both (ectendomycorrhizae). The host gets benefit in such relationship in many ways: by more pronounce extension of root system for maximum uptake of mineral nutrients and water, resistance to drought stress, prevention from disease attack, etc. (Smith and Read 2008). Still it is unclear that who gets more benefit from whom in each of these relationships (i.e., mycorrhizae and endophytic relationships), so there is some dispute as to the degree of parasitism in both cases. "Mycorrhizae" are the fungi that have a mutualistic association with plant roots. While "endophytes" are the fungal partners those live within above-ground healthy plant parts and don't harm it. Among vast array of benefits provided by AMF fungi, the most significant one is to improve phosphorus nutrition of the host plant with low phosphate levels that is achieved by the large surface area of their hyphae and their high affinity P mobilization mechanisms. Previous studies have highlighted the significant role of mycorrhizal fungi in fulfilling plant P demand. This is evident from the fact that the P-solubilization zone around the roots not colonized with mycorrhizal fungi spans to only 1–2 mm, whereas the extended hyphae of AM fungi extends to around 8 cm or more beyond the root hair surface making the phosphorus from more volume of soil available to the plant (van der Heijden et al. 2006).

Quite a lot of mycorrhizal and endophytic fungi are able to mineralize organic phosphorus and solubilize inorganic phosphorus. While merits of P-solubilizing endophytic fungi has been concerned, they are more competitive than non-endophytes since the endophytes have gained the ability to invade host cells during evolutionary process, and this interaction is regulated by genes of both the partners (Rosenblueth and Martínez-Romero 2006). Moreover, the endophytic microbes have the ability to colonize plant cells without triggering host defense response. Thus, the distinction between the bulk soil microbial population, the rhizospheric and the endosymbiotic microbial population of a host plant may represent a true continuum, where microbes are able to colonize between the soil particles, the rhizosphere, and inside the root environment (Farrar et al. 2014).

9.3 P-Solubilizing Fungus as Plant Growth Promoter

P-solubilizing fungi are known to promote plant growth by an array of mechanisms those include (1) providing macronutrients by solubilization/mobilization reactions, (2) produce biocontrol active metabolites, i.e., siderophores and antibiotics, (3) phytoprotection against various phytopathogens (biocontrol), and (4) production of growth hormones, i.e., auxins and gibberellins (Fig. 9.1).

The P-solubilizing fungus is being used by many agronomists as soil inoculants for enhancement of plant growth (Khan et al. 2010). Good quality of microbial inoculants is the basic need to achieve high productivity in a sustainable way. The fungi showing greater solubilization of phosphorus under laboratory conditions are selected for bulk inoculum production and ultimately supplied to the farmers. After proper selection of carrier and development of fungal inoculants, they are then tested for persistence of P-solubilization activity, fungal count/g of carrier, and are then stored for about 3 months at 28 °C. Few examples of commercial available inoculums, i.e., *Penicillium bilaiae* (JumpStart; Philom Bios, Saskatoon, Canada) and *Penicillium radicum* (PR-70 RELEASE; Bio-Care Technology, Somersby, Australia) have been demonstrated in large scale and showed an increase in plant P nutrition. Different methods used for production, multiplication, and application of P-solubilizing fungi have been depicted in Fig. 9.2.

Different studies showed the effect of monoculture and/or composite culture of P-solubilizing fungus on various plant growth parameters (Zaidi et al. 2003; Zaidi and Khan 2007). Mittal et al. (2008) observed the effect of composite inoculation of

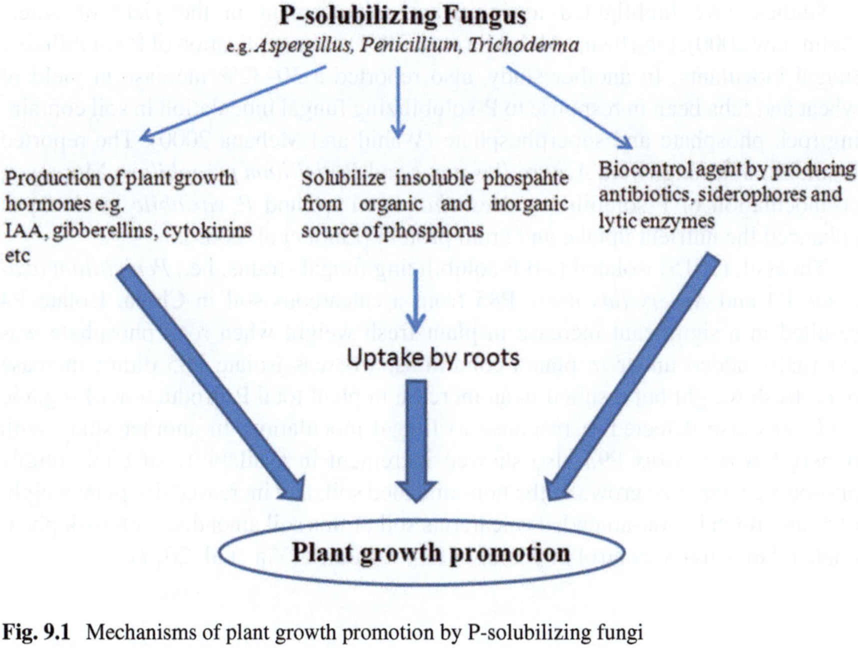

Fig. 9.1 Mechanisms of plant growth promotion by P-solubilizing fungi

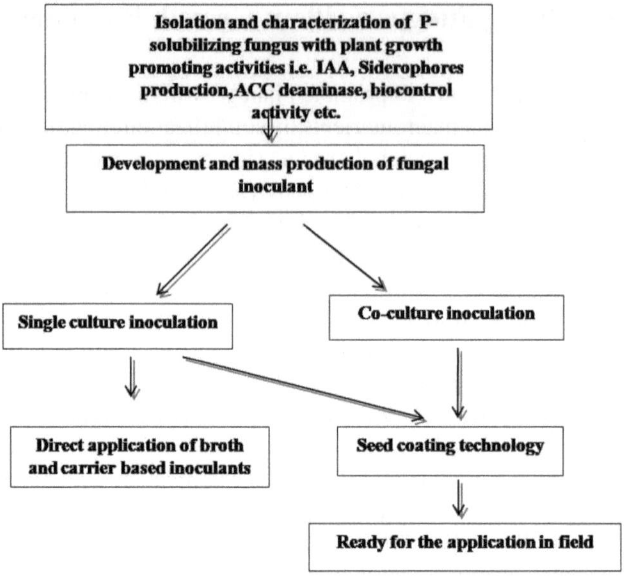

Fig. 9.2 Production and application of phosphate-solubilizing fungal inoculants

P-solubilizing fungal strains (two strains of *A. awamori*, and four of *P. citrinum*), on growth and seed production of chickpea in pot experiments. All isolates were showing synergistic effect and resulted in a significant enhancement of plant height, seed number, and weight compared to the uninoculated control.

Studies have highlighted a significant enhancement in the yield of wheat (Whitelaw 2000), faba bean (Abd-Alla et al. 2001) after application of P-solubilizing fungal inoculants. In another study, also reported a 30–32% increase in yield of wheat and faba bean in response to P-solubilizing fungal inoculation in soil containing rock phosphate and superphosphate (Wahid and Mehana 2000). The reported P-solubilizing fungi were *Aspergillus niger* and *Penicillium pinophilum*. Moreover, co-inoculation of P-solubilizing *Mesorhizobium* sp. and *P. variabile* in chickpea enhanced the nutrient uptake and grain protein (Zaidi et al. 2003).

Yin et al. (2015) isolated two P-solubilizing fungal strains, i.e., *Penicillium oxalicum* P4 and *Aspergillus niger* P85 from a calcareous soil in China. Isolate P4 resulted in a significant increase in plant fresh weight when rock phosphate was externally added under in planta conditions, whereas isolate P85 didn't increase plant fresh weight but resulted in an increase in plant total P. Production of organic acids was also detected in response to fungal inoculation. In another study with *Aspergillus aculeatus* P93 also showed increment in availability of total soluble phosphorus of maize grown in the non-amended soil, and increased dry plant weight of plant grown in non-amended calcareous soil or this soil amended with rock phosphate relative to the control in greenhouse pot studies (Yin et al. 2017).

A P-solubilizing fungi *Lecanicillium psalliotae* strain IISR-EPF-02 significantly increased plant growth parameters and leaf chlorophyll content of cardamom compared to uninoculated control (Senthil Kumar et al. 2018). It was concluded that the fungal strain exhibited the production IAA, ammonia, siderophores, and various hydrolytic enzymes in addition to P-solubilization which directly or indirectly enhanced the plant growth. In pot culture experiment with semi-sterile soil, the inoculation with *Penicillium* aculeatum significantly increased the shoot biomass and P content of wheat in biochar-amended treatments. These results led to the development of novel bioinoculants containing phosphate-solubilizing *Penicillium* fungi to increase the fertility value of P-rich biochar (Efthymiou et al. 2018).

9.4 Mechanism of Fungi for P-Solubilization

P-solubilizing fungus employs different mechanisms of P-solubilization based on the organic and inorganic type of P-source present in the soil. Thus, solubilization mechanisms can be categorized as inorganic and organic P-solubilization. Phosphate-solubilizing fungi are known to synthesize various organic acids protons, hydroxyl ions, carbon monoxide, siderophores, exopolysaccharides, and extracellular enzymes that have significant role in solubilization of insoluble phosphorus in soil (Fig. 9.3).

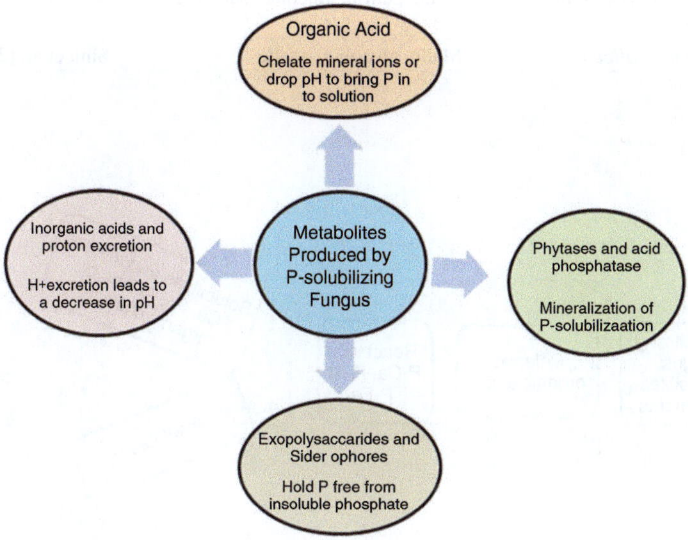

Fig. 9.3 Flow chart shows the metabolites produced by P-solubilizing fungi in different mechanisms of solubilization

9.4.1 Inorganic P-Solubilization

Secretion of various organic acids in soil solution seems to be most potent mechanism for inorganic P-solubilization. Primarily, organic acids like gluconic acid, maleic, acetic, citric, fumaric, glycolic, lactic, oxalic, propionic, succinic and tartaric, etc. (Table 9.1) (Ahmad and Shahab 2011). Further, secretion of organic acids by P-solubilizing fungus has also been well reported (Park et al. 2009; Kumar and Rai 2015). Among all, gluconic acid is most well known for mineralizing insoluble phosphate solubilization (Khan et al. 2010; Mehta et al. 2013b). The amount of organic acid varies with the producer fungal strain but also the type of organic acid produced greatly depends upon the insoluble P-source. Mendes et al. (2013) showed that *Aspergillus niger* FS1 primarily secrete oxalic acid with a major peak in treatments with $AlPO_4$ and $FePO_4$, whereas gluconic acid was produced in very less quantity in $AlPO_4$. On the contrary to this, *Penicillium canescens* FS23 produced gluconic acid and citric acid after treatments with $AlPO_4$, $Ca_3(PO4)_2$, and rock phosphate (Fig. 9.4).

Table 9.1 Prominent organic acids produced by P-solubilizing fungus

P-solubilizing fungus	Organic acids produced	References
Aspergillus flavus, A. candidus, A. niger, A. terreus, A. wentii, Fusarium oxysporum, Penicillium sp., *Trichoderma isridae, Ttrichoderma* sp.	Lactic, maleic, malic, Acetic, tartaric, citric, fumaric, Gluconic	Akintokun et al. (2007)
Aspergillus flavus, A. niger, P. canescens	Oxalic, citric, gluconic, succinic	Maliha et al. (2004)
Penicillium oxalicum	Malic, gluconic, oxalic	Shin et al. (2006)

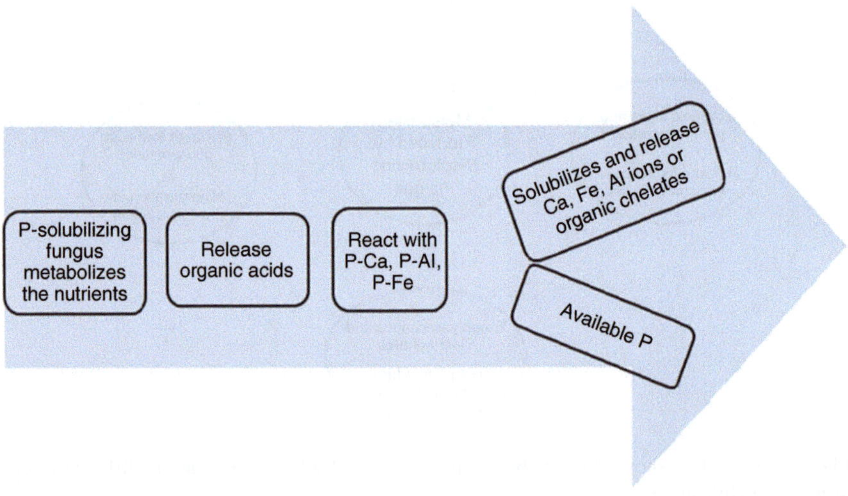

Fig. 9.4 Flow chart shows the P-solubilization by releasing organic acid as metabolites

P-solubilizing fungi release organic acid from the outer surface of cytoplasmic membrane which is the site of direct oxidation pathway. Organic acid forms a complex and chelate cations linked with P which releases soluble P into the soil through ligand exchange reactions with phosphate anion (Bolan et al. 1996). The efficiency of a given organic acid is related to its chemical characters, including number and placement of the carboxyl and hydroxyl groups, the stability constant of the metal–organic acid complex, types and quantity of organic acid as well as metal and pH of the soil solution (Kumar 2016) [Table 9.1]. The behavior of organic acid production by fungal isolates is their individual character and is also influenced by the culture conditions (Reyes et al. 2006).

As the concentration of free mineral phosphate and drop in pH of the solution do not always show positive correlation so release of organic acids could not be assumed as the only route for solubilization of inorganic phosphate (Gulati et al. 2008; Pei-Xiang et al. 2012). Some researchers believe that apart or along with organic acid production, some microorganisms release protons to the outer surface with the help of proton translocation ATPase accompanying NH_4^+ assimilation or respiration that promotes P-solubilization (Krishnaraj and Goldstein 2001). Organic acids were found to be absent in supernatant after P-solubilization by *Penicillium rugulosum* (Illmer and Schinner 1995). The authors suggested the release of protons that might be the possible means of phosphate solubilization in supernatant. *Trichoderma harzianum* Rifai has not produced any known organic acid but solubilized P by chelating and reducing molecules (Altomare et al. 1999).

Additionally, the other means of P-solubilization by fungi include the production of inorganic acids such as sulfuric, nitric, and carbonic acids, release of siderophores and exopolysaccharides (Yi et al. 2008; Sharma et al. 2013). Although, their role and mechanism are not yet much clear. Literature has well described the role of siderophores in P mineralization by phosphate-solubilizing microbes (Parker et al. 2005). On the other hand, releases of carbohydrate polymers (exopolysaccharides) have also been reported. The role of these carbohydrate polymers in solubilization of phosphate has been documented by Yi et al. (2008).

As can be seen from the above account, majority of the studies are relying on the organic acid concept of P-solubilization. Many genes are responsible for the production of organic acids. Among P-solubilizing fungus, most of the scientist considered gluconic acid as major organic acid. Goldstein and Liu (1987) studied the genetic basis of P-solubilization and observed the upregulation of glucose dehydrogenase (GDH) and pyrroloquinoline quinone (PQQ)(4,5-dihydro-4,5-dioxo-1H-pyrrolo-[2,3-] quinoline-2,7,9 tricarboxylic acid) genes in the P-solubilizing organisms. PQQ-dependent glucose dehydrogenase enzyme spans around cytoplasmic membrane and led to oxidation of glucose to gluconic acid. Due to the production of gluconic acid soil pH decrease which makes available the divalent and trivalent (HPO_4^{2-} and HPO_4^{3-}) soluble forms of phosphorus.

9.4.2 *Organic P-Solubilization*

Organic P mineralization is carried out with the help of different enzymes, i.e., phosphatases, phytases, and phosphonatases. Phosphatases carried out dephosphorylation or hydrolyze organic phospho-ester and phosphoanhydride bonds of organic matter. Among the phosphatases, phosphomonoesterases are the most abundant class (Nannipieri et al. 2011). Depending upon the pH optima, they are known as acid or alkaline phosphatase (Behera et al. 2014). Several genes encoding for acid and alkaline phosphatases with broad substrate specificity have been cloned and characterized (Nilgiriwala et al. 2008). However, a significant concentration of phosphatases is released by plant roots, but it has been reported that microbial phosphatases have greater affinity for substrate compared to plant-derived phosphatases (Chen et al. 2006).

Phosphorus is stored in plant seeds and pollen in the form of phytate. Phytate is naturally degraded by phytases which is a prime source of inositol phosphate and constitute about 50% or more of organic phosphate in soil (Rodríguez et al. 2006). Phytases hydrolyze phytate and make available free form of phosphate. Phosphonatases and C-P lyases hydrolyze C-P bond of organophosphates and release free phosphorus (Rodríguez et al. 2006). But due to the less availability, organophosphatases do not contribute much to the freely available phosphate in the soil solution.

It has been well documented that the concentration of organic acids released by P-solubilizing fungi is far greater than bacteria therefore exhibit greater P-solubilization activity. Srivastav et al. (2004) reported significantly higher solubilization of rock phosphate under in vitro conditions by fungal strains of *Aspergillus niger*, *Curvularia lunata*, *Rhizoctonia solani* and *Fusarium oxysporum*. In another study by Yadav and Tarafdar (2003), *Aspergillus*, *Emmericella*, and *Penicillium*, isolated from arid and semi-arid regions of India lysed organic phosphates like phytin and glycerophosphate.

9.5 Development of Phosphate-Solubilizing Fungal Inoculants

P-solubilizing microbes have been used for production of biofertilizers. These bioinoculants provide an eco-friendly substitute to the chemical fertilizers. Fungi have been reported to show higher phosphate solubilization and hence are gaining more importance especially from the agronomists (Khan et al. 2007). Several fungal biofertilizers have already been developed. For example, Indian Agricultural Research Institute (IARI) has developed a fungal bioinoculant comprising of *Aspergillus awamori, Aspergillus niger*, and *Penicillium digitatum*. Similarly, a biofertilizer containing *P. bilaii* has been commercialized by Novozymes Biologicals Ltd. (Canada) and a *P. radicum* based P-biofertilizer has been developed by Bio-Care

Technology (Australia) (Gupta and Rodriguez-Couto 2018). Similarly, Ambika Biotech & Agro Services, Madhya Pradesh, India produces a P-biofertilizer under the trade name of Ambiphos containing *A. niger* (Pal et al. 2015).

In order to produce phosphate-solubilizing fungal inoculants, there is requirement of bulk production of the desired fungal strains. The development of a biofertilizer can be broadly divided into three phases. The first phase includes the screening of a variety of potential phosphate solubilizers and followed by assessment and selection of fungal inoculants. The screening and selection is generally carried out with the help of media like Pikovskaya's medium (Pikovskaya 1948), modified Pikovskaya's Medium, NBRIP *medium* which contains tricalcium sulfate as the source of insoluble phosphorus (Chatli et al. 2008). The second phase includes the development of a proper fungal biofertilizer, and the third phase includes the quality checking of the biofertilizer as well as assessing the persistence of the PSMs and dissemination to the farmers (Khan et al. 2010). For production of the biofertilizer, the selected microbe is cultivated at a large scale under appropriate conditions in a suitable fermentation broth. For cultivation of fungal cultures, lower pH is generally preferred as acidic conditions promote fungal growth and the same time inhibits most of the bacterial contaminants (Nelofer et al. 2016). When the desired level of growth is attained, the biomass is usually mixed with generally pre-sterilized carrier material (in case of carrier-based biofertilizers), packed and usually stored for some time period under appropriate conditions, before being marketed. For example, the P-biofertilizers developed by IARI are packaged in polybags and stored for approx. Three months at 25 ± 2 °C. At each stage of the biofertilizer production, a strict quality checking regime is followed to assess the load of desired microbe as well as to check the presence of any contaminant.

One of the major drawbacks associated with microbial biofertilizers is the fact that often the added microbes are unable to survive under in vivo conditions. This may be due to the fact that the bioinoculants are either unable to cope up with the harsh environmental conditions or are outcompeted by the already existing microflora (Van Elsas and Van Overbeek 1993; Rodriguez-Navarro et al. 1991; Heijnen et al. 1992; Jones and Burges 1998; Walia et al. 2017). One of the possible strategies is to overcome this problem is to produce carrier-based biofertilizer. Kundu and Gaur (1981) have suggested peat, farmyard compost (FYM), soil and dairy animals waste cake powder as suitable carriers for P-based fertilizers.

An ideal carrier is desired to possess certain characteristics like it should be able to sustain the microbial and survival by maintaining the adequate level of moisture, pH, aeration etc. Thus, the carrier should have good moisture absorption capability, porousness, and pH buffering capacity. Also, the carrier should be eco-friendly and non-toxic to microbes, plants, animals, and humans. Apart from this, it is also expected to be sterlizable, and easy to handle, mix, and store. For keeping effective cost-benefit ratio, the carrier should also be cheap and easily available (Stephens and Rask 2000; Rebah et al. 2002; Rivera-Cruz et al. 2008). Several different carriers have been used for biofertilizer production. Smith (1995) has classified them into the following categories. The first group includes various types of soils, inorganic soil as well as coal and peat. The second group comprises of plant waste materials: spent mushroom manure, barnyard compost, soybean and shelled nut oil press-mud, etc. The last category contains various inert materials like vermiculite, perlite, ground rock phosphate, calcium sulfate, etc. These carriers have also been used in conjunction with each other by many workers. Wang et al. (2015) assessed the utilization of different carriers for developing a biofertilizer of *A. niger* 1107 and reported a mixture of wheat husk and 20% perlite to result into increased available phosphorus content.

Another very significant type of fungi which play significant role in the phosphorus nutrition of plants includes the arbuscular mycorrhizal fungi (AMF). These fungi not only are important in nutrient and water uptake of plants but also provide a host of other benefits to the host plant like disease resistance, tolerance to harsh environmental conditions, etc. (Pal et al. 2014). The production of AM-based biofertilizer presents an inherent difficulty because of the reason that these fungi are obligatory symbiotic. This behavior of the fungi makes it practically impossible to carry out their mass scale production using synthetic media (Pal et al. 2015). Berruti et al. (2016) have reported three strategies for utilization of AMF-based biofertilizer. As per the first strategy, root zone soil from the plant harboring AMF can be used as bio-inoculum. However, this is not a very reliable method and may result in generation of inefficient inoculants which may even carry propagules of weeds and pathogens. A better strategy involves isolation of AM spores and using them for generation of AM crude inoculum. For this purpose known AM fungal isolate is grown on a host trap plant usually in an inert medium. The trap plant is a host plant which is susceptible to massive AM fungal growth and hence is used for large-scale production of AM fungal biomass. This is the most commonly used type of inoculum for large-scale crop inoculation as it usually contains a more concentrated set of the same kind of propagules found in soil inocula. The third strategy involves the use of infected root fragments of a known AMF host obtained from a trap plant culture as a source of inoculum.

An ideal trap plant should not only be highly conducive to the desired AM fungus but ideally should also produce large-scale propagules of only the desired fungus. Also, the tap plant should show profuse root development within a shorter time span and tolerate the environmental conditions suitable for rapid fungal proliferation (Sadhana 2014). Suitable trap plants should be screened to ensure that mass scale propagation of AM fungi (Bagyaraj and Manjunath 1980). Sreenivasa and Bagyaraj (1988) reported Rhodes grass (*Chlorisgayana*) to be the best host for mass multiplication of *Glomus fasciculatum* while Selvakumar et al. (2016) reported maize to be a better host trap plant as compared to sudan grass for the propagation of *Claroideoglomus etunicatum.* Several trap plants including *Sorghum bicolor var. sudanense*, *Paspalum notatum*, *Ipomea batatas, Trifolium subterraneum*, *Fragaria* sp., *Sorghum vulgare*, *Zea mays*, *Allium cepa, Chloris gayana, Coleus* sp. etc. have been used for the purpose of mass scale culturing of various AM fungi (Sadhana 2014; Sreenivasa and Bagyaraj 1988; Hung and Sylvia 1998).

The AM fungi can be isolated from soil by wet sieving method and decantation (Singh et al. 2010) followed by microscopic observation. Large number of spores are selected and mass multiplication is undertaken by pot culture technique

(Wood 1984). Conventionally, the host trap plants and the AM fungi have been cultivated in solid growth medium containing soil, sand, peat, vermiculite, perlite, clay, various types of composted barks or their combinations (Tiwari and Adholeya 2002). As per the method described by Tamil Nadu Agricultural University (http://agritech.tnau.ac.in/org_farm/orgfarm_biofertilizers.html#mass), a trench (1 m × 1 m × 0.3 m) lined with polythene sheet to be used as a plant growth tub or pot. The trench is filled up to 20 cm height with 50 kg of vermiculite and 5 kg of sterilized soil. To this, 1 kg of AM inoculum is applied 2–5 cm below the surface of vermiculite. The surface sterilized seeds of trap plant (maize) are sown in the trench along with appropriate dosage urea, superphosphate, and muriate. The plants are grown till 2 months with quality testing in between. After the period of 2 months, the roots of the trap plants are cut and the mixture of vermiculite, spores, pieces of hyphae and infected root pieces thus obtained is used as AM fungal bioinoculant. Apart from this approach, some workers have also assessed soil-less culture system of hydroponics and aeroponics for cultivation of AM fungi-trap crops. The greatest advantage of these methods lies in the fact that they result into the production of pure clean spores (IJdo et al. 2011).

9.6 Mode of Application of Phosphate-Solubilizing Fungi

The bioinoculant is most commonly applied by treating the seed surface with the desired microbial inoculant prior to seeding. In fact, this has been reported to be the most common and popularly used method (Walia et al. 2013b; Khan et al. 2007). There are some few techniques which are extensively used for the delivery of endophytic microorganism's viz. seed treatment, soil application, bacterization of plant material, and even foliar spraying. In seed treatment process, carrier-based fungal inoculants are used for dipping the seeds into the liquid culture medium. In this method, a fungus attached sturdily to the surface of seed and confirms that each seed obtains the inoculated fungal culture. However, there may be some constraints in this method like (1) sometimes the amount of viable fungi onto the seed surface may not be sufficient, and (2) the survivability of inoculated strains adversely affected when seeds come in direct contact with applied seed chemicals. However, the usage of certain adhesive solutions like that of gum arabic, jaggery, and charcoal has been proven to increase the attachment property of the bioinoculant onto the seeds.

In case of those plant species which are cultivated commercially by vegetative propagation, the bioinoculant is usually applied onto the moist plant parts prior to their planting in the field (Panhwar et al. 2013). The shoots of such plants have been reported to be quite amenable towards bacterization by endophytic microorganisms. Soil application method is another mode of inoculant application involving direct supplementation of the soil with microbial culture. There are some inherent benefits associated with this technique, which include: (1) possibility of a higher population of the desired P-solubilizing fungi per unit area, (2) minimization of interactions between the bioinoculant and chemicals associated with chemically treated seeds, thus reducing chances of any adverse effect on the microbes, (3) in comparison to seed inoculation technique, this method is more rapid as since the stage of mixing the seeds with inoculants is not required, (4) inoculants can tolerate drier and desiccated conditions better as compared to carrier-based inoculants (Khan et al. 2010). In view of these aspects, two methodologies can be adopted for microbial inoculants applications: (1) the monoculture approach (MCA) which involves utilization of a single culture of P-solubilizing fungi, and (2) the co-culture or multiple culture approach (CCA), implicating the use of two or more different P-solubilizing fungal strains, which in turn are mixed together and then applied under natural field/pot house conditions.

9.7 Inoculation Effects of Phosphate-Solubilizing Fungi

Soil is a vast habitat for a multifarious community of intricately related naturally occurring soil microorganisms, whose activities essentially impacts the physico-chemical properties of the soil. Soil microorganisms including fungi execute several important roles in soil leading to direct as well as indirect effect. In the direct mechanism, microbes play a very crucial part in plant growth promotion through production of phytohormones and mineral nutrition (Mehta et al. 2014; Guleria et al. 2014) and in indirect mechanism it plays a role in biocontrol of phytopathogenic microbes (Walia et al. 2013a; Mehta et al. 2011). Microbes live in close association with plants starting from seed germination to plant maturity (Lynch 1983). Therefore, the rhizosphere which inhabits complex and different microbial communities with adequate P-solubilizing activities becomes an important member of soil habitat. There are several reports who clearly indicated a considerable enhancement in plant growth due to mono- or co-culture inoculation of rhizospheric microorganisms (Mehta et al. 2013c; Zaidi and Khan 2007). One report on growth promotion in groundnut was demonstrated by Malviya et al. (2011). They assessed the effectiveness of two fungi, i.e., *Aspergillus niger* and *Penicillium notatum* in tri-calcium phosphate (TCP) amended soils under pot culture conditions and reported that inoculation with fungal strains improved the height and dry weight of the plant. When dual culture of fungal strains were used for inoculation, then it substantially improved the height (81%) as well as plant dry weight (105%), respectively, as compared to control uninoculated TCP soil (Prasanna et al. 2011; Malviya et al. 2011). A noteworthy augmentation in number of plants and the weight of 50 seeds were also observed with the application of single or dual inoculation of the fungal strains. In an another experiment performed by Priyadharsini and Muthukumar in 2017 on pigeon pea where inoculation of fungus *Curvularia geniculata* positively impacted various plant growth parameters. *C. geniculata* inoculated seedlings of pigeon pea were taller (26.53%), showed greater shoot dry weight as well as root dry weight (16.67% and 33.33%), had greater number of leaves (21.33%), and also possessed greater total root lengths (15.82%) as compared to uninoculated control seedlings. Apart from phosphate solubilization, the plant growth promoting activity of endophytic fungi is also attributed to phytohormone production (Chanclud and Morel 2016). Thus, the utilization of P-solubilizing fungi is suggested as an eco-friendly and sustainable approach for enhancing crop yield in a better way, under all types of experimental conditions. There are so many reports that had indicated the enhancement in plant growth using P-solubilizing fungi.

9.8 Factors Affecting the Survival of Fungal Strains and Phosphate Solubilization

Environmental changes affect the survival of fungal strains in soil and microorganisms have to adapt to these new conditions with time. Plant–microbe ecology is an intricately interconnected system and is always exposed to biotic and abiotic factors which in turn impact their growth, interactions as well as development. Thus, it is significant not only from the ecological point of view but also from the economical aspect as these factors greatly influence the development of root and hence crop production and yield. As a result, we require a much better comprehension of how these varied factors impact the fungal populations in soil environments. Additionally, it is very important to ascertain the rate of these variations which occurs in the environment and how these factors influence the fungal population and their phosphate-solubilizing activity. In this perspective, different factors have been proposed that impact the constitution, structure, and roles of fungal population in soil environments. Based on different plant species, there are so many studies that showed that the quantity as well as the make-up or composition of exudates released by different plants impacts the rhizosphere population. These exudates contain a wide range of molecules and incorporate on the one hand certain simple substrates like sugars, phenolics, amino acids, organic acids, and certain secondary metabolites and higher molecular weight compounds like proteins and mucilage on the other hand. Thus, the comprehending interactions among microbe as well as with plants along with factors impacting the longevity, survivability, and residual activity of the bioinoculants in soils and could result into better guessing the suitability and activity of the microbial inoculants in various agro-ecological regions of the world. The age and life stage of plant age, their genotypes, composition, and structure of root exudates, and the presence as well as concentration of environmental pollutants like pesticides and heavy metals are the important factors which affect the colonization, development, and survivability of the inoculated fungal strain in soil. However, use of molecular approaches in phosphate-solubilizing fungi, we could genetically manipulate the fungi to increase their ability/host range to maximize plant growth by cloning those genes which are directly involved in phosphate solubilization and other plant growth promoting traits (Krishnaraj and Goldstein 2001).

9.9 Techniques to Detect Colonization

For developing an endophytic fungal biofertilizer, it is imperative that one should be able to assess the colonization of plants by these microbes. The colonization of *Curvularia geniculata* was assessed by microscopic visualization of the host plant (*Parthenium hysterophorus*) roots stained with tryptan blue (0.05%) in lactoglycerol was carried out by Priyadharsini and Muthukumar (2017).

Conventionally, microscopy-based techniques have been used to perceive and quantify AMF in the host plant roots. Such techniques include observations of key fungal structures in living host plant roots, non-vital staining methods, and vital root staining methods (Vierheilig et al. 2005). Many workers like Schmitz et al. (1991), Frey et al. (1992), and Bothe et al. (1994) have suggested the use of biochemical methods relying on the quantification of characteristic AM fungal markers. Also, nowadays many molecular methods have been developed (Sanders 2002). For example, Alkan et al. (2004) proposed a real-time PCR-based method for assessment of *Glomus intraradices* in host plant tissue. The non-destructive methods like observation of stained or unstained living roots, epifluorescence microscopy are relatively simple but have not been found to be very reliable for quantitative estimation (Vierheilig et al. 2001, 2005).

Most of the microscopic techniques used for visualization of AM harboring roots rely on an initial stage of clearing the roots. This step allows removal of the unwanted plant cell components thus making the roots more transparent and hence allowing the development of a clear image of AM fungi and its components. Several methods have been suggested for carrying out root clearing. The methods include autoclaving of samples (Bevege 1968; Deguchi et al. 2017), heating in a water bath or hot air oven (60–90 °C), bleaching with alkaline hydrogen peroxide (Bevege 1968, Kormanik and McGraw 1982), and heating with KOH. Most commonly, the hot alkali treatment (heating with KOH) is carried out. This process is usually followed by rinsing the roots with water several times followed by acidification with dilute acid (like HCl). Some workers have also suggested sectioning of root samples to ensure better visualization Dickson et al. (2003).

The root samples thus prepared are subjected to staining. Various non-vital dyes like acid fuchsin in lactoglycerol (Kormanik and McGraw 1982, Dickson and Kolesik 1999; Dickson et al. 2003), chlorazol black E in lactoglycerol (Cunningham 1972; Koske and Tessier 1983; Widden 2001), trypan blue and cotton blue (Bevege 1968; Kormanik and McGraw 1982; Phillips and Hayman 1970), and ink (5% ink diluted in vinegar (Vierheilig et al. 1998) have been used by different workers. Bonfante-Fasolo et al. (1990) have reported a method using fluorescein-labeled wheat germ agglutinin for observation of AM fugal arbuscles. Similarly, Lum et al. (2002) have also described a method using confocal laser scanning microscope based on a molecular probe containing labeled wheat germ agglutinin (WGA-Alexa Fluor_488 conjugate).

In case of destructive methods with non-vital staining, various stains which react with certain fungal specific metabolites or enzymes are used. For example, nitro blue tetrazolium which detects the presence of succinate dehydrogenase has been used by many workers for visualization of all types of fungal structures in the host plant roots (MacDonald and Lewis 1978; Vierheilig et al. 2001). Similarly diaminobenzidine, which detects hydrogen peroxide, has been found to be helpful in observation of clumped mycorrhizal structures (Salzer et al. 1999). Also, alkaline phosphatase activity of AM fungi has been assessed by using fluorescent as well as non-fluorescent staining methods (Van Aarle et al. 2001; Tisserant et al. 1993).

For estimation of colonization rate of AMF, most commonly the grid-line intersect method of Giovannetti and Mosse (1980) is used. Although it is a simple method and requires only a dissecting microscope and a Petri plate with grid lines, its accuracy depends on the skills and experience of the worker. Another commonly used method is the magnified intersections method developed by McGonigle et al. (1990). Apart from this, many researchers also use the five-class system of Trouvelot et al. (1986). Both these methods quantify different AM fungal structures like arbuscles, vesicles, spores, etc. (Füzy et al. 2015). With the aim of enhancing the accuracy and efficacy of such methods, computerized analysis of AM fungi containing root images have been developed (Smith and Dickson 1991). Deguchi et al. (2017) have proposed the use of WinRHIZO Pro software for accurate and reproducible assessment of root colonization by AM fungi.

9.10 Genetics of Phosphate-Solubilizing Endophytic Fungi

The initiation and subsequent interaction of the endophytic fungal partner with the host plant is quite a complicated process. Some aspects of the genes involved in such interaction including phosphorus and other nutrient exchanges between the AM fungi and the host plant tissues have been elucidated but still a lot more research efforts are needed to better understand the genetic aspect of such interactions. An improved comprehension of the genes and their regulation involved in the plant–fungal interaction will help the researchers to undertake genetic manipulation of the fungi, which in turn can be utilized for better phosphorus uptake and improved plant growth and yield.

The molecular techniques have played a very crucial role in study of the genetic aspects of host–fungal interactions. The AM fungi are completely dependent on the host plants; however, the fungal spores represent the only plant-independent stage of the AM fungi. They have been reported to contain up to 2000 nuclei per spore (Bécard and Pfeffer 1993). Burggraaf and Beringer (1989) have reported at least 1000 nuclei per spore for *Glomus caledonium* while Viera and Glenn (1990) have reported 35,000 nuclei in case of *Gigaspora decipiens.* However, some researchers (Bianciotto et al. 1996; Hijri et al. 2002) have pointed out that these values might have got inflated due to possible contamination by other microbes or the inherent problems associated with cultivation of AM fungi. Hosny et al. (1998) have reported spores of *Scutellospora pellucida* to contain 0.14 pg of DNA, while the spores of *Scutellospora gregaria* to contain 1.15 pg of DNA, thus indicating high quantity of

DNA in the AM fungal spores. A lot of variation in the genome size from 16.5 Mb (Hijri and Sanders 2004) up to 1058.4 Mb (Hosny et al. 1998) has also been reported. Another significant feature which has been reported by Hosny et al. (1997) is the fact that many of these fungi like *G. intraradices* DNA has low G + C content (30–35%).

With the arrival and popularization of molecular techniques, extensive analyses of the fungal rDNA sequences have been undertaken. From such analysis, many researchers have concluded that these fungi are heterokaryotic and can possess different nuclei simultaneously (Sanders et al. 1995, 1996; Kuhn et al. 2001).

Researchers have used both targeted and non-targeted approach for analyzing various fungal genes and their expression. Harrison and van Buuren (1995) have reported the cloning of a high affinity phosphate transporter gene from *G. versiforme*, which showed expression only in extraradical hyphae. Ferrol et al. (2000) isolated partial genomic clones of P-type ATPase (designated as GmHA1 to GmHA5) from *G. mosseae* and analyzed their expression. Some researchers have been able to identify various chitin synthesis genes in various AM fungi, with the help of primers designed to amplify the conserved regions of the gene (Lanfranco et al. 1999a, b). Many techniques like differential display, suppressive subtractive hybridization, differential screening etc. have been used by many workers. Such techniques allow analyzing those genes, whose sequence and functions are unknown (Harrier 2001). A novel gene, designated as GmGin1, was identified by Requena et al. (2002) from cDNA library of *Glomus mosseae.* The gene was found to encode for a protein with a possible role in signaling between the mycobiont and the host plant. Similarly, Delp et al. (2000) were able to isolate three partial cDNA molecules from *G. intraradices* which showed similarity to various eukaryotic proteins. Although, many aspects of host–fungal interactions have been deciphered, yet many aspects still remain obscure (Garg and Chandel 2010). It is expected that with better understanding of such interactions, we will be in a better position to utilize the full benefits of such fungal endophytes.

9.11 Conclusions and Future Applications

Due to the ever increasing human population and ever diminishing land resources, our agricultural lands are becoming overburdened. The modern agricultural techniques are resulting in development of various high yielding plant varieties. However, such varieties generally have higher requirements for various nutrients including phosphorus. So, the farmers tend to apply chemical fertilizers leading to greater economic investments. Also, a major part of this phosphorus tends to become converted into insoluble and unavailable form apart from creating environmental problems. Hence, there is a need to look for alternatives of chemical fertilizers. The root endophytic fungi offer one of the very promising avenues of solving this problem. These fungi are not only capable of increasing the phosphate availability for the plant growth but also offer a plethora of other advantages and thus play a very crucial role in the overall growth and development of plants. Also, from the point of view of environmental sustainability, these fungi can be a very promising and potent solution. In fact, there are many instances of commercial application of these fungi for plant growth promotion across the globe. The only major challenge being encountered in the mass scale application of these fungi is the fact that it is difficult to cultivate these microbes without appropriate host plants. However, efforts are being made in this direction also to look for appropriate substitutes for host plants like using plant root exudates, etc. and in the near future it is expected that we will be in a better position to utilize these fungi for the benefit of mankind as well as nature.

References

Abd-Alla MH, Omar SA, Omar SA (2001) Survival of rhizobia/bradyrhizobia and a rockphosphate-solubilizing fungus *Aspergillus* niger on various carriers from some agroindustrial wastes and their effects on nodulation and growth of faba bean and soybean. J Plant Nutr 24:261–272

Ahmad N, Shahab S (2011) Phosphate solubilization: their mechanism genetics and application. Int J Microbiol 9:4408–4412

Akintokun AK, Akande GA, Akintokun PO, Popoola TOS, Babalola AO (2007) Solubilization of insoluble phosphate by organic acid producing fungi isolated from Nigerian soil. Int J Soil Sci 2:301–307

Alkan N, Gadkar V, Coburn J, Yarden O, Kapulnik Y (2004) Quantification of arbuscular mycorrhizal fungus *Glomus intradices* in host tissue using real-time polymerase chain reaction. New Phytol 161:877–885

Altomare C, Norvell WA, Björkman T, Harman GE (1999) Solubilization of phosphates and micronutrients by the plant growth-promoting and biocontrol fungus *Trichoderma harzianum* Rifai 1295-22. Appl Environ Microbiol 65:2926–2933

Azevedo JL, Maccheroni Junior W, Pereira JO, Araújo WL (2000) Endophytic microrganisms: a review on insect control and recent advances on tropical plants. Electron J Biotechnol 3:40–65

Bagyaraj DJ, Manjunath A (1980) Selection of a suitable host for mass production of arbuscular mycorrhizal inoculum. Plant Soil 55:495–498

Bécard G, Pfeffer PE (1993) Status of nuclear division in arbuscular mycorrhizal fungi during in vitro development. Protoplasma 174:62–68

Behera BC, Singdevsachan SK, Mishra RR, Dutta SK, Thatoi HN (2014) Diversity, mechanism and biotechnology of phosphate solubilizing microorganisms in mangrove—a review. Biocatal Agric Biotechnol 3:97–110

Berruti A, Lumini E, Balestrini R, Bianciotto V (2016) Arbuscular mycorrhizal fungi as natural biofertilizers: Let's benefit from past successes. Front Microbiol 6:1559

Bevege DI (1968) A rapid technique for clearing tannins and staining intact roots for detection of mycorrhizas caused by *Endogone* spp. & some records of infection in Australasian plants. Trans Br Mycol Soc 51:808–810

Bianciotto V, Bandi C et al (1996) An obligately endosymbiotic mycorrhizal fungus itself harbors obligately intracellular bacteria. Appl Environ Microbiol 62:3005–3010

Bolan NS, Currie LD, Baskaran S (1996) Assessment of the influence of phosphate fertilizers on the microbial activity of pasture soils. Biol Fertil Soils 21:284–292

Bonfante-Fasolo P, Faccio A, Perotto S, Schubert A (1990) Correlation between chitin distribution and cell wall morphology in the mycorrhizal fungus Glomus versiforme. Mycol Res 94:157–165

Bothe H, Klingner A, Kaldorf M, Schmitz O, Esch H, Hundeshagen B, Kernebeck H (1994) Biochemical approaches to the study of plant–fungal interactions in arbuscular mycorrhiza. Experientia 50:919–923

Burggraaf AJP, Beringer JE (1989) Absence of nuclear-DNA synthesis in vesicular arbuscular mycorrhizal fungi during *in vitro* development. New Phytol 111:25–33

Chanclud E, Morel JB (2016) Plant hormones: a fungal point of view. Mol Plant Pathol 17:289–1297

Chanway CP (1996) Endophytes: they're not just fungi. Can J Bot 74:321–322

Chatli AS, Beri V, Sidhu BS (2008) Isolation and characterisation of phosphate solubilizing microorganisms from the cold desert habitat of *Salix alba* Linn. in trans Himalayan region of Himachal Pradesh. Indian J Microbiol 48:267–273

Chauhan A, Guleria S, Walia A, Mahajan R, Verma S, Shirkot CK (2014) Isolation and characterization of *Bacillus* sp. with their effect on growth of tomato seedlings. Indian J Agric Biochem 27(2):193–201

Chauhan A, Guleria S, Balgir P, Walia A, Mahajan R, Mehta P, Shirkot CK (2016) Tricalcium phosphate solubilization and nitrogen fixation by newly isolated *Aneurinibacillus aneurinilyticus* CKMV1 from rhizosphere of *Valeriana jatamansi* and its growth promotional effect. Braz J Microbiol 48(2):294–304

Chen YP, Rekha PD, Arun AB, Shen FT, Lai WA, Young CC (2006) Phosphate solubilizing bacteria from subtropical soil and their tricalcium phosphate solubilizing abilities. Appl Soil Ecol 34:33–41

Cunningham JL (1972) A miracle mounting fluid for permanent whole-mounts of microfungi. Mycologia 64:906–911

Deguchi S, Matsuda Y, Takenaka C, Sugiura Y, Ozawa H, Ogata Y (2017) Proposal of a new estimation method of colonization rate of arbuscular mycorrhizal fungi in the roots of *Chengiopanax sciadophylloides*. Mycobiology 45(1):15–19

Delp G, Smith SE, Baker SJ (2000) Isolation by differential display of three partial cDNAs potentially coding for proteins from the VA mycorrhizal *Glomus intraradices*. Mycol Res 104:293–300

Deshmukh S, Hückelhoven R, Schäfer P, Imani J, Sharma M, Weiß M, Waller F, Kogel KH (2006) The root endophytic fungus *Piriformospora indica* requires host cell death for proliferation during mutualistic symbiosis with barley. Proc Natl Acad Sci USA 103:18450–18457

Dickson S, Kolesik P (1999) Visualization of mycorrhizal structures and quantification of their surface area and volume using laser scanning confocal microscopy. Mycorrhiza 9:205–213

Dickson S, Schweiger P, Smith SA, Söderström B, Smith S (2003) Paired arbuscules in the Arum-type arbuscular mycorrhizal symbiosis with *Linum usitatissimum*. Can J Bot 81:457–463

Duponnois R, Kisa M, Plenchette C (2006) Phosphate solubilizing potential of the nematofungus *Arthrobotrys oligospora*. J Plant Nutr Soil Sci 169:280–282

Efthymiou A, Grønlund M, Müller-Stover DS, Jakobsen I (2018) Augmentation of the phosphorus fertilizer value of biochar by inoculation of wheat with selected *Penicillium* strains. Soil Biol Biochem 116:139–147

Farrar K, Bryant D, Cope-Selby N (2014) Understanding and engineering beneficial plant–microbe interactions: plant growth promotion in energy crops. Plant Biotechnol J 12(9):1193–1206

Ferrol N, Miguel-Barea J, Azcon-Aguilar C (2000) The plasma membrane H+-ATPase gene family in arbuscular mycorrhizal fungus *Glomus mosseae*. Curr Genet 37:112–118

Franken P (2012) The plant strengthening root endophyte *Piriformospora indica*: potential application and the biology behind. Appl Microbiol Biotechnol 96:1455–1464

Frey B, Buser HR, Schuepp H (1992) Identification of ergosterol in vesicular-arbuscular mycorrhizae. Biol Fert Soils 13:229–234

Füzy A, Biró I, Kovács R, Takács T (2015) Estimation of am fungal colonization—comparability and reliability of classical methods. Acta Microbiol Immunol Hung 62:435–451

Gaiero JR, McCall CA, Thompson KA, Day NJ, Best AS, Dunfield KE (2013) Inside the root microbiome: bacterial root endophytes and plant growth promotion. Am J Bot 100(9):1738–1750

Gallaud I (1905) Etudes sur les mycorrhizas endotrophs. Rev Gen Bot 17:5–500

Garg N, Chandel S (2010) Arbuscular mycorrhizal networks: process and functions. A review. Agron Sustain Dev 30:581–599

Gehring CA, Mueller RC, Whitham TG (2006) Environmental and genetic effects on the formation of ectomycorrhizal and arbuscular mycorrhizal associations in cottonwoods. Oecologia 149:158–164

Giovannetti M, Mosse B (1980) An evaluation of techniques for measuring vesicular-arbuscular infection in roots. New Phytol 84:489–500

Goldstein AH, Liu ST (1987) Molecular cloning and regulation of a mineral phosphate solubilizing gene from *Erwinia herbicola*. Nat Biotechnol 5:72–74

Gulati A, Rahi P, Vyas P (2008) Characterization of phosphate-solubilizing fluorescent Pseudomonads from the rhizosphere of seabuckthorn growing in the cold deserts of Himalayas. Curr Microbiol 56:73–79

Guleria S, Sharma K, Walia A, Chauhan A, Shirkot CK (2014) Population and functional diversity of phosphate solubilizing bacteria from apricot (*Prunus Armeniaca*) of mid and high regions of Himachal Pradesh. Bioscan 9(2):1435–1443

Gupta VG, Rodriguez-Couto S (2018) New and future developments in microbial biotechnology and bioengineering: Penicillium system properties and applications. Elsevier, New York, pp 270–271

Harman GE, Howell CR, Viterbo A, Chet I, Lorito M (2004) Trichoderma species-opportunistic, avirulent plant symbionts. Nat Rev Microbiol 2:43–56

Harrier LA (2001) The arbuscular mycorrhizal symbiosis: a molecular review of the fungal dimension. J Exp Bot 52:469–478

Harrison MJ, van Buuren ML (1995) A phosphate transporter from mycorrhizal fungus *Glomus versiforme*. Nature 378:626–629

Heijnen CE, Hok-A-Hin CH, van Veen JA (1992) Improvements to the use of bentonite clay as a protective agent, increasing survival levels of bacteria introduced into soil. Soil Biol Biochem 24:533–538

Hijri M, Sanders IR (2004) The arbuscular mycorrhizal fungus *Glomus intraradices* is haploid and has a small genome size in the lower limit of eukaryotes. Fungal Genet Biol 41(2):253–261

Hijri M, Redecker D et al (2002) Identification and isolation of two ascomycete fungi from spores of the arbuscular mycorrhizal fungus *Scutellospora castanea*. Appl Environ Microbiol 68:4567–4573

Hinsinger P, Bengough AG, Vetterlein D, Young IM (2009) Rhizosphere: biophysics, biogeochemistry and ecological relevance. Plant Soil 321:117–152

Hosny M, Pais de Barros JP, Gianinazzi-Pearson V, Dulieu H (1997) Base composition of DNA from glomalean fungi: high amounts of methylated cytosine. Fungal Genet Biol 22:103–111

Hosny M, Gianinazzi-Pearson V, Dulieu H (1998) Nuclear DNA contents of eleven fungal species in Glomales. Genome 41:422–428

Hung LL, Sylvia DM (1998) Production of vesicular-arbuscular mycorrhizal fungus inoculum in aeroponic culture. Appl Environ Microbiol 54:353–357

Igual JM, Valverde A, Cervantes E, Velazquez E (2001) Phosphate solubilizing bacteria as inoculants for agriculture: use of update molecular techniques in their study. Agron Sustain Dev 21:561–568

IJdo M, Cranenbrouck S, Declerck S (2011) Methods for large-scale production of AM fungi: past, present,and future. Mycorrhiza 21:1–16

Illmer PA, Schinner F (1995) Solubilization of inorganic calcium phosphates solubilization mechanisms. Soil Biol Biochem 27:257–263

Jones KA, Burges HD (1998) Technology of formulation and application. In: Burges HD (ed) Formulation of microbial biopesticides: beneficial microorganisms, nematodes and seed treatments. Kluwer Academic Publishers, Dordrecht, pp 7–30

Khan MR, Khan SM (2002) Effects of root-dip treatment with certain phosphate solubilizing microorganism on the Fusarial wilt of tomato. Bioresour Technol 85:213–215

Khan MS, Zaidi A, Wani PA (2007) Role of phosphate solubilizing microorganisms in sustainable agriculture. A review. Agron Sustain Dev 27:29–43
Khan MS, Zaidi A, Ahemad M, Oves M, Wani PA (2010) Plant growth promotion by phosphate solubilizing fungi—current perspective. Arch Agron Soil Sci 56(1):73–98
Khan AL, Hussain J, Al-Harrasi A, Al-Rawahi A, Lee IJ (2015) Endophytic fungi: resource for gibberellins and crop abiotic stress resistance. Crit Rev Biotechnol 35:62–74
Kormanik PP, McGraw AC (1982) Quantification of vesiculararbuscular mycorrhizae in plant roots. In: Schenck NC (ed) Methods and principles of mycorrhizal research. The American Phytopathological Society, St. Paul, pp 37–45
Koske RE, Tessier B (1983) A convenient permanent slide mounting medium. Mycol Soc Am Newsletter 34:59
Krishnaraj PU, Goldstein AH (2001) Cloning of a *Serratia marcescence* DNA fragment that induces quinoprotein glucose dehydrogenase mediated gluconic acid production Escherichia coli in the presence of stationary phase *Serratia marcescence*. FEMS Microbiol Lett 205(2):215–220
Kucey RMN (1983) Phosphate solubilizing bacteria and fungi in various cultivated and virgin Alberta soils. Can J Soil Sci 63:671–678
Kuhn G, Hijri M et al (2001) Evidence for the evolution of multiple genomes in arbuscular mycorrhizal fungi. Nature 414:745–748
Kumar A (2016) Phosphate solubilizing bacteria in agriculture biotechnology: diversity, mechanism and their role in plant growth and crop yield. Intern J Adv Res 4(4):116–124
Kumar A, Rai LC (2015) Proteomic and biochemical basis for enhanced growth yield of *Enterobacter* sp. LCR1 on insoluble phosphate medium. Microbiol Res 170:195–204
Kumar A, Guleria S, Mehta P, Walia A, Chauhan A, Shirkot CK (2015) Plant growth promoting traits of phosphate solubilizing rhizobacteria isolated from seabuckthorn growing in cold desert region of trans-Himalayas and evaluating their potential on growth of tomato seedlings. Acta Physiol Plantarum 37(3):1–12
Kundu BS, Gaur AC (1981) Effect of single and composite cultures on rock phosphate solubilization. Haryana Agric Univ J Res 11:559–562
Lanfranco L, Garnero L, Bonfante P (1999a) Chitin synthase genes in the arbuscular mycorrhizal fungus *Glomus vbrsiforme*: full sequence of a gene encoding a class IV chitin synthase. FEMS Microbio Lett 170:59–67
Lanfranco L, Vallino M, Bonfante P (1999b) Expression of chitin synthase genes in the arbuscular mycorrhizal fungus *Gigaspora margarita*. New Phytol 142:347–354
Lin TF, Huang HI, Shen FT, Young CC (2006) The protons of gluconic acid are the major factor responsible for the dissolution of tricalcium phosphate by *Burkholderia cepacia* CC-A174. Bioresour Technol 7:957–960
Lum MR, Li Y, Larue TA, David-Schwartz R, Kapulnik Y, Hirsch A (2002) Investigation of four classes of non-nodulating white clover (*Melilotus alba* annua Desr.) mutants and their responses to arbuscular-mycorrhizal fungi. Integr Comp Biol 42:295–303
Lynch JM (1983) Soil biotechnology: microbiological factors in crop productivity. Blackwell Scientific Publications, Oxford
MacDonald RM, Lewis M (1978) The occurrences of some acid phosphatases and dehydrogenases in the vesicular arbuscular mycorrhizal fungus *Glomus mosseae*. New Phytol 80:135–141
Maliha R, Samina K, Najma A, Sadia A, Farooq L (2004) Organic acids production and phosphate solubilization by phosphate solubilizing microorganisms under *in vitro* conditions. Pakistan J Biol Sci 7:187–196
Malviya J, Singh K, Joshi V (2011) Effect of phosphate solubilizing fungi on growth and nutrient uptake of ground nut (*Arachis hypogaea*) plants. Adv Biores 2(2):110–113
Mandyam K, Loughin T, Jumpponen A (2010) Isolation and morphological and metabolic characterization of common endophytes in annually burned tallgrass prairie. Mycologia 102:813–821
Martinazzo R, Santos DR, Gatiboni LC, Brunetto G, Kaminski J (2007) Fósforo microbiano no solo sob sistema plantio direto em resposta à adição de fosfato solúvel. Rev Bras Cienc Solo 31:563–570

Mayerhofer MS, Kernaghan G, Harper KA (2012) The effects of fungal root endophytes on plant growth: a meta-analysis. Mycorrhiza 23:119–128

McGonigle TP, Miller MH, Evans DG, Fairchild GL, Swan JA (1990) A new method which gives an objective measure of colonization of roots by vesicular-arbuscular mycorrhizal fungi. New Phytol 115:495–501

Mehta P, Walia A, Chauhan A, Shirkot CK (2011) Accelerated solubilization of inorganic phosphate and production of antifungal activity in soil by plant growth promoting rhizobacteria isolated from apple rhizosphere. J Mycol Plant Pathol 41(3):342–349

Mehta P, Walia A, Chauhan A, Kulshrestha S, Shirkot CK (2013a) Phosphate solubilization and plant growth promoting potential by stress tolerant *Bacillus* sp. isolated from rhizosphere of apple orchards in trans Himalayan region of Himachal Pradesh. Ann Appl Biol 163:430–443

Mehta P, Walia A, Chauhan A, Shirkot CK (2013b) Plant growth promoting traits of phosphate-solubilizing rhizobacteria isolated from apple trees in trans Himalayan region of Himachal Pradesh. Arch Microbiol 195:357–369

Mehta P, Walia A, Kulshrestha S, Chauhan A, Shirkot CK (2013c) Efficiency of plant growth-promoting P-solubilizing Bacillus circulans CB7 for enhancement of tomato growth under net house condition. J Basic Microbiol 53:1–12

Mehta P, Walia A, Kakkar N, Shirkot CK (2014) Tricalcium phosphate solubilisation by new endophyte Bacillus methylotrophicus CKAM isolated from apple root endosphere and its plant growth-promoting activities. Acta Physiol Plantarum 36:2033–2045

Mehta P, Walia A, Shirkot CK (2015) Functional diversity of phosphate solubilizing plant growth promoting Rhizobacteria isolated from apple trees in the trans Himalayan Region of Himachal Pradesh, India. Biol Agri Hort 31(4):265–288

Mendes GO, Moreira de Freitas AL, Pereira OL, da Silva IR, Vassilev NB, Costa MD (2013) Mechanisms of phosphate solubilization by fungal isolates when exposed to different P sources. Ann Microbiol 64:239. https://doi.org/10.1007/s13213-013-0656-3

Mittal V, Singh O, Nayyar H, Kaur J, Tewari R (2008) Stimulatory effect of phosphatesolubilizing fungal strains (*Aspergillus awamori* and *Penicillium citrinum*) on the yield of chickpea (*Cicer arietinum* L. cv. GPF2). Soil Biol Biochem 40:718–727

Nannipieri P, Giagnoni L, Landi L, Renella G (2011) Role of phosphatase enzymes in soil. In: Bunemann E, Oberson A, Frossard E (eds) Phosphorus in action: biological processes in soil phosphorus cycling, soil biology, vol 26. Springer, Heidelberg, pp 251–244

Nelofer R, Syed Q, Nadeem M, Bashir F, Mazhar S, Hassan A (2016) Isolation of phosphorus-solubilizing fungus from soil to supplement biofertilizer. Arab J Sci Eng 41:2131–2138

Neubert K, Mendgen K, Brinkmann H, Wirsel SGR (2006) Only a few fungal species dominate highly diverse mycofloras associated with the common reed. Appl Environ Microbiol 72:1118–1128

Nilgiriwala KS, Alahari A, Rao AS, Apte SK (2008) Cloning and overexpression of an alkaline phosphatase phoK from *Sphingomonas* sp.BSAR-1 for uranium bioprecipitation from alkaline solution. Appl Environ Microbiol 74:5516–5523

Pal S, Singh HB, Rakshit A (2014) The arbuscular mycorrhizal symbiosis: an underground world wide web. In: Singh DP, Singh HB (eds) Microbial communities for sustainable soil health and ecosystem productivity. Studium Press LLC, Houston, pp 219–253

Pal S, Singh HB, Farooqui A, Rakshit A (2015) Fungal biofertilizers in Indian agriculture: perception, demand and promotion. J Eco-friendly Agric 10:101–113

Panhwar QA, Jusop S, Naher UA, Othman R, Razi MI (2013) Application of potential phosphate-solubilizing bacteria and organic acids on phosphate solubilization from phosphate rock in aerobic rice. Sci World J 2013:272409

Park KH, Lee CY, Son HJ (2009) Mechanism of insoluble phosphate solubilization by *Pseudomonas fluorescens* RAF15 isolated from ginseng rhizosphere and its plant growth-promoting activities. Lett Appl Microbiol 49:222–228

Parker DR, Reichmann SM, Crowley DE (2005) Metal chelation in the rhizosphere. In: Zobel RW (ed) Roots and soil management: interactions between roots and the soil. Agronomy monograph, vol 48. American Society of Agronomy, Madison, pp 57–93
Pei-Xiang Y, Li MA, Ming-Hui C, Jia-Quin X, Feng HE, Chang-Qun D, Ming-He M, Dun-Huang F, Yan-Qing D, Fa-Xiang Y (2012) Phosphate solubilizing ability and phylogenetic diversity of bacteria from phosphorus rich soils around Dianchi lake drainage area of China. Pedosphere 22:707–716
Phillips JM, Hayman DS (1970) Improved procedures for clearing roots and staining parasitic and vesicular arbuscular mycorrhizal fungi for rapid assessment of infection. Trans Br Mycol Soc 55:158–161
Pikovskaya RI (1948) Mobilization of phosphorus in soil in connection with the vital activity of some microbial species. Mikrobiologiya 7:362–370
Prasanna A, Deepa V, Murthy PB, Deecaraman M, Sridhar R, Dhandapani P (2011) Insoluble phosphate solubilization by bacterial strains isolated from rice rhizosphere soils from southern India. Int J Soil Sci 6(2):134–141
Priyadharsini P, Muthukumar T (2017) The root endophytic fungus *Curvularia geniculata* from *Parthenium hysterophorus* roots improves plant growth through phosphate solubilization and phytohormone production. Fungal Ecol 27:69–77
Radhakrishnan R, Khan AL, Kang SM, Lee IJ (2015) A comparative study of phosphate solubilization and the host plant growth promotion ability of *Fusarium verticillioides* RK01 and *Humicola* sp. KNU01 under salt stress. Ann Microbiol 65:585–593
Rebah FB, Tyagi RD, Prevost D (2002) Wastewater sludge as a substrate for growth and carrier for rhizobia, the effect of storage conditions on survival of *Sinorhizobium meliloti*. Bioresour Technol 831:45–51
Requena N, Mann P, Hampp R, Franken P (2002) Early developmentally regulated genes in the arbuscular mycorrhizal fungus *Glomus mosseae*: identification of GmGin1 a novel gene with homology to the C-terminus of metazoan hedgehog proteins. Plant Soil 244:129–139
Reyes I, Valery A, Valduz Z (2006) Phosphate-solubilizing micro-organisms isolated from rhizospheric and bulk soils of colonizer plants at an abandoned rock phosphate mine. Plant Soil 287:69–75
Rivera-Cruz MC, Narcía AT, Ballona GC, Kohler J, Caravaca F, Roldán A (2008) Poultry manure and banana waste are effective biofertilizer carriers for promoting plant growth and soil sustainability in banana crops. Soil Biol Biochem 40:3092–3095
Rodríguez H, Fraga R, Gonzalez T, Bashan Y (2006) Genetics of phosphate solubilization and its potential applications for improving plant growth-promoting bacteria. Plant Soil 287:15–21
Rodriguez-Navarro DN, Temprano F, Orive R (1991) Survival of *Rhizobium* sp. (*Hedysarum coronarium* L.) on peat-based inoculants and inoculated seeds. Soil Biol Biochem 23:375–379
Rosenblueth M, Martínez-Romero E (2006) Bacterial endophytes and their interactions with hosts. Mol Plant-Microbe Interact 19(8):827–837
Sadhana B (2014) Arbuscular mycorrhizal fungi (AMF) as a biofertilizer—a review. Int J Curr Microbiol App Sci 3:384–400
Salzer P, Corbiere H, Boller T (1999) Hydrogen peroxide accumulation in *Medicago truncatula* roots colonized by the arbuscular mycorrhiza-forming fungus *Glomus intraradices*. Planta 208:319–325
Sanders IR (2002) Specificity in the mycorrhizal symbiosis. In: Van der Heijden MGA, Sanders IR (eds) Mycorrhizal ecology. Studies in ecology. Springer, Heidelberg, pp 415–437
Sanders IR, Alt M, Groppe K, Boller T, Wiekman T (1995) Identification of ribosomal DNA polymorphisms among and within spores of the Glomales—application to studies on the genetic diversity of arbuscular mycorrhizal fungal communities. New Phytol 130:419–427
Sanders IR, Clapp JP et al (1996) The genetic diversity of arbuscular mycorrhizal fungi in natural ecosystems—a key to understanding the ecology and functioning of the mycorrhizal symbiosis. New Phytol 133:123–134

Schmitz O, Danneberg G, Hundeshagen B, Klingner A, Bothe H (1991) Quantification of vesicular-arbuscular mycorrhiza by biochemical parameters. J Plant Physiol 139:106–114

Selvakumar G, Kim K, Walitang D, Chanratana M, Kang Y, Chung B, Sa T (2016) Trap culture technique for propagation of arbuscular mycorrhizal fungi using different host plants. Korean J Soil Sci Fertil 49:608–613

Senthil Kumar CM, Jacob TK, Devasahayam S, Thomas S, Geethu C (2018) Multifarious plant growth promotion by an entomopathogenic fungus *Lecanicillium psalliotae*. Microbiol Res 207:153–160

Sharma SB, Sayyed RZ, Trivedi MH, Gobi TA (2013) Phosphate solubilizing microbes: sustainable approach for managing phosphorus deficiency in agricultural soils. Springerplus 2:587

Sharma R, Walia A, Chauhan A, Shirkot CK (2015) Multi-trait plant growth promoting rhizobacteria from tomato rhizosphere and evaluation of their potential as bioinoculants. App Biol Res 17(2):1–12

Sharma R, Sharma P, Chauhan A, Walia A, Shirkot CK (2017) Plant growth promoting activities of rhizobacteria isolated from *Podophyllum hexandrum* growing in north-west region of Himalayas. Proc Natl Acad Sci India Sect B Biol Sci 87(4):1443–1457. https://doi.org/10.1007/s40011-016-0722-2. ISSN: 0369-8211.

Shin W, Ryu J, Kim Y, Yang J, Madhaiyan M, Sa T (2006) Phosphate solubilization and growth promotion of maize (Zea mays L.) by the rhizosphere soil fungus *Penicillium oxalicum*. In: 18th World Congress of Soil Science July 9–15, Philadelphia, Pennsylvania, USA

Singh BK, Munro S, Potts JM, Millard P (2007) Influence of grass species and soil type on rhizosphere microbial community structure in grassland soils. Appl Soil Ecol 36:147–155

Singh SR, Singh U, Chaubey AK, Bhat MI (2010) Mycorrhizal fungi for sustainable agriculture—a review. Agric Rev 31:93–104

Smith RS (1995) Inoculant formulations and applications to meet changing needs. In: Tikhonovich IA, Provorov NA, Romanov VI, Newton WE (eds) Nitrogen fixation: fundamentals and applications. Kluwer Academic Publishers, Dordrecht, pp 653–657

Smith SE, Dickson S (1991) Quantification of active vesicular-arbuscular mycorrhizal infection using image analysis and other techniques. Aust J Plant Physiol 18:637–648

Smith S, Read D (2008) Mycorrhizal symbiosis. Academic Press, New York

Sreenivasa MN, Bagyaraj DJ (1988) *Chloris gayana* (Rhodes grass), a better host for the mass production of *Glomus fasciculatum*. Plant Soil 106:289–290

Srivastav S, Yadav KS, Kundu BS (2004) Prospects of using phosphate solubilizing *Pseudomonas* as biofungicide. Indian J Microbiol 44:91–94

Stephens JHG, Rask HM (2000) Inoculant production and formulation. Field Crops Res 65:249–258

Tisserant B, Gianinazzi-Pearson V, Gianinazzi S, Gollotte A (1993) In planta histochemical staining of fungal alkaline phosphatase activity for analysis of efficient arbuscular mycorrhizal infections. Mycol Res 97:245–250

Tiwari P, Adholeya A (2002) *In vitro* co-culture of two AMF isolates Gigaspora margarita and *Glomus intraradices* on RiT-DNA transformed roots. FEMS Microbiol Lett 206:39–43

Trouvelot A, Kough JL, Gianinazzi-Pearson V (1986) Mesure du taux de mycorhization VA d'un systeme radiculaire. Recherche de methodes d'estimation ayant une signifi cantion fonctionnelle. In: Gianinazzi-Pearson V, Gianinazzi S (eds) Physiological and genetic aspects of mycorrhizae. INRA, Paris, pp 217–221

Van Aarle IM, Olsson PA, Soderstrom B (2001) Microscopic detection of phosphatase activity of saprophytic and arbuscular mycorrhizal fungi using a fluorogenic substrate. Mycologia 93:17–24

Van der Heijden MGA, Streitwolf-Engel R, Riedl R, Siegrist S, Neudecker A, Ineichen K, Boller T, Wiemken A, Sanders IR (2006) The mycorrhizal contribution to plant productivity, plant nutrition and soil structure in experimental grassland. New Phytol 172:739–752

Van Elsas JD, Van Overbeek LS (1993) Bacterial responses to soil stimuli. In: Kjelleberg S (ed) Starvation in bacteria. Plenum Press, New York, pp 55–79

Varma A, Singh A, Sudha S, Sharma J, Roy A, Kumari M, Rana D, Thakran S, Deka D, Bharti K, Hurek T, Blechert O, Rexer KH, Kost G, Hahn A, Maier W, Walter M, Strack D, Kranner I (2001) *Piriformospora indica*—an axenically culturable mycorrhiza-like endosymbiotic fungus. In: Hock B (ed) Mycota IX. Springer, Berlin, pp 123–150

Venkateswarlu B, Rao AV, Raina P, Ahmad N (1984) Evaluation of phosphorus solubilization by microorganisms isolated from arid soil. J Indian Soc Soil Sci 32:273–277

Verma S, Varma A, Rexer K, Hassel A, Kost G, Bisen P, Bütehorn B, Franken P (1998) *Piriformospora indica*, gen. et sp. nov., a new root-colonizing fungus. Mycologia 90:896–903

Viera A, Glenn MG (1990) DNA content of vesicular-arbuscular mycorrhizal fungal spores. Mycologia 82:263–267

Vierheilig H, Coughlan AP, Wyss U, Piche Y (1998) Ink and vinegar, a simple staining technique for arbuscular mycorrhizal fungi. Appl Environ Microbiol 64:5004–5007

Vierheilig H, Knoblauch M, Juergensen K, van Bel A, Grundler MW, Piche Y (2001) Imaging arbuscular mycorrhizal structures in living roots of *Nicotiana tabacum* by light, epifluorescence and confocal laser scanning microscopy. Can J Bot 79:231–237

Vierheilig H, Schweigerb P, Brundrett M (2005) An overview of methods for the detection and observation of arbuscular mycorrhizal fungi in roots. Acta Physiol Plantarum 125:393–404

Vitorino C, Bessa LA, Carvalho LG, Silva FG (2016) Growth promotion mediated by endophytic fungi in cloned seedlings of *Eucalyptus grandis* x *Eucalyptus urophylla* hybrids. Afr J Biotechnol 15:2729–2738

Wahid OA, Mehana TA (2000) Impact of phosphate solubilizing fungi on the yield and phosphorus uptake by wheat and faba bean plants. Microbiol Res 155:221–227

Wakelin SA, Warren RA, Harvey PR, Ryder MH (2004) Phosphate solubilization by *Penicillium* sp. closely associated with wheat roots. Biol Fertil Soils 40:36–43

Walia A, Mehta P, Chauhan A, Shirkot CK (2013a) Antagonistic activity of plant growth promoting rhizobacteria isolated from tomato rhizosphere against soil borne fungal plant pathogens. Int J Agri Environ Biotechnol 6(4):587–595

Walia A, Mehta P, Chauhan A, Shirkot CK (2013b) Effect of *Bacillus* sp. strain CKT1 as inoculum on growth of tomato seedlings under net house conditions. Proc Natl Acad Sci India Sect B Biol Sci 84(1):144–155

Walia A, Guleria S, Chauhan A, Mehta P (2017) Endophytic bacteria: role in phosphate solubilization. In: Maheshwari DK, Annapurna K (eds) Endophytes: crop productivity and protection, sustainable development and biodiversity. Springer, Berlin, pp 1–33

Wang H, Liu S, Zhai L, Zhang J, Ren T, Fan B, Liu H (2015) Preparation and utilization of phosphate biofertilizers using agricultural waste. J Int Agric Adv 14:158–167

Whitelaw MA (2000) Growth promotion of plants inoculated with phosphate solubilizing fungi. Adv Agron 69:99–151

Widden P (2001) The use of glycerin jelly for mounting stained roots for the observation and quantification of endomycorrhizal fungi. Mycologia 93:1026–1027

Wood T (1984) Commercialization of arbuscular mycorrhizal fungi inoculum: the reclamation market. In: Williams SE, Allen MF (eds) Arbuscular mycorrhizae and reclamation of arid and semiarid lands. University of Wyoming, Laramie, pp 21–27

Yadav RS, Tarafdar JC (2003) Phytase and phosphatase producing fungi in arid and semi-arid soils and their efficiency in hydrolyzing different organic P compounds. Soil Biol Biochem 35:745–751

Yi Y, Huang W, Ge Y (2008) Exopolysaccharide: a novel important factor in the microbial dissolution of tricalcium phosphate. World J Microbiol Biotechnol 24:1059–1065

Yin Z, Shi F, Jiang H, Roberts DP, Chen S, Fan B (2015) Phosphate solubilization and promotion of maize growth by *Penicillium oxalicum* P4 and *Aspergillus niger* P85 in a calcareous soil. Can J Microbiol 61(12):913–923

Yin Z, Fan B, Roberts DP, Chen S, Shi F, Buyer JS, Jiang H (2017) Enhancement of maize growth and alteration of the rhizosphere microbial community by phosphate-solubilizing fungus *Aspergillus aculeatus* P93. J Agric Biotechnol 2(2):1–10

Zaidi A, Khan MS (2007) Stimulatory effects of dual inoculation with phosphate solubilizing microorganisms and arbuscular mycorrhizal fungus on chickpea. Austr J Exp Agric 47:1016–1022

Zaidi A, Khan MS, Amil M (2003) Interactive effect of rhizotrophic microorganisms on yield and nutrient uptake of chickpea (*Cicer arietinum* L.). Eur J Agron 19:15–21

Zaidi A, Khan MS, Ahemad M, Oves M, Wani PA et al (2009) Recent advances in plant growth promotion by phosphate-solubilizing microbes. In: Khan MS (ed) Microbial strategies for crop improvement. Springer, Berlin, pp 23–50

Chapter 10
Endophytic Fungi: The Desired Biostimulants for Essential Oil Production

Hesham A. El Enshasy, Siti Zulaiha Hanapi, Roslinda Abd Malek, Soad A. Abdelgalil, and Ong Mei Leng

10.1 Introduction

Agarwood resin is a precious commodity extracted from the xylem tissue of *Aquilaria* and can be defined as a fragrant dark resinous heartwood of the diseased trunk and timber of *Aquilaria* and *Gyrinops* trees (Blanchette 2006). Agarwood goes by different names around the world as a result of varieties of the languages, cultures, and philosophies of the countries in which it is planted. These names include aloeswood, eaglewood, calambac, gharuwood, and oud (Fig. 10.1) (Blanchette 2006).

Since ancient times, agarwood resin has been used in many countries based on its vital role in social, cultural, and economic practices all over the world (Donovan and Puri 2004). It can be used as an incense, perfume, and multifunctional pharmaceutical product (Eurlings et al. 2010). It has also been utilized in

H. A. El Enshasy (✉)
Institute of Bioproduct Development, Universiti Teknologi Malaysia, Johor Bahru, Malaysia

Faculty of Chemical Engineering and Energy, Universiti Teknologi Malaysia, Johor Bahru, Malaysia

City of Scientific Research and Technological Applications, New Burg Al Arab, Alexandria, Egypt
e-mail: henshasy@ibd.utm.my

S. Z. Hanapi · R. A. Malek
Institute of Bioproduct Development, Universiti Teknologi Malaysia, Johor Bahru, Malaysia

S. A. Abdelgalil
Institute of Bioproduct Development, Universiti Teknologi Malaysia, Johor Bahru, Malaysia

City of Scientific Research and Technological Applications, New Burg Al Arab, Alexandria, Egypt

O. M. Leng
Harita Go Green Sdn. Bhd., Johor Bahru, Johor, Malaysia

B. P. Singh (ed.), *Advances in Endophytic Fungal Research*, Fungal Biology,
https://doi.org/10.1007/978-3-030-03589-1_10

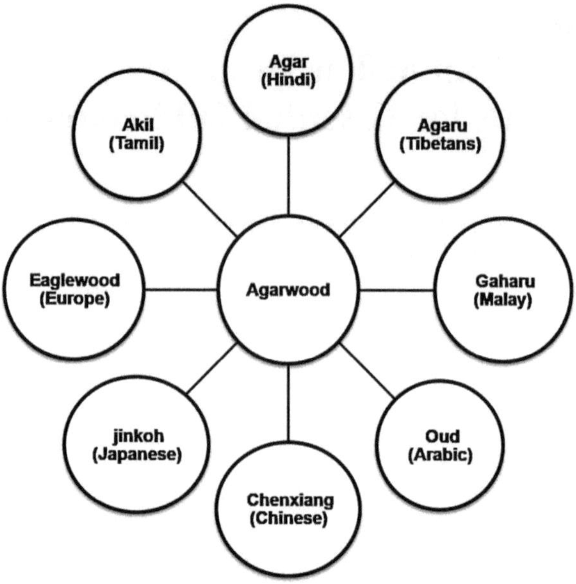

Fig. 10.1 Schematic chart of agarwood names in different countries, languages, and cultures

religious rituals around the world. Islamic culture uses agarwood incense in mosques, shops, and houses during ceremonies or rituals (Karimi et al. 2012). Ancient Egyptians used agarwood incense for embalming their dead starting 3000 years ago (Gerard 2007). Moreover, agarwood resin possesses medicinal value based on therapeutic properties like anticancer, anti-inflammatory, antimicrobial, immunomodulatory, and wound healing properties. Thus, it can be used as an antiemetic, tranquilizer, sedative, analgesic, digestive, or laxative agent (Bhore et al. 2013; Huang et al. 2013; Yang et al. 2013). The aromatic, fumigatory, and medicinal properties of agarwood have spurred demand for it around the world. In general, the current selling value of agarwood chips starts as low as USD 30 and goes up to USD 10,000/kg based on the quality and resin content. Pure agarwood oil, however, can sell for as high as USD 40,000/kg. The global market for agarwood oil and related products had a total value of around USD 6–8 billion in 2013 and was estimated to exceed USD 36 billion by 2017 (Chowdhury et al. 2017). Therefore, the production of agarwood and its derivatives has attracted considerable scientific interest in recent decades as a result of their increased market growth. To meet the increased global demand for agarwood, measures must be taken to prevent the extinction of wild *Aquilaria* trees. Thus, different methods for *Aquilaria* plantations have been studied in many countries, and various strategies have been implemented to spur biological induction using endophytic fungi to enhance the production of agarwood resin.

10.1.1 Aquilaria *Trees*

The tree of *Aquilaria* spp. is among the precious trees in the world and serves as the main source of precious nontimber forest products (NTFPs), agarwood tree which is an angiosperm within the order Malvales and Thymelaeaceae family in the subfamily Thymelaeoideae according to the latest classification as updated by Herber (2003). The number of species identified as agarwood resin producers is 13 out of a total of 21 total accepted species, named as follows: *A. malaccensis, A. crassna, A. sinensis, A. baillonii, A. beccariana, A. filaria, A. hirta, A. khasiana, A. microcarpa, A. rostrata, A. rugosa, A. subintegra*, and *A. yunnanensis* (Compton and Zich 2002; Kiet et al. 2005).

10.2 Distribution of *Aquilaria* Trees

The regions of South and Southeast Asia, such as Bhutan, Cambodia, Indonesia, China, India, the Philippines, Singapore, Laos, Malaysia, Myanmar, Thailand, Bangladesh, and Vietnam, are widely populated by *Aquilaria* spp. Indonesia and Malaysia are considered the two main countries for plantations of *Aquilaria* trees (Liu et al. 2013). On the other hand, United Arab Emirates, Saudi Arabia, and Japan are considered the main importing countries of agarwood resin (Regula 2010). Overexploitation and improper harvesting of natural agarwood have led to a scarcity of agarwood-producing tree stocks in natural populations, so the genus is currently conserved under the Convention on International Trade in Endangered Species of Wild Fauna and Flora (CITES) treaty and the International Union for Conservation of Nature Red List (Lee and Rozi 2016). Because of this, alternative strategies have been developed involving the plantation of agarwood trees in an attempt to produce agarwood resin artificially through the induction process, using various mechanical, physical, and biological methods (Okudera and Ito 2009).

10.3 Agarwood Resins

The heartwood of the healthy *Aquilaria malaccensis* tree is characterized by white, soft and free of scented resins as agarwood resin in absent in healthy intact plants. External factors like lightning strikes, fires, animal grazing, insect infestations, chemical applications, physical damage like injuries or cutting, and microbial invasions (endophytic fungi) contribute to the formation of a dark aromatic resin embedded in the heartwood of agarwood essential oil (Pojanagaroon and Kaewrak 2006; Dai et al. 2009). Agarwood essential oil is of a complex composition and is found as a mixute of more than 150 compounds including hydrocarbons like sesquiterpenes, sesquiterpene alcohols, oxygenated compounds (alcohols, aldehydes, ketones, acids, oxides, lactones, acetals, ethers, and esters, chromone derivatives and resin. All of these constituents contribute to the aroma and odor characteristics of the oil (Chang et al. 2002).There are many hypotheses to explain the reasons of agarwood production in this type of trea. The immunological response of the host tree to wounds or infections is an interesting topic of research and a key issue for the sustainable production of essential oil of the plant (Wong et al. 2015).

10.4 Agarwood Development in *Aquilaria* Trees

Naturally, when a plant is injured by a physical or biological method, plants respond to wounds by synthesizing self-protective compounds known as agarwood oil (Suharti et al. 2011; Xu et al. 2013; Zhang et al. 2014a, b; Chhipa and Kaushik 2017; Faizal et al. 2017). The development, the formation of agarwood resin in the plant is started by either natural or artificial induction is illustrated in Fig. 10.2.

There are four well-known methods namely: natural, physical, chemical and biological have been described for producing resinous oil in the *Aquilaria* tree (Table 10.1). Natural method take years for the formaiton of agarwood and depend on many factors are actively being explored. In addition, there is therefore some degree of overlap between the role of natural fungal growth in host plants and artificial fungus induction that needs further investigation.

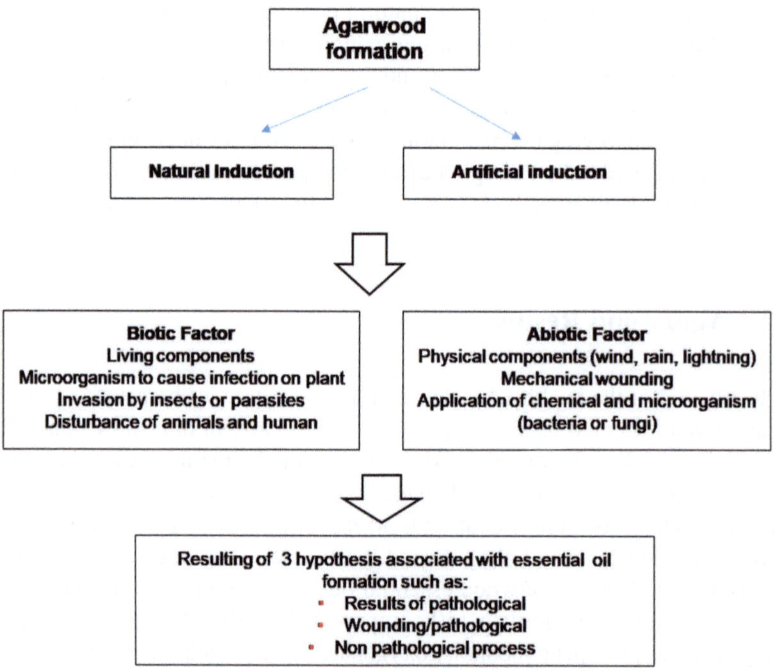

Fig. 10.2 Summary of hypothesis of essential oil/agarwood production

Table 10.1 Summary of different methods used for essential oil induction in agarwood tree

	References
Chemical induction	
Sulfuric acid, acetic acid, alcohol, sodium chloride, hydrogen peroxide, methyl jasmonate, soybean oil, brown sugar, formic acid, salicylic acid, sodium bisulfate, yeast extract, iron powder, jasmonic acid, formic acid	Sen et al. (2017), Faizal et al. (2017), Mohamed et al. (2014a, b), Tian et al. (2013), Zhang et al. (2012), Chen et al. (2011), Novriyanti et al. (2010)
Physical induction	
Burning, cauterizing, nailing, inflicting nail, Az wounds, wounding using chisels/knife, pulling a hole with screws, pinhole, burning chisel, drilling, agar-wit, drilling, aeration, partial trunk pruning	Putri et al. (2017), Chhipa and Kaukshik (2017), Chowdry et al. (2016), Chong et al. (2015), Zhang et al. (2012, 2014a, b), Liu et al. (2013), Tian et al. (2013)
Biological induction	
By different microorganisms including: *Acremonium* sp., *Aspergillus* sp., *Botryodiplodia* sp., *Cytosphaera* sp., *Fusarium* sp., *Libertella* sp., *Phytium* sp., *Penicillium* sp., *Rhizoctonia* sp., *Scytalidium* sp., *Fusarium solani*, *Melanotus flavour, Livens, Diplodia* sp., *Botryodylodis* sp., *Lasiodiplodis* sp.	Putri et al. (2017), Chhipa and Kaushik (2017), Chowdhury et al. (2016), Chong et al. (2015), Zhang et al. (2012, 2014a, b), Liu et al. (2013), Tian et al. (2013)
Natural induction	
Ants, snails, fungi	Ng et al. (1997)

Detailed information on the different methods used for infection or invasion of *Aquilaria* tree for essential oil production or agarwood is illustrated in Table 10.2. A natural method is where the injury is developed by microorganism intrusion, aggravating of insects, lightning bolts, and wind (Chong et al. 2015; Mohamed et al. 2014a, b). The sign of agarwood formation is typically the appearance of dark color around the wounded or rotting parts of the tree (Zhang et al. 2010; Chhipa and Kaushik 2017). However, this process is time-consuming and not frequently applied in old tree. It has been reported that only 10% of agarwood is produced in mature *Aquilaria* trees (about 25- to 30-year-old plants) with natural induction (Zhang et al. 2010; Chhipa and Kaushik 2017). Due to the low yield obtained using this method, it is not suitable for commercial production to meet the needs of a the growing market of agarwood resin. Therefore, artificial methods have been developed and introduced to agarwood farmers to obtain high quality and quantity of agarwood. Each method introduced shall combine the concept of agarwood production while at the same time conserve the *Aquilaria* stock.

10.4.1 Physical Approach

Physical method are among the most popular practices to induce resinous oil in *Aquilaria* tree. The injury modes by physical methods include removal, burning, cauterizing, nailing, nails, axe wounds, the use of chisels or knives, drilling a hole

Table 10.2 Summary of different procedures used for the infection or invasion of *Aquilaria* trees for essential oil production or agarwood

Technique/method	Type of method	Methodology	Advantages	Disadvantages	Reference
Nailing method	Physical	– Putting nails in tree trunk (hammering)	– Easy to apply – Low cost	Low yield Low quality labor intensive	Chowdhury et al. (2016)
Drilling method	Physical biological (*Fusarium solani*)	– Make a dimple (drilling holes) at set interval spiral design around a tree trunk, from base to highest point of tree – Drill holes 0.8 mm in diameter and their depth – Fungal species are introduced into hollows then plugged with wooden sticks – Sonic tomography is used to detect the presence of resinous wood	– Easy to apply – Low cost – High resin yield in shorter time – The resin concentration within the tree can be quantitied using tomography method	Height of estimation does not impact propagation speed of sound waves or tomographic results	Putri et al. (2017)
Drilling method	Physical biological	– Drill several small holes in a spiral in trunk/whole body – Apply agarwood inducer into xylem part of *Aquilaria* tree	Resin is produced and spread throughout xylem cell from trunk, environmentally friendly	–	Chong et al. (2015)
Burn-chisel-drill	Physical Chemical (jasmonic acid)	– Heat shock applied for treated cell suspension of *A. sinensis* – Reproducing burn-chisel-drill technique used on trees – Jasmonic acid affects accumulation of sesquiterpene compounds	Jasmonic acid act as signal transducer in this process to induce agarwood formation	–	Xu et al. (2016)

with screws, pinholes, burning–chisel drilling, and agar-wit (Table 10.2). Nevertheless, Chowdhury et al. (2016) have reported that the nailing method using hammering of nails in the trunk yielded poor quality of agarwood oil. Another method proposed by Zhang et al. (2014a, b) was including the use of a burning-chisel-drilling method (BCD) at high temperature (iron drill bit at 600 °C). This method induced higher sesquiterpene content after causing injury to the plant part, while at the same time preventing the infection of external microbes. Currently, a method known as agarwood kits (CA-kits) developed by Blanchette (2006) is known for its effectiveness. In most cases, physical methods demonstrate higher risks and take more time, produce poorer yields, and are very speculative with the unpredictable and low quality of agarwood (Liu et al. 2013). For this reason, the physical method cannot be used to meet consumers quality requirements.

10.4.2 Chemical Approach

A wide range of chemicals have been applied to induce agarwood production (Chen et al. 2011; Chong et al. 2015). Different chemicals has been inserted in a different range of concentrations, including the application of synthetic resin into the xylem of trees. Chemicals are usually transported to different plant location in the tree using water transportation system to reach the entire plant.

Chemical inducers such as sulfuric acid, acetic acid, alcohol, sodium chloride, hydrogen peroxide, methyl jasmonate, soybean oil, brown sugar, formic acid, and salicylic acid have additionally been useful to induce agarwood development (Table 10.1). Sulfuric acid and acetic acid have been used widely in several countries, however, it caused damage to trees. It was also reported that sodium chloride was used as an inducing agent, however, the yield results were in regard to the quantity and quality of essential oil production (Chen et al. 2011; Mohamed et al. 2014a, b). To improve agarwood resin formation and induce plant protective reaction, chemicals including methyl jasmonate, hydrogen peroxide, and salicylic corrosive were applied (Wijitphan 2009; Kumeta and Ito 2010).

Another study reported by Zhang et al. (2012) demonstrated the application of a whole-tree (agar-wit) method to induce agarwood and within 20 months, a high quality of agarwood with the major content of sesquiterpene compounds such as agarospirol and eudesmol was obtained. It was proposed that this strategy could fulfill the significant requirement for agarwood while conserving the plant itself. An aeration method proposed by Liu et al. (2013) that combined physical (aeration holes made by plastic, bamboo, or metal) and chemical methods (sodium bisulfate, yeast extract and iron powder) enhanced the defense mechanisms of the host plants and thus induced agarwood formation. The use of the chemical method also influenced the quality of the fragrance compound of agarwood through inoculation with artificial induction of salicylic and methyl jasmonate (Okudera and Ito 2009). This induction led to improving the production of sesquiterpenoid and chromanones, important aromatic compounds (Okudera and Ito 2009).

In another study, Xu et al. (2016) used jasmonic acid (JA), a well-known molecule which is responsible for plant defense and secondary metabolism, to induce sesquiterpene production in *A. sinensis*. This rapid method transforms the injured area from white to blackish after injection. Nevertheless, there is debate about this application due to the negative impacts of the solutions of JA when released back to the surroundings as it will be harmful to the surrounding organisms and be toxic to humans. Eventually, safety measures of the contents of the essential oil produced shall be taken if used for cosmetics, perfumes, incense, tea, and medicine to avoid exposure to allergens and cytotoxic compounds.

10.4.3 Biological Approach

The biological method is typically a method that uses fungi as inoculum to stimulate and induce biotic stress to host plants. The artificial injection of fungi was first introduced in late 1920s as reviewed by (Gibson 1977; Chhipa and Kaushik 2017). This natural method, with or without chemical inducer, was reported to develop a high yield of agarwood oil. In this method, agarwood resin was induced after the plant suffering from infection by phytoalexin including terpenoids, glycosteroids, and alkaloids compound secrete as a result of plant adaptation to the wound caused by pathogen infection. Moreover, natural plant terpene was classified as phytoalexins which exhibit bioactive function as an antimicrobial or anti-insect agents (Banerjee et al. 2006).

Novriyanti et al. (2010) reported on the potential use of biostimulants for the efficient development of fungi that can infected in regular base and result in the constant production of agarwood resins. Infection with fungi as biological inducer in *Aquilaria* tree can spread into the whole plant body and thus resulted in high essential oil yield. This is also considered to be an added value of this method compared to the limited spread of injury when using chemical and physical methods. Fungal mycelium usually spreads within plant tissues and can enter into the stem through holes. This results in the continuous spreading of infection and thus greatly increases the number of cells and uses large portions of plant tissues as a biofactory for resin production.

Artificial induction have been carrid out by a large number of fungi such as *Melanotus* sp., *Phytium* sp., *Penicillium* sp., *Diplodia* sp., *Botryodyplodis* sp., *Lasiodiplodis* sp., *and Fusarium* sp. The role of fungi in the biological artificial injection is to spead the biostimulant mechanism into different part of the tree for higher agarwood resin yield of high quality. In the work of Chhipa and Kaushik (2017), induction was done by inoculation of *Penicillium polinicum* using syringe injection method after holing (drill machine) in a zigzag approach in 4- to 5-year-old tree. After 3 months, agarospiril was detected by gas chromatography-mass spectrometry (GC-MS). A study by Tian et al. (2013) proposed a combination of chemical and biological methods called the pinhole infusion method for enhancing agarwood production. At the beginning, a hole is made using a drill in the trunk. Then a chemical inducer (formic acid) is injected into each hole after 1–2 years, followed by inoculation of *Botryosphaeria dothidea*, to accelerate resin formation. Another technique was the fermentation of resinous agarwood chips with associated *Fusarium* and juvenile plants to induce the formation of terpenoid precursor and agarwood sesquiterpenes (Sen et al. 2017). A tissue culture strategy has been investigated for the construction of agarwood scented in callus and cell suspension culture of *Aquilaria* (Okudera and Ito 2009; Jayaraman and Mohamed 2015).

10.5 Chemical Compositions of Agarwood Oil Produced by Biological Method

Essential oil or agarwood is composed of a complex combination of organic compounds, mainly sesquiterpenes and chromones that are responsible for the unique fragrance of agarwood. Agarol, a type of sesquiterpene, was the first chemical isolated from *A. Agallocha* species (Konishi et al. 2002). Practically, the grading system of the agarwood oils was examined physically through oil color, odor, and long-lasting aroma by an expert grader (Ismail et al. 2013). In Malaysia, agarwood is classified under various grades based on their color, density, and aroma (Nor Azah et al. 2013). However, researchers are looking for a scientifically based and quantitative grading system for the chemical constituents and the composition of the compounds (Liu et al. 2017).

To assess the quality of an essential oil, the content of 2-(2 phenylethyl)chromone and 2-[2-4-methoxyphenylethyl] chromone was evaluated (Fig. 10.3) (Liu et al. 2017). Out of 66 different sesquiterpenoids found in different agarwood types, the high-quality agarwood contains about 66.47% more 2-(2 phenylethyl)chromone and 2-[2-4-methoxyphenylethyl] chromones compared to the low quality. Most importantly, the grading of essential oil is anticipated with the aromadendrene content of the agarwood (Pasaribu et al. 2013). It is derived from the different grade of agarospirol and guaiol content, which might be different from one tree to another. γ-eudesmol and baimuxinal are just obtained in agarwood from *A. sinensis* and *A. crassna*. While *epi*-γ-Eudesmol, Jinkoh-eremol, and β-agarofuranare do not widely exist but can only be found in *A. malaccensis* (Fig. 10.4) (Naef 2011; Chen et al. 2012; Hashim et al. 2016). The other "main indicator" of gaharu oil is 2-(2-phenylethyl) chromium, which has been identified in different forms of 39 2-(2-phenylethyl) from different agarwood (Naef 2011; Lancaster and Espinoza 2012; Espinoza et al. 2014).

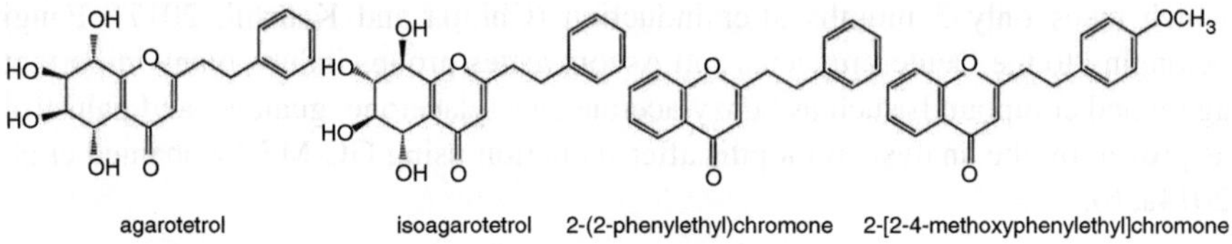

Fig. 10.3 Typical 2(2-phenylethyl) chromone compounds

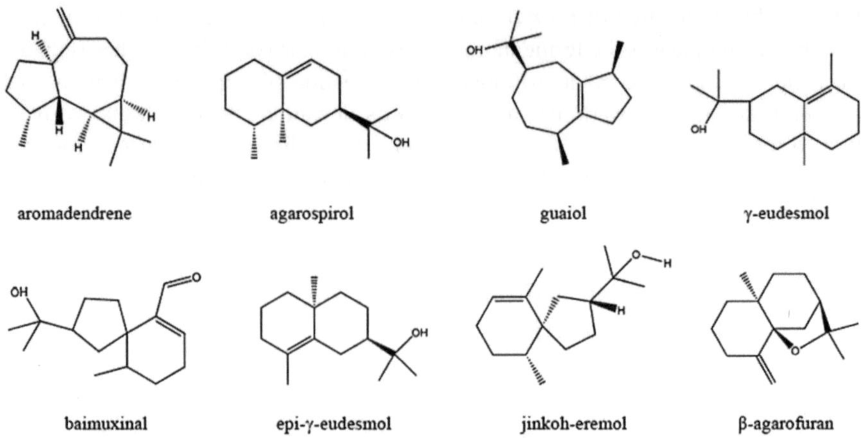

Fig. 10.4 Chemical structure of main sesquiterpenoids in agarwood resin

Notably, the method of induction can directly affect the quality of agarwood. It is also significant that the high quality of agarwood oil depends on the sesquiterpenes and aromatic content in agarwood. This is verified with the finding by Naef (2011) who reported that sesquiterpenes and benzyl acetone are key compounds for the determination of agarwood quality. Different types of endophytic fungi were reported previously as biological agents to induce the formation of the chemical compound. In other research of Zhang et al. (2014a, b), fungal induction by *Lasiodiplodia theobromae* resulted in the formation of 34 sesquiterpenes and 4 fragrant mixes with agarwood arrangement from *A. sinensis*. A recent study also reported on the exclusive isolation of 33 endophytic fungal strains belonging to the genera *Colletotrichum*, *Pestalotiopsis*, *Fusarium*, *Russula*, *Athrinum*, *Diaporthe*, and *Cladosporium* from *A. subintegra* (Monggoot et al. 2017). In this study, four potential strains, *Arthrinium* sp. MFLUCC16-0042, *Colletotrichum* sp. MFLUCC16-0047, *Colletotrichum* sp. MFLUCC16-0048, and *Diaporthes* sp. MFLUCC16-0051, generated a wide range of volatile compounds, comprising β-agarofuran, α-agarofuran, δ-eudesmol, oxo-agarospirol, and β-dihydro agarofuran.

However, among all endophytic fungi studied, *Fusarium solani* is considered one of the most potent essential oil inducer strains. This fungus typically can colonize in phloem boundaries within xylem tissues to induce high agarwood compounds such as tridecanoic acid, α-santalol, and spathulenol (Faizal et al. 2017). Agarospirol production was induced successfully using *Penicillium polonicum*, which takes only 3 months after induction (Chhipa and Kaushik 2017). Fungi belonging to the Deuteromycetes and Ascomycetes groups induce many important agarwood compounds such as benzylacetone, acetylacetone, guaiene, and palustrol as proven by the analysis 6 months after induction using GC-MS (Mohamed et al. 2014a, b).

10.6 Role of Different Endophytic Fungi in Induction of Phytochemical Compounds

In this area of study, fungi were categorized into two types, endophytes or pathogens. Endophytes are known as microorganisms acting to attack living plant tissue and cause asymptomatic infections inside plants. During invasion the microorganisms may establish an endosymbiotic relationship with the host plant by enhancing and helping to promote plant development and diminish ailment indications caused by pathogens or environmental stress. Nevertheless, endophytic fungi can be transformed into pathogenic fungi under specific conditions and are referred to as latent pathogens. Development of different biological compounds by endophytic fungi as a fresh source of such compounds has led to their high value in bioprospecting. Novel strains and their bioactive components have been identified by several researchers in academia, agencies, and private companies. Studies have reported that a biological agent of basidiomycete fungi, *Colletotrichum gloesporioides*, and *Botrysphaeria* sp. can produce high-grade agarwood oil within 8–18 months compared with the natural activity the oil obtained between 10 and 20 years.

A wide range of endophytic fungal strains was reported by several researchers as biostimulants in *Aquilaria*, including *Melanotus flavolivens*, *Epicoccum granulatum*, *F. lateritium*, *F. oxysporum*, *F. bulbiferum*, *F. lateritium*, *Diplodia* sp., *botryodylodia*, *Aspergillus* sp., *Chaetomium globosum*, *Botryoshaeria dothidea*, *Acremonium* sp., *Colletotrichum* sp., *xylaria* sp., *Lasiodiplodia theobromae*, and others types of basidiomycetes (Tamuli et al. 2005; Cui et al. 2013; Tian et al. 2013; Zhang et al. 2012, 2014a, b; Peng et al. 2015). However, *Fusarium* sp. is the most widely isolated followed by *Cladophialophora* sp. from *Aquilaria* sp. (Table 10.3) (Chhipa and Kaushik 2017).

10.6.1 Fungal Endophytes Associated with Agarwood Production

Endophytes can be defined as endosymbiotic microbes that establish mutualistic symbiosis relationships within plant tissues in one stage of their life cycle through which the mutually beneficial relationships were obtained for both plant and fungi (Behie and Bidochka 2014). It has been reported that fungal endophytes played a vital role in supporting plant growth, enhancing stress tolerance, and improving animal resistance of host plants (Cheplick and Faeth 2009). In general, *Alternaria, Cladosporium, Curvularia, Fusarium, Phaeoacremonium,* and *Trichoderma* are genera of known endophytic fungi (Mishra et al. 2017a, b; Turjaman et al. 2016).

Aquilaria trees in natural conditions are associated with a highly diversified cluster of microorganisms, whether in the rhizosphere through the soil microflora (Nimnoi et al. 2011) or located inside stems (Zhang et al. 2014a, b). *Aquilaria* is naturally associated with a wide variety of fungal agents involved in the agarwood

Table 10.3 Different species of fungi used as biostimulants for essential oil production and their chemical composition

Endophytic fungi	Plant species and type	Chemical composition	Reference
Fusarium isolate strain H15	*A. malaccensis* callus	Aroma compopunds	Sen et al. (2017)
Diaporthe sp. MFLUCC16-0051	*Aquilaria subintegra*	Antioxidant activity Agarwood volatiles	Monggoot et al. (2017)
F. solani	*A. malaccensis*	Agarwood compound	Faizal et al. (2017)
Penicillium polonicum	*A. malaccensis*	Production of Agarospirol	Chhipa and Kaushik (2017)
Fusarium oxysporum	*A. sinensis*	Sesquiterpenes (agarospirol), aromatic compounds	Zhang et al. (2014a, b)
Nigrospora oryzae	*Aquilaria sinensis*	Eudesmane-type sesquiterpene, 11-hydroxycapitulatin B (**1**), sesquiterpene, capitulatin B (**2**)	Li et al. (2014)
Fusarium sp.	*A. sinensis*	Sesquiterpenes, fatty acids, chromanone	Tian et al. (2013)
Mycosphaerella	*A. sinensis*	Sesquiterpenes, 2-(2-phenylethyl), chromanone, aromatics, fatty acid and esters	Tian et al. (2013)
Botryosphaeria sp.	*A. sinensis*	Sesquiterpenes, 2-(2-phenylethyl), chromanone, aromatics	Tian et al. (2013)
Trichoderma spirale	*A. sinensis*	Trichodermic acid A Trichodermic acid B Trichodermic acid C	Li et al. (2012)

production process like *Aspergillus* sp., *Botryodiplodia* sp., *Diplodia* sp., *Fusarium bulbiferum, F. laterium, F. oxysporum, F. solani, Penicillium* sp., and *Pythium* sp. (Sitepu et al. 2011). The biotechnological approach has been exploited for improving the artificial induction of agarwood production using different fungal strains. Earlier studies discovered that fungi act as infecting agent by entering the host plant through wounds and accelarate the process of agarwood resin production (Mohamed et al. 2014a, b). To induce agarwood production, tree trunks are wounded deliberately by large knives and the hammering of nails into tree trunks. The exposure of *Aquilaria* to wounding results in the weakness of the tree and its becoming susceptible to fungal invasion, which facilitates the induction process of agarwood resin production (Nobuchi and Siripatanadilok 2008).

A result of an endophytic fungal invasion is a high resin with volatile organic compounds produced by the host's defense system to suppress and retard the fungal infection by a process called tylosis. The presence of fungal injury in the tree trunk resulting in the various biochemical effects takes place and a white milky element known as oleoresin will be produced by the tree consequently. The resin intensely improves the mass and density of the affected wood, it darkens with time and is easily identified against the otherwise whitewood. The darkened area finally gets larger and indicates agarwood presence. Moreover, after the production of aromatic agarwood resin the whole tree gradually dries up, signaling its readiness to be harvested. For this, the accumulation of oleoresin causes the formation of agarwood resins as a response to fungal infection.

The harvesting time of agarwood can be 3 years or more after artificial inoculation and depends on the maturity of the agarwood resin. To screen agarwood from nonagarwood parts, cleaning and carving processes are required, which is a labor-intensive process. This step is followed by the extraction of essential oil from the agarwood resin. However, there are many methods used for the extraction of agarwood oil in the laboratory and on an industrial scale such as hydrodistillation, microwave-assisted solvent extraction, and supercritical fluid extraction. Hydrodistillation is considered the most applicable extraction method on a commercial scale and is considered safe and environmentally friendly. However, the main disadvantage of this method is the length of time of the extraction process and low yield. Nowadays, different methods in combination with different solvent systems are widely applied to increase the yield of oil production. The quality of the produced oil is usually determined by full chemical analysis using a gas chromatography flame-ionization detector (GC-FID) and GC-MS (Mucharommah 2011).

10.7 Mechanism of Agarwood Production

So far, little information is available on the association of fungal enzymatic activities and signaling pathway with the formation of agarwood (Turjaman et al. 2016). The intensity and success of efforts have been channeled into the investigation of the basic mechanism of agarwood production. The invasion of fungus into a plant host is accomplished by enzymatic machinery that enables fungi to penetrate and colonize inside plant tissues via specific mycotoxin, which is considered a secondary metabolite (Adam et al. 2015). As a result of this sequential infection mechanism, the plant produces terpene containing resins as a chemical defense to restrict the growth of fungal mycelia and inhibit spore formation (Yamada 1992).

The efforts to understand the basic mechanism of agarwood formation so far have been focused around terpenoid biosynthesis, particularly sesquiterpenes (Chang et al. 2002). There is a large research need in agarwood to take into account the diversity of interactions between the plant and the biological inducers, which affect the quality of the produced oil. This includes an in-depth study of the interactions at different levels (proteomics, transcriptomics, and metabolomics) to understand fungus–agarwood interaction, which is yet to be explored. Two previous studies conducted on these aspects included transcriptomics and metabolomics analyses of health and resinous agarwood (Behie and Bidochka 2014). However, these studies did not focus in depth on the plant–fungus interaction mechanism.

More recently, this interaction was studied using metabolomic approaches (GC-MS, multivariate statistics, and correlation network analysis) across three different platforms of interaction (involving *A. malaccensis* and an associated *Fusarium* sp.), viz. callus, juvenile plant, and fermentation of resinous wood chips (Sen et al. 2017). The study was designed to find chemometric signatures that could explain agarwood formation and scan the interactive landscape for new aroma compounds. This, apart from enhancing the basic understanding of agarwood formation, can potentially broaden the base of agarwood aromatics for the future.

Due to the fact that only wounded *Aquilaria* trees can produce agarwood, understanding the role of the signaling pathway in wound-induced agarwood formation is crucial. Xu et al. (2016) have reported on the role of indigenous JA in mediating a plant's defense response to produce secondary metabolites and agarwood sesquiterpene biosynthesis. They found that heat shock could stimulate JA production, thereby increasing the buildup of agarwood sesquiterpenes, while the presence of inhibitors such as nordohydroguaiaretic acid (NDGA) could block the JA signaling pathway and affect resin production.

10.8 Factors Affecting the Performance of Endophytic Fungi as Biostimulants for Agarwood Production

Typically, agarwood formation is influenced by several factors, mainly by the interaction between endophytic fungi and plant hosts. The effect of this interaction significantly influences the quality and quantity of metabolic products in plants (Rodriguez et al. 2009; Jia et al. 2016). The study of Santoso (2013) illustrated these factors as a triangle model between the genetic makeup of the host tree, type of pathogenic endophytic fungi, and environmental conditions. During an infection, by either natural or inoculated *A. lalaccensis*, in most cases fungi create pathogenic side effects and related enzymes such as polyphenol, oxidase, peroxidase, pectinase, and cellulase.

10.8.1 Natural Induction Using Indigenous Fungi

The potential use of indigenous fungi in essential oil formation was first discovered by Bhattacharyya in 1952 after the successful isolation of *Epicoccum* granulated from the infected tree. This finding attracted researchers to further investigate the connection between potential fungi and associated tree in the agarwood formation process. Biotrophic fungi in the internal plant tissue of the rhizosphere or phyllosphere could be triggered to become endophytic as a result of many transduction mechanisms within the host (Deka et al. 2017). The population of coexisting endophytic fungi can be greatly affected by physical elements, for example, temperature, humidity, geographic location (Mohamed et al. 2014a, b), genotypes of the applied strain including genetic background, age of strain (Liu et al. 2013). The tree of

Aquiliria species has been naturally infected by *Aspergillus* sp., *Botryodiplodia* sp., *Diplodia* sp., *Fusarium bulbiferum, F. laterium, F. osysporum, F. solani, Penicillium* sp., and *Pythium* sp. (Sitepu et al. 2011). Other studies revealed that the colonization of *Aspergillus, Lasiodiploidia, Chaetomium, Fusarium*, and *Penicillium* caused the accumulation of oleoresin from the tree (Sangareswari@Nagajothi et al. 2016).

Deka et al. (2017) stated that there could be a strong interaction between inter- and intraspecies of associated endophyte–endophyte within a plant, which is unavoidable to stimulate secondary metabolites production. In addition, abiotic environmental stress seems to confer on plants a greater capability to survive with the stress thanks to the balanced relationship between nutrient conditions of the plant and soils and the degree of endophyte colonization (Murphy 2013). On top of that, a period of stress has been determined to be a critical factor in agarwood resin induction in connection with fragrance quality (Tamuli et al. 2005; Faizal et al. 2017).

Little knowledge is available related to fungi and enzymatic activities that could be responsible for the development of agarwood. The concomitant production of enzymes from pathogenic or saprophytic fungi may occur during the decay of the plant, while at the same time the plant's defense mechanism may ultimately induce or inhibit the process. Tamuli et al. (2008) reported on the differentiations in the activity of several enzymes in a naturally infected tree by *Chaetomium globosum* and *F. oxysporum*as compared to a healthy tree, which could be the responsible agent for natural induction. Later, Sangareswari@Nagajothi et al. (2016) also reported on the high activities of cellulose and ligninolytic and laccase enzyme of *Aspergilus* AR13 isolated from the *Aquilaria* tree. They suggested that these enzymes play a role in the infection of the tree and higher agarwood resin production.

10.8.2 Artificial Treatment of Fungi

In general, the natural formation of agarwood in *Aquilaria malaccensis* is a time-consuming process characterized by low yield and quality. Therefore, noncustomary induction using agents was proposed, and these works were reviewed by Ng et al. (1997). However, before using fungi as biostimulants for agarwood formation, the technique needs to be optimized and proven. To prove any induction method, two directions need to be considered, the delivery mode and inducer microbe (Rasool and Mohamed 2016).

In an artificial fungal treatment technique, choosing the proper endophyte should be considered first to obtain the desired chemical constituents, the desired chemical constituents of the agarwood. The chosen strain must be able to adapt and colonize in a new environment. The technology uses endophytic fungi to enhance the formation of agarwood in a natural way, including using insects as a vector to transmit fungi (Turjaman et al. 2016).

The type of strain used in resin induction largely affects the type of metabolites produced and, subsequently, the oil quality. For example, it has been reported that tree inoculated with *F. solani* produced 20% more of the compounds of interest in agarwood such as sesquiterpenes, which contributed to a pleasant fragrance compared to a naturally wounded tree (Faizal et al. 2017). The length of the period between the induction of endophytic fungi and the time of harvest contributed a principal factor to the quality and quantity of the agarwood produced. Endophytic fungi induced in host plants require a certain period to present an infectious effect before the development of defense mechanisms from the plants. Santoso (2013) reported that the collection of agarwood requires no less than 3 months after induction to produce a significant amount of agarwood. While other work reported that 18 months were required to attain maximal yield (Turjaman et al. 2016). In another study, Rasool and Mohamed (2016) reported that at least 2 years are needed when using biological agents for induction in order to obtain higher chemical compounds such as sesquiterpenes to produce high-grade agarwood. Therefore, artificial induction using microbes is a challenge and needs further research to shorten the production process. Some works have even achieved high-grade agarwood, but many of these studies are restricted in the form of patent or are unpublished (Turjaman et al. 2016). A recent study by Sen et al. (2017) reported on the morphological changes in plants caused by plant–*Fusarium* interaction, which might be considered during intensive study at different levels of phenotypes (transcriptome, proteome, and metabolome).

10.9 Other Advantages/Impacts of Endophytic Fungi as Biostimulants for Essential Oil Production

The factors that determine the beneficence of plant–endophytic fungus relationships have been reported on by many authors (Chowdhary and Kaushik 2015; Du Jardin 2015; Hardoim et al. 2015; Jia et al. 2016). These mainly involve the specific arrangement of partner genotypes or strains (Liu et al. 2013), developmental phase, and the ecological and environmental fixing (Murphy 2013). Generally, endophytic fungi actions in host plants can be considered negative, positive, or neutral in their effects on the host plants. The positive effect, called mutualism, means that there will always be a positive effect on plant growth rate, whereby endophytic fungi provide fitness benefits for interacting partners, which are the plant hosts. A "sign change" happens if the interaction goes from positive (mutualism) to neutral (commensalism and neutralism), in which either endophytic fungi that reside inside the healthy tissues of plants obtain benefits from the host plants, but the latter are not affected (or vice versa), or the association between host plants and endophytic fungi occupy the same environment without affecting each other. However, most well-known endophytes are commensals, with known or unknown yet function in plants (Hardoim et al. 2015). Indeed, negative effects (predation, parasitism, amensalism, and competition) on host plants may lead to a reduction or cessation of its growth and quality of the secondary metabolite produced. However, it remains a matter of debate that negative effects could be transformed into positive or neutral effects once endophytes have lost their virulence.

Previous reports showed that endophytic fungi have significant effects on their host plants by expanding their development, enhancing their fitness, and fortifying their resistance to abiotic and biotic stresses (Jia et al. 2016). Furthermore, endophytic fungi are also able to influence the production of secondary metabolites in their plant host (Hardoim et al. 2015). At present, the exact relationship between endophytic fungi underlying the formation of agarwood for the production of essential oils has been less studied. Research conducted by Mohamed et al. (2014a, b) demonstrated that several endophytic fungi from *A. malaccensis*, such as *Cunninghamella bainieri, Fusarium solani*, and *Lasiodiplodia theobromae*, were involved in the formation of agarwood. Chen et al. (2011) found that the essential oils obtained from wild agarwood were rich in sesquiterpenes and aromatic constituents as well as higher antimicrobial activities toward *Bacillus subtilis* and *Staphyloccus aureus* compared to healthy trees.

However, in natural forests, the formation of agarwood occurs randomly and at low frequency. It was determined that only 7–10% of trees produce agarwood as a result of infections (Ng et al. 1997). Based on the fact that fragrant resin will not be produced by healthy trees unless they should be first infected by fungi, thus biological induction is considered as part of production process of resin (Liu et al. 2013). Current research also indicates that synthetic induction using *Fusarium solani* strains GSL1–4 to *Aquilaria malaccensis*, comparing wounded and unwounded trees, expedites the process of agarwood formation with a high quality of substances such as α-santalol, spathulenol, tridecanoic acid, and stigmasterol, which are the major elements in agarwood (Faizal et al. 2017). Other components such as sesquiterpenes were also found in abundance in other fungus-inoculated trees. This clearly indicates the varied responses of trees to pathogen attacks.

10.10 Future Perspective

Based on the increased market demand for agarwood, many studies have been carried out both upstream and downstream of this industry. Nowadays, the main research focus is on increasing the yield and quality of agarwood, decreasing production time, and enhancing oil recovery. In addition, the sustainability of this industry is of high concern and has spurred a search for good agricultural practices in connection with *Aquilaria* and *Gyrinops* trees. As is well known, the quality and quantity of agarwood resin are governed by various factors such as tree type and age and the type and age of fungal cells used in the induction process. To ensure the production of desired compounds in resin, a new approach could involve the inoculation of a certain fungal cocktail with the aim of stimulating specific plant responses to produce an agarwood resin with a desired aroma and chemical structure. This may require an in-depth understanding of the induction and biosynthesis of key chemicals in host plants at the genomic and metabolomic levels. Another challenge in the agarwood industry is to produce high-quality resin in a shorter amount of time. This may include improvements in agriculture practice for faster plant growth in a preinduction phase and accelerated maturation in a postfungal infection phase. Furthermore, research on using a plant cell culture system as an alternative to a whole plant for oil production may foster a greater sustainability of this industry, shorter production times, ease of regulation of pathways for the production of targeted chemical compounds, and ease of oil extraction process in a few steps. On the other hand, researchers have paid considerable attention to the downstream aspect of improving the extraction process of agarwood oil using a combination of environmentally friendly technologies of solvent-free extraction systems. All these together will not only increase our understanding of the unique process of oil production but also contribute to improving the economy of the agarwood oil industry.

References

Adam G, Wiesenberger G, Guldener U (2015) Fusarium mycotoxins and their role in plant–pathogen interactions. In: Zeilinger S, Martin JF, Garcia-Estrada C (eds) Biosynthesis and molecular genertics of fungal secondary metabolies, Fungal biology, vol 2. Springer, New York, pp 199–233

Banerjee NR, Furnes H, Muehlenbachs K, Staudigel H, de Wit MJ (2006) Preservation of ca. 3.4–3.5 Ga microbial biomarkers in pillow lavas and hyaloclastites from the Barberton Greenstone Belt, South Africa. Eart Planet Sci Lett 241:707–722

Behie SW, Bidochka MJ (2014) Nutrient transfer in plant-fungal symbioses. Trends Plant Sci 19:734–740

Bhore SJ, Preveena J, Kandasamy KI (2013) Isolation and identification of bacterial endophytes from pharmaceutical agarwood-producing *Aquilaria* species. Pharm Res 5(2):134–137

Blanchette RA (2006) The genus *Gyrinops*, is closely related to *Aquilaria* and in the past all species were considered to belong to *Aquilaria*. "Cultivated Agarwood—Training programs and Research in Papua New Guinea", Forest Pathology and Wood Microbiology Research Laboratory, Department of Plant Pathology, University of Minnesota

Chang YS, Nor Azah MA, Abu Said A, Lok EH, Reader S, Spiers A (2002) Gaharu. FRIM Tech Inf 69:1–7

Chen H, Yang Y, Xue J, Wei J, Zhang Z, Chen H (2011) Comparison of compositions and antimicrobial activities of essential oils from chemically stimulated agarwood, wild agarwood and healthy *Aquilaria sinensis* (Lour.) Gilg trees. Molecules 16(6):4884–4896

Chen HQ, Wei JH, Yang JS, Zhang Z, Yang Y, Gao ZH, Sui C, Gong B (2012) Chemical constituents of agarwood originating from the endemic genus *Aquilaria* plants. Chem Biodivers 9(2):236–250

Cheplick GP, Faeth SH (2009) Ecology and evolution of the grass-endophyte symbiosis. Oxford University Press, New York, p 241

Chhipa H, Kaushik N (2017) Fungal and bacterial diversity isolated from *Aquilaria malaccensis* tree and soil, induces agarospirol formation within 3 months after artificial infection. Front Microbiol 8:1286

Chong SP, Osman MF, Bahari N, Nuri EA, Zakaria R, Abdul-Rahim K (2015) Agarwood inducement technology: a method for producing oil grade agarwood in cultivated *Aquilaria malaccensis* Lamk. J Agrobiotech 6:1–16

Chowdhary K, Kaushik N (2015) Fungal endophyte diversity and bioactivity in the Indian medicinal lant *Ocimum sanctum Linn.* PLoS One 10(11):e0141444. https://doi.org/10.1371/journal.pone.0141444

Chowdhury M, Hussain D, Chung S, Kabir H, Rahman A (2016) Agarwood manufacturing: a multidisciplinary opportunity for economyof Bangladesh—a review. Agric Eng Int CIGR J 18:171–177

Chowdhury M, Rahman A, Hussain MD, Kabir E (2017) The economic benefit of agarwood production through aeration method into the *Aquilaria malaccensis* tree in Bangladesh. Bangladesh J Agric 42:191–196

Compton JGS, Zich FA (2002) Gyrinops ledermannii (Thymelaeaceae), being an agarwood-producing species prompts call for further examination of taxonomic implications in the generic delimitation between Aquilaria and Gyrinops. Flora Malesiana Bull 13:61–65

Cui J, Guo S, Fu S, Xiao P, Wang M (2013) Effects of inoculating fungi on agilawood formation in *Aquilaria sinensis*. Chin Sci Bull 58:3280–3287

Dai HF, Liu J, Zen YB, Wang H, Mei WI (2009) Two new 2-(2-phenylethyl) chromones from Chinese eaglewood. J Asian Nat Prod Res J 12:134–137

Deka D, Tayung K, Jha DK (2017) Harnessing fungal endophytes for plant and human health. In: Maheshwari DK (ed) Endophytes: biology and biotechnology. Springer, Cham, pp 59–98

Donovan DG, Puri RK (2004) Learning from traditional knowledge of non-timber forest products: Penan Benalui and the autecology of *Aquilaria* in Indonesian Borneo. Ecol Soc 9(3):3

Du Jardin P (2015) Plant biostimulants : Definition, concept, main categories and regulation. Sci Hort 196:3–14

Espinoza EO, Lancaster CA, Kreitals NM, Hata M, Cody RB, Blanchette RA (2014) Distinguishing wild from cultivated agarwood (*Aquilaria* spp.) using direct analysis in real time and time of-flight mass spectrometry. Rapid Commun Mass Spect 28(3):281–289

Eurlings MC, van Beek HH, Gravendeel B (2010) Polymorphic microsatellites for forensic identification of agarwood (*Aquilaria crassna*). Forensic Sci Int 197(3):30–34

Faizal A, Esyanti RR, Aulianisa EN, Iriawati, Santoso E, Turjaman M (2017) Formation of agarwood from *Aquilaria malaccensis* in response to inoculation of local strains of *Fusarium solani*. Trees 31:189–197

Gerard AP (2007) Agarwood: the life of a wounded tree. IIAS Newsletter 45: 24–25

Gibson I (1977) The role of fungi in the origin of oleoresin deposits (Agaru) in the wood of *Aquilaria agallocha Roxb*. Bano Biggan Patrika 6:16–26

Hardoim PR, van Overbeek LS, Berg G, Pirttilä AM, Compant S, Campisano A et al (2015) The hidden world within plants: ecological and evolutionary considerations for defining functioning of microbial endophytes. Microbiol Mol Biol Rev 79:293–320

Hashim YZH-Y, Kerr PG, Abbas P, Salleh HM (2016) *Aquilaria* spp. (Agarwood) as source of health beneficial compound: a review of traditional use, phytochemistry and pharmacology. J Ethnopharmacol 189:331–360

Herber BE (2003) Thymelaeaceae. In: The families and genera of vascular plants. Flowering plants dicotyledons (malvales, capparales and non betalain caryophyllales). Berlin: Springer, pp 373–438

Huang JQ, Wei JH, Zhang Z, Yang Y, Liu YY, Meng H, Zhang XL, Zhang JL (2013) Historical records and modern studies on agarwood production method and overall agarwood production method. Zhongguo Zhong Yao Za Zhi 38:302–306

Ismail N, Mohd Ali NA, Jamil M, Fazalul Rahiman MH, Tajuddin SN, Tain MN (2013) Analysis of high quality agarwood oil chemical compounds by means of spme/gc-ms and z-score technique. Malaysian J Anal Sci 17(3):403–413

Jayaraman S, Mohamed R (2015) Crude extract of *Trichoderma* elicits agarwood substances in cell suspension culture of the tropical tree, *Aquilaria malaccensis* Lam. Turk J Agric For 39:163–173

Jia M, Chen L, Xin H-L, Zheng C-J, Rahman K, Han T (2016) A friendly relationship between endophytic fungi and medicinal plants: a systematic review. Front Microbiol 9:906–920

Karimi I, Becker LA, Chalechale A, Ghashghaii A (2012) Biochemical plasma profile of male rats exposed to smoke of agarwood (*Aquilaria* spp.). Comp Clin Pathol 21:1053–1058

Kiet LC, Kessler PJA, Eurlings MCM (2005) A new species of *Aquilaria* (Thymelaeaceae) in Vietnam. Blumea-Biodivers Evol Biogeogr Plants 50(1):135–141

Konishi T, Konoshima T, Shimada Y, Kiyosawa S (2002) Six new 2-(2-phenylethyl) chromones from agarwood. Chem Pharm Bull 50:419–422

Kumeta Y, Ito M (2010) Characterization of delta-guaiene synthases from cultured cells of *Aquilaria*, responsible for the formation of the sesquiterpenes in agarwood. Plant Physiol 154:1998–2007

Lancaster C, Espinoza E (2012) Evaluating agarwood products for 2-(2-phenylethyl) chromones using direct analysis in real time time-of-flight mass spectrometry. Rapid Commun Mass Spectrom 26(23):2649–2656

Lee SY, Rozi M (2016) The origin and domestication of *Aquilaria*, an important agarwood-producing genus. In: Mohamed R (ed) Agarwood. Springer, Singapore, pp 1–20

Li DL, Chen YC, Tao MH, Li HH, Zhang WM (2012) Two new octahydro naphthalene derivatives from *Trichoderma* spirale, an endophytic fungus derived from *Aquilaria sinensis*. Helv Chim Acta 95(5):805–809

Li D, Chen Y, Pan Q, Tao M, Zhang W (2014) A new eudesmane sesquiterpene from *Nigrospora oryzae*, an endophytic fungus of *Aquilaria* sinensis. Rec Nat Prod 8:330–333

Liu Y, Chen H, Yang Y, Zhang Z, Wei J, Meng H, Gao Z (2013) Whole-tree agarwood-inducing technique: an efficient novel technique for producing high-quality agarwood in cultivated *Aquilaria sinensis* trees. Molecules 18(3):3086–3106

Liu YY, Wei JH, Gao Z, Zhang Z, Lyu J-C (2017) A review of quality assessment and grading for agarwood. Chin Herbal Med 9:22–30

Mishra VK, Passari AK, Chandra P, Leo VV, Kumar B, Gupta VK, Singh BP (2017a) Determination and production of antimicrobial compounds by *Aspergillus clavatonanicus* strain MJ31, an endophytic fungus from *Mirabilis jalapa* L. using UPLC-ESI-MS/MS and TD GC-MS. PLoS One 12(10):1–24. https://doi.org/10.1371/journal.pone.0186234

Mishra VK, Passari AK, Leo VV, Singh BP (2017b) Molecular diversity and detection of endophytic fungi based on their antimicrobial biosynthetic genes. In: Singh BP, Gupta VK (eds) Molecular markers in mycology, fungal biology. Springer, Basel, pp 1–35. https://doi.org/10.1007/978-3-319-34106-4_1

Mohamed R, Jong PL, Kamziah AK (2014a) Fungal inoculation induces agarwood in young *Aquilaria malaccensis* trees in the nursery. J For Res 25:201–204

Mohamed R, Jong PL, Nurul Irdayu I (2014b) Succession patterns of fungi associated to wound-induced agarwood in wild *Aquilaria malaccensis* revealed from quantitative PCR assay. World J Microbiol Biotechnol 30:2427–2436

Monggoot S, Popluechai S, Gentekaki E, Pripdeevech P (2017) Fungal endophytes: an alternative source for production of volatile compounds from agarwoodoil of *Aquilaria subintegra*. Microb Ecol 5:1–8

Mucharommah (2011) Development of eaglewood (gaharu) in Bengkulu, Sumatera. In: Turjaman M (eds) Proceeding of gaharu workshop: development of gaharu production technology, a forest community based empowerment. Indonesia's Work Programme for ITTO PD425/06 Rev.1 (I), Indonesia, pp 134

Murphy BR (2013) Fungal endophytes of barley roots—friend and foe? In: Endophytes for plant protection: the state of art. Braunschweig, Deutsche Phytomed Gesellschaft, pp 102–116

Naef R (2011) The volatile and semi-volatile constituents of agarwood, the infected heartwood of *Aquilaria* species: a review. Flavour Frag J 26(2):73–89

Ng LT, Chang YS, Kadir AA (1997) A review on agar (gaharu) producing *Aquilaria* species. J Trop Forest Prod 2:272–285

Nimnoi P, Pongsilp N, Lumyong S (2011) Actinobacterial community and diversity in rhizosphere soils of *Aquilaria crassna* Pierre ex Lec assessed by RT-PCR and PCR-DGGE. Biochem Syst Ecol 39:509–519

Nobuchi T, Siripatanadilok SA (2008) The formation of wood in tropical forest trees: a challenge from the perspective of functional wood anatom. In: Zimmerman MH (ed) Cytological observations of *Aquilaria crassna* wood associated with the formation of aloes-wood. UPM Press, Serdang, pp 147–160

Nor Azah MA, Husni SS, Mailina J, Sahrim L, Abdul Majid J, Mohd Faridz Z (2013) Classification of agarwood (gaharu) by resin content. J Trop Forest Sci 25:213–219

Novriyanti E, Santosa E, Syafii W, Turjaman M, Sitepu IR (2010) Anti-fungal activity of wood extract of *Aquilaria* crassna Pierre ex Lecomte against agarwood-inducing fungi, *Fusarium solani*. Indones J Res 7:155–165
Okudera Y, Ito M (2009) Production of agarwood fragrant constituents in *Aquilaria* calli and cell suspension cultures. Plant Biotechnol 26(3):307–315
Pasaribu G, Waluyo TK, Pari G (2013) Analisys of chemical compound in some of agarwood quality by gas chromatography mass spectrometry. J For Res Prod Res 31(3):181–185
Peng CS, Osman MF, Bahar N, Nuri EAK, Zakaria R, Rahim KA (2015) Agarwood inducement technology: a method for producing oil grade agarwood in cultivated Aquilaria malaccensis Lamk. J Agrobiotechnol 6:1–16
Pojanagaroon S, Kaewrak C (2006) Mechanical methods to stimulate aloes wood formation in *Aquiliria crassna* Pierre ex H Lec (kritsana) trees. ISHS Acta Hort 676:161–166
Putri N, Karlinasari L, Turjaman M, Wahyudi I, Nandika D (2017) Evaluation of incense-resinous wood formation in agarwood (*Aquilaria malaccensis* Lam.) using sonic tomography. Agric Nat Res 51:84–90
Rasool S, Mohamed R (2016) Understanding agarwood formation and its challenges. In: Mohamed R (ed) Agarwood: tropical forestry. Springer, Singapore, pp 39–56
Regula N (2010) The volatile and semi-volatile constituents of agarwood, the infected heartwood of *Aquilaria* species: a review. Flavour Fragr J 26:73–87
Rodriguez RJ, White JF Jr, Arnold AE, Redman RS (2009) Fungal endophytes: diversity and functional roles. New Phytol 182(2):314–330
Sangareswari@Nagajothi M, Thangamuthu Parthiban K, Umesh Kanna S, Karthiba L, Saravanakumar D (2016) Fungal microbes associated with agarwood formation. AJPS 7:1–8
Santoso E (2013) Agarwood formation by fungal bioinduction technology. In: Susmianto A, M Turjaman, Setio P (eds). Track record: agarwood inoculation technology. FORDA Press, Bogor Indonesia, pp 22-66
Sen S, Dehingia M, Talukdar NC, Khan M (2017) Chemometric analysis reveals links in the formation of fragrant bio-molecules during agarwood (*Aquilaria malaccensis*) and fungal interactions. Sci Rep 7:44406. https://doi.org/10.1038/srep44406
Sitepu IR, Santoso E, Siran SA, Turjaman M (2011) Fragrant wood gaharu: when the wild can no longer provide. ITTO and FORDA, Bogor, pp 1–67
Suharti SP, Pratiwi E, Santosa, Turjaman M (2011) Feasibility study of business in garwood inoculation at diffrent stem diameters and inoculation periods. Indon J Res 8:114–129
Tamuli P, Boruah P, Nath SC, Leclercq P (2005) Essential oil of eaglewood tree: a product of pathogenesis. J Ess Oil Res 17(6):601–604
Tamuli P, Baruah P, Samanta R (2008) Enzyme activities of agarwood (*Aquilaria malaccensis* Lamk.) stem under pathogenesis. J Spices Arom Crop 17(3):240–243
Tian JJ, Gao XX, Zhang WM, Wang L, Qu LH (2013) Molecular identification of endophytic fungi from *Aquilaria sinensis* and artificial agarwood induced by pinholes-infusion technique. Afr J Biotechnol 12(21):3115–3131
Turjaman M, Hidayat A, Santoso E (2016) Development of agarwood induction technology using endophytic fungi. In: Mohamed R (ed) Agarwood. Springer, Singapore, pp 57–71
Wong YF, Chin ST, Perlmutter P, Marriott PJ (2015) Evaluation of comprehensive two-dimensional gas chromatography with accurate mass time-of-flight mass spectrometry for the metabolic profiling of plant-fungus interaction in *Aquilaria malaccensis*. J Chromatogr A 1387:104–115
Xu Y, Zhang Z, Wang M, Wei J, Chen H, Gao Z et al (2013) Identification of genes related to agarwood formation: transcriptome analysis of healthy and wounded tissues of *Aquilariasinensis*. BMC Genom 14:227
Xu Y-H, Liao Y-C, Zhang Z, Liu J, Sun P-W, Gap Z-H et al (2016) Jasmonic acid is a crucial signal transducer in heat shock-induced sesquiterpene formation in *Aquilaria sinensis*. Sci Rep 6:21843. https://doi.org/10.1038/srep21843

Yamada T (1992) Biochemistry of gymnosperm xylem responses to fungal invasion. In: Blanchette RA, Biggs AR (eds) Defense mechanisms of woody plants against fungi. Springer, Heidelberg, pp 199–233

Yang L, Qiao C, Xie D, Gong NB, Lu Y, Zhang J, Dai J, Guo S (2013) Antidepressant abietane diterpenoids from chinese eaglewood. J Nat Prod 76:216–222

Zhang Z, Yang Y, Meng H, Sui CH, Wei JH, Chen HQ (2010) Advances in studies on mechanism of agarwood formation in *Aquilaria sinensis* and its hypothesis of agarwood formation induced by defense response. Chin Trad Herbal Drugs 41:156–159

Zhang XL, Liu YY, Wei JH, Yang Y, Zhang Z, Huang JQ et al (2012) Production of high-quality agarwood in *Aquilaria sinensis* trees via whole-tree agarwood-induction technology. Chin Chem Lett 23:727–730

Zhang Z, Wei J, Han X, Liang L, Yang Y, Meng H, Xu Y, Gao Z (2014a) The sesquiterpene biosynthesis and vessel-occlusion formation in stems of *Aquilaria sinensis* (Lour.) Gilg trees induced by wounding treatments without variation of microbial communities. Int J Mol Sci 15:23589–235603

Zhang Z, Zhang X, Yang Y, Wei JH, Meng H, Meng ZH et al (2014b) Hydrogen peroxide induces vessel occlusions and stimulates sesquiterpenes accumulation in stems of *Aquilaria sinensis*. Plant Growth Regul 72:81–87

Chapter 11
Current Perspectives on the Novel Structures and Antioxidant Properties of Mangrove Endophytic Fungal Exopolysaccharides

A. M. V. N. Prathyusha, Ganugula Mohana Sheela, Chanda V. Berde, and P. V. Bramhachari

11.1 Introduction

Mangroves are salt-tolerant coastal wetland forests confined to intertidal zones of deltas, estuaries, backwaters, lagoons, creeks, and marshes of tropical and subtropical latitudes (Thatoi et al. 2013). Mangroves harbor the second most important ecologically diverse group of fungi among marine ecosystems, in terms of productivity and sustained tertiary yield (Kobayashi and Tsuda 2004a, b; Sridhar 2004). Mangrove forests are "hotspots" for the biodiversity of marine fungi (Shearer et al. 2007) since bases of trunks and aerating roots are partially water-logged. At sea edges there is an overlap among marine and terrestrial fungi (Sarma and Hyde 2001). Mangrove fungi comprise lower fungi (Oomycetes, Thraustochytrids) and higher fungi (Ascomycetes, Basidiomycetes). The latest estimate of marine fungi is about 1500 species, and most of them are newly or inadequately described species (Hyde et al. 1998). Mangrove fungal communities play a key role in the nutritive cycle and in supporting mangrove ecosystems. Mangrove fungi usually occur as saprophytes (decomposing organic matter), symbionts (in plants and animals), and parasites (for plants) in mangrove ecosystems. Because of their exceptional ecosystems that bestride both the land and sea, mangroves originated as a rich source of mutualistic fungal endophyte microbiomes (Liu et al. 2007). Fungi associated with mangroves are recognized as manglicolous fungi. Nonetheless, most marine fungi are novel and scantily characterized species in mangrove ecosystems (Raghukumar 2008). A synergetic association among fungi and photosynthetic organisms is ancient and ubiquitous

A. M. V. N. Prathyusha · P. V. Bramhachari (✉)
Department of Biotechnology, Krishna University, Machilipatnam, Andhra Pradesh, India

G. Mohana Sheela
Department of Biotechnology, Vignan University, Guntur, Andhra Pradesh, India

C. V. Berde
Department of Microbiology, Gogate Jogalekar College, Ratnagiri, Maharashtra, India

B. P. Singh (ed.), *Advances in Endophytic Fungal Research*, Fungal Biology,
https://doi.org/10.1007/978-3-030-03589-1_11

(Muller and Krauss 2005). Nearly 200 species of endophytic fungi have been isolated and recognized from mangroves, of which the main genera are *Aspergillus, Alternaria, Clolletotrichum, Cladosporium, Penicillium, Trichodema, Paecilamyces, Phoma, Fusarium, Phomopsis, Pestalotiopsis* and *Phyllosticta* (Liu et al. 2007).

The production of commercially important hormones, volatile compounds, pigments, extracellular polysaccharides (EPSs), cross-kingdom signal molecules, and quorum sensing are characteristic aspects of endophytic fungi, which support proliferation and survival in the callous habitat of the phyllosphere, exposed to environmental extremes. A few interesting fungi, bacteria, and some cyanobacteria synthesize a high-molecular-weight EPS during their metabolic process. EPSs function in the formation of biofilm and the localization of biogeochemical processes in sediments and aggregates. Several marine microbial EPSs and their protein complexes possess unique chemical compositions, unusual structures with various biological activities, natural absorbents, and antioxidants owing to their unambiguous aquatic environment (Chi and Zhao 2003; Mancuso Nichols et al. 2004). In recent years, increasing attention has been devoted to the exploration and application of microbial polysaccharides for their potent industrial applications (Liu et al. 2017). Interestingly, microbial EPSs have numerous advantages over plant- or animal-derived EPSs, such as defined and reproducible production parameters to evade environmental conditions, along with the high yield and quality of end product.

EPSs of different microbial origins (bacteria and fungi) exhibit diverse structural combinations that furnish them with unique properties (Moscovici 2015; Delbarre-Ladrat et al. 2014). Owing to their beneficial properties in physicochemical and biological activities, microbial EPSs display a wide variety of commercial applications in the feed, food, chemical, pharmaceutical, textile, packaging, and cosmetics industries (Mahapatra and Banerjee 2013). However, over the last 5 years, several researchers have discovered that mangrove fungi are good-quality manufacturers of bioactive EPSs. These novel EPS molecules not only play an essential role in plant–endophyte interactions but also possess several biological functions, viz. antioxidant (Leung et al. 2009; Chen et al. 2011), antitumor, anti-inflammatory (Yan et al. 2011), and antiallergic activities (Mahapatra and Banerjee 2012). Therefore, mangrove fungi might be considered crucial sources for structurally unique and biologically potent EPSs with striking applications in different fields. In this review we try to refine a conceptual insight from previous reports, which may drive the scientific fraternity to address current impediments in fungal endophyte research. We emphasize the significance of endophytic fungi and their association with host plants. In addition, numerous biological associations of endophytic fungi with an exceptional reference to the production of mangrove fungal EPSs, its structure–function composition, and antioxidant potential are discussed in detail.

11.2 Significance of Mangrove Endophytic Fungi

Mangrove endophytic fungi possess wide scientific interest, for two key reasons. First, they comprise the second largest part of marine fungi. Second, these fungi may contribute to the overall performance of host plants by improving their fitness, photosynthetic efficiency, nutrient and water use, growth rate, and reproductive success, by acting as chemical defenses against herbivores, pathogens, or competitors (Strobel 2006; Singh et al. 2011), or by sharing genes and secondary metabolites that allow plants to tolerate abiotic or biotic stresses and, thus, adapt to changing environmental conditions (Singh et al. 2011). They might accordingly have a significant influence on plant biogeography, evolution, and community structure in terrestrial ecosystems (Rodríguez et al. 2009). As a result, scientists have learned that mangrove fungi are central for the adaptation of mangrove plants to extreme environmental conditions and for signifying that these fungi are promising bioresources for screening novel products, particularly endophytic mangrove fungi (Strobel et al. 2004). Endophytic fungi modulate plants at physiologic, genetic, and ecologic levels. These alterations induce severe changes in plant responses to their milieus, with impending consequences related to spatial differences in vegetation dynamics (Weishampel and Bedford 2006). Furthermore, transferring such endophytic fungi from host plants to nonhost plants would present enormous challenges in the form of a revolutionary biotechnological approach to building up new and improved cultivars for potential agricultural applications. Therefore, it is outlined how different host environments and biological pressures in unfavorable environments can perhaps be responsible for stimulating fungal endophytes to produce different significant bioactive molecules and to perform various other eco-friendly biological activities. It is known that metabolites may act as chemical defenses for the acclimatization of mangrove fungi (Rodriguez et al. 2004). Endophytic fungal species may number around one million, dispersed in plants, and represent a global potential source of novel and various undescribed species (Chandra 2012). Previous reports revealed that endophytic fungi that reside in plant tissues help them without causing any harm or disease. Furthermore, endophytic fungi support plant growth and synthesize secondary metabolites necessary for plant defense mechanisms (Tan and Zou 2001).

11.3 Natural Product Perspective of Mangrove Endophytic Fungi

Marine-derived fungi were reported in recent years to synthesize a number of novel biologically active secondary metabolites; however, few of them has new carbon frameworks and are unique in nature yet. Marine fungal endophytes may survive in metabolically hostile environments, constantly combating host defense compounds (Schulz et al. 1999). Such a hostile environment may acclaim for evolution of the potentially augmented synthetic ability of endophytes. This possibly explains the obvious difference detected when a species of endophytes isolated from a host plant produced a bioactive compound but was not successful in doing so when isolated from a different plant species (Li et al. 1996). Mangrove endophytic fungi have attracted the attention of several researchers because of their significance in ecology (Mayer and Hamann 2004; Bourguet–Kondracki and Kornprobst 2005). With notable progress in the modern fields of molecular separation, spectroscopic techniques, and microplate-based ultrasensitive in vitro and in vivo assays, the study of ecological products from marine endophytic fungi has advanced with increased attention devoted to unique chemical scaffolding (Kobayashi and Tsuda 2004a, b). Nonetheless, the occurrence and significance of endophytic fungi as they relate to the diversity of plant species are not understood yet. Therefore, there is a strong possibility of the discovery of unique biologically active molecules from these fungi linked in the company of therapeutically valuable plants (Nisa et al. 2015). In contrast, endophytic fungi are underexplored due to their inability to synthesize large amounts of bioactive molecules necessary for the discovery of drugs. Endophytic fungal species are documented to synthesize a vast variety of antimicrobial, antioxidant, and anticancer compounds that are biologically active (Chandra 2012; Nisa et al. 2015; Demain 2014).

11.4 Methods to Evaluate Endophytic Fungal EPSs

The structures of fungal EPSs are usually determined by monosaccharide composition, molecular weight, glycosidic linkages, branch structure, and chain lengths. The structural features of EPSs were determined by different tools of experimental analysis, including, for example, High-performance liquid chromatography (HPLC), Gas Chromatography-Mass Spectroscopy (GC-MS), Electronspray Ionization-Mass Spectroscopy (ESI-MS), and ^{1}H and ^{13}C Nuclear Magnetic Resonance (NMR) spectroscopy (Fig. 11.1) (Krcmar et al. 1999). The monosaccharide composition was analyzed by subjecting EPSs to acid hydrolysis followed by derivatization to alditolacetates or trimethylsilyl derivatives, which were analyzed by GC-MS (Ruiz-Matute et al. 2011; Chen et al. 2012). Some researchers determine monosaccharide composition using HPLC attached to a pulsed amperometric detector or UV detector (Chen et al. 2013a, b). Additional analytic approaches, viz. paper and Thin-layer chromatography (TLC), are rarely used for the compositional analysis of monosaccharides (Yang et al. 2005; Yan et al. 2006). The type of glycosidic bond in monosaccharides and functional groups of EPSs were evaluated by Fourier-transform infrared (FT-IR) spectroscopy (Fig. 11.1) (Chen et al. 2012).

Glycosidic linkage sequences of EPSs are generally established by methylation analysis of free hydroxyl (OH) groups. Methylated OH groups of polysaccharides are hydrolyzed to the corresponding monosaccharides, followed by reduction to alditols, followed by acetylation to form partly methylated alditol acetates. The newly formed methylated alditol acetates are finally analyzed by GC–MS. Glycosidic linkage patterns of EPSs can be deduced using periodic acid oxidation and Smith degradation methods (Fig. 11.1) (Ruiz-Matute et al. 2011; Sun et al. 2011). Moreover, using a nonspecific partial acid hydrolysis technique, the molecular weight and number of side chains are reduced, while low-molecular-weight portions of EPSs are obtained by partial hydrolysis. These fragments can be analyzed using various techniques, viz. electron spray ionization (ESI), tandem mass spectrometry (Sun et al. 2014), and NMR spectroscopy (Chen et al. 2011).

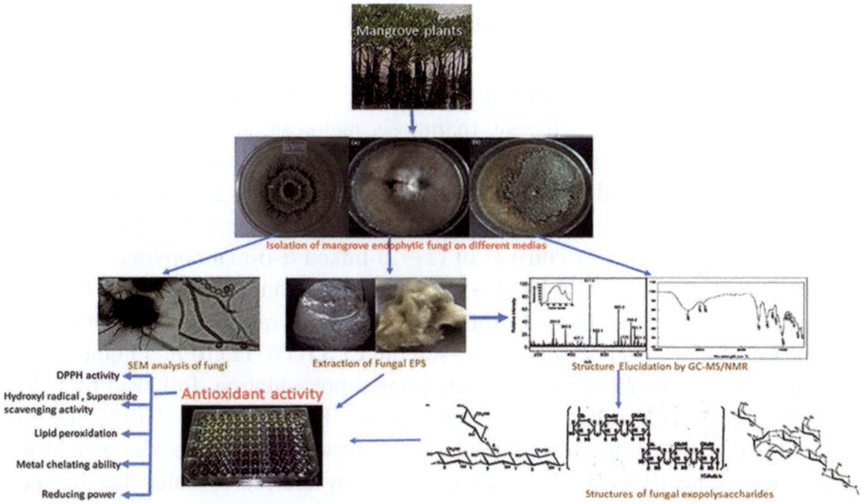

Fig. 11.1 Schematic overview of important steps involved in isolation of fungi from mangrove sources and in the isolation and identification of exopolysaccharides

NMR spectroscopy is an very beneficial and important tool for the structural description of EPSs. In ^{1}H NMR spectroscopy, ^{1}H signals for EPSs lie in a range of 3.5–5.5 ppm. However, in the case of the ^{13}C NMR spectrum, EPSs exhibit moderately broader chemical shifts than the ^{1}H NMR spectrum, typically ranging between 60 and 110 ppm. The structural patterns of EPSs comprising monosaccharide compositions and different glycosidic bonds can be studied using two-dimensional (2D) NMR technologies. Nowadays, 2D NMR analysis technology includes ^{1}H–^{1}H correlated spectroscopy (COSY), ^{1}H–^{1}H total correlation spectroscopy (TCOSY), ^{1}H–^{13}C heteronuclear multiple quantum coherence (HMQC) (Chen et al. 2012), ^{1}H–^{13}C heteronuclear multiple-bond correlation (HMBC), ^{1}H–^{13}C heteronuclear single-quantum coherence (HSQC), rotating frame Overhauser enhancement spectroscopy (ROESY), and nuclear Overhauser effect spectroscopy (NOESY), which have been effectively used for the structural description of EPSs (Fig. 11.1) (Chen et al. 2011, 2013a, b; Li et al. 2018). In addition, branching chains of EPSs can be identified using atomic force microscopy (AFM), transmission electron microscopy (TEM), scanning electron microscopy (SEM), and helix–coil transition assay (Mahapatra and Banerjee 2012). However, the detailed structural composition and functional applications of just a few fungal EPSs from diverse mangrove sources have been investigated.

11.5 Molecular Composition of Endophytic Fungal EPSs

Compositionally, an EPS produced by a marine filamentous fungus *Keissleriella* sp. YS contains galactose, glucose, rhamnose, mannose, and glucuronic acid in a proportion of 50:8:1:1:0.4 (Sun et al. 2004). Chen and coworkers showed that *Aspergillus* sp. Y16 produces a homogeneous EPS (As1-1) galactomannan with a molecular weight of about 15 kDa. 1D and 2D NMR spectroscopy analyses revealed that As1-1 mostly consists of (1→2)-linked α-D-mannopyranose units as backbone, substituted at C-6 by (1→6)-linked α-D-mannopyranose, (1→6)-linked β-D-galactofuranose, and (1→6)-linked β-D-mannopyranose units (Chen et al. 2011). Yan et al. (2014) purified water-soluble EPS (GW-12) from a liquid culture of mangrove fungus *P. solitum* through ethanol precipitation, followed by anion-exchange and size-exclusion chromatography. Reverse-phase HPLC analysis showed that GW-12 primarily contained D-mannose, and its molecular weight was expected to be nearly 11.3 kDa. NMR spectroscopy analysis revealed that the backbone of GW-12 was composed of (1→2)-linked α-D-mannopyranose and (1→6)-linked α-D-mannopyranose. Their branches are composed of terminal α-D-mannopyranose units with (1→6) linkages. Chen et al. (2015) extracted an EPS from mangrove fungus *F. oxysporum*, and its structure was characterized by FT-IR and ^{1}H and^{13}C NMR analysis. Results indicated that EPSs have a backbone with (1→6)-linked β-D-galactofuranose units, with multiple branches at the C-2 position composed of α-D-Glc*p*(1→2), β-D-Man*p*(1→2)-β-DMan*p*(1→2)-α-D-Glc*p*(1→2), and β-D-Man*p*(1→2)-α-DGlc*p*(1→2).

11.6 Antioxidant Activity of Mangrove Endophytic Fungal EPSs

It is noteworthy that the antioxidant activities of EPSs from mangrove fungi are most significantly analyzed among the numerous biological functions of EPS. Sun et al. (2004) reported free-radical-scavenging activities of an EPS (EPS2) isolated from the marine filamentous fungus *Keissleriella* sp. YS 4108. Radical elimination and other antioxidant actions of EPS2 (glycan) were evidenced in various in vitro systems showing that EPS2 demonstrated first-rate superoxide radical scavenging activity (Sun et al. 2004). EPS antioxidant activity was estimated by several in vitro assays, including by its metal chelating capability (Yang et al. 2005), reducing power, lipid peroxidation inhibition assay (Chen et al. 2011), and hydroxyl, superoxide, and DPPH (2,2-diphenyl-1-picrylhydrazyl) radical scavenging activity (Sun et al. 2009).

The antioxidant activity of EPSs in vitro is strongly associated with the source, composition of monosaccharides, protein molecular weight, and uronic acid residues. Chen et al. (2011) reported on the antioxidant activity of As1-1 (EPS) in vitro, isolated from the mangrove endophytic fungus *Aspergillus* sp. Y16. The strong

Table 11.1 Exopolysaccharides from mangrove fungi and their potential applications

S. no.	Species	Exopolysaccharide	Activity	Reference
1.	*A. pullulans* var. *melanogenium* P16	Pullulan	Thickener, a viscosity stabilizer, preparation of nontoxic, biodegradable, edible plastic materials	Ma et al. (2014)
2.	*Aspergillus* sp. Y16	Galactomannan	Antioxidant	Chen et al. (2011)
3.	*F. oxysporum*	Mannoglucogalactan	Antioxidant	Chen et al. (2015)
4.	*P. solitum*	Mannopyranose	Antioxidant	Yan et al. (2014)

DPPH and superoxide radical-scavenging activity of As1-1 was reported using EC_{50} values (1.45 mg/mL for DPPH and 3.4 mg/mL for superoxide radicals). Conversely, the As1-1 inhibition effect on lipid peroxidation was very slight. Chen et al. (2015) showed that EPSs produced by *F. oxysporum* (Fw-1) exhibited strong scavenging ability on hydroxyl radicals, DPPH activity, and higher EC_{50} values than As1-1. Nonetheless, the antioxidant mechanisms of EPSs are complex. Moreover, studies on EPSs with different structures and their antioxidant properties will play a significant part in understanding the mechanisms of antioxidant activity (Chen et al. 2015).

Although some alterations in the factors affecting antioxidant activity among in vitro and in vivo assays, the chemical composition of culture media and quantity of EPSs are major factors that can affect antioxidant activity in vitro (Anderson and Phillips 1999). Nonetheless, several elements, including bioavailability, digestibility, and polysaccharide metabolism, may also affect antioxidant activity in vivo. Henceforth, in vivo assays on antioxidant activity will be more critical for the assessment of the actual antioxidant potential of EPSs (Fardet et al. 2008). Nevertheless, very little attention has been paid to the in vivo antioxidant activities of EPSs that are synthesized by mangrove fungi (Table 11.1).

11.7 Conclusion and Future Perspectives

Although several mangrove fungal endophytes have been studied and reported for their important biological properties so far, in view of their diversity, they are seen comparatively as a new ecological source. Their exceptional biochemical and physiological properties direct the production of niche-explicit EPS molecules that are industrially potent. The discovery of specific EPS molecules from undesirable environments may help in understanding several biological mechanisms in fungi. Metabolites include micro- and macromolecules that are produced by the least studied and yet surprising pathways like hormones, pigments, volatile compounds, EPSs, cross-kingdom signals, and quorum sensing. Moreover, future research should focus on elucidating all the fundamental molecular pathways along with the effects of biochemical and physical inhabitants of specific hosts on endophytic function and metabolite elicitation. The currently most pressing need is to shift the research focus on endophytes and scientifically channel their ecological role toward potential industrial, biomedical, biological, and eco-friendly applications. Further research should focus on the mechanistic life of fungal endophytes in different ecoregions to make them suitable for their prospective biological functions.

Acknowledgment The author gratefully acknowledges UGC, Government of India, for its financial support in the form of UGC-CSIR-SRF under Grant F.No. 19-1/2015(SA-1).

References

Anderson D, Phillips BJ (1999) Comparative *in vitro* and *in vivo* effects of antioxidants. Food Chem Toxicol 37(9–10):1015–1025

Bourguet–Kondracki ML, Kornprobst JM (2005) Marine pharmacology: potentialities in the treatment of infectious diseases, osteoporosis and Alzheimer's disease. Adv Biochem Eng Biotechnol 97:105–131. https://doi.org/10.1007/b135824

Chandra S (2012) Endophytic fungi: novel sources of anticancer lead molecules. Appl Microbiol Biotechnol 95(1):47–59

Chen YL, Mao WJ, Tao HW, Zhu WM, Yan MX, Liu X et al (2015) Preparation and characterization of a novel extracellular polysaccharide with antioxidant activity, from the mangrove-associated fungus *Fusarium oxysporum*. Mar Biotechnol 17(2):219–228

Chen Y, Mao W, Tao H, Zhu W, Qi X, Chen Y et al (2011) Structural characterization and antioxidant properties of an exopolysaccharide produced by the mangrove endophytic fungus *Aspergillus* sp. Y16. Bioresour Technol 102(17):8179–8184

Chen Y, Mao W, Wang B, Zhou L, Gu Q, Chen Y et al (2013a) Preparation and characterization of an extracellular polysaccharide produced by the deep-sea fungus *Penicillium griseofulvum*. Bioresour Technol 132:178–181

Chen Y, Mao W, Wang J, Zhu W, Zhao C, Li N et al (2013b) Preparation and structural elucidation of a glucomannogalactan from marine fungus *Penicillium* commune. Carbohydr Polym 97(2):293–299

Chen Y, Mao W, Yang Y, Teng X, Zhu W, Qi X et al (2012) Structure and antioxidant activity of an extracellular polysaccharide from coral-associated fungus, *Aspergillus versicolor* LCJ-5-4. Carbohydr Polym 87(1):218–226

Chi Z, Zhao S (2003) Optimization of medium and cultivation conditions for pullulan production by a new pullulan-producing yeast strain. Enzyme Microb Technol 33(2):206–211

Delbarre-Ladrat C, Sinquin C, Lebellenger L, Zykwinska A, Colliec-Jouault S (2014) Exopolysaccharides produced by marine bacteria and their applications as glycosaminoglycan-like molecules. Front Chem 2:85

Demain AL (2014) Importance of microbial natural products and the need to revitalize their discovery. J Ind Microbiol Biotechnol 41:185–201

Fardet A, Rock E, Rémésy C (2008) Is the in vitro antioxidant potential of whole-grain cereals and cereal products well reflected in vivo? J Cereal Sci 48(2):258–276

Hyde KD, Jones EBG, Leano E, Pointing SB, Poonyth AD, Vrijmoed LLP (1998) Role of fungi in marine ecosystems. Biodivers Conserv 7:1147–1161

Kobayashi J, Tsuda M (2004a) Bioactive products from Okinawan marine micro-and macroorganisms. Phytochem Rev 3(3):267–274

Krcmar P, Novotny C, Marais MF, Joseleau JP (1999) Structure of extracellular polysaccharide produced by lignin-degrading fungus *Phlebia radiata* in liquid culture. Int J Biol Macromol 24(1):61–64

Leung PH, Zhao S, Ho KP, Wu JY (2009) Chemical properties and antioxidant activity of exopolysaccharides from mycelial culture of *Cordyceps sinensis* fungus Cs-HK1. Food Chem 114(4):1251–1256

Kobayashi J, Tsuda M (2004b) Bioactive products from Okinawan marine micro– and macro–organisms. Phytochem Rev 3:267–274

Li JY, Strobel GA, Sidhu R, Hess WM, Ford EJ (1996) Endophytic taxol-producing fungi from bald cypress, *Taxodium distichum*. Microbiology 142:2223–2226

Li H, Cao K, Cong P, Liu Y, Cui H, Xue C (2018) Structure characterization and antitumor activity of the extracellular polysaccharide from the marine fungus *Hansfordia sinuosae*. Carbohydr Polym 190:87–94

Liu AR, Wu XP, Xu T (2007) Research advances in endophytic fungi of mangrove. Chin J Appl Ecol 18:912–918

Liu J, Wang X, Pu H, Liu S, Kan J, Jin C (2017) Recent advances in endophytic exopolysaccharides: production, structural characterization, physiological role and biological activity. Carbohydr Polym 157:1113–1124

Ma ZC, Fu WJ, Liu GL, Wang ZP, Chi ZM (2014) High-level pullulan production by *Aureobasidium pullulans* var. *melanogenium* P16 isolated from mangrove system. Appl Microbiol Biotechnol 98(11):4865–4873

Mahapatra S, Banerjee D (2012) Structural elucidation and bioactivity of a novel exopolysaccharide from endophytic *Fusarium solani* SD5. Carbohydr Polym 90(1):683–689

Mahapatra S, Banerjee D (2013) Fungal exopolysaccharide: production, composition and applications. Microbiol Insights 6:1–16. https://doi.org/10.4137/MBI.S10957

Mancuso Nichols CA, Garon S, Bowman JP, Raguenes G, Guezennec J (2004) Production of exopolysaccharides by Antarctic marine bacterial isolates. J Appl Microbiol 96(5):1057–1066

Mayer AMS, Hamann MT (2004) Marine pharmacology in 2000: marine compounds with antibacterial, anticoagulant, antifungal, anti-inflammatory, antimalarial, antiplatelet, antituberculosis, and antiviral activities; affecting the cardiovascular, immune, and nervous systems and other miscellaneous mechanisms of action. Mar Biotechnol 6:37–52

Moscovici M (2015) Present and future medical applications of microbial exopolysaccharides. Front Microbiol 6:1012

Muller CB, Krauss J (2005) Symbiosis between grasses and asexual fungal endophytes. Curr Opin Plant Biol 8(4):450–456

Nisa H, Kamili AN, Nawchoo IA, Shafi S, Shameem N, Bandh SA (2015) Fungal endophytes as prolific source of phytochemicals and other bioactive natural products: a review. Microb Pathog 82:50–59

Raghukumar C (2008) Marine fungal biotechnology: an ecological perspective. Fungal Divers 31:19–35

Rodriguez RJ, Redman RS, Henson JM (2004) The role of fungi symbioses in the adaptation of plants to high stress environments. Mitig Adap Strat Glob Change 9:261–272

Rodríguez RJ, White JRJF, Arnold AE, Redman RS (2009) Fungal endophytes: diversity and functional roles. New Phytol 182:314–330

Ruiz-Matute AI, Hernandez-Hernandez O, Rodriguez-Sanchez S, Sanz ML, Martinez-Castro I (2011) Derivatization of carbohydrates for GC and GC–MS analyses. J Chromatogr B 879(17–18):1226–1240

Sarma VV, Hyde, KD (2001) A review on frequently occurring fungi in mangroves. Fungal Divers 8:1–34

Schulz B, Römmert A-K, Dammann U, Aust H-J, Strack D (1999) The endophyte–host interaction: a balanced antagonism. Mycol Res 103:1275–1283

Shearer CA, Descals E, Kohlmeyer B, Kohlmeyer J, Marvanová L, Padgett D et al (2007) Fuangal diversity in aquatic habitats. Biodivers Conserv 16:49–67

Singh LP, Gill SS, Tuteja N (2011) Unraveling the role of fungal symbionts in plant abiotic stress tolerance. Plant Signal Behav 6:175–191

Sun C, Wang JW, Fang L, Gao XD, Tan RX (2004) Free radical scavenging and antioxidant activities of EPS2, an exopolysaccharide produced by a marine filamentous fungus *Keissleriella* sp. YS 4108. Life Sci 75(9):1063–1073

Sun HH, Mao WJ, Chen Y, Guo SD, Li HY, Qi XH et al (2009) Isolation, chemical characteristics and antioxidant properties of the polysaccharides from marine fungus *Penicillium* sp. F23-2. Carbohydr Polym 78(1):117–124

Sun HH, Mao WJ, Jiao JY, Xu JC, Li HY, Chen Y et al (2011) Structural characterization of extracellular polysaccharides produced by the marine fungus *Epicoccum nigrum* JJY-40 and their antioxidant activities. Mar Biotechnol 13(5):1048–1055

Sun Y, Wang H, Guo G, Pu Y, Yan B (2014) The isolation and antioxidant activity of polysaccharides from the marine microalgae *Isochrysis galbana*. Carbohydr Polym 113:22–31

Sridhar KR (2004) Mangrove fungi in India. Curr Sci 86(12):1586–1587

Strobel GA, Daisy B, Castillo U, Harper J (2004) Natural products from endophytic microorganisms. J Nat Prod 67:257–268

Strobel GA (2006) *Muscodor albus* and its biological promise. J Ind Microbiol Biotechnol 33:514–522

Tan RX, Zou WX (2001) Endophytes: a rich source of functional metabolites. Nat Prod Rep 18(4):448–459

Thatoi H, Behera BC, Mishra RR (2013) Ecological role and biotechnological potential of mangrove fungi: a review. Mycology 4(1):54–71

Weishampel PA, Bedford BL (2006) Wetland dicots and monocots differ in colonization by arbuscular mycorrhizal fungi and dark septate endophytes. Mycorrhiza 16(7):495–502

Yan JK, Wang WQ, Li L, Wu JY (2011) Physiochemical properties and antitumor activities of two α-glucans isolated from hot water and alkaline extracts of *Cordyceps* (Cs-HK1) fungal mycelia. Carbohydr Polym 85(4):753–758

Yan J, Guo XQ, Li XG, Wu XY, Gou XJ (2006) TLC to fleetly analyze monosaccharide composition of polysaccharide. Food Sci 27(12):603–607

Yan M, Mao W, Chen C, Kong X, Gu Q, Li N et al (2014) Structural elucidation of the exopolysaccharide produced by the mangrove fungus *Penicillium solitum*. Carbohydr Polym 111:485–491

Yang XB, Gao XD, Han F, Tan RX (2005) Sulfation of a polysaccharide produced by a marine filamentous fungus *Phoma herbarum* YS4108 alters its antioxidant properties in vitro. Biochim Biophys Acta 1725(1):120–127

Chapter 12
Endophytic Fungi: Promising Source of Novel Bioactive Compounds

Fazilath Uzma, Chakrabhavi D. Mohan, Chandra N. Siddaiah, and Srinivas Chowdappa

12.1 Introduction

Microorganisms have been of considerable importance to mankind, both environmentally and economically. They possess unique traits and are substantiated by their ability to grow fast. Their enzyme systems can catalyze various chemical reactions which cannot be reproduced synthetically in chemical laboratories. They have been proven as potential drug candidates and are significant in processes such as fermentation and biotransformation. Also, bacterial and fungal microorganisms have fascinated researchers as producers of prospective bioactive compounds since they can be preserved indefinitely, assuring perennial availability of the biologically active compounds (Lam 2006). They can be manipulated genetically and physiochemically to obtain enhanced and higher yields of the desired bioactive compounds (Kharwar et al. 2011). Microorganisms produce secondary metabolites of biological interest for defense against invading pathogens (Vinale et al. 2014). The global scenario has witnessed several effective drugs of microbial origin, but a number of problems such as cancer, infections such as HIV, H1N1, cardiovascular diseases, and several microbial infections still remain to be cured. Hence, the search for novel compounds against infectious diseases is ever increasing and is a colossal task.

F. Uzma · S. Chowdappa
Fungal Metabolite Research Laboratory, Department of Microbiology and Biotechnology, Bangalore University, Jnanabharathi Campus, Bangalore, Karnataka, India

C. D. Mohan
Department of Studies in Molecular Biology, University of Mysore, Manasagangotri, Mysore, Karnataka, India

C. N. Siddaiah (✉)
Department of Studies in Biotechnology, University of Mysore, Manasagangotri, Mysore, Karnataka, India

B. P. Singh (ed.), *Advances in Endophytic Fungal Research*, Fungal Biology,
https://doi.org/10.1007/978-3-030-03589-1_12

12.1.1 *Natural Products as Basis for Potential New Compounds*

The invention of the antibiotic penicillin by Alexander Fleming in 1928 resulted in the discovery of numerous bioactive compounds for treating several infective diseases. Ever since, there have been persistent efforts towards novel discoveries to tackle the problem of increasing drug resistance in microorganisms (Davies and Davies 2010). Therefore, persistent search for novel therapeutic agents is essential to discover new structural compounds for novel drug development. The research on the novel bioactive metabolites from microorganisms has been anticipated to meet this demand and hence continues to be the enthralling area of investigation. The bioactive secondary metabolites are produced by microorganisms in response to external stimuli such as nutritional changes or foreign infection and have been identified as the source for the derivation of novel drugs (Pavarini et al. 2012). Presently, a minimum of 200,000 natural bioactive metabolites with novel biological activities have been identified (Demain and Sanchez 2009). Research on microbial metabolites has gained momentum due to less toxicity, low production costs, structural diversity, and broad spectrum activities with lesser amounts of compound administration to treat various diseases (Rout et al. 2009). They offer excellent prospects for finding potential biologically active compounds and are the key sources for the synthesis of effective drugs (Cragg and Newman 2013). Presently, the discovery of effective medicines with least side effects proves to be an alternative to traditional disease control and treatment methods (Kusari et al. 2009). Most of the antibacterial compounds presently used in therapeutic regimen are natural products or their derivatives. Research has been focused these days towards endophytic fungi from medicinal plants as they have traditionally been used for the treatment of several diseases. Bioactive products from fungal endophytes of medicinal and aromatic plants have been the continuous sources of bioactive metabolites which can have an abundant impact for novel drug discovery in modern medicine (Cragg et al. 2009). Several reports suggest that the biologically active compounds from endophytic fungi could be alternative approaches for novel drug discovery (Joseph and Priya 2011).

The plant kingdom is a potential source of bioactive compounds, and it is assessed that only 10–15% of existing higher plant species have been explored for their bioactive potentialities (Bisht et al. 2006). A very small number of these plants (about 6%) have been screened for their bioactivities (Demain and Vaishnav 2011). The usage of medicinal plants can be traced back to traditional practices where the indigenous populations have utilized them for treatment of human ailments (Okigbo et al. 2009). Medicinal plants are known to be a repository of various bioactive secondary metabolites and bioactive compounds which can be used as precursors for chemo-pharmaceutical semi-synthesis (Sasidharan et al. 2011). Nevertheless, overuse of medicinal plants for traditional medicinal practices in folklore medicine has resulted in environmental degradation and loss of biodiversity. Also, pathogens, forest fires, overgrazing, and overharvesting could substantially limit the abundance of these rare medicinal plant species. The present deteriorating condition of this valuable resource is very precarious and needs immediate attention for conservation and propagation. To combat the declining conditions of medicinal plants, bioactive compounds from fungal endophytes could play a lead role in providing constant supply of chemotherapeutics with high target specificity, least side effects, and lower cost, thereby reducing deforestation and extinction of ecologically important as well as valuable medicinal plants.

12.1.2 Fungal Endophytes

Medicinal plants could host vast number of microorganisms, commonly referred as endophytes (Akay et al. 2014). Endophytes reside in the internal host plant tissues symbiotically and spend most of their life cycle within the host plants. Approximately one million endophytic species are present in the plant kingdom (Kusari et al. 2009). Fungal endophytes are ubiquitous, highly diverse organisms but are generally asymptomatic. They are an unexplored group of microbial plant symbionts and usually occupy above-ground tissues such as petioles, leaves, stems, bark, and reproductive structures which discriminates them from mycorrhizal symbionts. This discrimination may not be ascertained as endophytic fungi also colonize the root tissues (Rodriguez et al. 2009). Nearly all plants investigated till date harbor at least one endophytic fungal species (Saikkonen 2007). The bioactive chemical metabolites produced by endophytes could be influenced by the defensive chemistry of their host plants. During the extensive coevolution period, some fungal endophytes produce related or identical biological active compounds similar to their host plants (Zhao et al. 2011). The endophyte could exist in a metabolically unfavorable atmosphere and could be subjected to encounter host defense chemicals continuously and their association could stimulate the secondary metabolite production by the host plant (Jalgaonwala and Mahajan 2011). Various studies have indicated that medicinal plants which host fungal endophytes could produce important natural products of pharmacological interest, thereby directing more studies on the plant–endophyte interactions for beneficial compounds. This interaction is regulated and controlled by the genetic machinery of the host and the endophyte and is modulated by the environment (Moricca and Ragazzi 2008). However, endophytic fungi remained unexploited in their ability to produce bioactive chemical constituents for drug development. Several endophytic fungi possess the distinctive ability of producing bioactive secondary metabolites which could be used as proficient therapeutic agents against various diseases (Uzma et al. 2018; Aly et al. 2010). Some examples include taxol (paclitaxel), podophyllotoxin, vincristine, etoposide, irinotecan, topotecan, and vinblastine presently in use for treatment of different types of human cancer (Balunas and Kinghorn 2005). Different endophytic fungal species producing a number of antimicrobial, antioxidant, and anticancer compounds have also been reported (Gouda et al. 2016).

12.1.3 Origin of Fungal Endophytes

The word "endophyte" was initially used to refer to any organism found inside the tissues of living autotrophs (Arnold and Lewis 2005). However, this term has undergone several redefinitions. In the year 1991, Petrini proposed a working definition for endophytes and has since been widely accepted by the scientific community which states, "Organisms that inhabit host plant organs at some time in their life can colonize internal plant tissues without causing apparent harm to their host." This definition is factual for endophytic beings that have a lengthy epiphytic phase and also for latent pathogens that live an asymptomatic life within their hosts (Petrini 1991). Endophytes are diverse microorganisms which include bacteria, fungi, and cyanobacteria which inhabit the vascular tissues and the reproductive structures of living plants. Predominantly, endophytes exist in mutual association with the host plants as they obtain nutrition from them, while the host plant enjoys enhanced resistance to herbivores, pathogens, increased competitive abilities to combat pathogens and resistance to various abiotic stresses due to endophytes residing within them. The interaction between the endophytes and host plant may differ from being an antagonistic to mutualistic association. It is also well-known fact that endophytes could influence the population dynamics, plant community diversity, and ecosystem functions as noted on a few grass endophytes (Hoffman and Arnold 2010).

Fungal endophytes, however, can share different relationships with different hosts and exhibit distinctive features which could possibly include symbiotic, mutualistic, commensalistic, and parasitic associations in response to host genotype and environmental factors (Singh et al. 2011). Fungal endophytes have the distinct ability to express multiple symbiotic lifestyles. However, it is understood that the life cycle of the pathogenic fungi could change depending on the host genotype and this switching can occur in genetically divergent or in the same species suggesting that the expression of the symbiotic lifestyles is dependent on the different host types. Also, single gene modifications can result in mutualistic existence of the pathogen. The fungal endophytes express a nonpathogenic life style and exist as latent pathogens in a disease free host (Schulz et al. 1999). The endophytic fungal origin may be from pathogens losing their virulence. Phytopathogens may get converted into an endophyte as a result of gene mutation which could have occurred during coevolution period in the host plant. Many of the known endophytic fungi are inactive phytopathogens that reside within a healthy host, but could cause disease when the host plant is aged or stressed. Schulz et al. (1999) developed a hypothesis that an antagonistic balance between endophytic virulence and plant defense response results in asymptomatic colonization. Furthermore, fungal endophytes may exhibit multiple lifestyles at different life stages and also can include pathogenic organisms that exhibit latency between infection and onset of disease symptoms. Hence, fungal endophytes are microorganisms that exist within plant tissues and comprise a diverse, polyphyletic group with the potential for myriad interactions with host plants (Petrini 1986). Interaction of endophytic fungi is host specific. Several fungal endophytes can be isolated from one host plant among which at least one fungal species demonstrates host specificity. For example, the species of *Neotyphodium* and *Epichloë* are the best known and intensively studied group of fungal grass endophytes (Márquez et al. 2010). The endophytic *Acremonium* which inhabits the tall fescue *Lolium perenne* L. is host specific to grasses and remains symptomless throughout its life cycle (Rudgers and Clay 2007).

12.1.4 Plant-Endophyte Symbiosis

The phenomenon of "endophytism" reveals the reciprocal or mutual association of a plant with a microbe, wherein the microbe exists within the host plant without causing an infection. Besides being a treasured biological resource, endophytes play varied roles and are essential for plant growth, development, stress tolerance, and adaptation. The first description of symbiotic relationship as "the living together of dissimilar organisms" was proposed by De Bary in 1879, based on which symbiosis has been defined depending on the impact to host and symbionts. These fungal symbionts have a solid impact on the ecology of the host plant, fitness, evolution, altering plant population and demonstrate major effects on the community structure and diversity of associated organisms. Symbiotic relationships may impact the response of the plants to any external environmental changes (Brundrett 2006). The term "endophyte" has become frequent with mutualism. During the asymptomatic phase of their life cycle, the ecological role of fungi may vary from antagonistic associations to mutualistic associations and hence these distinct plant–endophyte interactions have been referred to as continuum (Schulz and Boyle 2005).

Over the last decade, an array of diverse endophytic microflora including bacteria (Garbeva et al. 2001) and fungi (Saikkonen et al. 1998) have been isolated and identified by molecular methods from various asymptomatic plant tissues. This microflora includes latent pathogens, saprotrophs, and also symbionts (Faeth and Hamilton 2006). Most of the plants share a symbiotic relationship with the fungal endophytes in the natural environment (Petrini 1986). Endophytes produce diverse chemical entities that include phytohormones and can modulate the expression of genes in the bioactive secondary metabolism of the host plant (Wani et al. 2016). Although much information of the ecology, life history, and phylogeny of endophytic fungi has been progressively gathered in the past decades, major queries regarding the origin, evolution, speciation, and their ecological roles still remain to be answered. The most commonly studied mutualistic mechanism was identified in foliar grasses in which a small proportion of endophytes are symbiotic fungi that can form an asymptomatic and lifetime association in the host plant (Petrini 1991). However, in the twentieth century, Charles Bacon and his colleagues discovered the source of fescue toxicosis, a livestock disease when fed on the *Festuca arundinacea* pastures (Bacon et al. 1977). These investigators observed that though the grass pastures were symptomless, most plants of *F. arundinacea* had a systemic colonization in their leaves and stems by a fungus identified as *Neotyphodium coenophialum*, hence the cattle suffered intoxications. It was found that infected plants contained several toxic alkaloids which helped the *Neotyphodium* species to be beneficial to their host plants. They increased host plant tolerance to several biotic and abiotic stress factors, improved nutrient use and conferred anti-herbivore defense to their plant hosts (Schardl et al. 2004). The endophytic *Neotyphodium* species and their *Epichloë* teleomorphs are the best known and an extensively studied group of fungal endophytes. Some grass species are known to host more than a hundred different species of fungal endophytes (Márquez et al. 2007) and this number is larger for members of other plant families (Arnold and Lutzoni 2007). Endophytes are ubiquitous in the plant world. In addition, plant species without endophytes are rare. Grass endophytes systemically infect almost all plant–plant, plant–pathogen, and plant–herbivore interactions in grassland populations. However, two economically important introduced grass species—tall fescue and perennial ryegrass (*Lolium perenne* L.), and their artificially selected cultivars have been studied for their strong plant–endophyte mutualism (Saikkonen et al. 2010). Most endophytic surveys rely on the tissues of young and healthy plants. As a result, it aids in obtaining latent pathogens from such plants if the sampling for endophytes is done prior to the appearance of the disease symptoms (Photita et al. 2004). However, prior to the onset of disease, these dormant pathogens are not important constituent of the endophytic assemblages, as endophytes are symptomless in plants.

For example, the perennial grass *Dactylis glomerata* L. was studied for their endophytic assemblages. One hundred and nine fungal species were identified, amid which five species were recognized pathogens of that grass (Márquez et al. 2007). Sometimes, dormant saprophytes have been isolated as endophytic fungi in healthy tissues. These endophytic fungi behave as latent saprophytes which are spatially restricted in growth while the host is actively growing, but when the infected host plant senesces or dies, these fungi grow unrestricted and reproduce (Promputtha et al. 2007).

Many microorganisms including root endophytes establish symbiotic associations with plants and play a vital role in maintaining a better soil and plant health. One such unique member capable of colonizing roots of many plant species and establishing symbiotic relationships is *Piriformospora indica* (order Sebacinales of Basidiomycetes), known to benefit plants by colonizing the roots of several medicinal plants leading to their growth promotion (Varma et al. 2012). This endophyte, on root colonization, can upsurge the nutrient uptake; helps the plants to withstand salt, drought, and oxidative stress conditions; and confers toxin resistance, resistance to heavy metals as well as protection against root and foliar pathogens. Other positive effects of the endophyte *P. indica* include promotion of plant growth, enhanced nitrate and phosphate assimilation, adventitious root formation, enhanced secondary metabolite production, and higher seed yield (Varma et al. 2012). In addition, the root endophytic *Acremonium strictum* W. Gams in the grass *Ammophila arenaria* increased root biomass (Hol et al. 2007) and *Stagonospora species* enhanced the host growth (Ernst et al. 2003).

12.1.5 Alterations Between Mycorrhizal and Fungal Endophytic Associations Within the Host Plants

In the rhizosphere soil, many fungi including the pathogenic, necrotrophic, and beneficial fungi (endophytes which offer protection against pathogens) inhabit the root cortex of the host plant without causing any disease and it is a daunting task to precisely categorize these fungi based on their fungal-root associations. Hence, there was an imperative need to adequately define and separate the mycorrhizal associations from the other root-fungal associations. It was stated that endophytic associations are diverse from mycorrhiza by way of lack of localized interface of specialized hyphae, non-existence of synchronized plant fungus development, and the lack of plant benefits from nutrient transfer, which are the distinctive features of mycorrhizas. However, endophytes provide the distinct advantage to the plant by providing herbivore resistance, stress and pathogen protection. Mycorrhizal fungi are primarily different from other fungal organisms as they are efficient at influencing plant growth and nutrient uptake in soils and plants, whereas endophytes and pathogenic organisms are plant inhabitants which do not possess efficient means of nutrient uptake from soils and supply it to the plants. The multifunctional host fungal interface in mycorrhizal associations helps in rapid nutrient transfer in a relatively short span of time. Nevertheless, transfer of nutrients in endophytic associations is important to the fungus if it continues to persist over longer duration (months or years) than the active phase of mycorrhizal associations. Also, the plant-endophyte coevolution would not happen if the endophytes did not obtain benefits from the host plants (Brundrett 2009). It can also be probable that mycorrhizal fungi may have endophytic phases, involving longer periods of existence within the host plants without any nutrient transfer. For instance, old roots of the *Glomeromycete* fungi can exist in living roots as endophytes for a decade even after arbuscle termination. It is clear that fungi play different roles and the multifarious role of these multifunctional fungi is uncertain. Fungal versatility also possesses several benefits. Such multifunctional fungi are able to colonize novel habitats and breakdown a wide variety of substrates than specialized fungi. However, the role of endophytic fungi seems less common as compared to the mycorrhizal associations. More elaborate research would help to ascertain the exact nature and the relative importance of root-associated fungi, since it is unclear whether these are endophytic or mycorrhizal associates, or how these roles change with time. Further, the evolution of fungi is dependent on the host plant development and sometimes, the fungal species may be parasitic which appear at certain stages of their lifecycles. These parasitic fungal species depend on the host plant for a particular supply of nutrients which are imperative for their sustenance. Mainly two types of parasitism are recognized—nectrotrophic parasitism and biotrophic parasitism. The infection resulting in the destroying of the tissue and eventually to death of the plant is the nectrotrophic parasitism. The parasitic fungi can proliferate in the dead and withering plant material, whereas in the biotrophic parasitism, the parasite and host live together for longer periods of time. The parasite acquires nutrients and growth regulating substances such as vitamins from the host plant, but does not kill the host plant. Most of the biotrophic fungi are obligatory parasites. The transfer of nutrients in the parasitic fungi may be active or passive. Parasites alter the host metabolism and its biological functions while attaining specific benefits from the host.

12.1.6 Classification of Endophytes

The endophytes consist of a broad range of fungi which includes the latent pathogens and the dormant saprophytes as mentioned earlier. However, phylogenetic analysis suggests that some of the endophytes are genetically distinct than the parasites within the same host despite their morphological identity (Ganley and Newcombe 2006). In general, two major groups of endophytic fungi have been recognized—the clavicipitaceous endophytes (C-endophytes) and non-clavicipitaceous endophytes (NC-endophytes). Clavicipitaceous endophytes belong to the family *Clavicipitaceae* (*Hypocreales*; *Ascomycota*), many species produce bioactive molecules (mainly of the genera *Cordyceps*, *Balansia*, *Epichloe* or *Neotyphodium*, *Claviceps* and *Myriogenospora*). Contrastingly, non-clavicipitaceous endophytic fungi are a large group that have not been well defined taxonomically, but the majority of the species belong to the phyla *Ascomycota* and *Basidiomycota*, represented by the genera *Alternaria*, *Arthrobotrys*, *Aspergillus*, *Cladosporium*, *Colletotrichum*, *Coprinellus*, *Curvularia*, *Fusarium*, *Paecilomyces*, *Penicillium*, *Phanerochaete*, *Phoma*, and others (Spatafora and Bushley 2015).

C-endophytes are known to infect grasses, whereas NC-endophytes have been observed in healthy tissues of nonvascular plants, ferns, conifers, and angiosperms. Mostly, C- endophytes are systemic and vertically transmitted through seeds and exclusively infect grass. Whereas, non-systemic Class II endophytes are taxonomically diverse, horizontally transmitted from plants to plants and colonize almost all plants in ecosystems.

12.1.6.1 Clavicipitaceous Endophytes

Clavicipitaceous endophytes or Class I endophytes, also called Balansiaceous fungi represent a small number of phylogenetically related clavicipitaceous species that are fastidious in culture and limited to some cool- and warm-season grasses (Bischoff and White 2005). Distinctively these fungal endophytes spend their entire life cycle within the above-ground portion of the host grass, forming nonpathogenic, systemic, and usually intercellular associations. They confer drought resistance, herbivore resistance and are known to increase plant biomass. However, these benefits conferred by Class I endophytic fungi are reliant on the host species, host genotype, and environmental conditions (Faeth and Hamilton 2006). Class I endophytes follow primarily vertical transmission, wherein fungi are passed through maternal plants on to offspring through seed infections (Saikkonen et al. 2010).

Usually one fungal isolate or genotype dominates the colonized plants. The endophytic infection frequency upsurges in grass populations over a time period suggesting the adaptive advantage on the host species. The growth increase of perennial ryegrass and tall fescue was observed due to endophyte contagion by way of osmoregulation and stomatal regulation. They defend the plants against nitrogen starvation and water stress. Numerous studies conducted on single cultivars and natural ecotypes of tall fescue, meadow fescue, and perennial ryegrass in controlled environments suggest that their *epichloë* endophytes (*Neotyphodium coenophialum*, *N. uncinatum*, and *N. lolii*, respectively) have fixed positive effects on plant growth. Besides plant growth, increases in biomass production, tiller numbers, seed production, and root growth have been reported (Singh et al. 2011). Also, endophytic fungi in tall fescue and meadow fescue have presented with enhanced root growth, extended root hairs, and decreased root diameter. Grass endophytes express a wide range of adaptations to biotic and abiotic stress which include drought and soil acidity. Hence, the endophyte infected grasses are able to survive in limited resources and are more competitive than non- endophyte infected grasses (Malinowski and Belesky 2006). Though the systemic grass endophytes have been well studied, the interactions between host plants and endophytes in natural populations have not been comprehended fully.

12.1.6.2 Non-Clavicipitaceous Endophytes

Non-Clavicipitaceous (NC) endophytes or non-balansiaceous endophytes are highly diverse representing a polyphyletic assemblage of primarily ascomycetous fungi and their ecological roles are not entirely defined. Non-clavicipitaceous endophytes have been identified from all terrestrial biomes including both agroecosystems and biomes ranging from the tropical to the tundra regions (Arnold and Lutzoni 2007). NC-endophytes are differentiated into three functional classes (Class 2, 3, and 4) depending on their host colonizing patterns, transmission between host generations, *in planta* biodiversity levels, and ecological function. Even though the three classes have wide host ranges, Class 2 endophytes may propagate in both above- and below-ground tissues. By contrast, Class 3 and 4 endophytes are limited to above-ground tissues and roots, respectively. Colonization of host tissues also differs among the classes. Class 3 endophytes form highly localized infections, while Class 2 and 4 endophytes possess the capability of extensive tissue colonization.

(a) ***Class 2 Endophytes***

The members of the Class 2 endophytes comprise *Ascomycota* or *Basidiomycota*, but most belong to the *Ascomycota a*nd are different from the other NC-endophytes as they inhabit plant tissues like roots, stems, and leaves and are competent of forming far-reaching infections inside the plant. They spread via the seed coats or rhizomes and are sparse in the rhizosphere region. They typically possess high infection frequencies in plants growing in high stress habitats.

They receive considerable attention due to their vast species diversity, and possess lifestyle switching abilities between endophytic and free living existence and their various fungal ecological modes with prospective applications (Rodriguez et al. 2009). This endophytic class has been considered as the largest fungal symbiotic group. They can be cultured on artificial media and are able to colonize majority of the plants in natural ecosystems. The host-endophytic association occurs both in aerial and below-ground tissues. The first description of a Class 2 endophyte was in *Phoma* sp. of *Calluna vulgaris* (Rayner 1915). Although it was misunderstood as mycorrhizal association, actually it was a fungal endophyte that colonized all plant parts including the seed coat, and did not produce intracellular mycorrhizal structures. The characteristic feature of the class 2 endophytes is that they colonize plants via appressoria formation or penetrate the plant tissues directly through the hyphae. Although healthy plants have low levels of appressoria formation, these fungi are able to sporulate rapidly during host senescence (Rodriguez et al. 2009). Class 2 endophytes are mutualistic and are responsible in providing fitness benefits to hosts while they obtain nutrition from host tissues for growth and reproduction, and avoid abiotic stress via symbiosis.

(b) ***Class 3 Endophytes***

Most of the class 3 endophytic fungi belong to the Phylum *Ascomycota* with a small number belonging to *Basidiomycota* which propagate via hyphal fragmentation and produce sexual or asexual spores on dead tissues. Spores and hyphal fragments may be released passively, by herbivores or by physical factors such as wind or rain (Herre et al. 2007).

These endophytic fungi are differentiated based on their occurrence mainly above-ground tissues (shoots of the plant) with horizontal transmission. These endophytic fungi form highly localized infections and possess high *in planta* biodiversity. They also include the high diverse endophytic fungi existing in the leaves of tropical trees and above-ground tissues of nonvascular plants, seedless vascular plants, woody and herbaceous angiosperms in the tropical forest biomes (Davis and Shaw 2008). Besides their occurrence within photosynthetic and herbaceous tissues, these fungal endophytes are also found in flowers, fruits, inner bark and are highly diverse within individual host tissues and plants. Most healthy leaves in moist tropical forests contain numerous independent infections rather than systemic infections (Arnold 2007).

(c) ***Class 4 Endophytes***

Class 4 endophytes are characterized for the presence of darkly melanized septa. They are associated with the angiosperm root tissues comprising dark pigmented fungi, hence called "dark septate endophytes" (DSE) and are limited to plant roots. These endophytic classes are predominantly ascomycetous fungi, which are conidial or sterile and form melanized structures such as inter- or intracellular hyphae and microsclerotia within the roots. The presence of DSE in soils and plant roots confirms the horizontal mode of transmission. Under *in vitro* conditions, the major means of transmission is the mycelial fragmentation and conidial dispersal. Root colonization by *Phialocephala fortinii* is a classic illustration of DSE. The process of colonization starts with superficial and/or runner hyphae that form a loose hyphal network on the root surface. Root systems play a vital role in the growth of the individual hyphae which can grow between cortical cells, along the main axis of the root and also within the depressions between epidermal cells (Mandyam et al. 2012). They have been reported from plant systems from Antarctic, Arctic, temperate, and tropical ecosystems. These endophytes are frequent in high stress environments and have ubiquitous occurrence which is suggestive of the role of DSE in ecophysiology of plants. However, the role of these mysterious fungal symbionts remains uncertain. The biological and ecological roles of DSE are important in plant physiology owing to the abundant plant-root associations from different habitats.

12.1.7 Endophytes as Treasured Sources of Secondary Bioactive Metabolites

Secondary bioactive metabolites are the "extrolite" compounds which are produced during differentiation of a living organism (Frisvad et al. 2008). Secondary metabolites can be defined as "the metabolites which are produced after the active growth phase of the organisms and have no direct involvement in the growth and development of the organism." The primary metabolites that are essentially same for all living systems are produced during the dynamic phase of the organism. Contrastingly, the secondary metabolites are typically species specific and are the intermediate derivatives of primary metabolism. They have been isolated from endophytes and characterized for the production of numerous biologically important compounds.

Endophytes are potential agents for drug discovery. Plant endophytes are asymptomatic, subtle and coexist with their hosts under most situations. They are commonly nonpathogenic in nature, but may produce secondary metabolites which facilitate them to survive in the plant interstitial space. The endophytic fungi which exist within the host plants produce secondary metabolites which are similar or identical to, or with higher activity than that of their own hosts (Strobel 2002). Plant endophytes are a relatively less-studied niche in microbial drug discovery (Kharwar et al. 2011). The bioactive metabolite classes obtained from endophytic fungi include alkaloids, cytochalasins, polyketides, terpenoids, flavonoids, quinones, peptides, xanthones, flavonoids, phenols and phenolic compounds, and steroids (Zaferanloo et al. 2012). These bioactive metabolites play significant role in fungal ecology since these metabolites are important for interactions between fungi and plant hosts such as signaling, defense, and regulation of symbiosis. However, not much research has been carried out to comprehend the role of bioactive metabolites in host–endophyte interaction on balansiaceous endophytes or grass endophytes (Schulz and Boyle 2005). The key challenge in drug discovery is to develop efficient and effective strategies to identify and recover potential strains for production of novel bioactive compounds. Secondary metabolites from special habitats or unusual niches may enhance chances of finding new compounds for use in drug discovery. Numerous secondary metabolites have been tested for their use in agriculture, pharmaceuticals, immunosuppressive agents, anticancer and antitumor agents (Huang et al. 2007a, b). Fungal endophytes modify the evolutionary dynamics by impacting the growth and survival of the host. Endophytic fungi improve plant growth, provide resistance against several plant diseases, and confer host plant ability to withstand abiotic stress. In addition, they possess the ability to degrade environmental pollutants and improve the soil conditions (Xiao et al. 2010). The endophytic association with host plants resulted in the discovery of the clinically used anticancer drug, taxol from the Western Yew plant, *Taxus brevifolia* which became a breakthrough in the search for novel drugs (Stierle et al. 1993). This remarkable achievement has resulted in the identification of various new bioactive metabolites and novel bioactive compounds from endophytic fungi of different host plants.

Several reports have highlighted the use of endophytic bioactive secondary fungal metabolites in the treatment of various human diseases due to their high biofunctional activities. Different fungal endophytes from the selected medicinal plants of Western Himalayas have been reported to show antimicrobial and immune modulatory activity (Qadri et al. 2013). The endophytic metabolites of *Artemisia annua* have depicted anti- acetylcholinesterase activity and can be used to treat Alzheimer's disease. Fungal endophytes from *Kigelia Africana* are reported to possess antibacterial activity (Idris et al. 2013) and the mycoendophytes of *Aralia elata* are potent antibacterial agents (Wu et al. 2012). Also, certain endophytes produce the same metabolites as that of the host plant such as hypericin, camptothecin, podophyllotoxin, diosgenin, and paclitaxel (Zhao et al. 2010).

12.1.8 Biological Activities of Endophytic Fungi

Endophytes are promising sources of potential secondary metabolic products that demonstrate a range of biological activities, viz. antimicrobial, antiparasitics, anticancer, neuroprotective, antioxidant, insulin imitative, enzyme inhibitive, immunosuppressive, and many more. Also, endophytic fungi are known to inhibit fungal growth, bacterial growth and produce effective cytotoxic metabolites (Wang et al. 2007). Recently, endophytic fungi have been researched for diverse applications due to their vast extracellular enzyme production (Pimentel et al. 2011). The possibility of utilizing fungal endophytes as effective biotechnological sources for the industrial enzyme production has been barely exploited, hence necessitating a vital need to discover and utilize diverse novel enzymes with high stability which finds important commercial applications with high industrial potency (Desire et al. 2014). Many reports have revealed the use of these fungal endophytic compounds for treatment of human diseases such as cancer, due to their highly functionalized activity (Aly et al. 2011; Kusari et al. 2012). Therefore, ardent research is a prerequisite for the commercial and economical production of these bioactive compounds as their production from host plant is expensive, not environment-friendly, and time consuming.

12.1.9 Endophytic Fungi as Potential Sources of Anticancer Compounds

Bioactive compounds obtained from endophytic fungal strains are capable of killing cancer cells due to cytotoxicity, apoptosis, and disruption of cellular microfilaments, and thereby possess efficient cancer inhibitory activities (Joseph and Priya 2011). Several endophytic fungal strains produce bioactive secondary metabolites proficient in cytotoxic assays and have potential anticancer activity. Taxol, vincristine, etoposide, irinotecan, topotecan, and vinblastine are some of the plant-derived anticancer drugs presently in clinical use to treat various human cancers (Balunas and Kinghorn 2005). Endophytes due to their association with host plants led to the discovery of the clinically used anticancer drug, taxol from the Western Yew plant, *Taxus brevifolia* (Stierle et al. 1993). This extraordinary accomplishment led to the identification of several new taxol producing endophytic fungi from different host plants. The world's first billion dollar anticancer drug, taxol ($C_{47}H_{51}NO_{14}$) an FDA approved source of paclitaxel, is a highly functionalized, diterpenoid especially targeted to treat breast, lung, ovarian, head and neck cancers, as well as advanced forms of Kaposi's sarcoma. The breakthrough in the discovery of taxol came from the endophytic *Taxomyces andreanae* from *Taxus brevifolia* and *Pestalotiopsis microspora* from *Taxus wallichiana* (Aly et al. 2011). Much research is directed towards search for novel paclitaxel derivatives including the fungal endophytes of *Taxus* species and non *Taxus* species (Kaul et al. 2013). Similarly, podophyllotoxin and its analogs are pharmaceutically important as they possess cytotoxic and antiviral activities (Abd-Elsalam and Hashim 2013). The natural sources of podophyllotoxin have become limited due to overexploitation and deforestation of medicinal plants. Also, the synthetic production of compounds is still underway which has prompted for newer methods of production with measurable yields. As an alternate source of podophyllotoxin, etoposide was developed in 1966 and FDA approved in 1983. However, novel approaches for total podophyllotoxin synthesis have been discovered. Plant tissue culture offers to be a sustainable production system which has been consistently improved upon relative to reliability and capacity (Majumder and Jha 2009).

12.1.10 Endophytic Cytotoxic Compounds from Terrestrial Plants

Fungal endophytes are widely used as potential anticancer agents for novel drug discovery approach. The secondary metabolic extracts or compounds from endophytic fungi are screened for anti-proliferation with a group of cell lines as the primary screening procedure for cytotoxicity. This method helps in the rapid screening of extracts or compounds and can be used to identify compounds which demonstrate specific/selective cytotoxicity towards the cancerous cells. Cytotoxic studies have to be precisely specific and are imperative in cancer therapy as compounds if non-specifically toxic, would make the chemotherapy more destructive to the patient. Also, a small number of anticancer agents have been carcinogenic (Blagosklonny 2005). Hence, toxicity screening is the primary step used to identify anticancer compounds capable of destroying tumor cells selectively and specifically. A new phomoxanthone A was obtained from the endophytic *Phomopsis sp.* BCC 1323 of *Tectona grandis*. This compound exhibited a significant cytotoxic activity against KB cells, BC-1 cells, and non-malignant Vero cells. Phomoxanthone A displayed IC_{50} values of 0.99, 0.51, and 1.4 mg/mL as compared to the standard compound ellipticine (Isaka et al. 2010). A novel *Acremonium* species obtained from *Knema laurina* (Thai medicinal plant) produced the compound brefeldin A which showed potent activity against KB (epidermoid cancer of the mouth), BC-1 (breast cancer), and NCI-H187 (small-cell lung cancer), with IC_{50} values of 0.18, 0.04, and 0.11 mM, respectively (Chinworrungsee et al. 2008). Altersolanol, a hydroxylated tetrahydro anthraquinone, macrosporin and 1,2,4,5-tetrahydroxy-7-methoxy-2-methyl-1,2,3,4-tetrahydroanthracene-9,10-dione from an endophytic *Alternaria* sp. from *Erythrina variegata* possesses potent anti-angiogenic activity and proves to be a favorable cancer therapy agent for other pro-angiogenesis related diseases (Pompeng et al. 2013). Four novel isoprenylated chromone derivatives—pestaloficiol I, pestaloficiol J, pestaloficiol K, and pestaloficiol L—were obtained from the endophytic *Pestalotiopsis fici* of *Camellia sinensis*. Pestaloficiol L exhibited the most potent cytotoxicity with IC_{50} values of 8.7 mM for HeLa and 17.4 mM for MCF7 cell lines. The IC_{50} values of the remaining compounds ranged between 8.7 and >136.1 mM for HeLa cells and between 17.4 and >153.8 mM for MCF7 cells. The 5-fluorouracil used as the positive control gave with IC_{50} values of 10.0 and 15.0 mM, correspondingly (Ling et al. 2009). Recently, the endophytic *Eupenicillium sp.* LG41, isolated from the Chinese medicinal plant *Xanthium sibiricum*, produced two new decalin containing compounds eupenicinicol C and eupenicinicol D with the use of epigenetic modulators such as nicotinamide which lead to the novel compound production that were not produced without epigenetic modulation. Eupenicinicol D demonstrated cytotoxicity against THP-1 cell line with IC_{50} of 8 μM (Li et al. 2017). The novel alkaloids—6, 7-dehydropaxilline, spirotryprostatin F and N-demethylmelearoride A—were isolated from the endophytic cultures of *Penicillium brefeldianum*. Spirotryprostatin F demonstrated cytotoxicity against HepG2 and MDA-MB-231 cells with IC_{50} values of 14.1 μmol/L and 35.5 μmol/L, respectively. Also, the N-demethylmelearoride A demonstrated moderate cytotoxicity against HepG2 cells with IC_{50} values of 36.6 μmol/L (Gao et al. 2017).

12.1.11 Endophytic Fungi as Potential Sources of Antioxidants

Medicinal plants possess a wide variety of free scavenging molecules such as phenolic acids, flavonoids, quinones, coumarins, lignans, tannins, alkaloids, amines, vitamins, terpenoids, and other endogenous metabolites. The bioactive metabolites isolated from medicinal plants are a prolific source of novel bioactive compounds with antioxidant properties. Antioxidant compounds are efficient against ROS (reactive oxygen species) damage and oxygen-derived free radicals which can result in cellular degradation, DNA damage, and carcinogenesis (Huang et al. 2007a, b) and hence are promising in the prevention and treatment of ROS-linked diseases such as cancer, hypertension, atherosclerosis, cardiovascular diseases, diabetes mellitus, neurodegenerative diseases, and ageing (Rukachaisirikul et al. 2008). Recent research has focused on antioxidant studies from several endophytes of medicinal plants (Pimentel et al. 2011). The compounds pestacin and isopestacin were acquired from the endophytic *Pestalotiopsis microspora* from Papua New Guinea. The antioxidant activity of pestacin was found to be higher as compared to standard Trolox (a vitamin E derivative) (Harper et al. 2003). Investigations in the alpine plants *Rhodiola rosea* gave five endophytic fungal strains with good antioxidant potential (Cui et al. 2015). One hundred and twelve Chinese medicinal plants were surveyed for their antioxidant components and phenolic compounds were reported as the dominant component signifying positive linear correlations between the total antioxidant capacities and phenolic contents (Cai et al. 2004). Despite numerous studies on antioxidant activities and phenolic contents in plants, however, no comparative investigations have been accomplished for their endophytic assemblages. Hence, this can be researched further for the endophytic antioxidant component. The preliminary phytochemical investigations of the endophytic fungi of *Eugenia jambolana* showed the presence of alkaloids, phenols, flavonoids, saponins, and terpenes. Phenols and terpenes are the main chemical constituents responsible for lipid peroxidation reduction and hence act as primary and secondary antioxidants. Also, the endophytic extracts with higher phenolic content also showed good antioxidant activity (Yadav et al. 2014).

12.1.12 Endophytic Fungi as Plant Growth Enhancers

Endophytes promote the plant growth either actively or passively through different mechanisms. They produce phytohormones to enhance the host plant growth without any apparent facilitation of host nutrient uptake (Schulz and Boyle 2005). Endophytes aid in promoting plant growth by nitrogen fixation, P solubilization, indole acetic acid (IAA) production, and siderophore secretion. The benefits of the endophytic interaction include improved plant growth, higher nutrient content, insect pest and herbivore resistance, resistance or disease tolerance, increased competitiveness, enhanced tolerance to stressful factors such as heavy metals, low pH, and high salinity (Kumar et al. 2014; Siddaiah et al. 2017). A plant-root interacting fungus *Piriformospora indica* has been reported to possess growth promoting effects on a range of plants such as *Zea mays*, *Nicotiana tobaccum*, *Bacopa monnieri*, *Artemisia annua*, *Petroselinum crispum*, *Populus tremula*, *Oryza sativa*, *Sorghum vulgare*, *Triticum sativum*, *Glycine max*, *Cicer arietinum*, *Solanum melongena*, and terrestrial orchids like *Dactylorhiza purpurella*, *D. incarnata*, *D. majalis*, and *D. fuchsia*

(Varma et al. 1999). Hence, it is known to have commercial applications in agroforestry and floriculture. Also, a seedborne endophyte *Stagonospora* species from *Phragmites australis* (Common reed) enhances host vitality (Ernst et al. 2003). Endophytes could be utilized as biofertilizers for plant growth promotion (Pandya et al. 2013). The endophytic *Sebacina vermifera* increased the growth of *Nicotiana attenuate* by decreasing ethylene production (Barazani et al. 2007). The endophyte *Acremonium strictum* occurring in roots increased the plant biomass of *Ammophila arenaria* (Marram grass) through antagonist effects on root-feeding nematodes (Hol et al. 2007). The endophytic *Trichoderma* sp. was assessed for its biocontrol potential against *F. solani,* causing root rot disease and it was observed that there was a significant *F. solani* growth inhibition and spore germination (Toghueo et al. 2016).

12.1.13 Endophytic Fungi as Nanoparticle Synthesizers

Microbes such as bacteria, fungi, and yeast play a vital role in the remediation of toxic metals through metal ion reduction and act as interesting nanofactories (Durán et al. 2007). Recently, endophytic fungi have been used as potential nanoparticle synthesizers which have promising applications in diagnostics, biomarkers, cell labeling, antimicrobial agents, drug delivery systems, and nano preparations for treatment of various diseases (Singh et al. 2016). Several endophytic fungi have been studied for gold and silver nanoparticle synthesis. The endophytic fungus *Penicillium* sp. isolated from healthy leaves of *Curcuma longa* was subjected to extracellular silver nanoparticle biosynthesis and demonstrated effective antibacterial activity against Multi Drug Resistant *E. coli* and *S. aureus* (Singh et al. 2014). Also, the endophytic *Pestalotia* sp. from the leaves of *Syzygium cumini* has been used for the extracellular synthesis of silver nanoparticles and demonstrated efficient antibacterial activity (Raheman et al. 2011). The endophytic fungus *Aspergillus clavatus* from *Azadirachta indica* was synthesized for silver nanoparticles and revealed effective antimicrobial potential against *Candida albicans*, *Pseudomonas fluorescens*, and *E. coli* (Verma et al. 2010). Geranium leaves (*Pelargonium graveolens*) and its endophytic *Colletotrichum sp.* were utilized for the gold nanoparticle synthesis. In both cases, rapid metal ion reduction was observed which resulted in stable gold nanoparticle formation (Shankar et al. 2003). The endophytic fungi *Aspergillus tamari, A. niger*, and *Penicillium ochrochloron* isolated from *Potentilla fulgens* L. were assessed for silver nanoparticle synthesis, thus offering promising scope as fungal nano factories for their sustainable production (Devi et al. 2014). The endophytic *Aspergillus versicolor* from the medicinal plant *Centella asiatica* was evaluated for silver nanoparticle synthesis. The synthesized silver nanoparticles were highly toxic against bacteria but possessed good antifungal activity (Netala et al. 2016). There is a huge interest to develop risk free, economical and environment-friendly technologies for nanomaterial synthesis. Nanoparticles find wide applications in various fields as catalysts, sensors, and medicines. The nanoparticle synthesis via biological systems offers new routes to develop nanoparticles with desired properties for making their exploitation possible in diverse fields (Medina et al. 2007). The harmless and ecofriendly behavior of endophytes prove to be good nanoparticle synthesizers.

12.1.14 Antimicrobials from Endophytic Fungi

The need for effective antimicrobials to provide relief for human ailments and deal with the problem of drug resistance in bacteria is increasing. Endophytes have been identified as potential source of new bioactive products for use in therapeutic regime and most importantly a significant substitute to defeat the problem of drug resistance by phytopathogens and human pathogens (Pongcharoen et al. 2008). The antimicrobial compound equisetin was isolated from endophytic *Fusarium* sp. from invasive plant *Opuntia dillenii* of the arid zone possessed efficient antibacterial activities and the endophytic fungal population assisted the host to survive adverse environmental conditions (Ratnaweera et al. 2015). The endophytic *Phoma sp.* from *Taraxacum mongolicum* was evaluated for antimicrobial activities. The antimicrobial compound 2-hydroxy-6-methylbenzoic acid showed effective antimicrobial activity (Zhang et al. 2013). Also, the endophytic fungus *Emericella quadrilineata* from the medicinal plant *Pteris pellucid* was evaluated for antimicrobial potential. The fungus produced benzyl benzoate which is used in the treatment of scabies and other skin diseases (Goutam et al. 2014). Phomoxanthone A is obtained from a *Phomopsis* sp. from the stem of *Costus* sp. growing in the rain forest of Costa Rica. It has depicted antibacterial activity against *Bacillus megaterium* at a concentration of 10 mg/mL (Elsässer et al. 2005). The compounds cycloepoxylactone and cycloepoxytriol B were identified from *Phomopsis* sp. of *Laurus azorica* which were moderately active against *B. megaterium* (Mousa and Raizada 2013).

The antibacterial compound Epoxydine B, Epoxydon, (4R,5R,6S)-6-acetoxy-4,5-dihydroxy-2-(hydroxymethyl)cyclohex-2-en-1-one,2-chloro-6-(hydroxymethyl)benzene-1,4-diol, and the antibiotic ES-242-1 were identified from an endophytic *Phoma* sp. of *Salsola oppositifolia* (Qin et al. 2010). The effective antibacterials acremonisol A, semicochliodinol A, and cochliodinol were isolated from *C. globosum* SNB-GTC2114 and simultaneously pyrrocidine A, B, C and alterperylenol from *Lewia infectoria* SNB-GTC2402 obtained from *Besleria insolita* belonged to Amazon Rainforest biome of Cayenne and Roura, French Guiana (Casella et al. 2013). The polyketide, cryptosporioptide was from a *Cryptosporiopsis* sp., from the shrub *Viburnum tinus* inhibited *B. megaterium* and hence depicted antibacterial activity (Saleem et al. 2013). The compounds monocerin, (12*S*)-12-hydroxymonocerin, and isocoumarin were isolated from *Microdochium bolleyi*, an endophyte from *Fagonia cretica* and were active against *E. coli* and *B. megaterium* (Zhang et al. 2008). Dothideomycetide A from an endophytic *Dothideomycete* sp., of a Thai medicinal plant, *Tiliacora triandra*, showed antibacterial activity against *S. aureus* ATCC 25923 and MRSAATCC 33591 (Senadeera et al. 2012). The bioactive secondary alkaloids, fumigaclavine C and pseurotin were isolated from the endophyte *Aspergillus* sp. EJC08, of the medical plant *Bauhinia guianensis*. Fumigaclavine C possessed activity against *B. subtilis*, *E. coli*, *P. aeruginosa*, and *S. aureus*, respectively, while Pseurotin A has activity against *B. subtilis, E. coli, P. aeruginosa*, and *S. aureus*, respectively (Pinheiro et al. 2013).

12.2 Conclusion

Even though large numbers of drugs have been discovered, screening techniques are extensively required to isolate new biologically active metabolites from nature. The fungal endophytes from the therapeutic plants have not received much exploration of their secondary metabolites towards bioactivity. Since that only a lesser number of endophytic fungi have been studied in these medicinal plants, researchers are focused to evaluate and elucidate the probable and potential secondary metabolites produced by these fungi.

References

Abd–Elsalam KA, Hashim AF (2013) Hidden fungi as microbial and nano–factories for anticancer agents. Fungal Genom Biol 3:e115

Akay S, Ekiz G, Kocabas F, EE H–K, Korkmaz KS, Bedir E (2014) A new 5, 6-dihydro-2-pyrone derivative from *Phomopsis amygdali*, an endophytic fungus isolated from hazelnut (*Corylusavellana*). Phytochem Lett 7:93

Aly AH, Debbab A, Kjer J, Proksch P (2010) Fungal endophytes from higher plants: a prolific source of phytochemicals and other bioactive natural products. Fungal Divers 41(1):1–16

Aly AH, Debbab A, Proksch P (2011) Fungal endophytes: unique plant inhabitants with great promises. Appl Microbiol Biotechnol 90(6):1829–1845

Arnold AE (2007) Understanding the diversity of foliar endophytic fungi: progress, challenges, and frontiers. Fungal Biol Rev 21(2):51–66

Arnold AE, Lewis LC (2005) Ecology and evolution of fungal endophytes, and their roles against insects. In: Vega FE, Blackwell M (eds) Insect–Fungal Associations: Ecology and evolution. Oxford University Press, New York, pp 74–96

Arnold AE, Lutzoni F (2007) Diversity and host range of foliar fungal endophytes: are tropical leaves biodiversity hotspots? Ecology 88:541–549

Bacon CW, Porter JK, Robins JD, Lutrell EJ (1977) Epichloë typhina from toxic tall fescue grass. Appl Environ Microbiol 34:576–581

Balunas MJ, Kinghorn AD (2005) Drug discovery from medicinal plants. Life Sci 78(5):431–441

Barazani O, von Dahl CC, Baldwin IT (2007) *Sebacina vermifera* promotes the growth and fitness of *Nicotiana attenuata* by inhibiting ethylene signaling. Plant Physiol 144(2):1223–1232

Bischoff JF, White JF (2005) Evolutionary development of the Clavicipitaceae. Mycology Series, 23:505

Bisht GS, Awasthi AK, Dhole TN (2006) Antimicrobial activity of *Hedychium spicatum*. Fitoterapia 77(3):240–242

Blagosklonny MV (2005) Carcinogenesis, cancer therapy and chemoprevention. Cell Death Differ 12:592–602

Brundrett MC (2006) Understanding the roles of multifunctional mycorrhizal and endophytic fungi. In: Schulz B, Boyle C, Sieber TN (eds) Microbial root endophytes, Soil biology, Vol 9, Springer Verlag, Berlin, Heidelberg, Germany, pp 281–298.

Brundrett MC (2009) Mycorrhizal associations and other means of nutrition of vascular plants: understanding the global diversity of host plants by resolving conflicting information and developing reliable means of diagnosis. Plant Soil 320(1–2):37–77

Cai Y, Luo Q, Sun M, Corke H (2004) Antioxidant activity and phenolic compounds of 112 traditional Chinese medicinal plants associated with anticancer. Life Sci 74(17):2157–2184

Casella TM, Eparvier V, Mandavid H, Bendelac A, Odonne G, Dayan L, Duplais C, Espindola LS, Stien D (2013) Antimicrobial and cytotoxic secondary metabolites from tropical leaf endophytes: isolation of antibacterial agent pyrrocidine C from *Lewia infectoria* SNB–GTC2402. Phytochemistry 96:370–377

Chinworrungsee M, Wiyakrutta S, Sriubolmas N, Chuailua P, Suksamrarn A (2008) Cytotoxic activities of trichothecenes isolated from an endophytic fungus belonging to order hypocreales. Arch Pharm Res 31:611

Cragg GM, Newman DJ (2013) Natural products: a continuing source of novel drug leads. Biochim Biophys Acta 1830(6):3670–3695

Cragg GM, Grothaus PG, Newman DJ (2009) Impact of natural products on developing new anticancer agents. Chem Rev 109(7):3012–3043

Cui JL, Guo TT, Ren ZX, Zhang NS, Wang ML (2015) Diversity and antioxidant activity of culturable endophytic fungi from alpine plants of *Rhodiola crenulata, R. angusta,* and *R. sachalinensis*. PLoS One 10(3):e0118204

Devi LS, Joshi SR (2014) Ultrastructures of silver nanoparticles biosynthesized using endophytic fungi. J Microsc Ultrastruct 3(1):29–37.

Davies J, Davies D (2010) Origins and evolution of antibiotic resistance. Microbiol Mol Biol Rev 74(3):417–433

Davis EC, Shaw AJ (2008) Biogeographic and phylogenetic patterns in diversity of liverwort-associated endophytes. Am J Bot 95(8):914–924

Demain AL, Sanchez S (2009) Microbial drug discovery: 80 years of progress. J Antibiot 62(1):5

Demain AL, Vaishnav P (2011) Natural products for cancer chemotherapy. Microb Biotechnol 4(6):687–699

Desire MH, Bernard F, Forsah MR, Assang CT, Denis ON (2014) Enzymes and qualitative phytochemical screening of endophytic fungi isolated from *Lantana camara* Linn. Leaves. J Appl Biol Biotechnol 2(06):001–006

Durán N, Marcato PD, De S, Gabriel IH, Alves OL, Esposito E (2007) Antibacterial effect of silver nanoparticles produced by fungal process on textile fabrics and their effluent treatment. J Biomed Nanotechnol 3:203–208

Elsässer B, Krohn K, Flörke U, Root N, Aust HJ, Draeger S, Schulz B, Antus S, Kurtán T (2005) X-ray structure determination, absolute configuration and biological activity of phomoxanthone A. Eur J Org Chem 21:4563–4570

Ernst M, Mendgen KW, Wirsel SG (2003) Endophytic fungal mutualists: seed–borne *Stagonospora* spp. enhances reed biomass production in axenic microcosms. Mol Plant-Microbe Interact 16(7):580–587

Faeth SH, Hamilton CE (2006) Does an asexual endophyte symbiont alter life stage and long–term survival in a perennial host grass? Microb Ecol 52(4):748–755

Frisvad JC, Andersen B, Thrane U (2008) The use of secondary metabolite profiling in chemotaxonomy of filamentous fungi. Mycol Res 112(2):231–240

Ganley RJ, Newcombe G (2006) Fungal endophytes in seeds and needles of *Pinus monticola.* Mycol Res 110(3):318–327

Gao N, Shang ZC, Yu P, Luo J, Jian KL, Kong LY, Yang MH (2017) Alkaloids from the endophytic fungus *Penicillium brefeldianum* and their cytotoxic activities. Chinese Chem Lett 28:109

Garbeva P, Van Overbeek LS, Van Vuurde JWL, Van Elsas JD (2001) Analysis of endophytic bacterial communities of potato by plating and denaturing gradient gel electrophoresis (DGGE) of 16S rDNA based PCR fragments. Microb Ecol 41(4):369–383

Gouda S, Das G, Sen SK, Shin HS, Patra JK (2016) Endophytes: a treasure house of bioactive compounds of medicinal importance. Front Microbiol 7:1538

Goutam J, Sharma VK, Verma SK, Singh DK, Kumar J, Mishra A, Kumar A, Kharwar RN (2014) Optimization of culture conditions for enhanced production of bioactive metabolites rich in antimicrobial and antioxidant activities isolated from *Emericella quadrilineata* an endophyte of *Pteris pellucida*. J Pure Appl Microbiol 8(3):2059–2073

Harper JK, Arif AM, Ford EJ, Strobel GA, Porco JA, Tomer DP, Oneill KL, Heider EM, Grant DM (2003) Pestacin: a 1,3–dihydro isobenzofuran from *Pestalotiopsis microspora* possessing antioxidant and antimycotic activities. Tetrahedron 59(14):2471–2476

Herre EA, Mejía LC, Kyllo DA, Rojas E, Maynard Z, Butler A, Van Bael SA (2007) Ecological implications of anti-pathogen effects of tropical fungal endophytes and mycorrhizae. Ecology 88(3):550–558

Hoffman MT, Arnold AE (2010) Diverse bacteria inhabit living hyphae of phylogenetically diverse fungal endophytes. Appl Environ Microbiol 76(12):4063–4075

Hol WG, de la Pena E, Moens M, Cook R (2007) Interaction between a fungal endophyte and root herbivores of *Ammophila arenaria*. Basic Appl Ecol 8(6):500–509

Huang WY, Cai YZ, Hyde KD, Corke H, Sun M (2007a) Endophytic fungi from *Nerium oleander* L (Apocynaceae): main constituents and antioxidant activity. World J Microbiol Biotechnol 23(9):1253–1263

Huang WY, Cai YZ, Xing J, Corke H, Sun M (2007b) A potential antioxidant resource: endophytic fungi from medicinal plants. Econ Bot 61(1):14–30

Idris AM, Al–tahir I, Idris E (2013) Antibacterial activity of endophytic fungi extracts from the medicinal plant *Kigelia africana*. Egypt Acad J Biol Sci 5(1):1–9

Isaka M, Chinthanom P, Boonruangprapa T, Rungjindamai N, Pinruan U (2010) Eremophilane type sesquiterpenes from the fungus *Xylaria* sp. BCC 21097. J Nat Prod 73:683

Jalgaonwala RE, Mahajan RT (2011) Evaluation of hydrolytic enzyme activities of endophytes from some indigenous medicinal plants. J Agric Technol 7(6):1733–1741

Joseph B, Priya RM (2011) Bioactive compounds from endophytes and their potential in pharmaceutical effect: a review. Am J Biochem Mol Biol 1(3):291–309

Kaul S, Ahmed M, Zargar K, Sharma P, Dhar MK (2013) Prospecting endophytic fungal assemblage of *Digitalis lanata* Ehrh.(foxglove) as a novel source of digoxin: a cardiac glycoside. 3 Biotech 3(4):335–340

Kharwar RN, Mishra A, Gond SK, Stierle A, Stierle D (2011) Anticancer compounds derived from fungal endophytes: their importance and future challenges. Nat Prod Rep 28(7):1208–1228

Kumar PG, Mir Hassan Ahmed SK, Desai S, Leo Daniel Amalraj E, Rasul A (2014) *In vitro* screening for abiotic stress tolerance in potent biocontrol and plant growth promoting strains of *Pseudomonas* and *Bacillus* spp. Int J Bacteriol 2014:1–6

Kusari S, Lamshöft M, Spiteller M (2009) *Aspergillus fumigatus* Fresenius, an endophytic fungus from *Juniperus communis* L. Horstmann as a novel source of the anticancer pro-drug deoxypodophyllotoxin. J Appl Microbiol 107(3):1019–1030

Kusari S, Hertweck C, Spiteller M (2012) Chemical ecology of endophytic fungi: origins of secondary metabolites. Chem Biol 19(7):792–798

Lam KS (2006) Discovery of novel metabolites from marine actinomycetes. Curr Opin Microbiol 9(3):245–251

Li G, Kusari S, Golz C, Laatsch H, Strohmann C, Spiteller M (2017) Epigenetic modulation of endophytic *Eupenicillium sp.* LG41 by a histone deacetylase inhibitor for production of decalin-containing compounds. J Nat Prod 80:983–988

Ling L, Liu S, Niu S, Guo L, Chen X, Che Y (2009) Isoprenylatedchromone derivatives from the plant endophytic fungus, *Pestalotiopsis fici*. J Nat Prod 72:1482

Majumder A, Jha S (2009) Biotechnological approaches for the production of potential anticancer leads podophyllotoxin and paclitaxel: an overview. J Biol Sci 1(1):46–69

Malinowski DP, Belesky DP (2006) Ecological importance of *Neotyphodium* spp. grass endophytes in agroecosystems. Grassl Sci 52(1):1–14

Mandyam K, Fox C, Jumpponen A (2012) Septate endophyte colonization and host responses of grasses and forbs native to a tallgrass prairie. Mycorrhiza 22(2):109–119

Márquez LM, Redman RS, Rodríguez RJ, Roossinck MJ (2007) A virus in a fungus in a plant: three-way symbiosis required for thermal tolerance. Science 315:513–515
Márquez SS, Bills GF, Acuña LD, Zabalgogeazcoa I (2010) Endophytic mycobiota of leaves and roots of the grass *Holcus lanatus*. Fungal Divers 41(1):115–123
Medina C, Santos-Martinez MJ, Radomski A, Corrigan OI, Radomski MW (2007) Nanoparticles: pharmacological and toxicological significance. Br J Pharmacol 150:552–558
Moricca S, Ragazzi A (2008) Fungal endophytes in Mediterranean oak forests: a lesson from *Discula quercina*. Phytopathology 98:380–386
Mousa WK, Raizada MN (2013) The diversity of anti-microbial secondary metabolites produced by fungal endophytes: an interdisciplinary perspective. Front Microbiol 4:65
Netala VR, Bethu MS, Pushpalatha B, Baki VB, Aishwarya S, Rao JV, Tartte V (2016) Biogenesis of silver nanoparticles using endophytic fungus *Pestalotiopsis microspora* and evaluation of their antioxidant and anticancer activities. Int J Nanomedicine 11:5683–5696
Okigbo RN, Anuagasi CL, Amadi JE (2009) Advances in selected medicinal and aromatic plants indigenous to Africa. J Med Plants Res 3(2):086–095
Pandya M, Naresh Kumar G, Rajkumar S (2013) Invasion of rhizobial infection thread by non-rhizobia for colonization of Vignaradiata root nodules. FEMS Microbiol Lett 348(1):58–65
Pavarini DP, Pavarini SP, Niehues M, Lopes NP (2012) Exogenous influences on plant secondary metabolite levels. Anim Feed Sci Technol 176(1):5–16
Petrini O (1986) Taxonomy of endophytic fungi of aerial plant tissues. In: Fokkema NJ, van den Huevel J (eds) Microbiology of the phyllosphere. Cambridge University Press, Cambridge, pp 175–187
Petrini O (1991) Fungal endophytes of tree leaves. In: Andrew IA, Hirano SS (eds) Microbial ecology of leaves. Springer, New York, pp 179–197
Photita W, Lumyong S, Lumyong P, McKenzie EHC, Hyde KD (2004) Are some endophytes of *Musa acuminata* latent pathogens? Fungal Divers 16:131–140
Pimentel MR, Molina G, Dionísio AP, Maróstica MR Jr, Pastore GM (2011) The use of endophytes to obtain bioactive compounds and their application in biotransformation process. Biotechnol Res Int 2011:576286
Pinheiro EA, Carvalho JM, dos Santos DC, Feitosa Ade O, Marinho PS, Guilhon GM, de Souza AD, da Silva FM, Marinho AM (2013) Antibacterial activity of alkaloids produced by fungus Aspergillus sp. EJC08 isolated from medical plant *Bauhinia guianensis*. Nat Prod Res 27:1633–1638
Pompeng P, Sommit D, Sriubolmas N, Ngamrojanavanich N, Matsubara K, Pudhom K (2013) Antiangiogenetic effects of anthranoids from *Alternaria* sp. an endophytic fungus in a Thai medicinal plant *Erythrina variegata*. Phytomedicine 20:918
Pongcharoen W, Rukachaisirikul V, Phongpaichit S, Kühn T, Pelzing M, Sakayaroj J, Taylor WC (2008) Metabolites from the endophytic fungus Xylaria sp. PSU–D14. Phytochemistry 69(9):1900–1902
Promputtha I, Lumyong S, Dhanasekaran V, McKenzie EH, Hyde KD, Jeewon R (2007) A phylogenetic evaluation of whether endophytes become saprotrophs at host senescence. Microb Ecol 53(4):579–590
Qadri M, Johri S, Shah BA, Khajuria A, Sidiq T, Lattoo SK, Riyaz–Ul–Hassan S (2013) Identification and bioactive potential of endophytic fungi isolated from selected plants of the Western Himalayas. Springerplus 2(1):8
Qin S, Hussain H, Schulz B, Draeger S, Krohn K (2010) Two new metabolites, epoxydinea and b, from *Phoma* sp. Helv Chim Acta 93(1):169–174
Raheman F, Deshmukh S, Ingle A, Gade A, Rai M (2011) Silver nanoparticles: novel antimicrobial agent synthesized from a endophytic fungus *Pestalotia* sp. isolated from leaves of *Syzygium cumini*(L). Nano Biomed Eng 3:174–178
Ratnaweera PB, de Silva ED, Williams DE, Andersen RJ (2015) Antimicrobial activities of endophytic fungi obtained from the arid zone invasive plant *Opuntia dillenii* and the isolation of equisetin, from endophytic *Fusarium* sp. BMC Complement Altern Med 15(1):220
Rayner MC (1915) Obligate symbiosis in *Calluna vulgaris*. Ann Bot 29(113):97–133

Rodriguez RJ, White Jr JF, Arnold AE, Redman RS (2009) Fungal endophytes: diversity and functional roles. New Phytol 182(2):314–330

Rout SP, Choudary KA, Kar DM, Das L, Jain A (2009) Plants in traditional medicinal system–future source of new drugs. Int J Pharm Pharm Sci 1(1):1–23

Rudgers JA, Clay K (2007) Endophyte symbiosis with tall fescue: how strong are the impacts on communities and ecosystems? Fungal Biol Rev 21(2):107–124

Rukachaisirikul V, Sommart U, Phongpaichit S, Sakayaroj J, Kirtikara K (2008) Metabolites from the endophytic fungus *Phomopsis sp.* PSU-D15. Phytochemistry 69(3):783–787

Saikkonen K (2007) Forest structure and fungal endophytes. Fungal Biol Rev 21(2):67–74

Saikkonen K, Faeth SH, Helander M, Sullivan TJ (1998) Fungal endophytes: a continuum of interactions with host plants. Annu Rev Ecol Syst 29(1):319–343

Saikkonen K, Saari S, Helander M (2010) Defensive mutualism between plants and endophytic fungi? Fungal Div 41:101–113

Saleem M, Tousif MI, Riaz N, Ahmed I, Schulz B, Ashraf M, Nasar R, Pescitelli G, Hussain H, Jabbar A, Shafiq N, Krohn K (2013) Cryptosporioptide: a bioactive polyketide produced by an endophytic fungus *Cryptosporiopsis* sp. Phytochemistry 93:199–202

Sasidharan S, Chen Y, Saravanan D, Sundram KM, Latha LY (2011) Extraction, isolation and characterization of bioactive compounds from plants' extracts. Afr J Trad Comp Alt Med 8(1):1–10

Schardl CL, Leuchtmann A, Spiering MJ (2004) Symbioses of grasses with seedborne fungal endophytes. Annu Rev Plant Biol 55:315–340

Schulz B, Boyle C (2005) The endophytic continuum. Mycol Res 109(6):661–686

Schulz B, Römmert AK, Dammann U, Aust HJ, Strack D (1999) The endophyte–host interaction: a balanced antagonism? Mycol Res 103(10):1275–1283

Senadeera SP, Wiyakrutta S, Mahidol C, Ruchirawat S, Kittakoop P (2012) A novel tricyclic polyketide and its biosynthetic precursor azaphilone derivatives from the endophytic fungus Dothideomycete sp. Org Biomol Chem 10:7220–7226

Shankar SS, Ahmad A, Parischa R, Sastry M (2003) Bioreduction of chloroaurate ions by geranium leaves and its endophytic fungus yields gold nanoparticles of different shapes. J Mater Chem 13:1822–1826

Siddaiah CN, Satyanarayana NR, Mudili V, Gupta VK, Gurunathan S, Rangappa S, Huntrike SS, Srivastava RK (2017) Elicitation of resistance and associated defense responses in Trichoderma hamatum induced protection against pearl millet downy mildew pathogen. Sci Rep 7:43991

Singh LP, Gill SS, Tuteja N (2011) Unraveling the role of fungal symbionts in plant abiotic stress tolerance. Plant Signal Behav 6(2):175–191

Singh D, Rathod V, Ninganagouda S, Hiremath J, Singh AK, Mathew J (2014) Optimization and characterization of silver nanoparticle by endophytic fungi *Penicillium* sp. isolated from *Curcuma longa* (turmeric) and application studies against MDR *E. coli* and *S. aureus*. Bioinorg Chem Appl 8

Singh P, Kim YJ, Zhang D, Yang DC (2016) Biological synthesis of nanoparticles from plants and microorganisms. Trends Biotechnol 34(7):588–599

Spatafora JW, Bushley KE (2015) Phylogenomics and evolution of secondary metabolism in plant–associated fungi. Curr Opin Plant Biol 26:37–44

Stierle A, Strobel G, Stierle D (1993) Taxol and taxane production by *Taxomyces andreanae,* an endophytic fungus of Pacific yew. Science 260:214–214

Strobel GA (2002) Microbial gifts from the rain forest. J Phytopathol 24:14–20

Toghueo RMK, Eke P, Zabalgogeazcoa Í, de Aldana BRV, Nana LW, Boyom FF (2016) Biocontrol and growth enhancement potential of two endophytic *Trichoderma* spp. from *Terminalia catappa* against the causative agent of Common Bean Root Rot (*Fusarium solani*). Biol Conserv 96:8–20

Uzma F, Mohan CD, Hashem A, Konappa NM, Rangappa S, Kamath PV, Singh BP, Mudili V, Gupta VK, Siddaiah CN, Chowdappa S, Alqarawi AA, Abd Allah EF (2018) Endophytic fungi—alternative sources of cytotoxic compounds: a review. Front Pharmacol 9(309):1–37. https://doi.org/10.3389/fphar.2018.00309

Varma A, Verma S, Sahay N, Bütehorn B, Franken P (1999) *Piriformospora indica*, a cultivable plant-growth-promoting root endophyte. Appl Environ Microbiol 65(6):2741–2744
Varma A, Bakshi M, Lou B, Hartmann A, Oelmueller R (2012) *Piriformospora indica*: a novel plant growth-promoting mycorrhizal fungus. Agric Res 1(2):117–131
Verma VC, Kharwar RN, Gange AC (2010) Biosynthesis of antimicrobial silver nanoparticles by the endophytic fungus *Aspergillus clavatus*. Nanomedicine 5:33–40
Vinale F, Manganiello G, Nigro M, Mazzei P, Piccolo A, Pascale A, Ruocco M, Marra R, Lombardi N, Lanzuise S, Varlese R (2014) A novel fungal metabolite with beneficial properties for agricultural applications. Molecules 19(7):9760–9772
Viswambari Devi R, Doble M, Verma R (2015) Nanomaterials for early detection of cancer biomarker with special emphasis on gold nanoparticles in immunassays/sensors. Biosens Bioelectron 68:688–698
Wang FW, Jiao RH, Cheng AB, Tan SH, Song YC (2007) Antimicrobial potentials of endophytic fungi residing in *Quercus variabilis* and brefeldin A obtained from *Cladosporium* sp. World J Microbiol Biotechnol 23(1):79–83
Wani K, Shweta S, Pooja S, Pooja T (2016). Production of novel secondary metabolites from endophytic fungi by using fermentation process. Indo Am J Pharm Res 6(03)
Wu H, Yang H, You X, Li Y (2012) Isolation and characterization of saponin-producing fungal endophytes from *Aralia elata* in Northeast China. Int J Mol Sci 13(12):16255–16266
Xiao X, Luo S, Zeng G, Wei W, Wan Y, Chen L, Xi Q (2010) Biosorption of cadmium by endophytic fungus (EF) *Microsphaeropsis* sp. LSE10 isolated from cadmium hyperaccumulator *Solanum nigrum* L. Bioresour Technol 101(6):1668–1674
Yadav M, Yadav A, Yadav JP (2014) *In vitro* antioxidant activity and total phenolic content of endophytic fungi isolated from *Eugenia jambolana* Lam. Asian Pac J Trop Med 7:S256–S261
Zaferanloo B, Mahon PJ, Palombo EA (2012) Endophytes from medicinal plants as novel sources of bioactive compounds. Medicinal plants: diversity and drugs. Science Publisher: USA, 355–411
Zhang Q, Wu J, Nguyen A, Wang BD, He P, Laurent GS, Su YA (2008) Molecular mechanism underlying differential apoptosis between human melanoma cell lines UACC903 and UACC903 (+6) revealed by mitochondria-focused cDNA microarrays. Apoptosis 13(8):993–1004
Zhang Q, Wang SQ, Tang HY, Li XJ, Zhang L, Xiao J, Gao YQ, Zhang AL, Gao JM (2013) Potential allelopathic indole diketopiperazines produced by the plant endophytic *Aspergillus fumigatus* using the one strain–many compounds method. J Agric Food Chem 61(47):11447–11452
Zhao J, Zhou L, Wang J, Shan T, Zhong L, Liu X, Gao X (2010) Endophytic fungi for producing bioactive compounds originally from their host plants. Curr Res Technol Educ Trop Appl Microbiol Microbiol Biotechnol 1:567–576
Zhao J, Shan T, Mou Y, Zhou L (2011) Plant-derived bioactive compounds produced by endophytic fungi. Mini Rev Med Chem 11(2):159–168

Chapter 13
Endophytic Fungi: Recent Advances in Identification and Explorations

Vinay Kumar, Ravindra Soni, Lata Jain, Biplab Dash, and Reeta Goel

13.1 Introduction

Fungal endophytes have reflective impressions on plants, including advantageous effects on agriculturally important crops (Lugtenberg et al. 2016; Busby et al. 2016; Cocq et al. 2017; Doty 2017). Endophytic fungi reside in the internal tissues of almost all living plants and are involved in the production of various beneficial traits and are thus known for enhancing plant growth, increasing their fitness, and strengthening their tolerances to various abiotic and biotic stresses. These endophytes have the ability to colonize internal plant tissues of healthy root, stem, petiole, leaves, and flower (Park et al. 2012), twigs (Sanz-Ros et al. 2015), bark, fruit (Verma et al. 2011; Fouda et al. 2015), and seeds (Parsa et al. 2016), without causing any apparent harm or pathogenic infection to their host plants. Considering the presence of endophytes in almost all known plant species, such characteristics make fungal endophytes as one of the diverse components of biomass which is dynamically being modified to adjust to the ecological changes and to host physiology (Aly et al. 2011). It is estimated that there are approximately more than one million endophytic fungal species worldwide (Ganley et al. 2004) which represent a significant genetic resource for biotechnology.

Furthermore, endophytic fungi can have countless effects on host plant fitness, with the outcome of various interactions including both beneficial and antagonistic effects. This fungi-plant association is always accompanied by numerous chemical

V. Kumar · L. Jain
ICAR-National Institute of Biotic Stress Management, Raipur, Chhattisgarh, India

R. Soni · B. Dash
Department of Agril. Microbiology, Indira Gandhi Krishi Vishwavidhyalaya, Raipur, Chhattisgarh, India

R. Goel (✉)
Department of Microbiology, GB Pant University of Agriculture and Technology, Pantnagar, Uttarakhand, India

B. P. Singh (ed.), *Advances in Endophytic Fungal Research*, Fungal Biology,
https://doi.org/10.1007/978-3-030-03589-1_13

and physical interactions, thereby establishing them either in localized and/or systemic manner (Kusari et al. 2014). The multifaceted interactions demonstrated by the fungal endophytes against the host phytopathogens (Katoch and Pull 2017) indicated that their efficiencies are either due to production of natural compounds, secondary metabolites, or intermediates in the biosynthetic pathway of those metabolites, activated upon pathogen-attack (Pusztahelyi et al. 2015). This reveals that these endophytes are capable of producing various enigmatic metabolites when provoked under certain selective interacting conditions apart from the common metabolites produced under normal/natural conditions (Kumara et al. 2014).

Increasing interest in cryptic work of internal tissues of healthy plants by endophytic micro fungi has led to an increasing consciousness that higher plants most likely harbor a reservoir of uncharacterized fungi (Arnold et al. 2007; Rodriguez et al. 2009). It is assumed that the structure of fungal endophytes assemblages within the same species may vary not only due to the geographical differences, but also due to changes in agro-climatic conditions in the region (Arnold and Herre 2003). Understanding of the multitude of endophytic relationships with host plants needs more attention, investigation, and exploration in numerous aspects, namely the fungal-plant interactions, multispecies crosstalk, and linkages with herbivores and predators.

13.2 Importance/Functional Significance of Fungal Endophytes

Fungi play complex and important ecological roles in the ecosystem, as they continue the cycle of nutrients through ecosystems by decomposing dead organic material, and providing nutrients to plants. All plants in natural ecosystems appear to be symbiotic with fungal endophytes (Rodriguez et al. 2009). A highly diverse group of endophytic fungi may offer significant benefits to their host plants by producing secondary metabolites that provide protection and survival value, such as conferring abiotic and biotic stress tolerance, increasing biomass and decreasing water consumption, enhancing insect and disease resistance, secretion of source for bioactive and pharmacological compounds.

13.2.1 Agriculture

Fungal endophytes can have myriad effects on host plant fitness, with the outcome of interactions ranging from beneficial to antagonistic. Benefits include protection against damage by pathogens, herbivores, and salt or water stress (Arnold and Herre 2003; Obledo et al. 2003; Donoso et al. 2008; Miller et al. 2008). Arnold and Herre (2003) compared the ability of endophyte infected and uninfected leaves of Theobroma cacao to resist damage caused by a common foliar pathogen, and found that leaves infected with a community of endophytes had significantly less visible evidence of disease. Some fungal endophyte species provide their host with elevated tolerance to extreme environmental conditions. The grass species *Dichanthelium lanuginosum* is able to withstand high soil temperatures only when in association with the fungal endophyte *Curvularia protuberata* (Redman et al. 2002).

Fungal endophytes exhibit a range of symbiotic relationships with their host plant and are well known to contribute for plant fitness, which helps the host for better adaptation towards stress (both biotic and abiotic stress) conditions (Aly et al. 2011; Franken 2012; Johnson et al. 2013; Card et al. 2016; Potshangbam et al. 2017). Many of these fungal endophytes are also involved in mineral nutrition (nitrogen fixation, phosphate solubilization, and uptake of phosphorus) and siderophores production besides producing several plant growth hormones like IAA (indole acetic acid), auxin, gibberellins, ethylene, and abscisins (Hamayun et al. 2009; Waqas et al. 2015). Some endophytic fungi like *Penicillium chrysogenum* and *Alternaria alternata* along with some sterile fungal hyphae isolated from a medical plant *Asclepias sinaica* produced IAA in very high quantity which leads to significant increase in the root length of maize plants (Fouda et al. 2015). Fungal endophytes are capable of protecting the plant against pathogens through various strategies, such as competition with pathogens for colonization and nutrients, the production of antibiotics, and the induction of resistance in the host plant (Clarke et al. 2006; Ownley et al. 2010; Yuan et al. 2017). Ambuic acid, a cyclohexenone, reported in some strains of endophyte *P. microspora* exhibited activity against an important plant pathogen *Pythium ultimum* (Oomycete) (Li et al. 2001). They also protect plants against attacks from herbivorous insects through the production of toxins (Gange et al. 2012; Jia et al. 2016). Some of the endophytes reported from various important agricultural crops are summarized in Table 13.1.

13.2.2 Bioactive Compound

Fungal endophytes have proved to be an important source for bioactive antimicrobial compounds such as phenols, steroids, alkaloids, and peptides, which have a wide range of applications in the medical field (Mishra et al. 2017; Strobel et al. 2004; Joseph and Priya 2011). However, endophytic fungal isolates have been investigated for their biological applications including their ability for antimicrobial activity towards pathogenic microbes. For example, naturally bioactive chemicals produced by the endophytic fungi *Penicillium chrysogenum*, *Alternaria alternata* (Fouda et al. 2015), *Fusarium* spp., *Acremonium* spp. (Powthong et al. 2013), and *Pestalotiopsis* sp. (Bagyalakshmi et al. 2012) show antimicrobial activity against gram positive and gram negative bacteria by inhibiting their growth. In several cases, it has been found out that entophytic microbe functions like a biological defense shield for host plants against the foreign phytopathogens. Endophytic fungi produces or secretes many biological compounds and metabolites (like antibiotics and hydrolytic enzymes), which prevents the colonization of pathogenic microbes in plants (Strobel 2003;

Table 13.1 List of some major agricultural crops and their reported endophytes

Plant	Major fungal endophytes	References
Maize	*Fusarium, Aspergillus, Penicillium, Acremonium, Botryodiplodia* sp., *Alternaria alternata, Aureobasidium pullulans* var. *melanigeru, Colletotrichum graminicola, Saccharomyces cerevisiae, Trichoderma koningii, Epicocum nigrum, Cladosporium* sp., *Bipolariszeicola, Phomasorghina*	Nur Amin (2013), Fisher et al. (1992), Orole and Adejumo (2011); Szilagyi-Zecchin et al. (2016)
Rice/paddy	*Fusarium, Aspergillus, Penicillium, Acremonium, Gilmaniella, Arthrobotrys foliicola. Curvularia,* Chaetomium spp., Colletotrichum spp., Curvularia spp., Phytophthora spp., Rhizopus spp. and Trichoderma spp. *Curvularia, Gibberella fujikuro*i	Potshangbam et al. (2017); Zakaria et al. (2010) Leewijit et al. (2016); Pili et al. (2016)
Glycine max (Soybean)	*Fusariumoxysporum, Fusariumsolani and Fusarium* sp. *Ampelomyces* sp., *Cladosporiumcladosporioides, Colletotrichumgloeosporioides, Diaporthehelianthi, Guignardiamangiferae and Phoma* sp. *Boeremia* sp., *Cadophora* sp., *Coniothyrium* sp., *Corynespora cassiicola, Peyronellaea* sp., *Phaeosphaeria* sp., *Scopulariopsis brevicaulis* *Alternaria, Epicoccum, Cladosporium, Alternaria, Diaporthe, Epicoccum* *Penicillium, Rhizoctonia, Phoma* sp., *Guignardia mangiferae*	Fernandes et al. (2015) Rothen et al. (2017) Russo et al. (2016) Impullitti and Malvick (2013) Dalal and Kulkarni (2014)
Wheat	*Alternaria* spp., *Cladosporium, Stemphylium, Chaetomium* sp., *Aspergillus and Phaeosphaeria, Epicoccum nigrum, Cryptococcus* sp., *Rhodotorula rubra, Penicillium* sp., *Fusarium graminearum* *Darksidea* spp., *Cochliobolus* spp., *Periconia* spp., *Embellisia* spp., *Cochliobolus* spp.	Ofek-Lalzar et al. (2016) Larran et al. (2007) Bokati et al. (2016)
Common bean	*Aureobasidium pullulans, Fusarium oxysporum, Xylaria* sp., and *Cladosporium cladosporioides*	Parsa et al. (2016)
Barley	*Cladosporium, Penicillium glabrum, Lophiostoma corticola, Penicillium brevicompactum, Metarhizium, Neotyphodium gen. nov.*	Murphy et al. (2015) Clement et al. (1997)

Berg and Hallmann 2006) and also simultaneously reduces insect (Azevedo et al. 2000; Strobel et al., 2007; Kusari et al. 2012) and nematode (Hol et al. 2016) infestation. Since the onset of report that endophyte *Taxomyces andreanae* produces the same bioactive compound taxol (paclitaxel) as its host plant *Taxus brevifolia* in 1993 (Stierle et al. 1993), several other studies have found out that plant-derived secondary metabolites like huperzine A, diosgenin, α-irone, β-irone, toosendanin, and hypericin are also produced by endophytes (Zhao et al. 2010, 2011). Similarly, bioactive compounds produced by the plants as well as their associated endophytes also include various types of immunosuppressants, antiparasitics, antioxidants, and anticancer drug (Uzma et al. 2018; Strobel 2003; Puri et al. 2005, 2006; Gangadevi and Muthumary 2008; Pasut and Veronese 2009; Giridharan et al. 2012).

Many essential oils have been reported as secondary metabolites of endophytic fungi. Within the various groups of sesquiterpenes occurring in essential oils, the

Table 13.2 Some important bioactive compound produced by endophytic fungi

S. no.	Bioactive compound	Nature/application	Source endophytes (fungi)	References
1.	Taxol	Anticancer	Taxomyces andreanae	Stierle et al. (1993)
2.	Camptothecin	Anticancer	Fusarium solani	Kusari et al. (2009)
3.	Ergoflavin	Anticancer	Sterile mycelium	Deshmukh et al. (2009)
4.	Phenylpropanoids	Anticancer	Penicillium brasilianum	Fill et al. (2010)
5.	Podophyllotoxin	Anticancer	Trametes hirsute	Puri et al. (2006)
6.	Cytochalasins	Anticancer	Rhinocladiella sp.	Wagenaar et al. (2000)
7.	Cytoskyrins	Anticancer	Cytospora sp. CR200	Singh et al. (2007)
8.	Phomoxanthones A and B	Anticancer	Phomopsis species	Isaka et al. (2001)
9.	7-amino-4-methylcoumarin	Antimicrobial compounds	Xylaria sp. YX-28	Liu et al. (2008)
10.	Griseofulvin	Antimicrobial compounds	Xylaria sp.	Park et al. (2005)
11.	Chaetomugilin A and D	Antimicrobial compounds	Chaetomium globosum	Qin et al. (2009)
12.	(−)-mycorrhizin A (+)-cryptosporiopsin	Antimicrobial compounds	Pezicula	Schulz et al. (1995)
13.	Hypericin	Antimicrobial compounds	Aspergillus niger and Candida albicans	Kusari et al. (2008)
14.	Pestacin	Antioxidant compounds	Pestalotiopsis microspora	Harper et al. (2003)
15.	Graphislactone A	Antioxidant compounds	Cephalosporium sp. IFB-E001	Song et al. (2005)

eremophilanes are most common among endophytic fungi (Nicoletti and Fiorentino 2015). A *Xylariaceous* endophytic fungus was isolated from leaves of *Cupressus lusitanica* which also produce essential oil (Filho et al. 2011). Reports suggest that tomato plant which harbors the endophyte *Acremonium strictum* influences the composition of volatile compounds released by the host plant. Some of the bioactive compounds produced by the endophytic fungi are enlisted in Table 13.2.

13.2.3 Bioremediation

Endophytic fungi are known to degrade or deteriorate wide variety of compounds and materials and are a useful source of biodiversity with potential application for bioremediation (Oses et al. 2006; Verma et al. 2015; Sudha et al. 2016; Wang et al. 2017). Investigation of Russell et al. (2011) established the robust polyurethane degradation under anaerobic conditions in which the synthetic polymer served as the only carbon source for the fungus *Pestalotiopsis microspora*. Wood-inhabiting fungal endophytes of Chilean tree species *Drimys winteri* (fungi *Bjerkandera* sp.) and *Prumnopitys andina* (*Mycelia sterilia*) were also reported for lignocellulolytic enzymes production and wood biodegradation (Oses et al. 2006). Similarly, a diuron-degrading endophyte *Neurospora intermedia* was isolated by Wang et al. (2017) from sugarcane root grown in diuron-treated soil. Phytoremediation is another key bioremediation aspect of endophytic fungi in soils contaminated with heavy metals and hydrocarbons. It was also reported that bioaugmentation of two grass species (*Festuca pratensis* Huds and *Festuca arundinacea* Schreb.) with endophytic fungi *Neotyphodium uncinatum* and *Neotyphodium coenophialum* resulted in significant removal of polycyclic aromatic hydrocarbons and total petroleum hydrocarbons from the plant rhizosphere of up to 84 and 72%, respectively, compared with 56 and 31% in control plants, respectively (Soleimani et al. 2010). Endophytic fungi intriguingly exhibit excellent metal-binding capacity and provide more advantages over bacterial bioaugmentation (Gadd 1990; D'Annibale et al. 2006; Sani et al. 2017). The endophyte *Penicillium funiculosum* acts against the copper stress condition and this property can be used in bioremediation of pollution in cultivated area by stress-mediating endophytes (Khan and Lee 2013). Similarly, cadmium stress tolerance on seed germination and seedling growth of *Elymus dahuricus* increases due to the infection with the *Neotyphodium* endophyte (Zhang et al. 2012).

13.3 Techniques for Characterization of Fungal Endophytes

13.3.1 Traditional Methods

The identification and characterization of fungal endophytes based on the traditional methodologies for morphological characters by culture is as follows: Surface sterilization of plant tissues, and maceration of the plant tissue and streaking the macerate onto culture agar plate, or plating small sterilized segments onto potato dextrose agar or any other medium (Hallmann et al. 2006). The information generated from these existing techniques is not sufficient to provide a complete overview of the endophytes-plant associations. These methods are often time consuming, lack sensitivity and specificity, are slow, labor intensive and difficult to interpret (Hyde and Soytong 2008). Many fungi have uncultivable characteristics and could not grow in culture (Duong et al. 2008). Some fungi are very slow growing (Zhu et al. 2008), while others require specific media (Van Wyk et al. 2007). Culture based identification of endophytic fungi from natural environments has been limited due to non-culturable and non-sporulating nature of most of the endophytic fungi.

13.3.2 Molecular Tools

The advanced molecular techniques have also been routinely used by various scientific groups to estimate the occurrence of fungal endophytes inside the living plants (Raja et al. 2017). Advancement in molecular technique, namely ITS sequencing, pyrosequencing, DNA barcoding, fatty acid methyl ester analysis (FAME), and MALDI-TOF led to the rapid and effectual identification of both culture dependent and independent fungi. PCR based technique is designed to amplify a specific DNA sequence from a starting template to provide multiple copies from one single copy of particular gene. Further, fungal endophytes classification based on DNA sequence analysis using polymerase chain reaction (PCR) of internal transcribed spacer (ITS) region is also used as a universal DNA barcode marker (Schoch et al. 2012). The fungal ribosomal Intergenic Spacer region 1 (ITS1), 5.8S (where present) and ITS2 regions were also amplified using the ITS1 and ITS4 specific primers to identify endophytes (Gardes and Bruns 1993). Ofek-Lalzar et al. (2016) isolated endophytes from wheat using culture based and cultivation-independent methods with an analysis of total, 514 intergenic spacer region sequences from single culture. Nonetheless, 18S and 28S genes have generally been used for the identification of endophytic fungi at high taxonomic levels. A total of 59 morphologically unidentifiable strains were isolated and identified from healthy stems and pods of *Theobroma cacao* trees based on the sequence analyses of the 28S gene (Crozier et al. 2006). Further, amplicons for illumina MiSeq sequencing platform were produced with the primer combination NSI1 (Martin and Rygiewicz 2005) and ITS2 (White et al. 1990).

Furthermore, protein coding genes can also be used for identification; these genes usually occur as single copies in genomes, which may be valuable for quantitative comparison of taxon abundances but have certain disadvantages during PCR amplification. As a consequence of the non-conserved third nucleotide base, genes encoding for proteins contain more variation in terms of substitutions as compared with insertions and deletions enabling alignment across phylogenetically distant groups in spite of high variation in sequence. Furthermore, coding genes often contain introns with sufficient power for discrimination among species. A major drawback with numerous protein encoding genes is that they occur in gene families where within genome gene duplications often take place within the same time-frame as speciation, making the identification of orthologues gene problematic (Lindahl and Taylor 2004; Bödeker et al. 2009). Due to non-conserved third nucleotide base creates difficulty in the designing of primers which cover all possible sequence variants, even when highly conserved functional domains are targeted.

In addition, high throughput sequencing (HTS) methods outclass previous approaches in terms of magnitude and resolution. They allow identification and relative quantification of community and offer new insights into fungal ecology. These techniques are now taking over as the prime tool to evaluate plant-associated endophytic fungal communities including mycorrhizal symbionts, pathogens as well as free-living saprotrophs.

Table 13.3 List of Fungal endophyte isolated from different plants and their molecular characterization techniques

S. no.	Crop plants	Primers used	Name of fungal endophyte	Molecular techniques used	References
1.	*Tylophora indica* (Leaf and stem)	ITS1 and ITS4	*Colletotrichum truncatum, Dothediomycetes* sp., *Thielavia subthermophila, Alternaria tenuissima, Nigrospora oryzae, Alternaria* sp., *and Chaetomium* sp.	ITS region	Kumar et al. (2011)
2.	Pigeon pea (*Cajanus cajan*) (Root)	18S ribosomal RNA sequences and ITS	*Neonectria radicicola Fusarium spp Fusarium oxysporum Fusarium subglutinan Fusarium solani*	ITS and β-tubulin gene	Gao et al. (2011)
3.	Grapevine (*Vitis vinifera* L.)	ITS1/ITS4 nu-SSU-0817-59/ nu-SSU-1196-39	*Ampelomyces humuli, Neofusicoccum parvum, Xylaria* sp., *Davidiella tassiana, Gibberella pulicaris, Truncatella angustata,* and *Phoma herbarum*	Internal Transcribed Spacer sequence and Novel colony-PCR method	Pancher et al. (2012)
4.	*Capsicum annuum* (Roots)	ITS and Translation elongation factor gene	*Heterobasidion araucariae*	ITS and Translation elongation factor gene	Paul et al. (2012)
5.	*Acer ginnala*	ITS-rDNA sequences	*Alternaria species, Phomopsis species, Phoma* sp., and *Neurospora* sp. *were dominant endophytes while Pleosporales Cladosporium* sp., *Incertae Sedis* sp., *Epicoccum* sp. and *Trichoderma* sp. *were rare taxa.*	ITS-rDNA sequences	Qi et al. (2012)
6.	*Eucalyptus grandis*	ITS1F and ITS2	*Mycosphaerellaceae, Botryosphaeriaceae, Pleosporaceae, Nectriaceae, and Teratosphaeriaceae*	ITS1 nuclear encoded ribosomal RNA sequenced using Ion Torrent Personal Genome Machine (PGM)	Kemler et al. (2013)

(continued)

Table 13.3 (continued)

S. no.	Crop plants	Primers used	Name of fungal endophyte	Molecular techniques used	References
7.	*Aloe vera*	(V3–V4 regions)	*Proteobacteria, Firmicutes, Actinobacteria, and Bacteroidetes*	NGS illumina sequencing	Akinsanya et al. (2015)
8.	Common bean (*Phaseolus vulgaris*)	ITS4 and ITS5	*Xylaria* sp., *Fusarium oxysporum, and Cladosporium cladosporioides*	Ribosomal DNA internal transcribed spacer (ITS) region	Parsa et al. (2016)
9.	Quinoa plant *Chenopodium quinoa* Root	ITS1-F-ITS4	*Alternaria alternata, Fusarium acuminatumFusarium sambucinum Penicillium* sp. *Plectosphaerella* sp. *Rhinocladiella similis*	Internal transcribed spacer (ITS) region	González-Teuber et al. (2017)
10.	Rice (*Oryza sativa* L.) and maize (*Zea mays* L.)	ITS1 and ITS4	*Fusarium, Sarocladium, Aspergillus, Penicillium Acremonium* sp., and *Penicillium simplicisssum*	Internal transcribed spacer (ITS) region	Potshangbam et al. (2017)
11.	Lemon beebalm (*Monarda citriodora*)	ITS1 and ITS4	*Fusarium oxysporum Fusarium redolens Muscodor yucatanensis*	ITS-5.8 S rDNA sequencing	Katoch and Pull (2017)
12.	Populus trees *Populus trichocarpa*	ITS1 and ITS4	*Atractiella rhizophila*	Fungal-specific internal transcribed spacer (ITS)	Vélez et al. (2017)

Further, laser capture microdissection pressure catapulting (LMPC) is also an auspicious tool which allows selective procurement of the targeted cells under direct microscopic visualization and permits rapid one step procurement of populations from a section of complex, heterogeneous, e.g., plant tissue (Balestrini and Bonfante 2008). LMPC is basically based on wavelengths in the infrared or in the ultraviolet regions, admitting a high-energy laser that allows for very focused and precise cutting where the high concentration of photons destroys the chemical bonds in the tissue. It allows working on living cells or tissues without causing significant artifacts on DNA (Richard J. Howard 2004). A list of various fungal endophytes isolated from different plant species and their characterization using molecular approaches is summarized in Table 13.3.

13.4 Conclusion

Endophytic fungi are part of a plants' microbiome and are ubiquitously found across plant species and their ecosystems. The better understanding of their diversity led to their exploration for important sources of several bioactive natural products having enormous potential for the discovery of new molecules for drug discovery, industrial use, and agricultural practices.

References

Akinsanya MA, Goh JK, Lim SP, Ting ASY (2015) Metagenomics study of endophytic bacteria in Aloe vera using next-generation technology. Genom Data 6:159–163

Aly AH, Debbab A, Proksch P (2011) Fungal endophytes: unique plant inhabitants with great promises. Appl Microbiol Biotechnol 90(6):1829–1845

Amin N (2013) Diversity of endophytic fungi from root of Maize var. Pulut (waxy corn local variety of South Sulawesi, Indonesia). Int J Curr Microbiol Appl Sci 2(8):148–154

Arnold AE, Herre EA (2003) Canopy cover and leaf age affect colonization by tropical fungal endophytes: ecological pattern and process in *Theobroma cacao* (*Malvaceae*). Mycologia 95(3):388–398

Arnold AE, Henk DA, Eells RL, Lutzoni F, Vilgalys R (2007) Diversity and phylogenetic affinities of foliar fungal endophytes in loblolly pine inferred by culturing and environmental PCR. Mycologia 99(2):185–206

Bagyalakshmi TA, Ramesh V, Arivudainambi USE, Rajendran A (2012) A novel endophytic fungus Pestalotiopsis sp. Inhibiting Pinus caneriensis with antibacterial and antifungal potential. Int J Adv Life Sci 1:1–7

Balestrini R, Bonfante P (2008) Laser Microdissection (LM): applications to plant materials. Plant Biosyst 142(331):336

Berg G, Hallmann J (2006) Control of plant pathogenic fungi with bacterial endophytes. In: Schulz BJE, Boyle CJC, Sieber TN (eds) Microbial root endophytes. Springer, Berlin, pp 53–69

Bödeker ITM, Nygren CMR, Taylor AFS, Olson A, Lindahl BD (2009) ClassII peroxidase encoding genes are present in a wide phylogenetic range of ectomycorrhizal fungi. ISME J 3:1387–1395

Bokati D, Herrera J, Poudel R (2016) Soil influences colonization of root-associated fungal endophyte communities of maize, wheat, and their progenitors. J Mycol Article ID 8062073, 9 p

Busby PE, Ridout M, Newcombe G (2016) Fungal endophytes: modifiers of plant disease. Plant Mol Biol 90:645–655

Card SD, Johnson LJ, Teasdale S, Caradus J (2016) Deciphering microbial behaviour—the link between endophyte biology and efficacious biological control agents. FEMS Microbiol Ecol 92

Clarke BB, White JF Jr, Hurley RH, Torres MS, Sun S, Huff DR (2006) Endophyte-mediated suppression of dollar spot disease in fine fescues. Plant Dis 90:994–998

Clement SL, Wilson AD, Lester DG, Davitt CM (1997) Fungal endophytes of wild barley and their effects on Diuraphis noxia population development. Entomol Exp Appl 82:275–281

Cocq KL, Gurr SJ, Hirsch PR, Mauchline TH (2017) Exploitation of endophytes for sustainable agricultural intensification. Mol Plant Pathol 18:469–473

Crozier J, Thomas SE, Aime MC, Evans HC, Holmes KA (2006) Molecular characterization of fungal endophytic morphospecies isolated from stems and pods of Theobroma cacao. Plant Pathol 55:783–791

D'Annibale A, Rosetto F, Leonardi V, Federici F, Petruccioli M (2006) Role of autochthonous filamentous fungi in bioremediation of a soil historically contaminated with aromatic hydrocarbons. Appl Environ Microbiol 72(1):28–36

Dalal JM, Kulkarni NS (2014) Population variance and diversity of endophytic fungi in soybean *Glycine max* (L) Merril. RRJBS 3(4):33–39

Deshmukh SK, Mishra PD, Kulkarni-Almeida VSA, Sahoo MR, Periyasamy G, Goswami H et al (2009) Anti-inflammatory and anti-cancer activity of Ergoflavin isolated from an endophytic fungus. Chem Biodivers 6:784–789

Donoso EP, Bustamante RO, Caru M, Niemeyer HM (2008) Water deficit as a driver of the mutualistic relationship between the fungus Trichoderma harzianum and two wheat genotypes. Appl Environ Microbiol 74(5):1412–1417

Doty SL (2017) Functional importance of the plant endophytic microbiome: implications for agriculture, forestry, and bioenergy. In: Doty S (ed) Functional importance of the plant microbiome. Springer, Cham

Duong LM, McKenzie EHC, Lumyang S, Hyde KD (2008) Fungal sucession of senescent leaves of *Castaopsis diversifolia* in Doi Suthep-Pui National Park, Thiland. Fungal Divers 30:23–36

Fernandes E, Estadual U, Claros M, Liparini O, Universidade P, Silva C, Federal U, Queiroz M, Federal U (2015) Diversity of endophytic fungi in Glycine max. Microbiol Res 181:84–92

Filho FCS, da Silva Amaral L, Rodrigues-Filho E (2011) Composition of essential oils from *Cupressus lusitanica* and a Xylariaceous fungus found on its leaves. Biochem Syst Ecol 39:485–490

Fill TP, da Silva BF, Rodrigues-Fo E (2010) Biosynthesis of phenylpropanoid amides by an endophytic penicillium brasilianum found in root bark of Melia azedarach. J Microbiol Biotechnol 20(3):622–629

Fisher PJ, Petrini O, Lappin Scott HM (1992) The distribution of some fungal andbacterialendophytes in maize (Zea mays L.). New Phytol 122:299–305

Fouda AH, Hassan SED, Eid AM, Ewais EED (2015) Biotechnological applications of fungal endophytes associated with medicinal plant *Asclepias sinaica* (Bioss.). Ann Agric Sci 60(1):95–104

Franken P (2012) The plant strengthening root endophyte Piriformospora indica: potential application and the biology behind. Appl Microbiol Biotechnol 96(6):1455–1464

Gadd GM (1990) Fungi and yeast metal accumulation. In: Ehrlich HL, Brierley CL (eds) Microbiol mineral recovery. McGraw Hill, New York, pp 249–276

Gangadevi V, Muthumary J (2008) Taxol, an anticancer drug produced by an endophytic fungus *Bartalinia robillardoides* Tassi, isolated froma medicinal plant, *Aegle marmelos* Correaex Roxb. World J Microbiol Biotechnol 24(5):717–724

Gange AC, Eschen R, Wearn JA, Thawer A, Sutton BC (2012) Differential effects of foliar endophytic fungi on insect herbivores attacking a herbaceous plant. Oecologia 168(4):1023–1031

Ganley RJ, Brunsfeld SJ, Newcombe G (2004) A community of unknown, endophytic fungi in western white pine. Proc Natl Acad Sci U S A 101(27):10107–10112

Gao Y, Zhao JT, Zu YG, Fu YJ, Wang W et al (2011) Characterization of five fungal endophytes producing Cajaninstilbene acid isolated from pigeon pea [*Cajanus cajan* (L.) Millsp.]. PLoS One 6(11):e27589

Gardes M, Bruns T (1993) ITS primers with enhanced specificity for basidiomycetes—application to the identification of mycorrhizae and rusts. Mol Ecol 2:113 8

Giridharan P, Verekar SA, Khanna A, Mishra PD, Deshmukh SK (2012) Anticancer activity of sclerotiorin isolated from an endophytic fungus *Cephalotheca faveolata* Yaguchi, Nishim. & Udagawa. Indian J Exp Biol 50:464–468

González-Teuber M, Vilo C, Bascuñán-Godoy L (2017) Molecular characterization of endophytic fungi associated with the roots of *Chenopodium quinoa* inhabiting the Atacama Desert, Chile. Genom Data 11:109–112

Hallmann J, Berg G, Schulz B (2006) Isolation procedures for endophytic microorganisms. In: Schulz BJE, Boyle CJC, Sieber TN (eds) Microbial root endophytes. Soil biology, vol 9. Springer, Berlin/Heidelberg

Hamayun M, Khan SA, Ahmad N, Tang D, Kang S, Na C, Sohn E, Hwang Y, Shin D, Lee B, Kim J, Lee I (2009) *Cladosporium sphaerospermum* as a new plant growth-promoting endophyte from the roots of *Glycine max* (L.) Merr. World J Microbiol Biotechnol 25:627–632

Harper K, Arif AM, Ford EJ et al (2003) Pestacin: a 1,3-dihydro isobenzofuran from Pestalotiopsis microspora possessing antioxidant and antimycotic activities. Tetrahedron 59(14):2471–2476

Hol WHG, Raaijmakers CE, Mons I, Meyer KM, van Dam NM (2016) Root-lesion nematodes suppress cabbage aphid population development by reducing aphid daily reproduction. Front Plant Sci 7:111

Howard RJ, Bourett TM, Czymmek KJ (2004) Essential microscopy-based technology for studies of fungal pathogen-plant host interactions. Microsc Microanal 10(S02):218–219

Hyde KD, Soytong K (2008) The fungal endophyte dilemma. Fungal Divers 33:163–173

Impullitti AE, Malvick DK (2013) Fungal endophyte diversity in soybean. J Appl Microbiol 114:1500–1506

Isaka M, Jaturapat A, Rukseree K, Danwisetkanjana K, Tanticharoen M, Thebtaranonth Y (2001) Phomoxanthones A and B, novel xanthone dimers from the endophytic fungus Phomopsis species. J Nat Prod 64(8):1015–1018

Jia M, Chen L, Xin H-L, Zheng C-J, Rahman K, Han T, Qin L-P (2016) A friendly relationship between endophytic fungi and medicinal plants: a systematic review. Front Microbiol 7:906

Johnson LJ, De Bonth ACM, Briggs LR, Caradus JR, Finch SC, Fleetwood DJ, Fletcher LR, Hume DE, Johnson RD, Popay AJ, Tapper BA, Simpson WR, Voisey CR, Card SD (2013) The exploitation of epichloae endophytes for agricultural benefit. Fungal Divers 60:1–18

Joseph B, Mini Priya R (2011) Bioactive compounds from endophytes and their potential in pharmaceutical effect: a review. Am J Biochem Mol Biol 1(3):291–309

João Lúcio Azevedo, Walter Maccheroni Jr., Welington Luiz de Araújo, Endophytic microorganisms: a review on insect control and recent advances on tropical plants. EJB Electronic Journal of Biotechnology,Vol.3 No.1, Issue of April 15, 2000.Katoch M, Pull S (2017) Endophytic fungi associated with *Monarda citriodora*, an aromatic and medicinal plant and their biocontrol potential. Pharm Biol 55(1):1528–1535

Kemler M, Garnas J, Wingfield MJ, Gryzenhout M, Pillay K-A, Slippers B (2013) Ion torrent PGM as tool for fungal community analysis: a case study of endophytes in *Eucalyptus grandis* reveals high taxonomic diversity. PLoS One 8(12):e81718

Khan AL, Lee IJ (2013) Endophytic *Penicillium funiculosum* LHL06 secretes gibberellin that reprograms *Glycine max* L. growth during copper stress. BMC Plant Biol 13:86

Kumar S, Kaushik N, Edrada-Ebel R, Ebel R, Proksch P (2011) Isolation, characterization, and bioactivity of endophytic fungi of Tylophora indica. World J Microbiol Biotechnol 27(3):571–577. https://doi.org/10.1007/s11274-010-0492-6

Kumara PM, Shweta S, Vasanthakumari MM, Sachin N, Manjunatha BL, Jadhav SS, Ravikanth G, Ganeshaiah KN, Shaanker RU (2014) Endophytes and plant secondary metabolite synthesis: molecular and evolutionary perspective. In: Verma VC, Gange AC (eds) Advances in endophytic research. Springer, New Delhi, pp 177–190

Kusari S, Lamshöft M, Zühlke S, Spiteller M (2008) An endophytic fungus from Hypericum perforatumthat produces hypericin. J Nat Prod 71(2):159–162

Kusari S, Zühlke S, Spiteller M (2009) An endophytic fungus from *Camptotheca acuminata* that produces camptothecin and analogues. J Nat Prod 72(1):2–7

Kusari S, Verma VC, Lamshöft M, Spiteller M (2012) An endophytic fungus from *Azadirachta indica* A. Juss. that produces azadirachtin. World J Microbiol Biotechnol 28:1287–1294

Kusari S, Singh S, Jayabaskaran C (2014) Biotechnological potential of plant-associated endophytic fungi: hope versus hype. Trends Biotechnol 32:297–303

Larran S, Perelló A, Rosa Simón M, Moreno V (2007) The endophytic fungi from wheat (Triticum aestivum L.). World J Microbiol Biotechnol 23:565–572

Leewijit T, Pongnak W, Soytong K, Poeaim S (2016) Int J Agric Technol 12(7.2):2191–2202

Li JY, Harper JK, Grant DM, Tombe BO, Bashyal B, Hess WM, Strobel GA (2001) Ambuic acid, a highly functionalized cyclohexenone with antifungal activity from *Pestalotiopsis spp.* and *Monochaetia sp.* Phytochemistry 56(5):463–468

Lindahl BD, Taylor AFS (2004) N-acetylhexosaminidase coding genes in ectomycorrhizal basidiomycetes. New Phytol 164:193–199

Liu X, Dong M, Chen X, Jiang M, Lv X, Zhou J (2008) Antimicrobial activity of an endophytic Xylariasp.YX-28 and identification of its antimicrobial compound 7-amino-4-methylcoumarin. Appl Microbiol Biotechnol 78(2):241–247

Lugtenberg BJ, Caradus JR, Johnson LJ (2016) Fungal endophytes for sustainable crop production. FEMS Microbiol Ecol 92(12):194

Martin K, Rygiewicz P (2005) Fugal-specific primers developed for analysis of the ITS region of environmental DNA extracts. BMC Microbiol 5:28

Miller JD, Sumarah MW, Adams GW (2008) Effect of a rugulosinproducing endophyte in Picea glauca on Choristoneura fumiferana. J Chem Ecol 34:362e368

Mishra VK, Passari AK, Chandra P, Leo VV, Kumar B, Gupta VK, Singh BP (2017) Determination and production of antimicrobial compounds by *Aspergillus clavatonanicus* strain MJ31, an endophytic fungus from Mirabilis jalapa L. using UPLC-ESI-MS/MS and TD GC-MS. PLoS One 12(10):1–24. https://doi.org/10.1371/journal.pone.0186234

Murphy BR, Nieto LM, Doohan FM, Trevor R (2015) Hodkinsona Profundae diversitas: the uncharted genetic diversity in a newly studied group of fungal root endophytes. Mycology 6(3–4):139–150

Nicoletti R, Fiorentino A (2015) Plant bioactive metabolites and drugs produced by endophytic fungi of Spermatophyta. Agriculture 5(4):918–970

Obledo, E.N., Barragán-Barragán, L.B., Gutiérrez-González, P. , B. C. Ramírez-Hernández, B. C., José J. RamírezBenjamín Rodríguez-Garay (2003) Plant Cell, Tissue and Organ Culture 74: 237.

Ofek-Lalzar M, Gur Y, Ben-Moshe S, Sharon O, Kosman E, Mochli E, Sharon A (2016) Diversity of fungal endophytes in recent and ancient wheat ancestors Triticum dicoccoides and Aegilops sharonensis. FEMS Microbiol Ecol 92(10)

Orole OO, Adejumo TO (2011) Bacterial and fungal endophytes associated with grains and roots of maize. J Ecol Nat Environ 3:298–303

Oses R, Valenzuela S, Freer J, Baeza J, Rodríguez J (2006) Evaluation of fungal endophytes for lingo cellulolytic enzyme production and wood biodegradation. Int Biodeterior Biodegrad 57:129–135

Ownley BH, Gwinn KD, Vega FE (2010) Endophytic fungal entomopathogens with activity against plant pathogens: ecology and evolution. BioControl 55:113–128

Pancher M, Ceol M, Corneo PE, Longa CMO, Yousaf S, Pertot I, Campisano A (2012) Fungal endophytic communities in grapevines (*Vitis vinifera* L.) respond to crop management. Appl Environ Microbiol 78(12):4308–4317

Park J-H, Choi GJ, Lee HB et al (2005) Griseofulvin from Xylaria sp. strain F0010, an endophytic fungus of Abies holophylla and its antifungal activity against plant pathogenic fungi. J Microbiol Biotechnol 15(1):112–117

Park YH, Lee SG, Ahn DJ et al (2012) Diversity of fungal endophytes in various tissues of *Panax ginseng* meyer cultivated in Korea. J Ginseng Res 36(2):211–217

Parsa S, García-Lemos AM, Castillo K, Ortiz V, López-Lavalle LAB, Braun J, Vega FE (2016) Fungal endophytes in germinated seeds of the common bean, *Phaseolus vulgaris*. Fungal Biol 120(5):783–790

Pasut G, Veronese FM (2009) PEG conjugates in clinical development or use as anticancer agents: an overview. Adv Drug Deliv Rev 61(13):1177–1188

Paul NC, Deng JX, Shin KS, Yu SH (2012) Molecular and morphological characterization of endophytic *Heterobasidion araucariae* from roots of *Capsicum annuum* L. in Korea. Mycobiology 40(2):85–90

Pili NN, França SC, Kyndt T, Makumba BA, Skilton R, Höfte M, Mibey RK, Gheysen G (2016) Analysis of fungal endophytes associated with rice roots from irrigated and upland ecosystems in Kenya. Plant Soil 405(1–2):371–380

Potshangbam M, Devi SI, Sahoo D, Strobel GA (2017) Functional characterization of endophytic fungal community associated with *Oryza sativa* L. and *Zea mays* L. Front Microbiol 8:325

Powthong P, Thongmee A, Suntornthiticharoen P (2013) Antioxidant and antimicrobial activities of endophytic fungi isolated from Sesbania grandiflora (L.) Pers. Int J Phytomed 5(1):102–107. ISSN 0975-0185

Puri SC, Verma V, Amna T, Qazi GN, Spiteller M (2005) An endophytic fungus from *Nothapodytes foetida* that produces camptothecin. J Nat Prod 68:1717–1719

Puri SC, Nazir A, Chawla R et al (2006) The endophytic fungus Trametes hirsuta as a novel alternative source of podophyllotoxin and related aryl tetralin lignans. J Biotechnol 122(4):494–510

Pusztahelyi T, Holb IJ, Pócsi I (2015) Secondary metabolites in fungus-plant interactions. Front Plant Sci 6:573

Qi F, Jing T, Zhan Y (2012) Characterization of endophytic fungi from *Acer ginnala* Maxim. in an artificial plantation: media effect and tissue-dependent variation. PLoS One 7(10):e46785

Qin JC, Zhang Y-M, Gao J-M et al (2009) Bioactive metabolites produced by Chaetomium globosum, an endophytic fungus isolated from Ginkgo biloba. Bioorg Med Chem Lett 19(6):1572–1574

Raja HA, Miller AN, Pearce CJ, Oberlies NH (2017) Fungal identification using molecular tools: a primer for the natural products research community. J Nat Prod 80(3):756–770

Redman RS, Sheehan KB, Stout RG, Rodriguez RJ, Henson JM (2002) Thermotolerance conferred to plant host and fungal endophyte during mutualistic symbiosis. Science 298:1581

Rodriguez RJ, White JF Jr, Arnold AE, Redman RS (2009) Fungal endophytes: diversity and functional roles. New Phytol 182(2):314–330

Rothen C, Miranda V, Aranda-Rickert A, Fracchia S, Rodríguez MA (2017) Characterization of dark septate endophyte fungi associated with cultivated soybean at two growth stages. Appl Soil Ecol 120:62–69

Russell JR, Huang J, Anand P, Kucera K, Sandoval AG, Dantzler KW, Hickman D, Jee J, Kimovec FM, Koppstein D, Marks DH, Mittermiller PA, Núñez SJ, Santiago M, Townes MA, Vishnevetsky M, Williams NE, Vargas MPN, Boulanger LA, Slack CB, Strobel SA (2011) Biodegradation of polyester polyurethane by endophytic fungi. Appl Environ Microbiol 77:6076–6084

Russo ML, Pelizza SA, Cabello MN, Stenglein SA, Vianna MF, Scorsetti AC (2016) Endophytic fungi from selected varieties of soybean (Glycine max L. Merr.) and corn (Zea mays L.) grown in an agricultural area of Argentina. Rev Argent Microbiol 48(2):154–160

Sani A, Venkateswarulunagam V, Netala R, Tartte V (2017) Screening and identification of heavy metal-tolerant endophytic fungi *lasiodiplodia theobromae* from *boswellia ovalifoliolata* an endemic plant of tirumala hills. Asian J Pharm Clin Res 10(3):488–491

Sanz-Ros VA, Muller MM, Martin S, Diez JJ (2015) Fungal endophytic communities on twigs of fast and slow growing Scots pine (*Pinus sylvestris* L.) in northern Spain. Fungal Biol 119:870–883

Schoch CL, Seifert KA, Huhndorf S, Robert V, Spouge JL, Levesque CA, Chen W (2012) Nuclear ribosomal internal transcribed spacer (ITS) region as a universal DNA barcode marker for Fungi. Proc Natl Acad Sci 109(16):6241–6246

Schulz B, Sucker J, Aust HJ et al (1995) Biologically active secondary metabolites of endophytic Peziculaspecies. Mycol Res 99(8):1007–1015

Singh MP, Janso JE, Brady SF (2007) Cytoskyrins and cytosporones produced by Cytospora sp. CR200: taxonomy, fermentation and biological activities. Mar Drugs 5:71–84

Soleimani M, Afyuni M, Hajabbasi MA, Nourbakhsh F, Sabzalian MR, Christensen JH (2010) Phytoremediation of an aged petroleum contaminated soil using endophyte infected and non-infected grasses. Chemosphere 81:1084–1090

Song YC, Huang WY, Sun C, Wang FW, Tan RX (2005) Characterization of graphislactone A as the antioxidant and free radical-scavenging substance from the culture of Cephalosporium sp. IFB-E001, an endophytic fungus in Trachelospermum jasminoides. Biol Pharm Bull 28(3):506–509

Stierle A, Strobel G, Stierle D (1993) Taxol and taxane production by *Taxomyces andreanae*, an endophytic fungus of Pacific yew. Science 260:214–216

Strobel GA (2003) Endophytes as sources of bioactive products. Microbes Infect 5(6):535–544
Strobel G, Daisy B, Castillo U, Harper J (2004) Natural products from endophytic microorganisms. J Nat Prod 67(2):257–268
Strobel GA, Kluck K, Hess WM, Sears J, Ezra D, Vargas PN (2007) Muscodor albus E-6, an endophyte of Guazuma ulmifolia making volatile antibiotics: isolation, characterization and experimental establishment in the host plant. Microbiology 153(8):2613–2620
Sudha V, Baskar RGK, Al-Dhabi NA, Duraipandiyan V (2016) Biological properties of endophytic fungi. Braz Arch Biol Technol 59:e16150436
Szilagyi-Zecchin VJ, Adamoski D, Gomes RR, Hungria M, Ikeda AC, Kava-Cordeiro V, Glienke C, Galli-Terasawa LV (2016) Composition of endophytic fungal community associated with leaves of maize cultivated in south Brazilian field. Acta Microbiol Immunol Hung 63:449–466
Uzma F, Hashem A, Murthy N, Mohan HD, Kamath PV, Singh BP, Venkataramana M, Gupta VK, Siddaiah CN, Chowdappa S, Alqaeawi AA, Abd-Allah EF (2018) Endophytic fungi—alternative sources of cytotoxic compounds: a review. Front Pharmacol 9(309):1–37. https://doi.org/10.3389/fphar.2018.00309
Van Wyk M, Pegg G, Lawson S, Wingfield MJ (2007) Ceratocystis atrox sp. nov associated with Phoracanthta acanthocera infestations on Eucalyptus in Australia. Australas Plant Pathol 36:407–414
Vélez JM, Tschaplinski TJ, Vilgalys R, Schadt CW, Bonito G, Hameed K, Hamilton CE (2017) Characterization of a novel, ubiquitous fungal endophyte from the rhizosphere and root endosphere of Populus trees. Fungal Ecol 27:78–86
Verma VC, Gond SK, Kumar A, Kharwar RN, Boulanger LA, Strobel GA (2011) Endophytic fungal flora from roots and fruits of an Indian neem plant Azadirachta indica A. Juss., and impact of culture media on their isolation. Indian J Microbiol 51(4):469–476
Verma SK, Kumar A, Lal M, Das MD (2015) Biodegradation of synthetic dye by endophytic fungal isolate in *calotropis procera* root. Int J Appl Sci Biotechnol 3(3):373–380
Wagenaar MM, Corwin J, Strobel G, Clardy J (2000) Three new cytochalasins produced by an endophytic fungus in the genus Rhinocladiella. J Nat Prod 63(12):1692–1695
Wang Y, Li H, Feng G, Du L, Zeng D (2017) Biodegradation of diuron by an endophytic fungus *Neurospora intermedia* DP8-1 isolated from sugarcane and its potential for remediating diuron-contaminated soils. PLoS One 12(8):e0182556
Waqas M, Khana AL, Md Hamayun C, Shahzada R, Kanga S, Kime JG, Lee IJ (2015) Endophytic fungi promote plant growth and mitigate the adverse effects of stem rot: an example of *Penicillium citrinum* and *Aspergillus terreus*. J Plant Interact 10(1):280–287
White T, Bruns T, Lee S et al (1990) Amplification and direct sequencing of fungal ribosomal RNA genes for phylogenetics. In: Innis MA, Shinsky DG, Sninsky JJ (eds) PCR protocols: a guide to methods and applications, vol 315. Academic, San Diego, p 22
Yuan Y, Feng H, Wang L, Li Z, Shi Y, Zhao L et al (2017) Potential of endophytic fungi isolated from cotton roots for biological control against *Verticillium* Wilt disease. PLoS One 12(1):e0170557
Zakaria L, Yaakop AS, Salleh B, Zakaria M (2010) Trop Life Sci Res 21(1):101–107
Zhang X, Li C, Nan Z (2012) Effects of cadmium stress on seed germination and seedling growth of *Elymus dahuricus* infected with the Neotyphodium endophyte. Sci China Life Sci 55(9):793–799
Zhao J, Zhou L, Wang J, Shan T, Zhong L, Liu X, Gao X (2010) Endophytic fungi for producing bioactive compounds originally from their host plants. In: Vilas AM (ed) Current research, technology and education topics in applied microbiology and microbial biotechnology, vol 1. Formatex Research Center, pp 567–576
Zhao J, Shan T, Mou Y, Zhou L (2011) Plant derived bioactive compounds produced by endophytic fungi. Mini Rev Med Chem 11:159–168
Zhu GS, Yu ZN, Gui Y, Liu ZY (2008) A novel technique for isolating orchid mycorrhizal fungi. Fungal Divers 33:123–137

Chapter 14
Endophytic Fungi and Their Enzymatic Potential

Rashmi Mishra, J. S. Kushveer, P. Revanthbabu, and V. Venkateswara Sarma

14.1 Introduction

A complex network of interactions is shared between the hosts and their endophytic fungi. Evidences depicting the presence of endophytes in the fossils of leaves and stem tissues implicate their evolution with the first appearance of higher plants on the earth (Redecker et al. 2000). Through a co-evolution between the fungal endophytes and their hosts, a beneficial alliance has been established (Schulz and Boyle 2005). Endophytes are considered to be ubiquitous due to their presence in relatively all above ground flora including bryophytes, pteridophytes, gymnosperms, and angiosperms. Furthermore, this polyphyletic diverse group spans throughout the globe spanning the arctic to tropics, agronomical fields to forests of tropics (Arnold 2007). Endophytic fungi reside internally in the tissues, beneath the layers of epidermal cells, inter/intracellularly, without effecting any apparent infections to their hosts or causing any disease symptoms. They are ubiquitous organisms having symbiotic relationship with their hosts throughout their life cycle (Yan et al. 2015). The presence of endophytes is found in every host tissue, i.e., leaves, seeds, buds, stems, ovules, roots, tubers, fruits, xylem, and bark (Breen 1994). While the hosts bestow protective shelter and nutrients, the fungus yields secondary metabolites, organic acids, enzymes, etc. (Khan et al. 2016). Besides aiding growth to their hosts, these endophytes cater defense against pathogenic microbes as they secrete several enzymes (Sunitha et al. 2012). The endophytic fungi in addition to contributing to the host in physiology and metabolism, the mutualistic relation also provides the host the plant growth promoting hormones, solubilizing phosphatases, degradation of xenobiotics, antimicrobial properties, etc., (Yuan et al. 2017).

R. Mishra · J. S. Kushveer · P. Revanthbabu · V. V. Sarma (✉)
Department of Biotechnology, Pondicherry University, Kalapet, Puducherry, India
e-mail: sarmavv.dbt@pondiuni.edu.in

B. P. Singh (ed.), *Advances in Endophytic Fungal Research*, Fungal Biology,
https://doi.org/10.1007/978-3-030-03589-1_14

The significance of evolution and the dynamics of interactions between the endophytes and their hosts are often overlooked. The lack of such an acquaintance on their ecological importance has resulted in impeded exploitation of endophytes in industries, pharmaceutics, agriculture, human welfare, and environment. Endophytic fungal research, often entangled with their hosts and the environment, has flared up contributions in the areas of host and fungal ecology, fungal evolutionary biology, systematics and applied biotechnological research ranging from biocontrol to bioprospecting of endophytes (Arnold 2007). Besides being conspicuous producers of secondary metabolites that have potential as antibiotic, anticancer, antioxidant, and anti-parasitic compounds, used in the pharmaceutical industries, endophytic fungi are also reservoirs of a battery of enzymes. Different types of extracellular enzymes secreted by endophytic fungi have applications in food, textile, leather, confectionery, agriculture, beverage, and human health sectors (Mishra et al. 2017). Furthermore, around 60% of enzymes exploited in industries are a product of fungal origin (Suryanarayanan et al. 2012). Fungal endophytes are one of the most favored candidates in enzyme production. In spite of these facts, their exploitation in enzyme production for biotech, pharmaceutical, and food industries or in human welfare is sparse (Uzma et al. 2018; Mishra et al. 2016).

Endophytic fungi in association with the hosts secrete proteins which are presumed to aid in growth, nourishment, and perchance for defense. Availability of soluble sugars often determines the growth of fungal endophytes. In consequence, cellulose and lignin production depicts a strategic benefit to decompose the tissues and endure as saprobes after senescence (Oses et al. 2006). At present, endophytic fungi have a leading edge in the manufacture of industrially purposive enzymes like amylases, cellulases, chitinases, lipases, and proteases. The benefits of enzymes over chemical catalysts make them superior candidates for commercial and pharmaceutical uses as they function under comparatively moderate conditions of pH, temperature, and pressure, often stereoselective and specific (Tiwari 2015). Extracellular hydrolases, such as cellulases, pectinases, lipases, and xylanases, produced by endophytic fungi equip the hosts with resistance mechanisms against pathogenic infiltration (Mhatre et al. 2017).

From biotechnological point of view, there are few relevant enzymes that are discussed further, which are as follows:

14.1.1 Xylanases

The use of xylanases in biotech industry has increased remarkably. Xylanases have numerous applications in food, paper and pulp, pharmaceutical, and textile industries.

14.1.1.1 Paper and Pulp Industry

Xylanases are one of the favored enzymes in cellulose and paper industries. Bridges between the xylan and lignin are acted upon by xylanases that open cellulosic pulp structure, concomitantly guiding xylan fragmentation and fragment extraction subsequently (Paice et al. 1992). Use of xylanase in pre-bleaching has received considerable attention as much lower amount of chlorine compounds is used (up to 30%), resulting in 20% reduction of organochlorine composition of effluents. As otherwise formation of organochlorines, during the decomposition of lignin, with the use of chlorine, is hazardous to biotic as well as abiotic environmental factors as they are extremely mutagenic, noxious, and threatening.

14.1.1.2 Textiles

Xylanases also have been extensively used in the pretreatment of plant fibers. The xylanases employed in the process should be free from cellulolytic enzymes. For example, liberation of long cellulose fibers after the treatment of China grass stems. The need to use the strong bleaching step is no longer required, considering the lignin does not go through oxidation that leads to darkening of the fibers (Prade 1996; Brühlmann et al. 2000).

14.1.1.3 Food

Xylanases break down hemicelluloses present in wheat flour that result in a softer dough, increased volume of bread, higher water absorption, and resistance to fermentation process (Maat et al. 1992). Xylanases along with endoglucanases participate in arabinoxylan hydrolysis, separation of starch and gluten isolation from starch present in wheat flour (Wong et al. 1988; Gilbert et al. 1993). Xylanases are nowadays, used in combination with amylases, pectinases, and cellulases to improve juice quality of fruits by means of liquefaction.

14.1.1.4 Endophytic Fungi and Xylanases

Thermotolerant extracellular xylanase from an endophytic fungal strain, *Aspergillus terreus* residing inside *Memora peregrina*, a native plant to Brazil, was purified and characterized. The fungal strain was shown to have high levels of xylanase activity when grown on wheat bran as carbon source at 30 °C for 48 h, while cellulase activity was low. Xylanase was purified to 45-fold and 67% of recovery by carboxy methyl cellulose chromatography. A single band of 23 kD of purified band was observed in SDS PAGE. Optimal enzyme activity was observed at 55 °C and pH 4.5. Isolated enzyme was found to be thermotolerant at 45 °C with half-life of 55 min and at 50 °C for 36 min, respectively. Enzyme kinetic studies have revealed that K_m was 22 μg/mL and V_{max} 625 μg/mL. Metal ions K^+ and Mn^{+2} were involved in the xylanase activation (Sorgatto et al. 2012).

From the leaves of *Croton oblongifolius*, a Thai medicinal plant, 54 endophytic fungi were isolated. Primary screening for xylanase activity was investigated on solid xylan agar plates. Thirty fungal strains were shown positive and secondary screening was performed in xylan liquid culture. The highest xylanase was produced by PTRa9 strain. Further purification and characterization of xylanase was carried out. Optimum production of xylanase was accomplished after 96 h of culture, supplemented with 2% (w/v) rice bran and 0.1% (w/v) $(NH_4)_2SO_4$ (ammonium sulfate), as sources of carbon and nitrogen, respectively. The xylanase enzyme was purified by initial precipitation with ammonium sulfate followed by diethylaminoethyl cellulose ion exchange chromatography and Superdex 75 gel filtration chromatography techniques. The purified xylanase was found to be 54.8 KDa and 161.1 U/mg protein of specific activity was observed. More than 90% of activity of enzyme was maintained from −20 to 45 °C, and optimum activity was recorded at 45 °C. The xylanase enzyme was found to be active at a broad range of pH 3–11 and optimal pH was 5.0. The enzyme was sensitive to Cu^{2+}, Hg^{2+}, and EDTA. From the analysis of kinetic part, it had a K_m of 0.421 μg/mL and a V_{max} of 0.826 U/mg protein (Wipusaree et al. 2011).

A fungal strain (NRCF5) was isolated from the inner tissue of coral *Rhytisma* sp. (Egypt). Based on the morphological studies and ITS sequencing results, the species was confirmed as *Aspergillus* sp. The principles underlying in shuffling of genome to achieve cultivation of xylanase producing fungi, from marine source, *Aspergillus* sp. NRCF5 was exploited as starting strain and genetic variability was induced using different combinations and doses of mutagens. Ultraviolet irradiation (5 min) and N-methyl N-nitro-N-nitrosoguanidine (NTG, 100 mg/mL) for 30 (UNA) and 60 (UNB) min along with NTG (100 mg/mL) and ethidium bromide (250 mg/mL) for 30 (NEA) and 60 (NEB) min were used in combination as mutagens. This mutagenesis led to an increase in xylanase activity in five fungal strains and 0.25% (w/v) antimetabolite 2DG tolerance was also observed. There after recursive protoplast fusion was carried out. Four rounds of genome shuffling led to the rise of seven high xylanase producing fungal strains, and these were also shown tolerance to 1.0% (w/v) 2 DG. R4/31 strain was the best xylanase producing strain among the 7 strains. The recombinant xylanase has shown 6.13 higher folds of xylanase activity when compared with starting strain NCRF5 with 427.5/mL xylanase and 2.48 times more than that of the original (Mutant NEA51) (El-Bondkly 2012).

14.1.2 *Amylases*

14.1.2.1 Textiles

Starch is extensively used in sizing process to prevent yarn from breaking, as it is easily available and cheap. After weaving process, the starch is removed by amylases as they act only on starch and do not interfere with fabrics. The starch thus removed, from the fabrics due to desizing by amylases, which break starch into dextrins, can be washed off (Hendriksen et al. 1999).

14.1.2.2 Paper

Starch is used in paper industry to size the paper. Sizing gives mechanical strength to the paper which involves applying less viscous and high molecular weight starch to the paper. Usually, the starch is highly viscous. Upon sizing process, differences in viscosity of starch gives different grades of paper (Godfrey and West 1996).

14.1.2.3 Detergents

Approximately, 90% of liquid detergents contain amylases (Kottwitz et al. 1994). Amylases are stable at low temperatures, alkaline pH, and easily digest starch present in food particles that are attached to the clothes and form water-soluble oligosaccharides.

14.1.2.4 Bread and Baking Industry

Amylases are used in bread and baking industry for decades. Amylases give better color, higher volume, and softer crumb to the bread which improves the taste. Fungal amylases also got authorized as bread additives in the USA and in the UK since 1994 and 1963, respectively, after confirmation of their GRAS status (Pritchard 1992).

14.1.2.5 Endophytic Fungi and Amylases

Isolation of 14 endophytic fungi associated with *Acanthus ilicifolius* L., a mangrove angiosperm, and *Acrostichum aureum* L., a mangrove fern from Nethravathi Mangrove, southwestern coasts, India, was reported. Among these fungi, four of the fungal isolates from *A. aureum* host have shown positive activity when subjected to amylase plate assay (Maria et al. 2005).

Endophytic fungi isolated from flowers and leaves of *Alpinia calcarata* were studied for amylolytic activity. Out of the 30 fungal strains, one strain (*Cylindrocephalum* sp.- no. 7) has shown highest amylolytic activity. The influence of physical and chemical parameters that influence the production of amylase was additionally investigated with temperature, pH, carbon, and nitrogen sources as the parameters. Optimal amylase activity was observed at 30 °C and at pH 7.0. The concentration of 1.5% of maltose as carbon source and 0.3% sodium nitrate were recorded to be the best carbon and nitrogen sources, respectively, for the *Cylindrocephalum* sp. (Sunitha et al. 2012).

Adhatoda vasica, Coleus aromaticus, Costus aromaticus, and *Lawsonia inermis* were processed for endophytic fungi and 12 fungal isolates were isolated. Out of the 12 fungi, four hyphomycetes: *Cladosporium cladosporioides, Curvularia brachyspora, Drechslera hawaiiensis*, and *Nigrospora sphaerica*, along with three coelomycetes: *Colletotrichum crassipes, C. gloeosporioides*, and *Phyllosticta.* sp. were found to be positive for amylase in the plate assay (Amirita et al. 2012).

From *Acremonium* sp., an endophyte isolated from Malaysian forest trees, a novel sago starch degrading glucoamylase was purified and characterized. Optimum enzyme activity was found at 55 °C and at pH 5.5. Glucoamylase was stable at a broad range of pH ranging from 3.0 to 7.0 and was stable up to 60 °C. EDTA was found to strongly inhibit the enzyme. The enzyme catalyzed hydrolysis of amylopectin and amylose, displayed apparent V_{max} of 195 mmol/mL/min and 391 mmol/mL/min and K_m values of 10.0 and 3.8 mg/mL, respectively. The purified enzyme is an oligomeric enzyme containing 2 subunits of 22 and 34 kD. End product analysis of raw sago starch by purified glucoamylase has shown glucose alone, illustrating that the enzyme displayed an exo-action of starch-degrading activity (Marlida et al. 2000).

Amylase producing *Presumina minima*, an endophytic isolate of *Eremophila longifolia*, a native plant of Australia. Zymogram of α amylase has revealed it as a 70 kD enzyme. Its optimum activity was observed at 25 °C and optimum pH was 9.0. Additional study of metal ion after effect was carried out. From these studies it was confirmed that manganese and calcium promote and stabilize amylase activity. The impact of different nitrogen and carbon sources on amylase activity was observed. L-asparaginase and starch were confirmed as best nitrogen and carbon sources, respectively. Enzyme production during scale-up fermentation was encouraging as bioreactors revealed comparable production to that obtained in shaker cultures. The purified enzyme has shown similarity with the amylase of *Magnaporthe oryzae*, as confirmed from the partial sequence matching of the enzyme (Zaferanloo et al. 2014).

A foliar endophytic fungus was isolated from *Zea mays.* The highest rate of production of glucoamylase was observed during initial growth phase, i.e., after 24 h incubation period at pH 8.0. The glucoamylase activity was lost at 100 °C while stable at 70 °C and activity recorded was 158 U/mg protein and 100% residual activity was retained for 30 min. Saturation of 60% yielded 421 U/mg protein which was 74% with a 2.7-fold purification.

14.1.3 Cellulases

14.1.3.1 Textiles

Cellulases find various roles in textile industries such as stone washing of jeans. Stone washing gives faded look to the jeans. Cellulases are used in the place of pumice stone, a process known as biostoning that breaks minute fibrillar ends on the yarn surface. Stoning process involves more fiber damage, less machines productivity, more work intensive and not environment friendly. These drawbacks are overcome by biostoning process.

14.1.3.2 Pulp and Paper

The process of mechanical pulping such as grinding and refining of the raw wood material leading to pulps containing high fibers, bulk, and stiffness. During biomechanical pulping, employing cellulases has derived significant energy savings of 20–40% during grinding, improving, and refining of the raw wood material. Cellulases only, or in combination with xylanases, are profitable in the process of deinking waste papers. Various applications intended until now, in exploiting cellulases are for the extraction of ink from the surface of fiber by limited hydrolysis of carbohydrate molecule.

14.1.3.3 Food Industry

Cellulases have comprehensive applications in food biotechnology too. Enriched methods of clarification, extraction, and stabilization are needed for vegetable and fruit juice production, in macerating enzyme complexes (xylanases, cellulases, and pectinases) to elevate the yield of juices, enhance properties such as flavor, texture, and aroma of vegetables and fruits, diminishing bitterness by infusion of enzymes such as β-glucosidases and pectinases.

14.1.3.4 Detergent Industry

Employment of cellulases conjointly with lipases and proteases in the manufacturing of detergents has been a contemporary introduction in the industry. The alkaline cellulases find their use as a detergent additive as an industrial practice is duly trailed as an aspect of selective cellulose contact within the fibers and dig out dirt from the inter-fibrillary spaces.

14.1.3.5 Wine and Brewery Industry

Enzymes such as glucanases, pectinases, and hemi-cellulases play a determining aspect in wine formulation by correcting color, macerating skin, filtration, and clarification and conclusively, the wine peculiarity and stability. Aromaticity of wines can be enhanced by β-glucosidases as they remodel glycosylated precursors. The assistance of using such enzymes in course of wine preparation comprise reformed quality, finer maceration, apparent clarification, improved color extraction, elementary filtration, and improved stability.

14.1.3.6 Endophytic Fungi and Cellulases

Peng and Chen (Peng and Chen 2007) isolated 149 endophytic fungi from stems of 7 plants. About 69 (48.9%) out of the 149 fungal species showed the presence of intercellular lipid bodies when observed under microscope of which 26 endophytic fungal strains depicted larger lipid bodies. These strains were selected and cultivated on potato dextrose broth which improves the taste to accumulate 21.3% to 35% of lipid content of dry cell weight. When solid state fermentation was carried out, these isolates were found positive for cellulase production and microbial oil with the yields of 0.31~0.69 filter paper unit as well as 19~42 mg/g initial dry substrate, respectively. Thus, the above results indicate that some endophytic fungal isolates associated with the oleaginous host plants possess the attributes of oil accumulation and cellulase production simultaneously.

1,4,β-D endoglucanase, 1,4 β-D cellobiohydrolase, and β-D glucosidase are the three main members of cellulolytic enzymes. *Periconia* sp. (BCC2871), an endophytic fungus, produces thermotolerant β-glucosidase. The full length gene of β-glucosidase from this strain was cloned in *Pichia pastoris* KM71 strain. The recombinant enzyme was shown to have its optimal pH at 5 and 6 and the optimal temperature to be at 70 °C. High enzyme activity was observed even after longer incubation periods at higher temperatures thus retaining a 60% activity after 1.5 h at 70 °C. The enzyme was found to be stable at higher pH conditions also and retained 100% activity recorded even after 2 h at pH ≥ 8. The inclusion of β-glucosidase into the hydrolysis reaction of rice straw encompassing a commercial cellulase, Celluclast 1.5L (Novozyme, Denmark), resulted in the boosting of reducing sugars release upon hydrolysis. This recombinant enzyme is apt and suitable apt for treating lignocellulosic substrates in the production of biofuels and other chemicals (Harnpicharnchai et al. 2009).

Twelve endophytic fungal isolates were isolated from four medicinal plants, namely, *Adhatoda vasica, Costus igneus, Coletus aromatics*, and *Lawsonia inermis,* and were screened for cellulase activity. Among these twelve strains, *Cladosporium cladosporioides* and *Curvularia vermiformis* displayed a positive activity for cellulase assay (Amirita et al. 2012). Wood dwelling fungal endophytes were isolated from *Drimys winteri,* a Chilean tree, and *Prumnopitys andina*. *Bjerkandera* sp. isolated from *D. winteri* were positive to cellulase assay while *T. versicolor*, *G. australe* (A464), and *C. subvermispora* from *P. andina* were positive for cellulase activity (Oses et al. 2006).

Cellobiohydrolase (CBHI) encoding gene was isolated from a fungal endophyte *Fusiccoccum* sp. belonging to Ascomycota (Kanokratana et al. 2008). Expression of CBHI gene in *Pichia pastoris* KM71 was studied. The recombinant CBHI enzyme could degrade Avicel, filter paper, and 4-methyl umbelliferyl β-D-cellobioside (MUC) but could not hydrolyze carboxy methyl cellulose. CBHI was shown to have optimal activity at 40 °C at pH 5.0 K_m and V_{max} values were 0.57 mM and 3.086 nmol/min/mg, respectively, and the enzyme was stable at pH 3–11. CBH I has retained its

50% activity at 70–90 °C for 30 min. Since this enzyme has been observed to be stable in a wide range of pH, and moderately stable at high temperatures, it seems to have a promising potential in various biotechnological applications.

14.1.4 Lipases

14.1.4.1 Food Industry

Lipases have been extensively used in food industry to synthesize flavors or flavor precursors. Selective hydrolysis of triacyl glycerides releases free fatty acids that act as flavors or flavor precursors (Alves Macedo et al. 2003). Lipases are used in flavor development of dairy products such as cheese, margarine, chocolate milk, butter, and sweets to remove the fat from the meat to produce lean meat, fish meat processing.

14.1.4.2 Detergents

Lipases are used in detergents along with various other enzymes such as amylases, cellulases, and proteases (Benjamin and Pandey 1998). The ideal lipase should hydrolyze various lipids, withstand higher pH (pH 10–11), and temperatures (30–60 °C), resistant to proteases used in the same detergent with an ability to act on greasy substances like lipsticks, frying fats, butter, and sauces (Jaeger and Reetz 1998).

14.1.4.3 Pulp and Paper

Lipases are employed to remove the pitch, i.e., triacyl glycerides and waxes of wood. Lipases hydrolyze 90% of triacylglycerols into monoacylglycerols, making pulp-hydrophilic and less sticky and increase pulping rate, witness of the paper, decrease chemical usage, reduce pollution, save energy and prolong lifetime of equipment.

14.1.4.4 Leather

Proteins and lipids of collagen fibers of hides and skin should be totally removed before the tanning process of the leather. Non-fibril proteins are removed by proteases whereas fat is removed by lipases without damaging the leather.

14.1.4.5 Biodiesel Production

Biodiesel is produced from triacyl glycerides of animals, plants, algae, and fungi. Algae-based biodiesel production has been extensively used. Biodiesel synthesis involves removal of triacyl glycerides and digestion of the triacyl glycerides into fatty acids and the transesterification of fatty acids into fatty acid methyl. Lipases are required to break the triacyl glycerides into free fatty acids, which is the second step in the production of biodiesel.

14.1.4.6 Endophytic Fungi and Lipases

A total of 212 endophytic fungi residing in the leaves and fruits of Amazon plants were screened for lipase bound mycelium producing biocatalysts. About 87% of the isolates could hydrolyze tributyrin substrate and among these 30% of the isolates have shown commendable growth in lipase amended media. These isolates were evaluated for their ability of esterification and transesterification reactions in organic solvents. Among these fungal isolates best 9 isolates were investigated for enantioselectivity of mycelium bound lipase activity in (R,S)-2-octanol resolution reaction. The endophyte UEA_115 found to be the most ingenuous biocatalyst, demonstrating a splendid performance in esterification reactions (conversion >90%) and enhanced aptitude for the resolution of (R, S)-2-octanol (ees 29%; eep 99%; c 22%; E > 200). Thus, this investigation revealed the potent nature of the endophytic fungi as lipase suppliers in lipid biocatalysis (Zanotto et al. 2009).

Immobilization and enzyme activity of lipases secreted by *Crecospora kikuchii*, an endophytic fungus, associated with the host *Tithonia diversifolia,* was studied. The maximum production of lipase obtained was when the fungal strain was grown on 2% soya bean oil after 6 days and the yield was 9 U/mL. The stability of the lipase was studied with spray dried lipase with different adjuvants. The remaining enzymatic activity, once dried with 10% (w/v) of lactose maltodextrin, b-cyclodextrin, gum arabic, mannitol as well as trehalose ranged from 63 to 100%. No lipase activity was observed when adjuvants were absent. After 8 months, the enzyme activity at 5 °C and 25 °C was 50% and 40%, respectively. When 10% of β cyclodextrin was used to dry the lipase, the activity observed was 72% at 50 °C. Lipase separation was carried out by butyl sepharose column followed by partial purification (33.1%; 269.5 U/mg protein). Spray drying of this enzyme in maltodextrin DE10 retained 100% of its activity (Costa-Silva et al. 2011).

In a study, the seeds of sesame, pongamia, coconut, peanut, rubber, neem, and castor were sampled from different sites in Tamil Nadu state, India and then were processed for endophytic fungi isolation. Totally, 1279 endophytic fungal strains were obtained from above oil seeds. Out of the 1279 strains, 19 strains were positive for lipolytic activity, in addition to exhibiting cellulolytic, proteolytic, and amylolytic activities. These fungi belonged to the following genera: *Aspergillus, Alternaria, Chalaropsis, Cladosporium, Colletotrichum, Curvularia, Drechslera, Fusarium, Lasiodiplodia, Mucor, Penicillium, Pestalotiopsis, Phoma, Phomopsis,*

Phyllosticta, Rhizopus, Stachybotrys, Sclerotinia, and *Trichoderma.* Five fungal strains, viz., *Chalaropsis thielavioides, Lasiodiplodia theobromae, Colletotrichum gloeosporioides, Aspergillus niger*, and *Phoma glomerata*, displayed highest lipase activity. *Lasiodiplodia* sp. was the best producer of lipase and the activity observed was 108 U/ml. This fungal strain was characterized by ITS sequencing of 18s r RNA (Venkatesagowda et al. 2012).

Purification, characterization, stabilization, and enzyme kinetics of lipase production by fungal endophyte *Cercospora kikuchii* was studied (Costa-Silva et al. 2011). The lipase was purified up to 9.31 times, with a reported recovery of 26.6% whereas specific activity was found to be 223.6 U/mg. The optimal enzyme activity was observed at 35 °C of temperature and at pH of 4.6. Enzyme kinetic studies have shown that the K_m was 0.0324 mM and V_{max} 10.28 μmol/min/mg protein. The lipase possessed resistance attribute to Tween 80 and 20, Triton X, SDS, and to proteases. In the presence of oxidants also lipase retained 100% of activity. After spry drying of the purified lipase, it possessed 85.2% of enzymatic activity. The lipase enzyme obtained from *Cercospora kikuchii* has several properties in relevance with industrial applications and depicted sufficing stabilization and retention of its enzymatic activity after spray drying.

14.1.5 Pectinases

14.1.5.1 Wine Processing

Pectinolytic enzymes are added to the fruits before the addition of inoculum. This process improves the characteristics of wine. Pectinases support maximize juice yield, wine extraction, facilitate filtration, and intensify color and flavor of wine.

14.1.5.2 Tea and Coffee Processing

Pectinases break down pectins present in the tea leaves, accelerate the fermentation process, and act as antifoaming agents for tea as the pectin of tea leaves are destroyed by pectinases. Pectinases are generally used in combination with other enzymes such as cellulases, hemicellulases, and proteinases. Crude enzyme extract from *Aspergillus* sp. is more effective for tea fermentation when compared to purified pectin enzyme alone used in the tea leaves fermentation. This is because of the crude fungal enzymes that contain other enzymes as well.

14.1.5.3 Endophytic Fungi and Pectinases

Coffee beans are coated with mucilaginous coat. It is composed of 89% moisture, 9% protein, 4% sugar, and 2.8% pectin. Demucilation process is required to carry out the fermentation of coffee beans. The demucilation process is performed with pectinases and proteases. Demucilation process leads to the reduction in pH and increased release of sugars. Murthy and Naidu 2011 reported that crude pectinase obtained from *Aspergillus niger* (CFR) causes degradation of mucilaginous layer about 54% after 1 h and 71% after 2 h, present in coffee beans and complete decomposition of pectin was attained after 3.5 h.

14.1.6 L-asparaginases

L-asparaginase (LA) is one of the first enzymes with antileukemic properties and exploited as anti-tumor drugs. In combination with other chemotherapeutic agents, this enzyme is used to treat acute lymphoblastic leukemia (ALL), acute myelocytic leukemia, Hodgkin's disease, chronic lymphocytic leukemia, acute myelomonocytic leukemia, lymphosarcoma, melanosarcoma, and reticulosarcoma (Batool et al. 2016). In the past, bacterial sources for asparaginases, mainly from *Escherichia coli* and *Erwinia carotovora,* were exploited owing to their cost-effective nature (Theantana et al. 2009). Although the bacterial asparaginases have anti-leukemic properties, they also have after effects. Allergic reactions arise due to hypersensitivity leading to anaphylaxis thus making bacterial asparaginases inferior. Similarly, L-asparaginase causes hydrolysis of L-glutamine (L-Gln) producing ammonia and glutamic acid, which are neurotoxic (Suryanarayanan et al. 2012). Hence, a need has arisen to screen fungal endophytes as sources of L-asparaginases. Since animals are more related to fungi than bacteria, as per phylogenetic studies, the extent of hypersensitivity might be less conspicuous (Baldauf and Palmer 1993). Allergy development is another obstacle in the usage of bacterial L-asparaginases. Though, to address this allergy problem, one may switch from *Escherichia coli* to *Erwinia carotovora,* the latter is not as effective as compared to the former, due to its short life (Asselin et al. 1993). Another advantage of endophytic enzyme sources is the ease of purification and the cost-effective procedures, due to their extracellular production of enzymes.

Several marine algicolous endophytic fungal species belonging to the genera *Chaetomium, Alternaria, Cladosporium, Colletotrichum, Nigrospora, Curvularia, Phoma, Phaeotrichoconis, Paecilomyces*, and *Pithomyces* were also reported to produce L-asparaginases (Murali 2011). Nagarajan et al. (Nagarajan et al. 2014) qualitatively screened and recorded the selectivity towards L-Asn over L-Gln from several endophytes including *Colletotrichum acutatum, Phomopsis* sp., *Curvularia eragrostidis, C. lunata*, and *Nigrospora oryzae*. Chow and Ting (Chow and Ting 2015) screened four medicinal plants, viz., *Murraya koenigii, Oldenlandia diffusa, Pereskia bleo*, and *Cymbopogon citratus*, possessing anticancerous properties, from Malaysia for L-asparaginase production. They recorded four endophytic species: *Colletotrichum* sp., *Fusarium oxysporum, Penicillium simplicissimum*, and *Phoma* sp. for L-asparaginase production with 28% of total isolates positive for the production of this enzyme. They recorded mean activity of *Fusarium oxysporum* and

Penicillium simplicissimum as 0.013 and 0.019 μM/mL/min, which is lower than the previous studies (Nakahama et al. 1973). Pradeep et al. (2010) screened endophytes from *Ocimum sanctum* for L-asparaginase production and found that 18% of the recovered endophytes were positive for this enzyme. However, more efforts are needed to find novel sources of the enzyme that are free from glutaminase activity. Several studies have been carried out in this regard. Thangavel et al. (2013) made an attempt to isolate endophytes capable of producing glutaminase-free L-asparaginase. *Alternaria* sp., *Fusarium* sp., *Aspergillus* sp., and *Colletotrichum* sp. isolated from seaweeds *Gracilaria follifera, Amphiroa corallina, Sargassum wightii*, and *Amphiroa anceps* displayed positive results L-asparaginase (LA) activity.

Nagarajan et al. (2014) screened various trees from moist and evergreen forests of Western Ghats, India. Out of 33 isolates of endophytic fungi, 31 were positive for LA and 19 isolates were found to be positive for glutaminase-free L-asparaginase. Prominent among them were *Acremonium* sp., *Alternaria* sp. 2, *Aspergillus* sp., *Botrytis* sp., *Cylindrocladium* sp., *C. cladosporioides, Corynespora* sp., *Fusarium* sp., *Sordaria* sp., *L. theobromae,* and *Pestalotiopsis* sp. and particular mention should be made of *Alternaria* sp. for producing glutaminase-free L-asparaginase with specific activity of 1.65 U/mg. Apart from the medical importance, LAs find their use in food industry as well. Curtailing the compilation of acrylamide in baked or fried eatables, L-asparaginase diminishes the liabilities of carcinogenicity. During food processing, heat-induced reaction derives acrylamide formation. Free amino groups from asparagine and carbonyl groups of reducing sugars react to form acrylamine when heated or baked at 120 °C. Thus, a pretreatment with LA degrades asparagine and curtails acrylamide formation, which is neurotoxic in nature.

14.1.7 Tyrosinases

Tyrosinase enzyme is a monooxygenase that belongs to class-3 copper containing protein family. They catalyze the monophenolic oxidation (diphenolase or catecholase activity) or hydroxylation (monophenolase or cresolase activity) into respective o-quinones or catechols (Zaidi et al. 2014b). Tyrosinase also catalyzes the melanin synthesis, by oxidation, in mammalian systems. Tyrosinase causes the hydroxylation of L-tyrosine to 3,4-dihydroxyphenylalanine (L-DOPA) which is involved in melanin biosynthesis (Parvez et al. 2006). Tyrosinase which is present in mushrooms, fruits, vegetables, etc. is responsible for the blackening during long-term storage. In humans, abnormalities such as flecks and defects during hyperpigmentation of skin are attributed to tyrosinases. Quite a significant role is played by tyrosinases in agriculture and industry.

Tyrosinases are also used as biosensors to detect phenolic compounds in environment and thus find their applications in detoxification of wastewater and soils contaminated with phenols. Since tyrosinases are used as markers in melanoma patients and during the treatment of Parkinson's disease, they are of therapeutic importance too. In food industry also, they are used as food modifiers as they have the ability to crosslink proteins. Similarly, in cosmetic industry also inhibitors of these enzymes are screened for skin treatments such as dermatological disorders associated with melanin hyperpigmentation.

Zaidi et al. (2014b) described important aspects of tyrosinases including their defensive role in fungi and their spores, against stress, radiation, dehydration, free radicals, abnormal temperatures, and gaining stability; wound healing in plants through immune responses; target as prodrug activation, etc. *Agaricus bisporus*, *Neurospora crassa*, and *Streptomyces glaucescens* are known for best characterized tyrosinases. Endophytic fungi are well-known producers of tyrosinases. Pavithra et al. (2012) screened endophytes from Tulsi (*Ocimum sanctum*) and found that 7 out of 40 isolates produced tyrosinase, i.e., 27.5% were positive for its production. Zaidi et al. (Zaidi et al. 2014b) also purified tyrosinases from button mushroom, *Agaricus bisporus.* They obtained a K_m value of 0.933 mM which was higher than previous studies from *Pycnoporus sanguineus* and *Lentinula edodes*, respectively, have been reported. In continuation of search for treatment of dermatological disorders, various endophytes were screened for tyrosinase inhibitors.

An algicolous endophyte, *Pestalotiopsis* sp. Z233, associated with alga *Sargassum horneri,* was reported to secrete two novel sesquiterpenes, namely, 4α,5α-diacetoxy-9α-benzoyloxy-7βH-eudesman-1β,2β,11, 14-tetraol, and 1β,5α,6α,14-tetraacetoxy-9α-benzoyloxy-7β H-eudesman-2β,11-diol, that exhibited potent tyrosinase inhibitory potential. Similarly, a new tyrosine kinase inhibitor, chaetominedione (Zaidi et al. 2014a), had been isolated from an algicolous fungus *Chaetomium* sp. From terrestrial plants, *Azadirachta indica*, *Calotropis gigantea*, *Ocimum tenuiflorum*, and *Lantana camara* were screened for endophytic fungi producing tyrosinase enzymes. Of the total 50 endophytic fungi obtained from the four host plants, 27 were found to be positive for tyrosinase production of which one isolate of *Fusarium*, and two belonging to *Penicillium*, had highest activity of 2.8, 3.2, and 2.43 U/mL. Endophytes from *Ocimum tenuiflorum* and *Azadirachta indica* had highest activity as compared to *Lantana camara* and *Calotropis gigantea*. A similar study on endophytic fungi from *Azadirachta indica* had been carried out (Zaidi et al. 2014a). Two taxa, *Alternaria* sp., and *Colletotrichum truncatum* out of nine, have shown tyrosinase activity. Tyrosinases also possess the property to degrade lignin, thus the endophytic fungal production of tyrosinases implicates the characteristic association and recycling of natural ecosystem. Concomitantly, tyrosinase inhibitor production implicates host vegetative growth which represents a balance between the both. Several investigators have screened fungal endophytes for tyrosinase production on the hypothesis of change of ecological strategy of endophytes to saprobic lifestyle following senescence. Sun et al. (2011) screened endophytes from *Acer truncatum* for tyrosinase production and recorded that out of 21 strains, 10 endophytic taxa were positive.

14.1.8 Proteases

Apart from the destructive nature of proteases, they catalyze proteolytic processes with sharp accuracy and specificity yielding protein products (López-Otín and Bond 2008). In biological systems also, the roles of proteases are too many to enumerate including heat shock and unfolded protein responses, cell proliferation and differentiation, DNA replication and transcription, etc. Proteases are of therapeutic importance also as they show benefits in clinical studies in blood rheology control, inflammatory conditions, oncology, and immune regulation (Naidu 2011).

In biotechnological industry, proteases find their role as biochemical reagents in the manufacture of various products in brewing industry and in biscuit manufacturing by way of protein degradation to release required amino acids. Based on the nature of functional groups present in the enzyme active sites, the proteases are classified into aspartic proteases, cysteine proteases, serine proteases, and metalloproteases (Rao et al. 1998). More than two thirds of worldwide market is made up of proteases, which signifies their importance (Saran et al. 2007). Natural sources of proteases are plants, animals, and microbes. But microbial products are more preferred as they are economic, stable, and biologically diverse and can be easily manipulated genetically.

Fungal endophytes possess the ability to produce proteases. Morsy1, an unidentified endophytic fungus, isolated from a marine soft coral *Dendronephthya hemprichi*, has shown highest keratinase activity when grown on different agricultural and poultry wastes in solid state fermentation having maximum activity of 1600 U/g. The maximum enzyme production was observed with rice straw as the carbon source at 26 °C with moderate pH at 6 and the moisture content at 80%. Ahm1 and Ahm2, two types of keratinases, were purified and characterized by precipitation with diazanium sulfate (ammonium sulfate), DEAE sepharose, and gel exclusion chromatography in a sequential manner. Molecular weight of enzyme Ahm1 was 19 kDa and Ahm2 was 40 kDa. For purified keratinases, various kinetic parameters were optimized, to investigate the hydrolysis of azokeratin by Ahm1 (pH range 7.0–8.0, at 50 °C, stable in pH range of 6.0–8.0) and Ahm2 (pH range 10.0–11.0, at 60–65 °C, stable in pH range of 6.0 to 11.0). Both the keratinases, Ahm1 and Ahm2, were reported to undergo inhibition by chelating agents such as EGTA and EDTA. Serine protease inhibitor phenylmethylsulfonyl fluoride and cysteine protease inhibitor iodoacetamide, depicted inconsiderable consequences on keratinases (El-Gendy 2010).

Characterization and purification of novel fibrinolytic enzyme from an endophytic fungus, *Fusarium* sp. (Cpcc 480097), isolated from *Chrysanthemum* stems was investigated. This fungal strain could produce a highly active fibrinolytic enzyme. A novel fibrinolytic enzyme, named as Fu-P, was purified and characterized in a sequential manner, following ammonium sulfate precipitation of crude extract and purification by ion exchange and gel filtration chromatographies. The molecular weight of enzyme was found to be 28 kDa and the isoelectric point of enzyme was 8.1. Optimum activity of the Fu-P was found at 45 °C and pH 8.5. Purified Fu-P could degrade α-chain of fibrin efficiently, whereas its effect on β and γ chains was less. Fu-P had high affinity and specificity for the chymotrypsin substrates-2586. Therefore, it was described as a chymotrypsin like metalloprotease. N-terminal analysis, Q-A-S–S-G-T-P-A-T-I-R-V-L-V–V of the enzyme has revealed that it lacked any homology when compared to previously known fibrinolytic enzymes (Wu et al. 2009).

A novel endoprotease was produced by *Acremonium typhium*, an endophytic symbiotic fungus that infects the grass plant *Poa ampla.* The endoprotease was recorded to be highly stable when sodium dodecyl sulfate was present, which is an unusual feature of any protease. The enzyme was a thiol comprising serine protease and was found localized to a crude membrane fraction. *Acremonium typhium* that colonizes grass plants such as *Poa autumnalis* and *Poa sylvestris* also had shown similar protease activity, whereas endophytic fungi colonizing other grass plants *Festuca arundinacea* or *Lolium perenne* have not shown a similar activity. Therefore, protease expression in grasses belonging to *Poa* spp. is significant (Lindstrom and Belanger 1994). Rajput et al. (2016) screened a gymnosperm *Cupressus torulosa* for endophytic enzymatic potential. They reported four endophytes as protease positive out of eight isolates. The isolate with highest activity was *Alternaria alternata*.

Pavithra et al. (2012) investigated Tulsi for enzymatic potential of fungal endophytes and found that 50% of the isolates were positive for the protease production. Similarly, *Phoma herbarum, Alternaria alternata*, and an unclassified fungus, isolated from *Eremopholia*, a native plant to Australia, were assayed for production of proteases. The fungal media was lyophilized and investigated for the protease activity. Among the three strains, *Alternaria alternata* had highest protease activity. The protease of the fungus was active from pH 3 to 9. A wide range of activity from 9 to 50 °C was studied at pH 7.0 and implicated that it was a neutral protease. The optimum protease activity was found at 37 °C and pH 7 possessing an optimum specific activity having value of 69.86 BAFE units/mg. These characteristics suggesting that the protease can be used in dairy industry. Snitha et al. (2013) studied endophytic fungi associated with various medicinal plants: *Calophyllum inophyllum, Bixa orellana, Alpinia calcarata*, and *Catharanthus roseus* and explored their enzymatic potential. Maximum protease activity was attributed to *Aspergillus* sp., *Isaria* sp., and *Fusarium solani* from the host *Calophyllum inophyllum.* From the host *Alpinia calcarata, Aspergillus* sp. and a species under mycelia sterilia were positive for proteases. This was followed by *Cladosporium* sp. from the host *Catharanthus roseus* and; *Colletotrichum* sp., *Xylaria* sp., and *Alternaria* sp. from *Alpinia calcarata* were also found to produce proteases. Moderate activity was shown by *C. falcatum* from *Bixia orellana. Colletotrichum crassipes, C. gloeosporioides, C. falcatum, Drechslera hawaiiensis, Curvularia vermiformis*, and a Xylarialean fungus, reported by Amirita et al. (2012) as endophytes from four medicinal plants, viz., *Adhatoda vasica, Coleus aromaticus, Costus igneus,* and *Lawsonia inermis* from Tamil Nadu (India), were found to secrete extracellular protease enzymes. Endophytic fungi from *Bauhinia forficata* were screened for protease production and 13 out of 19 strains were found to be positive including *Curvularia australiensis, Cochliobolus lunatus, Myrmecridium schulzeri, Phoma putaminum, Penicillium commune, A. ochraceus C. lunatus, G. fujikuroi, A. chartarum, Nodulisporium* sp., *P. glabrum, A. ochraceus*, and *Pithomyces atro-olivaceus* (Bezerra et al. 2015). Interestingly, out of them, *Phoma putaminum* had the highest activity, which also is a potential herbicide. Similarly, 40 endophytic fungi inhabiting oil seeds were screened for protease activity and *Colletotrichum gloeosporioides* displayed the highest proteolytic activity along with *Penicillium citrinum, Pestalotiopsis palmarum*, and *Aspergillus niger*. Moderate activity was shown by *Lasiodiplodia theobromae.*

Jalgaonwala and Mahajan (2011) reported that twelve endophytic fungi had proteolytic activity of which a strain belonging to mycelia sterilia, from roots of *Catharanthus roseus*, showed a greater protease activity as compared to other fungal isolates tested. Alberto et al. (2016) examined four host trees, viz., *Luehea divaricata, Sapindus saponaria, Trichilia elegans, Piper hispidum*, and *Saccharum* spp., for production of various enzymes by fungal endophytes. Approximately 64% of endophytes were found to produce proteases. Isolation and partial purification of a protease produced by *Acremonium* sp., isolated from leaves of *Saraca asoca*, when grown on skim milk medium, has shown protease activity. Fructose and ammonium sulfate were found to be the prime carbon and nitrogen sources for the highest protease production and the optimum activity of the enzyme was recorded at pH 7.0. The enzyme activity was found to be 3.4 U/mL; the protein content 20 μg/mL and; the specific activity of enzyme 0.167 U/μg (Jain et al. 2012).

14.1.9 Phosphatases

Phosphorus present in soil is indispensable for plant growth, but is generally unavailable to plants in its native form. Phosphatases are the enzymes that make phosphorus available to the plants for enhancing their growth. Endophytic fungi are also known to produce phosphatases. Khan et al. (2016) screened various endophytes from frankincense tree for the secretion of extracellular enzymes. Species belonging to the genus *Preussia* exhibited higher enzyme secretion while a moderate activity was shown by *Penicillium citrinum* and *Aureobasidium pullulans.* Production of phosphatases also aids in the growth of hosts as monitored and reported in the above study with respect to the root/shoot length. Previous studies also support the fact that inoculation of endophytes has resulted in a distinct growth as easily assimilable nutrients are made available by the action of phosphatases.

Shubha and Srinivas (2017) reported phosphatase activity of endophytic fungi from *Cymbidium aloifolium. Colletotrichum truncatum* was reported to show the highest enzyme activity index of 1.58. They screened endophytes from flowers, roots, and leaves with 100%, 93.5%, and 77.7% of isolates positive for phosphatase activity, respectively. Except for two isolates, viz., *A. alternata* and *Curvularia* sp., all the other endophytic fungi were positive for phosphatase enzyme. Phosphate solubilization through phosphatase production or various organic acids is an attribute of plant growth promoting fungi (Illmer and Schinner 1995). Many reports clearly suggest that among different microbes, fungi are more capable of solubilizing phosphates (Nahas 1996). According to Wakelin et al. (2004), species belonging to *Penicillium* and *Aspergillus* are prominent endophytes known for competent phosphate solubilization. Endophytes thriving in extreme conditions such as high or low temperatures, nutrient deficient soils, lesser moisture are more potent in encouraging host growth (Chadha et al. 2015), and can applicate hosts by residing under such unfavorable conditions. Similar results were recorded by Chadha et al. (2015). All 12 isolates of endophytic fungi isolated from the roots of tomato plants from India, displayed phosphatase solubilizing activity. *Trichoderma pseudokoningii* showed the highest rate (37.45 ± 2.78 to 64.32 ± 2.87 μg/mL) of phosphate solubilization, with the rest showing lesser rate, viz., *Chaetomium globosum* (33.62 ± 5.92 to 69.32 ± 3.21 μg/mL), *Fusarium semitectum* (32.64 ± 1.89 to 57.63 ± 2.11 μg/mL), and *Aspergillus versicolor* (31.63 ± 2.02 to 63.72 ± 2.36 μg/mL).

14.1.10 Endophytes and Their Role in Phytoremediation/Biodegradation

The ability of microbes to detoxify certain hazardous substances is an integral part of the ecosystem functioning in the environment. Since endophytes secrete a host of extracellular enzymes such as xylanases, proteases, lipases, and tyrosinases, they are widely explored to treat wastes and clean environment (Mishra and Sarma 2017). Expression of enzymatic genes under environmental and nutritional stress conditions, remodel the fungal metabolic pathways. Endophytic basidiomycetes and soil mycoflora synergistically transform complex polymers into simpler compounds such as humus substances, CO_2, and glycoproteins (Grinhut et al. 2007; Mishra and Sarma 2017). In short, they influence the availability of nutrients to plant communities, quality of litter produced, and microflora associated with litter formation, also in a way acting as saprobes (Saikkonen et al. 2015).

During phytoremediation also, organic contaminants get degraded by enzymes produced by endophytes reducing phytotoxicity as well as evapotranspiration of volatile contaminants (Li et al. 2012). Bioconversion of organic matter and global mineral cycling is greatly modulated by enzymatic and non-enzymatic systems of fungal endophytes. *Neotyphodium coenophialum* and *N. uncinatum* isolated from two species of grass, *Festuca arundinacea* and *F. pratensis*, respectively, were found to be capable in the removal of rhizospheric polyaromatic hydrocarbons (PAH) and total petroleum hydrocarbons (TPH) (Soleimani et al. 2010). Tian and Schaich (2013), Dai et al. (2010) reported manganese peroxidases as dominant enzymes for degrading PAHs. Polyester polyurethane (PUR), a synthetic polymer was reported to get degraded by *Pestalotiopsis microspora,* an endophytic fungus, through extracellular secretion of serine hydrolases. *Pestalotiopsis microspora* survived under both aerobic and anaerobic conditions with PUR as the sole carbon source (Russell et al. 2011). Fungi play important roles in biotransformations (Abourashed and Hufford 1996). They can undertake reactions at remote regions and possess high substrate tolerance. Moreover, they can affect stereo-, chemo-, and region-selective processes resulting in intermediate compounds. Borges et al. (2009) reviewed the biotransformations effected by several endophytic fungi including *Chaetosphaeria* sp., an endophyte of *Aphelandra tetragon*, *Plectosporium tabacinum*, *Gliocladium cibotii*, etc. (Borges et al. 2009).

14.1.11 Ecological Relationships

Though the usage of the term "Endophytes" has been debated by some scientists ever since it came into the picture (Petrini 1991), the multitude of studies, nevertheless, still rely on the accustomed foresight of mutualistic behavior of endophytes towards their hosts (Borges et al. 2009). Improvizing a genetic system that grants the endophytes to establish information transfer with host could possibly be by the virtue of long-term synergy between them. This synergy often leads to barter biochemical synthetic pathways that causes production of common metabolites or enzymes with a wide range of applications (Strobel 2002). Fungal endophytes colonize host tissues in spite of the resistance mechanisms that hosts have such as secondary metabolite production against pathogens. Thus, the endophytes secrete a wide range of enzymes to degrade metabolites and gain entry into their hosts regardless of the host defense mechanisms. The fungal endophytes and the enzymes produced by them give an insight of their functional roles and patterns of substrate utilization (Snitha et al. 2013). The action of enzymes possibly explains the chances of genetic recombination between an endophyte and its host during the evolutionary time. Studies concerning the host and endophytes draw light on the ecological and evolution of endophytes with a symbiotic role towards the host, strength of connections established with the host, and factors influencing this partnership (Snitha et al. 2013). The magnitude and duration of enzymes produced by an endophyte also differs depending on the host selected indicating the need of endophytes to survive in the dwelling habitat, host age, biotic factors, etc. A few reports also suggest the role of endophytes in enhancing the host fitness and growth leading to the expression of certain genes and enzymes produced by them. The starch-degrading enzyme and nitrate reductase enzyme were found to be expressed by *Piriformospora indica,* dwelling in tobacco roots (Sherameti et al. 2005).

14.1.12 Genetic Manipulation in Enzyme Production

Endophytes are enriched with a reservoir of enzymes that are capable of directing the biosynthesis of structurally discrete and miscellaneous molecules that are challenging to synthesize artificially (Kusari and Spiteller 2011). Hence, establishing the existence of crucial enzymes, associated with specific pathways, could prove to be a molecular marker, for screening endophytes. Biosynthesis of secondary metabolites in endophytic fungi is encoded by genes arranged in clusters (Smith et al. 1990). Development of a formula for secondary metabolite production at a large scale by exploiting fermentation procedures is often limited by arrangement of these gene clusters (Wu and Chappell 2008), which encode the enzyme complexes, viz., the non-ribosomal peptide synthetases [NRPS] or polyketide synthases [PKS], containing several modules and domains having specific features and functions (Kusari and Spiteller 2011). Several tools of bioinformatics such as SMURF (Secondary Metabolite Unknown Regions Finder) (Khaldi et al. 2010), FungiFun (Priebe et al. 2011), and antiSMASH (Medema et al. 2011) are advantageous for predicting the genes involved. Multiple synonymous codons encode an amino acid, which reflects on the redundancy in the genetic code. Therefore, with several possible codons for a naturally occurring one provides a plot to genetically manipulate the heterologous expressions (Anbu et al. 2017). Genome mining of these clusters to predict the unidentified secondary products is a promising approach towards quest for new discoveries.

Anbu et al. (2017) also emphasized the need for regulatory mechanisms, which can possibly modulate the enzyme production through induction. In other words, mutagenesis to guide expression of genes can be exploited for obtaining an enzyme system of a microbe in favor of human welfare. Various enzymes of industrial, agricultural, pharmaceutical, and ecological importance could be made available in large amounts to meet the needs by implementing such strategies of recombination technology for expressing the enzymes needed. For example, endo-β-1,4-xylanase, a recombinant of *Aspergillus usamii,* was optimized and expressed in *Pichia pastoris*, as a result of improvement in gene expression (Wang et al. 2010).

14.1.13 Enzymatic Profiling of Endophytic Fungi

It is intriguing to notice that most of the work related to enzymatic profiling of endophytic fungi came from ethno-botanically important medicinal plants as a source of these fungi (Pavithra et al. 2012; Alberto et al. 2016; Mhatre et al. 2017). The rationale for the selection of such hosts lies with the concept that endophytes mimic their hosts in the production of secondary metabolites (Kusari et al. 2014). Hence, endophytes from such hosts produce bioactive molecules with antimicrobial activities, enzymes of varying importance and biodegradative abilities (Sudha et al. 2016).

A table is provided at the end that presents the enzymatic profiling of endophytic fungi from various hosts (Table 14.1). It lists down a few major enzymes for which endophytes have been screened for. Few enzymes such as phytase (Marlida et al. 2010), phenoloxidase (Oses et al. 2006), and chitosanase (Venkatachalam et al. 2015) have not been included as the information available on them in the literature is very less as few researchers would have concentrated on these enzymes. We have included information on 16 enzymes in Table 14.1 on which a number of researchers have worked. Those endophytic strains, which lacked a name, as, probably, they could not be identified, but displayed a positive activity for a particular enzyme, have not been included in the table. We know only the presence or absence of a particular enzyme activity (marked as plus or minus in Table 14.1) through plate assays, which only give qualitative assessment but not their quantitative proficiency. Hence, care should be taken while interpreting the results. On around 300 endophytic fungal species have information on their enzymatic potential. This reflects on the fact that more fungi need to be screened. It is difficult to say that some of the fungi included in the table lack the capability to produce particular enzymes. Only 150 fungal species have names known up to species level and the remaining up to genus level. A large number of fungi (40%) were found to be positive for amylase, and this could probably explain the mode of nutrition that the particular endophytic fungi have with reference to their hosts. Most of the fungi (36%) were also found to produce cellulases which reflect on their saprobic role after senescence and also capability to colonize the hosts during earlier stages of infestation. Enzyme pectinase was produced by 12% of the fungi listed while 22% secrete proteases (Table 14.1), thus implicating a strategy of these fungi in thriving inside the hosts. Lipases is another group of enzymes that most endophytic fungi (28%) secrete, which explains the way they might derive nutrition from the host. Since most of the fungi are from economically or medicinally important hosts, it is no longer a surprise that 12.7% of isolates possessed asparaginase enzyme producing ability, which is of therapeutic importance. Rest of the enzymes such as chitinases, esterases, gelatinases, glucosidases, laccases, ligninases, phosphatases, and tyrosinases were secreted by less than 10% of fungal population of endophytes. This observation indicates that either the endophytic fungi lack capability to produce some of these enzymes or that they were not studied by researchers in general, as could be seen from the table that many of them were listed as "Not Determined". Strains belonging to three endophytic fungal genera: *Aspergillus*, *Phoma*, and *Penicillium* were found to be positive for all the 15 enzymes listed in the table. This is followed by *Cladosporium* and *Curvularia* as they were positive for 14 enzymes production. Other species such as *Alternaria* (12), *Acremonium* (13), *Chaetomium* (13), *Colletotrichum* (12), *Fusarium* (12), and *Trichoderma* (12) are some of the genera that are positive for a wide range of enzymes production. These fungi are also commonly reported in many diversity studies on endophytic fungi on various plants from different regions.

It is very compelling to figure that several researchers have showed a bias towards the enzymes selected in their endophytic fungal screening programs. Amylases, cellulases, pectinases, proteases, and lipases are some of the enzymes predominantly surveyed, while the rest were less explored. Therefore, though we get some picture from the table, based on the information available in the literature, we still cannot definitely draw any decisive conclusions on the enzymatic potential of the endophytic fungi until a vast majority of other endophytic fungi are also screened for a spectrum of enzymes listed in the table.

Table 14.1 An overview of different enzymes produced by endophytic fungi

Fungi name	Amylase	Cellulase	Pectinase	Proteinase	Lipase	Laccase	Glucosidase	Phosphatase	Chitinase	Gelatinase	Asparaginase	Tyrosinase	Ligninase	Esterase	Polyurethanase	Xylanase	References
Acremonium zeae	ND	ND	ND	ND	ND	ND	ND	ND	ND	ND	ND	ND	ND	ND	ND	+	Corrêa et al. (2014)
Acremonium curvulum	ND	+	ND	+	+	ND	ND	ND	ND	ND	ND	ND	ND	ND	ND	+	Bezerra et al. (2015)
Acremonium implicatum	+	–	+	–	+	–	ND	ND	ND	ND	ND	ND	ND	ND	ND	ND	Snitha et al. (2013)
Acremonium spp.	+	+	+	–	+	+	ND	ND	+	+	+	+	+	+	ND	ND	Maria et al. (2005), Neema Job and Philip (2015), Marlida et al. (2000), Lisboa et al. (2013) and Prathyusha et al. (2015)
Acremonium terricola	ND	+	–	+	ND	ND	ND	ND	ND	ND	ND	ND	ND	ND	ND	+	Bezerra et al. (2012)
Acremonium typhinum	ND	ND	ND	+	ND	ND	ND	ND	ND	ND	ND	ND	ND	ND	ND	ND	Lindstrom et al. (1993) and Reddy et al. (1996)
Acrophialophora nainiana	ND	ND	ND	ND	ND	ND	ND	ND	ND	ND	ND	ND	ND	ND	ND	+	Polizeli et al. (2005)
Alternaria alternata	+	+	+	+	ND	–	ND	–	+	+	ND	+	–	ND	ND	+	Sheik et al. (2015), Wipusaree et al. (2011), Fouda et al. (2015), Zaferanloo et al. (2013), Rajulu et al. (2011), Rajput et al. (2016), Prathyusha et al. (2015) and Shubha and Srinivas (2017)
Alternaria dauci	ND	ND	ND	ND	ND	ND	ND	ND	ND	ND	ND	ND	ND	ND	+	ND	Russell et al. (2011)
Alternaria franaseriae	+	ND	ND	ND	ND	ND	ND	ND	ND	ND	+	ND	ND	ND	ND	ND	Masumi et al. (2014)
Alternaria porri	+	–	ND	–	+	ND	ND	ND	ND	ND	ND	ND	ND	ND	ND	ND	Katoch et al. (2014)

Alternaria spp.	+	+	–	+	+	–	+	ND	ND	ND	+	ND	–	ND	+	+	Robl et al. (2013), Snitha et al. (2013), Maria et al. (2005), Murali (2011), Corrêa et al. (2014), Masumi et al. (2014), Zaferanloo et al. (2013), Russell et al. (2011) and Uzma et al. (2016)
Alternaria tangelonis	+	ND	ND	ND	ND	ND	ND	ND	ND	ND	+	ND	ND	ND	ND	ND	Masumi et al. (2014)
Amanita muscaria	ND	ND	ND	+	ND	ND	ND	ND	ND	ND	ND	ND	ND	ND	ND	ND	Nygren et al. (2007)
Ampelomyces sp.	–	+	ND	ND	–	ND	ND	ND	ND	ND	ND	ND	ND	ND	ND	ND	Toghueo et al. (2017)
Annulohypoxylon stygium	ND	+	ND	+	ND	ND	+	ND	ND	ND	ND	ND	ND	ND	ND	+	Robl et al. (2013)
Aphanocladium sp.	ND	ND	ND	ND	ND	ND	ND	ND	ND	ND	ND	ND	ND	ND	ND	–	Thirunavukkarasu et al. (2015)
Arthrinium phaeospermum	+	–	+	+	–	–	ND	ND	ND	ND	ND	ND	ND	ND	ND	ND	Prathyusha et al. (2015)
Aspergillus aculeatus	+	–	ND	–	–	ND	ND	ND	ND	ND	ND	ND	ND	ND	ND	+	Katoch et al. (2014) and Polizeli et al. (2005)
Aspergillus awamori	+	ND	ND	ND	ND	ND	ND	ND	ND	ND	ND	ND	ND	ND	ND	+	Sundarram and Murthy (2014) and Polizeli et al. (2005)
Aspergillus chartarum	ND	+	ND	+	+	ND	ND	ND	ND	ND	ND	ND	ND	ND	ND	+	Bezerra et al. (2015)
Aspergillus fischeri	ND	ND	ND	ND	ND	ND	ND	ND	ND	ND	ND	ND	ND	ND	ND	+	Polizeli et al. (2005)
Aspergillus flavus	–	–	+	–	–	–	ND	ND	ND	ND	ND	ND	ND	ND	ND	ND	Snitha et al. (2013)
Aspergillus fumigatus	+	+	+	+	+	–	ND	+	ND	ND	ND	ND	ND	+	ND	+	Snitha et al. (2013), Tuppad and Shishupala (2014), Nath et al. (2015), Polizeli et al. (2005) and Venkatesagowda et al. (2012)
Aspergillus japonicas	+	+	+	+	–	–	ND	+	ND	ND	ND	ND	ND	ND	ND	+	Shubha and Srinivas (2017) and Bezerra et al. (2012)
Aspergillus kawachii	+	ND	ND	ND	ND	ND	ND	ND	ND	ND	ND	ND	ND	ND	ND	+	Sundarram and Murthy (2014) and Polizeli et al. (2005)

(continued)

Table 14.1 (continued)

Fungi name	Amylase	Cellulase	Pectinase	Proteinase	Lipase	Laccase	Glucosidase	Phosphatase	Chitinase	Gelatinase	Asparaginase	Tyrosinase	Ligninase	Esterase	Polyurethanase	Xylanase	References
Aspergillus nidulans	ND	ND	ND	ND	ND	ND	ND	ND	ND	ND	+	ND	ND	ND	ND	ND	Kaur and Saxena (2014)
Aspergillus nidulans	ND	ND	ND	ND	ND	ND	ND	ND	ND	ND	ND	ND	ND	ND	ND	+	Polizeli et al. (2005)
Aspergillus niger	+	+	–	+	+	+	+	+	ND	ND	+	ND	ND	+	ND	+	Muthezhilan et al. (2014), Snitha et al. (2013), Murali (2011), Robl et al. (2013), Bezerra et al. (2015), Nath et al. (2015), Rajput et al. (2016), Kaur and Saxena (2014), Sundarram and Murthy (2014) and Venkatesagowda et al. (2012)
Aspergillus ochraceus	–	+	+	+	+	–	ND	ND	ND	ND	ND	ND	ND	ND	ND	+	Bezerra et al. (2015) and Prathyusha et al. (2015)
Aspergillus oryzae	+	ND	ND	ND	ND	ND	ND	ND	ND	ND	ND	ND	ND	ND	ND	+	Sundarram and Murthy (2014) and Polizeli et al. (2005)
Aspergillus sojae	ND	ND	ND	ND	ND	ND	ND	ND	ND	ND	ND	ND	ND	ND	ND	+	Polizeli et al. (2005)
Aspergillus spp.	+	+	+	+	+	+	ND	+	+	+	+	+	+	+	ND	+	Venkatesagowda et al. (2012), Sheik et al. (2015), Murali (2011), Nygren et al. (2007), Neema Job and Philip (2015), Shubha and Srinivas (2017), Polizeli et al. (2005), Maria et al. (2005), Patil et al. (2015), Snitha et al. (2013), Uzma et al. (2016), Patel et al. (2013), Masumi et al. (2014) and Venkatachalam et al. (2015)

Aspergillus terreus	+	–	–	+	–	–	ND	+	ND	ND	+	ND	ND	ND	ND	+	Murali (2011), Kalyanasundaram et al. (2015), Thirunavukkarasu et al. (2015), Shubha and Srinivas (2017), Kaur and Saxena (2014), Polizeli et al. (2005) and Sorgatto et al. (2012)
Aspergillus versicolor	+	+	ND	+	+	ND	ND	ND	ND	ND	ND	ND	ND	+	ND	+	Polizeli et al. (2005) and Venkatesagowda et al. (2012)
Aspergillus terreus	+	+	ND	+	+	ND	ND	ND	ND	ND	ND	ND	ND	+	ND	ND	Venkatesagowda et al. (2012)
Aureobasidium pullulans	ND	+	ND	ND	ND	ND	+	+	+	ND	ND	ND	ND	ND	ND	+	Rajulu et al. (2011), Venkatachalam et al. (2015), Polizeli et al. (2005) and Khan et al. (2016)
Aureobasidium sp.	ND	+	ND	ND	ND	ND	+	+	ND	ND	ND	ND	ND	ND	ND	ND	Khan et al. (2016)
Basidiomycetes sp	+	–	+	–	+	–	ND	ND	ND	ND	ND	ND	ND	ND	ND	ND	Snitha et al. (2013)
Bionectria sp.	ND	ND	ND	ND	ND	ND	ND	ND	ND	ND	ND	ND	ND	ND	+	ND	Russell et al. (2011)
Biosporus sp.	–	+	ND	+	–	ND	ND	ND	ND	ND	ND	ND	ND	ND	ND	ND	Patil et al. (2015)
Bipolaris papendorfii	ND	ND	ND	ND	ND	–	ND	ND	ND	ND	ND	ND	ND	ND	ND	ND	Sheik et al. (2015)
Bipolaris sp.	+	+	+	+	+	–	ND	+	ND	ND	+	ND	ND	ND	ND	ND	Shubha and Srinivas (2017), Martín-Rodríguez et al. (2014) and Uzma et al. (2016)
Bjerkandera sp.	ND	+	ND	ND	ND	ND	ND	ND	ND	ND	ND	ND	ND	ND	ND	ND	Corrêa et al. (2014)
Boletus luridus	ND	ND	ND	+	ND	ND	ND	ND	ND	ND	ND	ND	ND	ND	ND	ND	Nygren et al. (2007)
Botrytis sp.	ND	ND	ND	ND	ND	ND	ND	ND	+	ND	ND	ND	ND	ND	ND	ND	Rajulu et al. (2011)
Candida sp.	+	+	+	ND	+	+	ND	ND	+	+	+	+	+	ND	ND	ND	Neema Job and Philip (2015)
Cenococcum geophilum	ND	ND	ND	+	ND	ND	ND	ND	ND	ND	ND	ND	ND	ND	ND	ND	Nygren et al. (2007)
Cephalosporium sp.	ND	+	ND	ND	ND	ND	ND	ND	ND	ND	ND	ND	ND	ND	ND	ND	Peng and Chen (2007)
Cercospora chrysanthemi	+	+	ND	ND	+	ND	ND	ND	ND	ND	ND	ND	ND	ND	ND	ND	Toghueo et al. (2017)

(continued)

Table 14.1 (continued)

Fungi name	Amylase	Cellulase	Pectinase	Proteinase	Lipase	Laccase	Glucosidase	Phosphatase	Chitinase	Gelatinase	Asparaginase	Tyrosinase	Ligninase	Esterase	Polyurethanase	Xylanase	References
Cercospora flagellaris	ND	+	ND	ND	+	ND	ND	ND	ND	ND	ND	ND	ND	ND	ND	ND	Toghueo et al. (2017)
Cercospora kikuchii	ND	ND	ND	ND	+	ND	ND	ND	ND	ND	ND	ND	ND	ND	ND	ND	Costa-Silva et al. (2011)
Cercospora olivascens	+	+	ND	ND	+	ND	ND	ND	ND	ND	ND	ND	ND	ND	ND	ND	Toghueo et al. (2017)
Cercospora sp.	+	+	ND	ND	+	ND	ND	ND	ND	ND	ND	ND	ND	ND	ND	ND	Toghueo et al. (2017)
Chaetomium cellulolyticum	ND	ND	ND	ND	ND	ND	ND	ND	ND	ND	ND	ND	ND	ND	ND	+	Polizeli et al. (2005)
Chaetomium crispatum	+	+	+	ND	ND	ND	ND	ND	ND	ND	ND	ND	ND	ND	ND	ND	Tuppad and Shishupala (2014)
Chaetomium globosum	+	+	ND	ND	–	+	ND	ND	ND	ND	ND	ND	ND	ND	ND	ND	Toghueo et al. (2017) and Corrêa et al. (2014)
Chaetomium spp.	+	+	–	+	+	+	+	+	+	ND	+	ND	ND	+	ND	ND	Patil et al. (2015), Murali (2011), Khan et al. (2016), Venkatachalam et al. (2015), Snitha et al. (2013), Lisboa et al. (2013) and Prathyusha et al. (2015)
Chalaropsis thielavioides	+	+	ND	+	+	ND	ND	ND	ND	ND	ND	ND	ND	+	ND	ND	Venkatesagowda et al. (2012)
Chlamydosporus sp.	ND	–	–	ND	ND	ND	ND	ND	ND	ND	ND	ND	ND	ND	–	–	Maria et al. (2005)
Cladosporioides spp.	ND	ND	ND	ND	ND	ND	ND	ND	ND	ND	ND	ND	ND	ND	ND	ND	Bezerra et al. (2012)
Cladosporium cladosporioides	+	+	–	+	+	–	ND	ND	ND	ND	+	ND	ND	+	ND	+	Amirita et al. (2012), Masumi et al. (2014), Venkatesagowda et al. (2012), Prathyusha et al. (2015), Rajput et al. (2016) and Bezerra et al. (2012)

Cladosporium spp.	+	+	+	+	+	+	ND	+	+	+	+	+	+	ND	ND	+	Patil et al. (2015), Murali (2011), Bezerra et al. (2012), Uzma et al. (2016), Shubha and Srinivas (2017), Venkatachalam et al. (2015), Thirunavukkarasu et al. (2015), Tuppad and Shishupala (2014), Zaferanloo et al. (2013) and Neema Job and Philip (2015)
Cladosporium sphaerospermum	ND	–	–	+	ND	ND	ND	ND	ND	ND	ND	ND	ND	ND	ND	+	Bezerra et al. (2012) and Rajput et al. (2016)
Cladosporium tenuissimum	+	ND	ND	ND	ND	ND	ND	ND	ND	ND	ND	ND	ND	ND	ND	ND	Toghueo et al. (2017)
Cochliobolus australiensis	ND	+	ND	+	+	ND	ND	ND	ND	ND	ND	ND	ND	ND	ND	+	Bezerra et al. (2015)
Cochliobolus lunatus	ND	+	ND	+	+	ND	ND	ND	ND	ND	ND	ND	ND	ND	ND	+	Bezerra et al. (2015)
Coelomycete species	+	+	–	ND	ND	ND	ND	ND	ND	ND	ND	ND	–	ND	ND	+	Choi et al. (2005)
Colletothericum sp.	+	ND	ND	ND	ND	ND	ND	ND	ND	ND	+	ND	ND	ND	ND	ND	Masumi et al. (2014)
Colletotrichum acutatum	ND	ND	ND	ND	ND	ND	ND	ND	+	ND	ND	ND	ND	ND	ND	ND	Rajulu et al. (2011)
Colletotrichum carssipes	+	–	ND	+	–	–	ND	ND	ND	ND	ND	ND	ND	ND	ND	ND	Amirita et al. (2012)
Colletotrichum clidemiae	–	–	ND	ND	–	ND	ND	ND	ND	ND	ND	ND	ND	ND	ND	ND	Toghueo et al. (2017)
Colletotrichum dematium	+	+	+	+	+	–	ND	ND	ND	ND	ND	ND	ND	+	ND	ND	Prathyusha et al. (2015) and Venkatesagowda et al. (2012)
Colletotrichum falcatum	+	+	–	+	+	–	ND	ND	ND	ND	ND	ND	ND	ND	ND	ND	Amirita et al. (2012) and Smitha et al. (2012)
Colletotrichum fructicola	+	+	ND	ND	+	–	ND	ND	ND	ND	ND	ND	ND	ND	ND	ND	Toghueo et al. (2017)

(continued)

Table 14.1 (continued)

Fungi name	Amylase	Cellulase	Pectinase	Proteinase	Lipase	Laccase	Glucosidase	Phosphatase	Chitinase	Gelatinase	Asparaginase	Tyrosinase	Ligninase	Esterase	Polyurethanase	Xylanase	References
Colletotrichum gloeosporioides	+	+	+	+	+	+	ND	+	+	ND	+	+	ND	+	ND	ND	Prathyusha et al. (2015), Katoch et al. (2014), Shubha and Srinivas (2017), Toghueo et al. (2017), Venkatesagowda et al. (2012), Amirita et al. (2012), Snitha et al. (2013), Lee et al. (2014) and Rabha et al. (2014)
Colletotrichum gossypiicola	ND	ND	ND	ND	ND	ND	ND	+	ND	ND	ND	ND	ND	ND	ND	ND	Nath et al. (2015)
Colletotrichum lindemuthianum	+	+	ND	+	+	ND	ND	ND	ND	ND	ND	ND	ND	+	ND	ND	Venkatesagowda et al. (2012)
Colletotrichum sp.	+	+	+	+	+	+	ND	ND	+	ND	+	ND	–	ND	ND	+	Tuppad and Shishupala (2014), Patil et al. (2015), Murali (2011), Chow and Ting (2015), Ayob and Simarani (2016), Sunitha et al. (2013), Rajulu et al. (2011), Manasa and Nalini (2014), Uzma et al. (2016), Choi et al. (2005), Theantana et al. (2009) and Lumyong et al. (2002)
Colletotrichum truncatum	+	+	–	+	+	+	ND	+	ND	ND	ND	ND	ND	+	ND	ND	Venkatesagowda et al. (2012), Snitha et al. (2013) and Shubha and Srinivas (2017)
Coniothyrium minitans	ND	ND	ND	ND	ND	ND	ND	ND	ND	ND	ND	ND	ND	+	ND	ND	Lisboa et al. (2013)
Coniothyrium sp.	+	+	+	–	–	–	ND	ND	ND	ND	ND	ND	ND	ND	ND	ND	Snitha et al. (2013)
Cortinarius glaucopus	ND	ND	ND	+	ND	ND	ND	ND	ND	ND	ND	ND	ND	ND	ND	ND	Nygren et al. (2007)

Cortinarius purpurascens	ND	ND	ND	+	ND	ND	ND	ND	ND	ND	ND	ND	ND	ND	ND	ND	Nygren et al. (2007)
Corynespora cassiicola	−	+	ND	ND	+	+	ND	ND	ND	ND	ND	ND	ND	ND	ND	ND	Toghueo et al. (2017)
Cryphonectria cubensis	ND	ND	ND	ND	ND	ND	ND	ND	ND	ND	ND	ND	ND	ND	ND	+	Suto et al. (2002)
Cryphonectria havanensis	ND	ND	ND	ND	ND	ND	ND	ND	ND	ND	ND	ND	ND	ND	ND	+	Suto et al. (2002)
Cryptococcus sp.	ND	ND	ND	ND	ND	ND	ND	ND	ND	ND	ND	ND	ND	ND	ND	+	Polizeli et al. (2005)
Cunninghamella echinulata	ND	ND	ND	ND	ND	+	ND	ND	ND	ND	ND	ND	ND	ND	ND	ND	Sheik et al. (2015)
Curvularia akaii	+	ND	ND	ND	ND	ND	ND	ND	ND	ND	+	ND	ND	ND	ND	ND	Masumi et al. (2014)
Curvularia brachyspora	+	−	ND	+	+	+	ND	ND	ND	ND	ND	ND	ND	ND	ND	ND	Amirita et al. (2012)
Curvularia clavata	ND	ND	ND	ND	ND	+	ND	ND	ND	ND	ND	ND	ND	ND	ND	ND	Sheik et al. (2015)
Curvularia lunata	+	+	+	+	+	−	ND	+	ND	ND	+	ND	ND	+	ND	ND	Sheik et al. (2015), Murali (2011), Tuppad and Shishupala (2014), Shubha and Srinivas (2017), Polizeli et al. (2005) and Venkatesagowda et al. (2012)
Curvularia pallescens	ND	ND	ND	ND	ND	+	ND	ND	ND	ND	ND	ND	ND	ND	ND	ND	Sheik et al. (2015)
Curvularia spp.	+	+	+	+	+	+	ND	−	+	+	+	+	+	ND	ND	ND	Patil et al. (2015), Murali (2011), Uzma et al. (2016), Shubha and Srinivas (2017), Neema Job and Philip (2015), Patel et al. (2013) and Prathyusha et al. (2015)
Curvularia tuberculata	ND	ND	ND	ND	ND	ND	ND	ND	ND	ND	+	ND	ND	ND	ND	+	Murali (2011) and Thirunavukkarasu et al. (2015)
Curvularia vermiformis	−	+	ND	+	+	−	ND	ND	ND	ND	ND	ND	ND	ND	ND	ND	Amirita et al. (2012)

(continued)

Table 14.1 (continued)

Fungi name	Amylase	Cellulase	Pectinase	Proteinase	Lipase	Laccase	Glucosidase	Phosphatase	Chitinase	Gelatinase	Asparaginase	Tyrosinase	Ligninase	Esterase	Polyurethanase	Xylanase	References
Cylindrocapron obtusisporum	ND	ND	ND	ND	ND	ND	ND	ND	ND	ND	+	ND	ND	ND	ND	ND	Kaur and Saxena (2014)
Cylindrocephalum sp.	+	+	+	–	+	–	ND	+	ND	ND	ND	ND	ND	ND	ND	ND	Shubha and Srinivas (2017) and Snitha et al. (2012, 2013)
Cylindrocladium sp.	ND	ND	ND	ND	ND	–	ND	ND	ND	ND	ND	ND	ND	ND	ND	ND	Sheik et al. (2015)
Cymodocea serrulata	ND	ND	ND	ND	ND	ND	ND	ND	ND	ND	ND	ND	ND	ND	ND	+	Thirunavukkarasu et al. (2015)
Daldinia eschscholzii	ND	+	ND	ND	ND	ND	ND	ND	ND	ND	ND	ND	ND	ND	ND	ND	Wu et al. (2017)
Debaryomyces hansenii	ND	ND	ND	ND	ND	–	ND	ND	ND	ND	ND	ND	ND	ND	ND	ND	Sheik et al. (2015)
Diaporthalean species	+	–	ND	ND	–	ND	ND	ND	ND	ND	ND	ND	ND	ND	ND	ND	Toghueo et al. (2017)
Diaporthe eucalyptorum	ND	ND	ND	ND	ND	–	ND	ND	ND	ND	ND	ND	ND	ND	ND	ND	Toghueo et al. (2017)
Diaporthe longicolla	–	–	ND	ND	–	ND	ND	ND	ND	ND	ND	ND	ND	ND	ND	ND	Toghueo et al. (2017)
Diaporthe sp.	+	+	+	+	–	ND	ND	ND	ND	ND	ND	ND	ND	ND	ND	ND	Rajput et al. (2016), Alberto et al. (2016) and Toghueo et al. (2017)
Discosia sp.	+	+	–	–	–	+	ND	ND	ND	ND	ND	ND	ND	ND	ND	ND	Snitha et al. (2013) and Corrêa et al. (2014)
Dothideomycetes sp.	ND	+	ND	ND	ND	ND	+	+	ND	ND	ND	ND	ND	ND	ND	ND	Khan et al. (2016)

Drechslera hawaiiensis	+	–	ND	–	+	–	ND	ND	ND	ND	ND	ND	ND	ND	ND	ND	Amirita et al. (2012)
Drechslera sp.	+	+	+	+	+	–	ND	ND	+	ND	+	ND	ND	ND	ND	–	Murali (2011), Snitha et al. (2013), Thirunavukkarasu et al. (2015), Uzma et al. (2016), Rajulu et al. (2011) and Prathyusha et al. (2015)
Edenia gomezpompae	ND	ND	ND	ND	ND	ND	ND	ND	ND	ND	ND	ND	ND	ND	+	ND	Russell et al. (2011)
Emericella nidulans	–	–	+	–	–	–	ND	ND	ND	ND	+	ND	ND	ND	ND	ND	Prathyusha et al. (2015) and Murali (2011)
Endothia gyrosa	ND	ND	ND	ND	ND	ND	ND	ND	ND	ND	ND	ND	ND	ND	ND	+	Suto et al. (2002)
Epichloe festucae	ND	ND	ND	+	ND	ND	ND	ND	ND	ND	ND	ND	ND	ND	ND	ND	Bryant et al. (2009)
Eupenicillium sp.	ND	ND	ND	ND	ND	ND	ND	ND	ND	ND	+	ND	ND	ND	ND	ND	Theantana et al. (2009)
Eurotium sp.	ND	ND	ND	ND	ND	ND	ND	ND	ND	ND	+	ND	ND	ND	ND	ND	Jalgaonwala and Mahajan (2014)
Eutypella sp.	+	–	ND	+	+	ND	ND	ND	ND	ND	ND	ND	ND	ND	ND	ND	Katoch et al. (2014)
Flavodon flavus	+	–	ND	–	+	ND	ND	ND	ND	ND	ND	ND	ND	ND	ND	ND	Katoch et al. (2014)
Fomitopsis cf. *meliae*	+	+	ND	+	+	ND	ND	ND	ND	ND	ND	ND	ND	ND	ND	ND	Katoch et al. (2014)
Fusarium acuminatum	+	ND	ND	ND	ND	ND	ND	ND	ND	ND	+	ND	ND	ND	ND	ND	Masumi et al. (2014)
Fusarium chlamydosporum	+	+	+	+	+	+	ND	+	ND	ND	ND	ND	ND	ND	ND	ND	Snitha et al. (2013) and Shubha and Srinivas (2017)
Fusarium decemcellulare	+	+	ND	ND	–	ND	ND	ND	ND	ND	ND	ND	ND	ND	ND	ND	Toghueo et al. (2017)
Fusarium equiseti	+	+	ND	ND	–	+	ND	ND	ND	ND	+	ND	ND	ND	ND	ND	Masumi et al. (2014) and Toghueo et al. (2017)
Fusarium fusaroides	ND	ND	ND	ND	ND	+	ND	ND	ND	ND	ND	ND	ND	ND	ND	ND	Sheik et al. (2015)

(continued)

Table 14.1 (continued)

Fungi name	Amylase	Cellulase	Pectinase	Proteinase	Lipase	Laccase	Glucosidase	Phosphatase	Chitinase	Gelatinase	Asparaginase	Tyrosinase	Ligninase	Esterase	Polyurethanase	Xylanase	References
Fusarium graminearum	ND	ND	ND	ND	ND	ND	ND	ND	ND	ND	+	ND	ND	ND	ND	ND	Manasa and Nalini (2014)
Fusarium javanicum	+	ND	ND	ND	ND	ND	ND	ND	ND	ND	+	ND	ND	ND	ND	ND	Masumi et al. (2014)
Fusarium lateritium	+	+	–	+	ND	ND	ND	ND	ND	ND	+	ND	ND	ND	ND	+	Masumi et al. (2014) and Bezerra et al. (2012)
Fusarium oxysporum	+	+	+	+	+	–	ND	+	ND	ND	+	ND	ND	+	ND	+	Masumi et al. (2014), Corrêa et al. (2014), Katoch et al. (2014), Nath et al. (2015), Thirunavukkarasu et al. (2015), Katoch et al. (2014), Shubha and Srinivas (2017), Toghueo et al. (2017), Polizeli et al. (2005), Venkatesagowda et al. (2012), Manasa and Nalini (2014) and Sheik et al. (2015)
Fusarium proliferatum	ND	+	ND	ND	ND	ND	+	+	ND	ND	ND	ND	ND	ND	ND	ND	Khan et al. (2016)
Fusarium reticulatum	+	ND	ND	ND	ND	ND	ND	ND	ND	ND	+	ND	ND	ND	ND	ND	Masumi et al. (2014)
Fusarium solani	+	+	+	+	–	–	ND	ND	ND	ND	ND	ND	ND	ND	ND	ND	Snitha et al. (2013), Ayob and Simarani (2016), Toghueo et al. (2017), Tuppad and Shishupala (2014) and Prathyusha et al. (2015)

Fusarium spp.	+	+	+	+	+	+	ND	ND	+	+	+	+	+	ND	ND	–	Muthezhilan et al. (2014), Masumi et al. (2014), Wu et al. (2009), Murali (2011), Maria et al. (2005), Chow and Ting (2015), Patil et al. (2015), Snitha et al. (2013), Muthezhilan et al. (2014), Masumi et al. (2014), Theantana et al. (2009), Choi et al. (2005), Manasa and Nalini (2014), Zaferanloo et al. (2013), Katoch et al. (2014), Uzma et al. (2016), Patel et al. (2013) and Neema Job and Philip (2015)
Fusarium sterilihyphosum	+	+	+	ND	ND	ND	ND	ND	ND	ND	ND	ND	ND	ND	ND	ND	Tuppad and Shishupala (2014)
Fusarium striatum	+	+	ND	ND	+	ND	ND	ND	ND	ND	ND	ND	ND	ND	ND	ND	Toghueo et al. (2017)
Fusarium subglotinans	+	ND	ND	ND	ND	ND	ND	ND	ND	ND	+	ND	ND	ND	ND	ND	Masumi et al. (2014)
Fusarium verticillioides	+	+	+	ND	ND	ND	ND	ND	ND	ND	+	ND	ND	ND	ND	ND	Tuppad and Shishupala (2014) and Manasa and Nalini (2014)
Fusicoccum sp.	+	+	+	+	+	–	ND	ND	ND	ND	ND	ND	ND	ND	ND	ND	Snitha et al. (2013) and Toghueo et al. (2017)
Gibberella baccata	ND	+	ND	+	+	ND	ND	ND	ND	ND	ND	ND	ND	ND	ND	+	Bezerra et al. (2015)
Gibberella fujikuroi	ND	+	ND	+	+	ND	ND	ND	ND	ND	ND	ND	ND	ND	ND	+	Bezerra et al. (2015)
Glomerella cingulata	–	–	+	–	–	–	ND	ND	ND	ND	ND	ND	ND	ND	ND	ND	Prathyusha et al. (2015)
Gonatophragmium mori	ND	ND	ND	ND	ND	ND	ND	ND	ND	ND	ND	ND	ND	ND	ND	+	Thirunavukkarasu et al. (2015)
Guignardia mangiferae	+	ND	ND	ND	+	–	ND	ND	ND	ND	ND	ND	ND	ND	+	ND	Russell et al. (2011), Toghueo et al. (2017) and Sheik et al. (2015)
Guignardia sp.	ND	ND	ND	ND	ND	–	ND	ND	ND	ND	ND	ND	ND	ND	ND	ND	Sheik et al. (2015)
Hebeloma incarnatulum	ND	ND	ND	+	ND	ND	ND	ND	ND	ND	ND	ND	ND	ND	ND	ND	Mayerhofer et al. (2015)

(continued)

Table 14.1 (continued)

Fungi name	Amylase	Cellulase	Pectinase	Proteinase	Lipase	Laccase	Glucosidase	Phosphatase	Chitinase	Gelatinase	Asparaginase	Tyrosinase	Ligninase	Esterase	Polyurethanase	Xylanase	References
Helminthosporium sp.	–	–	–	–	–	–	ND	+	ND	ND	ND	ND	ND	ND	ND	ND	Shubha and Srinivas (2017)
Hormonema sp.	ND	+	ND	ND	ND	+	ND	ND	ND	ND	ND	ND	ND	ND	ND	ND	Fillat et al. (2016)
Humicola sp.	ND	ND	ND	ND	ND	ND	ND	ND	+	ND	ND	ND	ND	ND	ND	ND	Venkatachalam et al. (2015)
Hydnum sp.	ND	ND	ND	+	ND	ND	ND	ND	ND	ND	ND	ND	ND	ND	ND	ND	Nygren et al. (2007)
Hyphomycetous species	+	+	–	ND	ND	ND	ND	ND	ND	ND	ND	ND	–	ND	ND	+	Choi et al. (2005)
Hypoxylon anthochroum	ND	ND	ND	ND	ND	+	ND	ND	ND	ND	ND	ND	ND	ND	ND	ND	Sheik et al. (2015)
Hypoxylon investiens	+	+	ND	ND	–	ND	ND	ND	ND	ND	ND	ND	ND	ND	ND	ND	Toghueo et al. (2017)
Hypoxylon sp.	ND	+	ND	ND	ND	ND	ND	ND	ND	ND	ND	ND	ND	ND	ND	ND	Wu et al. (2017)
Isaria sp.	+	+	–	+	+	+	ND	ND	ND	ND	ND	ND	ND	ND	ND	ND	Snitha et al. (2013)
Jania adherens	ND	ND	ND	ND	ND	ND	ND	ND	ND	ND	ND	ND	ND	ND	ND	–	Thirunavukkarasu et al. (2015)
Laccaria bicolor	ND	ND	ND	+	ND	ND	ND	ND	ND	ND	ND	ND	ND	ND	ND	ND	Mayerhofer et al. (2015)
Lasiodiplodia sp.	+	–	–	ND	ND	–	ND	ND	ND	ND	–	ND	ND	ND	+	ND	Russell et al. (2011) and Uzma et al. (2016)
Lasiodiplodia theobromae	+	+	ND	+	+	+	ND	ND	ND	ND	ND	ND	ND	+	ND	ND	Toghueo et al. (2017), Venkatesagowda et al. (2012), Amirita et al. (2012) and Sheik et al. (2015)
Leptosphaerulina Americana	+	ND	ND	+	ND	ND	ND	ND	ND	ND	ND	ND	–	ND	ND	–	Zaferanloo et al. (2013)
Leptosphaerulina sp.	+	ND	ND	+	ND	ND	ND	ND	ND	ND	ND	ND	–	ND	ND	–	Zaferanloo et al. (2013)

Leptosphaerulina trifolii	+	ND	ND	+	ND	ND	ND	ND	ND	ND	ND	ND	–	ND	ND	–	Zaferanloo et al. (2013)
Leptosphrulina chartarum	–	–	–	–	–	–	ND	ND	ND	ND	ND	ND	ND	ND	ND	ND	Prathyusha et al. (2015)
Leucostoma personii	ND	ND	ND	ND	ND	ND	ND	ND	ND	ND	ND	ND	ND	ND	ND	+	Suto et al. (2002)
Lignosus rhinoceros	ND	ND	ND	+	ND	ND	ND	ND	ND	ND	ND	ND	ND	ND	ND	ND	Noor et al. (2016)
Macrophomina phaseolina	–	+	ND	–	ND	ND	ND	ND	ND	ND	ND	ND	ND	ND	ND	ND	Ayob and Simarani (2016)
Meliniomyces variabilis	ND	ND	ND	+	ND	ND	ND	ND	ND	ND	ND	ND	ND	ND	ND	ND	Mayerhofer et al. (2015)
Meyerozyma caribbica	ND	ND	ND	ND	ND	–	ND	ND	ND	ND	ND	ND	ND	ND	ND	ND	Sheik et al. (2015)
Meyerozyma guilliermondii	ND	ND	ND	ND	ND	–	ND	ND	ND	ND	ND	ND	ND	ND	ND	ND	Sheik et al. (2015)
Microsphaeropsis sp.	ND	+	ND	ND	ND	ND	ND	ND	ND	ND	ND	ND	ND	ND	ND	ND	Peng and Chen (2007)
Monodictys castaneae	ND	–	–	–	ND	ND	ND	ND	ND	ND	ND	ND	ND	ND	ND	+	Bezerra et al. (2012)
Monotospora sp.	ND	ND	ND	ND	ND	+	ND	ND	ND	ND	ND	ND	ND	ND	ND	ND	Wang et al. (2006)
Mortierella hyaline	+	+	ND	+	+	ND	ND	ND	ND	ND	ND	ND	+	ND	ND	+	Corrêa et al. (2014) and Bhagobaty et al. (2010)
Mucor racemosus	+	+	ND	+	+	ND	ND	ND	ND	ND	ND	ND	ND	+	ND	ND	Venkatesagowda et al. (2012)
Mucor sp.	+	–	–	ND	ND	–	ND	ND	ND	ND	+	ND	ND	ND	ND	ND	Uzma et al. (2016) and Kaur and Saxena (2014)
Myceliophthora sp.	ND	ND	ND	ND	ND	ND	ND	ND	ND	ND	ND	ND	ND	ND	ND	+	Polizeli et al. (2005)
Mycosphaerella thailandica	+	ND	ND	ND	–	ND	ND	ND	ND	ND	ND	ND	ND	ND	ND	ND	Toghueo et al. (2017)
Myrmecridium schulzeri	ND	+	ND	+	+	ND	ND	ND	ND	ND	ND	ND	ND	ND	ND	+	Bezerra et al. (2012)

(continued)

Table 14.1 (continued)

Fungi name	Amylase	Cellulase	Pectinase	Proteinase	Lipase	Laccase	Glucosidase	Phosphatase	Chitinase	Gelatinase	Asparaginase	Tyrosinase	Ligninase	Esterase	Polyurethanase	Xylanase	References
Myrothecium sp.	+	+	+	+	–	–	ND	ND	ND	ND	ND	ND	ND	ND	ND	ND	Snitha et al. (2013)
Myrothecium verrucaria	+	+	ND	+	+	ND	ND	ND	ND	ND	ND	ND	ND	ND	ND	+	Bezerra et al. (2015) and Katoch et al. (2014)
Nectria rigidiuscula	+	+	ND	ND	–	ND	ND	ND	ND	ND	ND	ND	ND	ND	ND	ND	Toghueo et al. (2017)
Nectria sp.	ND	ND	ND	ND	ND	ND	ND	ND	ND	ND	ND	ND	ND	ND	+	ND	Russell et al. (2011)
Nemania bipapillata	+	ND	ND	ND	+	ND	ND	ND	ND	ND	ND	ND	ND	ND	ND	ND	Toghueo et al. (2017)
Neofusicoccum austral	ND	ND	ND	ND	ND	+	ND	ND	ND	ND	ND	ND	ND	ND	ND	ND	Fillat et al. (2016)
Neofusicoccum luteum	ND	ND	ND	ND	ND	+	ND	ND	ND	ND	ND	ND	ND	ND	ND	ND	Fillat et al. (2016)
Nigrospora oryzae	+	+	ND	ND	+	ND	ND	ND	+	ND	ND	ND	ND	ND	ND	ND	Toghueo et al. (2017) and Rajulu et al. (2011)
Nigrospora sp.	+	+	–	+	+	+	ND	+	ND	ND	+	ND	–	ND	ND	–	Zaferanloo et al. (2013), Patel et al. (2013), Thirunavukkarasu et al. (2015), Uzma et al. (2016), Shubha and Srinivas (2017), Peng and Chen (2007) and Murali (2011)
Nigrospora sphaerica	+	+	+	+	–	–	ND	ND	ND	ND	ND	ND	ND	ND	ND	+	Bezerra et al. (2012), Amirita et al. (2012), Prathyusha et al. (2015), Snitha et al. (2013) and Ayob and Simarani (2016)
Nodulisporium gregarium	ND	ND	ND	ND	ND	ND	ND	ND	+	ND	ND	ND	ND	ND	ND	ND	Rajulu et al. (2011)

Nodulisporium sp.	ND	+	ND	+	+	ND	ND	ND	ND	ND	ND	ND	ND	ND	ND	+	Bezerra et al. (2015) and Thirunavukkarasu et al. (2015)
Ophioceras leptosporum	+	+	ND	ND	+	ND	ND	ND	ND	ND	ND	ND	ND	ND	ND	ND	Toghueo et al. (2017)
Paecilomyces lilacinus	ND	ND	ND	ND	ND	+	ND	ND	ND	ND	ND	ND	ND	ND	ND	ND	Sheik et al. (2015)
Paecilomyces spp.	–	+	–	ND	ND	–	ND	ND	ND	ND	+	ND	ND	+	ND	ND	Murali (2011), Lisboa et al. (2013) and Uzma et al. (2016)
Paecilomyces variabilis	+	+	ND	+	+	ND	ND	ND	ND	ND	ND	ND	+	ND	ND	+	Corrêa et al. (2014) and Bhagobaty and Joshi (2012)
Paecilomyces variotii	+	+	+	+	+	–	ND	ND	ND	ND	ND	ND	ND	+	ND	ND	Snitha et al. (2013) and Venkatesagowda et al. (2012)
Papulaspora immerse	+	+	+	ND	ND	ND	ND	ND	ND	ND	ND	ND	ND	ND	ND	ND	Tuppad and Shishupala (2014)
Paraconiothyrium variabile	+	+	ND	ND	–	ND	ND	ND	ND	ND	ND	ND	ND	ND	ND	ND	Toghueo et al. (2017)
Penicillium aurantiogriseum	ND	+	–	+	ND	ND	ND	ND	ND	ND	ND	ND	ND	ND	ND	+	Bezerra et al. (2012)
Penicillium brasilianum	ND	ND	ND	ND	ND	ND	ND	ND	ND	ND	ND	ND	ND	ND	ND	+	Polizeli et al. (2005)
Penicillium capsulatum	ND	ND	ND	ND	ND	ND	ND	ND	ND	ND	ND	ND	ND	ND	ND	+	Polizeli et al. (2005)
Penicillium chermesinum	+	+	ND	ND	+	ND	ND	ND	ND	ND	ND	ND	ND	ND	ND	ND	Toghueo et al. (2017)
Penicillium chrysogenum	+	+	+	–	–	–	ND	+	ND	+	ND	+	ND	ND	ND	+	Sundarram and Murthy (2014), Shubha and Srinivas (2017), Fouda et al. (2015) and Nath et al. (2015)

(continued)

Table 14.1 (continued)

Fungi name	Amylase	Cellulase	Pectinase	Proteinase	Lipase	Laccase	Glucosidase	Phosphatase	Chitinase	Gelatinase	Asparaginase	Tyrosinase	Ligninase	Esterase	Polyurethanase	Xylanase	References
Penicillium citrinum	+	+	+	+	+	−	+	+	ND	ND	ND	ND	ND	+	ND	ND	Corrêa et al. (2014), Shubha and Srinivas (2017), Khan et al. (2016) and Venkatesagowda et al. (2012)
Penicillium commune	ND	+	ND	+	+	ND	ND	ND	ND	ND	ND	ND	ND	ND	ND	+	Bezerra et al. (2015)
Penicillium crustosum	ND	ND	ND	ND	ND	ND	ND	+	ND	ND	ND	ND	ND	ND	ND	ND	Nath et al. (2015)
Penicillium fellutanum	+	ND	ND	ND	ND	ND	ND	ND	ND	ND	ND	ND	ND	ND	ND	ND	Sundarram and Murthy (2014)
Penicillium frequentans	+	−	−	+	−	−	ND	ND	ND	ND	ND	ND	ND	ND	ND	ND	Prathyusha et al. (2015)
Penicillium funiuclas	+	+	ND	+	+	ND	ND	ND	ND	ND	ND	ND	ND	ND	ND	ND	Escudero et al. (2016)
Penicillium glabrum	ND	+	ND	+	+	ND	ND	ND	ND	ND	ND	ND	ND	ND	ND	+	Bezerra et al. (2015)
Penicillium glandicola	ND	+	+	+	ND	ND	ND	ND	ND	ND	ND	ND	ND	ND	ND	+	Bezerra et al. (2012)
Penicillium janthinellum	+	ND	ND	ND	ND	ND	ND	ND	ND	ND	ND	ND	ND	ND	ND	ND	Sundarram and Murthy (2014)
Penicillium megasporum	ND	ND	ND	+	ND	ND	ND	ND	ND	ND	ND	ND	ND	ND	ND	ND	Rajput et al. (2016)
Penicillium parvum	+	+	ND	ND	+	ND	ND	ND	ND	ND	ND	ND	ND	ND	ND	ND	Toghueo et al. (2017)
Penicillium purpurogenum	−	+	−	+	−	+	ND	+	ND	ND	ND	ND	ND	ND	ND	ND	Shubha and Srinivas (2017)
Penicillium roquefortii	+	ND	ND	ND	ND	ND	ND	ND	ND	ND	ND	ND	ND	ND	ND	ND	Sundarram and Murthy (2014)

Penicillium sclerotiorum	ND	ND	ND	ND	ND	ND	ND	+	ND	ND	ND	ND	ND	ND	ND	ND	Nath et al. (2015)
Penicillium simplicissimum	+	+	ND	ND	+	+	ND	ND	+	ND	+	+	ND	ND	ND	ND	Lee et al. (2014)
Penicillium spp.	+	+	+	+	+	+	ND	+	+	+	+	+	+	ND	ND	+	Govindappa et al. (2016), Neema Job and Philip (2015), Bhagobaty and Joshi (2012), Theantana et al. (2009), Muthezhilan et al. (2014), Rubini et al. (1980), Venkatachalam et al. (2015), Nath et al. (2015), Uzma et al. (2016), Kaur and Saxena (2014), Polizeli et al. (2005), Chow and Ting (2015), Snitha et al. (2013) and Murali (2011)
Penicillium variabile	+	+	ND	+	+	ND	ND	ND	ND	ND	ND	ND	ND	+	ND	ND	Venkatesagowda et al. (2012)
Periconia atropurpurea	ND	ND	ND	ND	ND	ND	ND	ND	ND	ND	ND	ND	ND	+	ND	ND	Lisboa et al. (2013)
Periconia sp.	ND	ND	ND	ND	ND	ND	+	ND	ND	ND	ND	ND	ND	ND	ND	ND	Harnpicharnchai et al. (2009)
Pestalotiopsis clavispora	ND	ND	ND	ND	ND	–	ND	ND	ND	ND	ND	ND	ND	ND	ND	ND	Sheik et al. (2015)
Pestalotiopsis disseminata	+	–	+	+	–	+	ND	ND	ND	ND	ND	ND	ND	ND	ND	ND	Snitha et al. (2013)
Pestalotiopsis glandicola	+	+	+	–	–	–	ND	ND	ND	ND	ND	ND	ND	ND	ND	ND	Prathyusha et al. (2015)
Pestalotiopsis guepinii	ND	+	–	–	ND	ND	ND	ND	ND	ND	ND	ND	ND	ND	ND	+	Bezerra et al. (2012)
Pestalotiopsis mangiferae	+	+	+	+	–	–	ND	ND	ND	ND	ND	ND	ND	ND	ND	ND	Prathyusha et al. (2015)
Pestalotiopsis microspora	+	–	+	–	–	–	ND	ND	ND	ND	ND	ND	ND	ND	+	ND	Russell et al. (2011) and Prathyusha et al. (2015)

(continued)

Table 14.1 (continued)

Fungi name	Amylase	Cellulase	Pectinase	Proteinase	Lipase	Laccase	Glucosidase	Phosphatase	Chitinase	Gelatinase	Asparaginase	Tyrosinase	Ligninase	Esterase	Polyurethanase	Xylanase	References
Pestalotiopsis palmarum	+	+	ND	+	+	ND	ND	ND	ND	ND	ND	ND	ND	+	ND	ND	Venkatesagowda et al. (2012)
Pestalotiopsis spp.	+	+	–	+	+	+	ND	ND	+	ND	+	ND	ND	ND	+	+	Murali (2011), Russell et al. (2011), Sheik et al. (2015), Maria et al. (2005), Lii et al. (2017), Lumyong et al. (2002), Poling et al. (2008), Snitha et al. (2013), Toghueo et al. (2017) and Uzma et al. (2016)
Pestalotiopsis versicolor	ND	ND	ND	+	ND	ND	ND	ND	ND	ND	ND	ND	ND	ND	ND	ND	Rajput et al. (2016)
Phaeosphaeria sp.	ND	ND	ND	ND	ND	ND	ND	ND	ND	ND	ND	ND	ND	ND	+	ND	Russell et al. (2011)
Phaeotrichoconis sp.	ND	ND	ND	ND	ND	ND	ND	ND	ND	ND	+	ND	ND	ND	ND	ND	Murali (2011)
Phanerochaete chrysosporium	ND	ND	ND	ND	ND	–	ND	ND	ND	ND	ND	ND	ND	ND	ND	ND	Sheik et al. (2015)
Phanerochaete sp.	ND	ND	ND	ND	ND	+	ND	ND	ND	ND	ND	ND	ND	ND	ND	ND	Sheik et al. (2015)
Phialemonium dimorphosporum	ND	ND	ND	ND	ND	–	ND	ND	ND	ND	ND	ND	ND	ND	ND	ND	Sheik et al. (2015)
Phialocephala fortinii	ND	ND	ND	+	ND	ND	ND	ND	ND	ND	ND	ND	ND	ND	ND	ND	Mayerhofer et al. (2015)
Phoma glomerata	+	+	ND	+	+	ND	ND	ND	ND	ND	ND	ND	ND	+	ND	ND	Venkatesagowda et al. (2012)
Phoma herbarum	+	ND	ND	+	ND	ND	ND	ND	ND	ND	ND	ND	–	+	ND	–	Zaferanloo et al. (2014) and Lisboa et al. (2013)
Phoma medicaginis	ND	+	ND	ND	ND	ND	+	+	ND	ND	ND	ND	ND	ND	ND	ND	Khan et al. (2016)

Phoma microchlamydospora	+	+	ND	ND	+	−	ND	ND	ND	ND	ND	ND	ND	ND	ND	ND	Toghueo et al. (2017)
Phoma moricola	+	ND	ND	+	ND	ND	ND	ND	ND	ND	ND	ND	−	ND	ND	−	Zaferanloo et al. (2013)
Phoma multirostrata	+	−	ND	−	+	ND	ND	ND	ND	ND	ND	ND	ND	ND	ND	ND	Katoch et al. (2014)
Phoma putaminum	ND	+	ND	+	+	ND	ND	ND	ND	ND	ND	ND	ND	ND	ND	+	Bezerra et al. (2015)
Phoma spp.	+	+	+	+	+	+	+	+	+	+	+	+	+	ND	ND	+	Lumyong et al. (2002), Zaferanloo et al. (2013), Neema Job and Philip (2015), Thirunavukkarasu et al. (2015), Uzma et al. (2016), Choi et al. (2005), Alberto et al. (2016), Katoch et al. (2014), Masumi et al. (2014), Snitha et al. (2013), Khan et al. (2016), Murali (2011) and Chow and Ting (2015)
Phoma tropica	ND	−	−	+	ND	ND	ND	ND	ND	ND	ND	ND	ND	ND	ND	+	Bezerra et al. (2012)
Phomopsis archeri	ND	−	−	+	ND	ND	ND	ND	ND	ND	ND	ND	ND	ND	ND	+	Bezerra et al. (2012)
Phomopsis cassiae	ND	ND	ND	ND	ND	ND	ND	ND	ND	ND	ND	ND	ND	+	ND	ND	Lisboa et al. (2013)
Phomopsis liquidambari	+	ND	ND	ND	ND	ND	ND	ND	ND	ND	ND	ND	ND	ND	ND	ND	Siddikee et al. (2016)
Phomopsis longicolla	+	+	+	+	+	+	ND	ND	ND	ND	ND	ND	ND	+	ND	ND	Venkatesagowda et al. (2012) and Snitha et al. (2013)
Phomopsis phyllanticolla	+	+	ND	ND	+	ND	ND	ND	ND	ND	ND	ND	ND	ND	ND	ND	Toghueo et al. (2017)
Phomopsis spp.	+	+	+	+	+	+	ND	ND	+	ND	+	+	−	ND	ND	+	Lumyong et al. (2002), Choi et al. (2005), Lee et al. (2014), Murali (2011), Toghueo et al. (2017), Peng and Chen (2007), Prathyusha et al. (2015), Thirunavukkarasu et al. (2015) and Uzma et al. (2016)

(continued)

Table 14.1 (continued)

Fungi name	Amylase	Cellulase	Pectinase	Proteinase	Lipase	Laccase	Glucosidase	Phosphatase	Chitinase	Gelatinase	Asparaginase	Tyrosinase	Ligninase	Esterase	Polyurethanase	Xylanase	References
Phomopsis stipata	+	+	ND	+	+	ND	ND	ND	ND	ND	ND	ND	ND	+	ND	ND	Lisboa et al. (2013) and Venkatesagowda et al. (2012)
Phomopsis theicola	–	+	ND	ND	–	ND	ND	ND	ND	ND	ND	ND	ND	ND	ND	ND	Toghueo et al. (2017)
Phyllosticta capitalensis	ND	ND	ND	ND	ND	–	ND	ND	ND	ND	ND	ND	ND	ND	ND	ND	Sheik et al. (2015)
Phyllosticta occulta	+	+	ND	+	+	ND	ND	ND	ND	ND	ND	ND	ND	+	ND	ND	Venkatesagowda et al. (2012)
Phyllosticta sp.	+	–	ND		+	–	ND	ND	ND	ND	ND	ND	ND	ND	ND	ND	Amirita et al. (2012)
Pithomyces atro-olivaceus	ND	+	ND	+	+	ND	ND	ND	ND	ND	ND	ND	ND	ND	ND	+	Bezerra et al. (2015)
Pithomyces chartarum	+	+	+	ND	ND	ND	ND	ND	ND	ND	ND	ND	ND	ND	ND	ND	Tuppad and Shishupala (2014)
Pithomyces sp.	ND	ND	ND	ND	ND	ND	ND	ND	ND	ND	+	ND	ND	ND	ND	ND	
Plectania milleri sp.	+	–	+	–	+	+	ND	ND	ND	ND	ND	–	ND	ND	ND	ND	Fatima et al. (2016)
Plectosphaerella sp.	ND	ND	ND	ND	ND	ND	ND	ND	ND	ND	ND	ND	ND	ND	+	ND	Russell et al. (2011)
Pleosporales species	+	+	ND	+	+	ND	ND	ND	ND	ND	ND	ND	ND	ND	+	ND	Russell et al. (2011), Katoch et al. (2014) and Toghueo et al. (2017)
Pochonia chlamydosporia	ND	ND	ND	+	ND	ND	ND	ND	ND	ND	ND	ND	ND	ND	ND	ND	Escudero et al. (2016)
Portieria hornemonii	ND	ND	ND	ND	ND	ND	ND	ND	ND	ND	ND	ND	ND	ND	ND	+	Thirunavukkarasu et al. (2015)
Preussia minima	+	ND	ND	+	ND	ND	ND	ND	ND	ND	ND	ND	–	ND	ND	–	Zaferanloo et al. (2013) and Corrêa et al. (2014)
Preussia sp.2	ND	+	ND	ND	ND	ND	+	+	ND	ND	ND	ND	ND	ND	ND	ND	Khan et al. (2016)

Pringsheimia smilacis	ND	ND	ND	ND	ND	+	ND	ND	ND	ND	ND	ND	ND	ND	ND	ND	Fillat et al. (2016)
Pseudocercospora sp.	+	–	ND	ND	+	ND	ND	ND	ND	ND	ND	ND	ND	ND	ND	ND	Toghueo et al. (2017)
Pseudofusicoccum kimberleyense	+	+	ND	ND	–	ND	ND	ND	ND	ND	ND	ND	ND	ND	ND	ND	Toghueo et al. (2017)
Pycnoporus sanguineus	+	ND	ND	ND	ND	ND	ND	ND	ND	ND	ND	ND	ND	ND	ND	ND	Sundarram and Murthy (2014)
Pyllosticta sp.	+	–	+	–	–	–	ND	ND	ND	ND	ND	ND	ND	ND	ND	ND	Snitha et al. (2013)
Rhizoctonia sp.	+	+	–	+	–	–	ND	+	ND	ND	ND	ND	ND	ND	ND	ND	Shubha and Srinivas (2017) and Patil et al. (2015)
Rhizopus sp.	+	+	+	ND	ND	–	ND	ND	ND	ND	+	ND	ND	ND	ND	ND	Uzma et al. (2016)
Rhizopus stolonifera	+	+	ND	+	+	ND	ND	ND	ND	ND	ND	ND	ND	+	ND	ND	Venkatesagowda et al. (2012)
Saccharicola sp.	+	+	+	+	ND	ND	ND	ND	ND	ND	ND	ND	ND	ND	ND	ND	Alberto et al. (2016)
Sarcosomataceous sp.	+	+	ND	ND	+	+	ND	ND	+	ND	+	+	ND	ND	ND	ND	Lee et al. (2014)
Sargassum sp.	ND	ND	ND	ND	ND	ND	ND	ND	ND	ND	ND	ND	ND	ND	ND	+	Thirunavukkarasu et al. (2015)
Sargassum wightii	ND	ND	ND	ND	ND	ND	ND	ND	ND	ND	ND	ND	ND	ND	ND	+	Thirunavukkarasu et al. (2015)
Sclerocystis sp.	ND	+	ND	ND	ND	ND	ND	ND	ND	ND	ND	ND	ND	ND	ND	ND	Peng and Chen (2007)
Sclerotinia sclerotiorum	+	+	ND	+	+	ND	ND	ND	ND	ND	ND	ND	ND	+	ND	ND	Venkatesagowda et al. (2012)
Sclerotium sp.	+	+	+	ND	ND	ND	ND	ND	ND	ND	ND	ND	ND	ND	ND	ND	Tuppad and Shishupala (2014)
Scopulariopsis canadensis	+	+	+	ND	ND	ND	ND	ND	ND	ND	ND	ND	ND	ND	ND	ND	Tuppad and Shishupala (2014)
Septoria sp.	–	+	ND	ND	+	ND	ND	ND	ND	ND	ND	ND	ND	ND	ND	ND	Toghueo et al. (2017)
Sordaria fimicola	+	–	–	–	–	–	ND	ND	ND	ND	ND	ND	ND	ND	ND	ND	Prathyusha et al. (2015)
Sordaria sp.	–	+	–	ND	ND	–	ND	ND	ND	ND	+	ND	ND	ND	ND	ND	Uzma et al. (2016)

(continued)

Table 14.1 (continued)

Fungi name	Amylase	Cellulase	Pectinase	Proteinase	Lipase	Laccase	Glucosidase	Phosphatase	Chitinase	Gelatinase	Asparaginase	Tyrosinase	Ligninase	Esterase	Polyurethanase	Xylanase	References
Sordariomycetes sp.	ND	+	ND	ND	ND	ND	+	+	ND	ND	ND	ND	ND	ND	ND	ND	Khan et al. (2016)
Sphaerospermum sp.	ND	ND	ND	ND	ND	ND	ND	ND	ND	ND	ND	ND	ND	ND	ND	ND	Bezerra et al. (2012)
Sporothrix schenckii	ND	ND	ND	ND	ND	ND	ND	ND	ND	ND	ND	ND	ND	ND	ND	+	Suto et al. (2002)
Stachybotrys chartarum	+	–	+	–	–	–	ND	ND	ND	ND	ND	ND	ND	ND	ND	ND	Prathyusha et al. (2015)
Stemphylium sp.	+	ND	ND	+	ND	ND	ND	ND	ND	ND	ND	ND	–	ND	ND	–	Zaferanloo et al. (2013)
Streptomyces rimosus	+	ND	ND	ND	ND	ND	ND	ND	ND	ND	ND	ND	ND	ND	ND	ND	Sundarram and Murthy (2014)
Syringodium sp.	ND	ND	ND	ND	ND	ND	ND	ND	ND	ND	ND	ND	ND	ND	ND	+	Thirunavukkarasu et al. (2015)
Talaromyces emersonii	+	+	+	–	–	–	ND	ND	ND	ND	ND	ND	ND	ND	ND	ND	Snitha et al. (2013)
Talaromyces flavus	+	+	ND	+	+	–	ND	ND	ND	ND	ND	ND	+	ND	ND	+	Corrêa et al. (2014), Bhagobaty and Joshi (2012) and Sheik et al. (2015)
Talaromyces rotundus	+	+	–	+	–	–	ND	+	ND	ND	ND	ND	ND	ND	ND	ND	Shubha and Srinivas (2017)
Talaromyces sp.	ND	ND	ND	ND	ND	ND	ND	ND	ND	ND	+	ND	ND	ND	ND	ND	Theantana et al. (2009)
Talaromyces wortmannii	ND	+	ND	+	ND	ND	+	ND	ND	ND	ND	ND	ND	ND	ND	+	Robl et al. (2013)
Tetraploa aristata	ND	–	–	+	ND	ND	ND	ND	ND	ND	ND	ND	ND	ND	ND	+	Bezerra et al. (2012)
Thalassia sp.	ND	ND	ND	ND	ND	ND	ND	ND	ND	ND	ND	ND	ND	ND	ND	+	Thirunavukkarasu et al. (2015)
Thermomyces lanuginosus	+	ND	ND	ND	ND	ND	ND	ND	ND	ND	ND	ND	ND	ND	ND	ND	Sundarram and Murthy (2014)

Thielavia arenaria	ND	+	ND	ND	ND	ND	+	+	ND	ND	ND	ND	ND	ND	ND	ND	Khan et al. (2016)
Torulomyces sp.	ND	ND	ND	ND	ND	ND	ND	ND	ND	ND	ND	ND	ND	ND	ND	+	Thirunavukkarasu et al. (2015)
Trametes versicolor	+	+	+	ND	ND	ND	ND	ND	ND	ND	ND	ND	+	ND	ND	+	Choi et al. (2005)
Tretopileus sphaerophorus	–	–	–	–	–	–	ND	ND	ND	ND	ND	ND	ND	ND	ND	ND	Prathyusha et al. (2015)
Trichoderma asperellum	+	–	–	–	–	+	ND	ND	ND	ND	ND	–	ND	ND	ND	ND	Fatima et al. (2016)
Trichoderma atroviride	ND	+	ND	+	ND	ND	+	ND	ND	ND	ND	ND	ND	ND	ND	+	Robl et al. (2013)
Trichoderma aureoviride	+	–	ND	–	+	ND	ND	ND	ND	ND	ND	ND	ND	ND	ND	ND	Katoch et al. (2014)
Trichoderma harzianum	+	+	ND	+	+	ND	ND	ND	ND	ND	ND	ND	ND	+	ND	+	Thirunavukkarasu et al. (2015), Polizeli et al. (2005) and Venkatesagowda et al. (2012)
Trichoderma piluliferum	ND	+	ND	+	+	ND	ND	ND	ND	ND	ND	ND	ND	ND	ND	+	Bezerra et al. (2015)
Trichoderma spp.	+	+	+	+	–	–	ND	+	+	ND	+	ND	ND	ND	ND	ND	Rajulu et al. (2011), Murali (2011), Shubha and Srinivas (2017) and Uzma et al. (2016)
Trichoderma viride	+	+	ND	+	+	ND	ND	ND	ND	ND	ND	ND	ND	+	ND	ND	Lisboa et al. (2013), Venkatesagowda et al. (2012)
Turbenaria sp.	ND	ND	ND	ND	ND	ND	ND	ND	ND	ND	ND	ND	ND	ND	ND	–	Thirunavukkarasu et al. (2015)
Ulocladium sp.	ND	ND	ND	ND	ND	+	ND	ND	ND	ND	ND	ND	ND	ND	ND	ND	Fillat et al. (2016)
Ulva lactuca	ND	ND	ND	ND	ND	ND	ND	ND	ND	ND	ND	ND	ND	ND	ND	–	Thirunavukkarasu et al. (2015)
Umbelopsis isabellina	ND	ND	ND	+	ND	ND	ND	ND	ND	ND	ND	ND	ND	ND	ND	ND	Mayerhofer et al. (2015)
Ustilaginoidea sp.	ND	ND	ND	ND	ND	ND	ND	ND	ND	ND	ND	ND	ND	+	ND	ND	Lisboa et al. (2013)

(continued)

Table 14.1 (continued)

Fungi name	Amylase	Cellulase	Pectinase	Proteinase	Lipase	Laccase	Glucosidase	Phosphatase	Chitinase	Gelatinase	Asparaginase	Tyrosinase	Ligninase	Esterase	Polyurethanase	Xylanase	References
Verticillium lecaniit	ND	ND	ND	ND	ND	ND	ND	ND	ND	ND	+	ND	ND	ND	ND	ND	Manasa and Nalini (2014)
Volutella sp.	ND	ND	ND	ND	ND	ND	ND	ND	ND	ND	+	ND	ND	ND	ND	ND	Manasa and Nalini (2014)
Xylaria adscendens	+	+	ND	ND	+	–	ND	ND	ND	ND	ND	ND	ND	ND	ND	ND	Toghueo et al. (2017)
Xylaria apiculata	+	–	ND	ND	–	–	ND	ND	ND	ND	ND	ND	ND	ND	ND	ND	Toghueo et al. (2017)
Xylaria castorea	+	+	ND	ND	+	–	ND	ND	ND	ND	ND	ND	ND	ND	ND	ND	Toghueo et al. (2017)
Xylaria persicaria	+	+	ND	ND	+	+	ND	ND	ND	ND	ND	ND	ND	ND	ND	ND	Toghueo et al. (2017)
Xylaria spp.	+	+	+	+	+	+	ND	+	ND	ND	ND	ND	+	+	–	+	Lumyong et al. (2002), Lisboa et al. (2013), Shubha and Srinivas (2017), Toghueo et al. (2017), Snitha et al. (2013), Urairuj et al. (2003), Toghueo et al. (2017) and Chow and Ting (2015)
Zopfiella karachiensis	ND	ND	ND	ND	ND	ND	ND	ND	ND	ND	ND	ND	ND	ND	+	ND	Russell et al. (2011)

ND not determined, + presence of enzyme activity, – absence of enzyme activity

14.1.14 Constraints

Based on the information presented in this chapter, we could find that very few endeavors have been made to exploit the endophytic fungi to take them for enzyme production at large scale. This not only reflects on the lack of concerted efforts on research in this direction but also several obstacles related to endophytic fungi. Despite the fact that endophytes, apparently, are also better candidates for enzyme production, often they lose the vigor of enzyme production in axenic cultures under laboratory conditions. Probable reason for this could be the absence of native environmental conditions and a web of network of biotic and abiotic factors in the artificial culture media and the vessels, which otherwise are provided by the hosts to the endophytic fungi (Khan et al. 2016).

Most of the research concerning the enzymatic potential of endophytic fungi is restricted to primary level, i.e., preliminary screening programs. Endophytes are screened for the production of a broad range of enzymes and are mostly not taken up further, for example, scaling up studies at industrial level. The research is mostly limited to laboratory level, and this is mainly due to lack of facilities. Since optimization, isolation, purification, and characterization of enzymes require more sophisticated facilities, most of the researchers find it difficult to carry out further studies even if the results were seemingly promising.

Studies on endophytic fungi with respect to their enzyme production are majorly dependent on the rationale of the host selection. In most of the cases discussed here, more than 80% of the work is with hosts that have medicinal properties, few with extreme environmental conditions and very few from a biodiversity aspect. Thus, the bias shown in host selection is again another limiting factor.

14.1.15 Conclusions

It is a well-established fact that fungal endophytes metabolize many substrates found in the hosts as they can produce a variety of enzymes such as amylases, laccases, proteases, phenol oxidases, xylanases, mannanases, lipases, and pectin lyases (Khan et al. 2015). Despite this fact, their potential to secrete important enzymes is often overlooked. The endophytes, in most of the cases, are enzymatically profiled, in the initial screening programs, to select the potent ones producing such proteins and provide clues on production of enzymes when in interaction with the hosts. Besides the preliminary screening of endophytes in laboratories, there is a grave need for taking them to large production. Along with the quantification of enzymes produced, desperate attempts are needed with advanced techniques to optimize, scale up, and exploit these enzymes for human as well as ecological well-being. Advanced techniques with higher sensitivity such as near-infra red (NIR), fluorescence spectrophotometry, and FTIR-based methods are required to bring up the applications restricted to laboratories only. One of the shortcomings is that most of the research on endophytic fungi does not find its way out of laboratories or applications in different sectors including industrial, environmental, and healthcare. There is a need to focus more on the application and optimization part to extract most out of the potential that the endophytic fungi have in diverse arena that impact human welfare. Further, high-throughput screening techniques coupled with genetic engineering for the processing of efficient endophytes may prove to be a promising approach towards the application of the precious enzymes in various realms.

Acknowledgements The Department of Biotechnology, Pondicherry University is thanked for providing the facilities.

References

Abourashed EA, Hufford CD (1996) Microbial transformation of artemether. J Nat Prod 59:251–253

Alberto RN, Costa AT, Polonio JC, Santos MS, Rhoden SA, Azevedo JL, Pamphile JA (2016) Extracellular enzymatic profiles and taxonomic identification of endophytic fungi isolated from four plant species. Genet Mol Res 15(4)

Alves Macedo G, Soberón Lozano MM, Pastore GM (2003) Enzymatic synthesis of short chain citronellyl esters by a new lipase from Rhizopus sp. Electron J Biotechnol 6:3–4

Amirita A, Sindhu P, Swetha J, Vasanthi NS, Kannan KP (2012) Enumeration of endophytic fungi from medicinal plants and screening of extracellular enzymes. World J Sci Technol 2:13–19

Anbu P, Gopinath SCB, Chaulagain BP, Lakshmipriya T (2017) Microbial enzymes and their applications in industries and medicine 2016. BioMed Res Int 2017:2195808

Arnold AE (2007) Understanding the diversity of foliar endophytic fungi: progress, challenges, and frontiers. Fungal Biol Rev 21:51–66

Asselin BL, Whitin JC, Coppola DJ, Rupp IP, Sallan SE, Cohen HJ (1993) Comparative pharmacokinetic studies of three asparaginase preparations. J Clin Oncol 11:1780–1786

Ayob FW, Simarani K (2016) Endophytic filamentous fungi from a Catharanthus roseus: identification and its hydrolytic enzymes. Saudi Pharm J 24:273–278

Baldauf SL, Palmer JD (1993) Animals and fungi are each other's closest relatives: congruent evidence from multiple proteins. Proc Natl Acad Sci 90:11558–11562

Batool T, Makky EA, Jalal M, Yusoff MM (2016) A comprehensive review on L-asparaginase and its applications. Appl Biochem Biotechnol 178:900–923

Benjamin S, Pandey A (1998) Candida rugosa lipases: molecular biology and versatility in biotechnology. Yeast 14:1069–1087

Bezerra JDP, Santos MGS, Svedese VM, Lima DMM, Fernandes MJS, Paiva LM, Souza-Motta CM (2012) Richness of endophytic fungi isolated from Opuntia ficus-indica Mill.(Cactaceae) and preliminary screening for enzyme production. World J Microbiol Biotechnol 28:1989–1995

Bezerra JDP, Nascimento CCF, Barbosa R do N, da Silva DCV, Svedese VM, Silva-Nogueira EB, Gomes BS, Paiva LM, Souza-Motta CM (2015) Endophytic fungi from medicinal plant Bauhinia forficata: Diversity and biotechnological potential. Braz J Microbiol 46:49–57

Bhagobaty RK, Joshi SR (2012) Enzymatic activity of fungi endophytic on five medicinal plant species of the pristine sacred forests of Meghalaya, India. Biotechnol Bioprocess Eng 17:33–40

Bhagobaty RK, Joshi SR, Kumar R (2010) Penicillium verruculosum RS7PF: a root fungal endophyte associated with an ethno-medicinal plant of the indigenous tribes of Eastern India. Afr J Microbiol Res 4:766–770

Borges W, Borges KB, Bonato PS, Said S, Pupo MT (2009) Endophytic fungi: natural products, enzymes and biotransformation reactions. Curr Org Chem 13:1137–1163

Breen JP (1994) Acremonium endophyte interactions with enhanced plant resistance to insects. Annu Rev Entomol 39:401–423
Brühlmann F, Leupin M, Erismann KH, Fiechter A (2000) Enzymatic degumming of ramie bast fibers. J Biotechnol 76:43–50
Bryant MK, Schardl CL, Hesse U, Scott B (2009) Evolution of a subtilisin-like protease gene family in the grass endophytic fungus Epichloë festucae. BMC Evol Biol 9:168
Chadha N, Prasad R, Varma A (2015) Plant promoting activities of fungal endophytes associated with tomato roots from central Himalaya, India and their interaction with Piriformospora indica. Int J Pharm Bio Sci 6:333–343
Choi YW, Hodgkiss IJ, Hyde KD (2005) Enzyme production by endophytes of Brucea javanica. J Agric Technol 1:55–66
Chow Y, Ting ASY (2015) Endophytic L-asparaginase-producing fungi from plants associated with anticancer properties. J Adv Res 6:869–876
Corrêa RCG, Rhoden SA, Mota TR, Azevedo JL, Pamphile JA, de Souza CG, Polizeli Mde L, Bracht A, Peralta RM (2014) Endophytic fungi: expanding the arsenal of industrial enzyme producers. J Ind Microbiol Biotechnol 41:1467–1478
Costa-Silva TA, Nogueira MA, Fernandes Souza CR, Oliveira WP, Said S (2011) Lipase production by endophytic fungus Cercospora kikuchii: stability of enzymatic activity after spray drying in the presence of carbohydrates. Dry Technol 29:1112–1119
Dai C, Tian L, Zhao Y, Chen Y, Xie H (2010) Degradation of phenanthrene by the endophytic fungus Ceratobasidum stevensii found in Bischofia polycarpa. Biodegradation 21:245–255
El-Bondkly AMA (2012) Molecular identification using ITS sequences and genome shuffling to improve 2-deoxyglucose tolerance and xylanase activity of marine-derived fungus, Aspergillus sp. NRCF5. Appl Biochem Biotechnol 167:2160–2173
El-Gendy MMA (2010) Keratinase production by endophytic Penicillium spp. Morsy1 under solid-state fermentation using rice straw. Appl Biochem Biotechnol 162:780–794
Escudero N, Ferreira SR, Lopez-Moya F, Naranjo-Ortiz MA, Marin-Ortiz AI, Thornton CR, Lopez-Llorca LV (2016) Chitosan enhances parasitism of Meloidogyne javanica eggs by the nematophagous fungus Pochonia chlamydosporia. Fungal Biol 120:572–585
Fatima N, Kondratyuk TP, Park E-J, Marler LE, Jadoon M, Qazi MA, Mehboob Mirza H, Khan I, Atiq N, Chang LC (2016) Endophytic fungi associated with Taxus fuana (West Himalayan Yew) of Pakistan: potential bio-resources for cancer chemopreventive agents. Pharm Biol 54:2547–2554
Fillat Ú, Martín-Sampedro R, Macaya-Sanz D, Martín JA, Ibarra D, Martínez MJ, Eugenio ME (2016) Screening of eucalyptus wood endophytes for laccase activity. Process Biochem 51:589–598
Fouda AH, Hassan SE-D, Eid AM, Ewais EE-D (2015) Biotechnological applications of fungal endophytes associated with medicinal plant Asclepias sinaica (Bioss.). Ann Agric Sci 60:95–104
Gilbert M, Yaguchi M, Watson DC, Wong KKY, Breuil C, Saddler JN (1993) A comparison of two xylanases from the thermophilic fungi Thielavia terrestris and Thermoascus crustaceus. Appl Microbiol Biotechnol 40:508–514
Godfrey T, West S (1996) Textiles. In: Industrial enzymology. Macmillan Press, London, pp 360–371
Govindappa M, Farheen H, Chandrappa CP, Rai RV, Raghavendra VB (2016) Mycosynthesis of silver nanoparticles using extract of endophytic fungi, Penicillium species of Glycosmis mauritiana, and its antioxidant, antimicrobial, anti-inflammatory and tyrokinase inhibitory activity. Adv Nat Sci Nanosci Nanotechnol 7:35014
Grinhut T, Hadar Y, Chen Y (2007) Degradation and transformation of humic substances by saprotrophic fungi: processes and mechanisms. Fungal Biol Rev 21:179–189
Harnpicharnchai P, Champreda V, Sornlake W, Eurwilaichitr L (2009) A thermotolerant β-glucosidase isolated from an endophytic fungi, Periconia sp., with a possible use for biomass conversion to sugars. Protein Expr Purif 67:61–69

Hendriksen HV, Pedersen S, Bisgard-Frantzen H (1999) A process for textile warp sizing using enzymatically modified starches. Patent application WO 99:35325
Illmer P, Schinner F (1995) Solubilization of inorganic calcium phosphates—solubilization mechanisms. Soil Biol Biochem 27:257–263
Jaeger K-E, Reetz MT (1998) Microbial lipases form versatile tools for biotechnology. Trends Biotechnol 16:396–403
Jain P, Aggarwal V, Sharma A, Pundir RK (2012) Isolation, production and partial purification of protease from an endophytic Acremonium sp. J Agric Technol 8:1979–1989
Jalgaonwala RE, Mahajan RT (2011) Evaluation of hydrolytic enzyme activities of endophytes from some indigenous medicinal plants. J Agric Technol 7:1733–1741
Jalgaonwala RE, Mahajan RT (2014) Production of anticancer enzyme asparaginase from endophytic Eurotium Sp. isolated from rhizomes of Curcuma longa. Eur J Exp Biol 4:36–43
Kalyanasundaram I, Nagamuthu J, Srinivasan B, Pachayappan A, Muthukumarasamy S (2015) Production, purification and characterisation of extracellular L-asparaginase from salt marsh fungal endophytes. World J Pharm Pharmac Sci 4:663–677
Kanokratana P, Chantasingh D, Champreda V, Tanapongpipat S, Pootanakit K, Eurwilaichitr L (2008) Identification and expression of cellobiohydrolase (CBHI) gene from an endophytic fungus, Fusicoccum sp.(BCC4124) in Pichia pastoris. Protein Expr Purif 58:148–153
Katoch M, Salgotra A, Singh G (2014) Endophytic fungi found in association with Bacopa monnieri as potential producers of industrial enzymes and antimicrobial bioactive compounds. Braz Arch Biol Technol 57:714–722
Kaur P, Saxena SG (2014) Screening of endophytic fungi for production of Asparaginase enzyme. Thapar University, Patiala
Khaldi N, Seifuddin FT, Turner G, Haft D, Nierman WC, Wolfe KH, Fedorova ND (2010) SMURF: genomic mapping of fungal secondary metabolite clusters. Fungal Genet Biol 47:736–741
Khan AL, Waqas M, Hussain J, Al-Harrasi A, Hamayun M, Lee I-J (2015) Phytohormones enabled endophytic fungal symbiosis improve aluminum phytoextraction in tolerant Solanum lycopersicum: an examples of Penicillium janthinellum LK5 and comparison with exogenous GA 3. J Hazard Mater 295:70–78
Khan AL, Al-Harrasi A, Al-Rawahi A, Al-Farsi Z, Al-Mamari A, Waqas M, Asaf S, Elyassi A, Mabood F, Shin J-H (2016) Endophytic fungi from Frankincense tree improves host growth and produces extracellular enzymes and indole acetic acid. PLoS One 11:e0158207
Kottwitz B, Upadek H, Carrer G (1994) Application and benefits of enzymes in detergents. Chim Oggi 12:21–24
Kusari S, Spiteller M (2011) Are we ready for industrial production of bioactive plant secondary metabolites utilizing endophytes? Nat Prod Rep 28:1203–1207
Kusari S, Singh S, Jayabaskaran C (2014) Biotechnological potential of plant-associated endophytic fungi: hope versus hype. Trends Biotechnol 32:297–303
Lee JM, Tan WS, Ting ASY (2014) Revealing the antimicrobial and enzymatic potentials of culturable fungal endophytes from tropical pitcher plants (Nepenthes spp.). Mycosphere 5:364–377
Li H-Y, Wei D-Q, Shen M, Zhou Z-P (2012) Endophytes and their role in phytoremediation. Fungal Divers 54:11–18
Lii SBW, Wong C, Al-Obaidi JR, Rahmad N, Mujahid A, Mueller M (2017) Ability of endophytic fungi isolated from Nepenthes ampullaria to degrade polyurethane. Malay J Microbiol 13:172–179
Lindstrom JT, Belanger FC (1994) Purification and characterization of an endophytic fungal proteinase that is abundantly expressed in the infected host grass. Plant Physiol 106:7–16
Lindstrom JT, Sun S, Belanger FC (1993) A novel fungal protease expressed in endophytic infection of Poa species. Plant Physiol 102:645–650
Lisboa HCF, Biasetto CR, de MJB, Araújo ÂR, Silva DHS, Teles HL, Trevisan HC (2013) Endophytic fungi producing of esterases: evaluation in vitro of the enzymatic activity using pH indicator. Braz J Microbiol 44:923–926

López-Otín C, Bond JS (2008) Proteases: multifunctional enzymes in life and disease. J Biol Chem 283:30433–30437

Lumyong S, Lumyong P, McKenzie EHC, Hyde KD (2002) Enzymatic activity of endophytic fungi of six native seedling species from Doi Suthep-Pui National Park, Thailand. Can J Microbiol 48:1109–1112

Maat J, Roza M, Verbakel J, da SH, da Silva MJS, Bosse M, Egmond MR, Hagemans MLD, Gorcom R, Hessing JGM (1992) Xylans and xylanases. In: Progress in biotechnology. Elsevier Science, Amsterdam, p 349

Manasa C, Nalini MS (2014) L-Asparaginase activity of fungal endophytes from Tabernaemontana heyneana wall (apocynaceae), endemic to the Western Ghats (India). Int Scholarly Res Notices 2014

Maria GL, Sridhar KR, Raviraja NS (2005) Antimicrobial and enzyme activity of mangrove endophytic fungi of southwest coast of India. J Agric Technol 1:67–80

Marlida Y, Saari N, Hassan Z, Radu S, Bakar J (2000) Purification and characterization of sago starch-degrading glucoamylase from Acremonium sp. endophytic fungus. Food Chem 71:221–227

Marlida Y, Delfita R, Adnadi P, Ciptaan G (2010) Isolation, characterization and production of phytase from endophytic fungus its application for feed. Pak J Nutr 9:471–474

Martín-Rodríguez AJ, Reyes F, Martín J, Pérez-Yépez J, León-Barrios M, Couttolenc A, Espinoza C, Trigos Á, Martín VS, Norte M (2014) Inhibition of bacterial quorum sensing by extracts from aquatic fungi: first report from marine endophytes. Mar Drugs 12:5503–5526

Masumi S, Mirzaei S, Kalvandi R, Zafari D (2014) Asparaginase and amylase activity of thyme endophytic fungi. J Crop Protect 3:655–662

Mayerhofer MS, Fraser E, Kernaghan G (2015) Acid protease production in fungal root endophytes. Mycologia 107:1–11

Medema MH, Blin K, Cimermancic P, de Jager V, Zakrzewski P, Fischbach MA, Weber T, Takano E, Breitling R (2011) antiSMASH: rapid identification, annotation and analysis of secondary metabolite biosynthesis gene clusters in bacterial and fungal genome sequences. Nucleic Acids Res 39:W339–W346

Mhatre A, Narwankar R, Rawat A, Tembadmani K, Mishra S (2017) Characterization of endophytic fungi from medicinal plants for application in therapeutic enzyme extraction. In: Current perspectives in sustainable environment management. SIES Indian Institute of Environment Management, Nerul, pp 230–239

Mishra R, Sarma VV (2017) Mycoremediation of heavy metal and hydrocarbon pollutants by endophytic fungi. In: Mycoremediation and environmental sustainability. Springer, Berlin, pp 133–151

Mishra VK, Passari AK, Singh BP (2016) In vitro antimycotic and biosynthetic potential of fungal endophytes associated with Schima Wallichii. In: Kumar P et al (eds) Current trends in disease diagnostics. Springer International, Cham, pp 367–381

Mishra VK, Passari AK, Chandra P, Leo VV, Kumar B, Gupta VK, Singh BP (2017) Determination and production of antimicrobial compounds by Aspergillus clavatonanicus strain MJ31, an endophytic fungus from Mirabilis jalapa L. using UPLC-ESI-MS/MS and TD GC-MS. PLoS One 12(10):1–24. https://doi.org/10.1371/journal.pone.0186234

Murali TS (2011) L-asparaginase from marine derived fungal endophytes of seaweeds. Mycosphere 2:147–155

Murthy PS, Naidu MM (2011) Improvement of robusta coffee fermentation with microbial enzymes. Eur J Appl Sci 3:130–139

Muthezhilan R, Vinoth S, Gopi K, Jaffar Hussain A (2014) Dye degrading potential of immobilized laccase from endophytic fungi of coastal sand dune plants. Int J ChemTech Res 6:4154–4160

Nagarajan A, Thirunavukkarasu N, Suryanarayanan TS, Gummadi SN (2014) Screening and isolation of novel glutaminase free L-asparaginase from fungal endophytes. Res J Microbiol 9:163

Nahas E (1996) Factors determining rock phosphate solubilization by microorganisms isolated from soil. World J Microbiol Biotechnol 12:567–572

Naidu KS (2011) Characterization and purification of protease enzyme. J Appl Pharmaceut Sci 1:107–112
Nakahama K, Imada A, Igarasi S, Tubaki K (1973) Formation of L-asparaginase by Fusarium species. Microbiology 75:269–273
Nath R, Sharma GD, Barooah M (2015) Plant growth promoting endophytic fungi isolated from tea (Camellia sinensis) shrubs of Assam, India. Appl Ecol Environ Res 13:877–891
Neema Job MS, Philip R (2015) Isolation and characterization of endophytic fungi from Avicennia officinalis. Int J Res Biomed Biotechnol 5:4–8
Noor ZM, Ahmad MS, Ariffin ZZ (2016) Purification and characterisation of fibrinolytic enzymes from endophytic fungi and Lignosus rhinocerus. Jurnal Teknologi 78:53–57
Nygren CMR, Edqvist J, Elfstrand M, Heller G, Taylor AFS (2007) Detection of extracellular protease activity in different species and genera of ectomycorrhizal fungi. Mycorrhiza 17:241
Oses R, Valenzuela S, Freer J, Baeza J, Rodríguez J (2006) Evaluation of fungal endophytes for lignocellulolytic enzyme production and wood biodegradation. Int Biodeter Biodegr 57:129–135
Paice MG, Gurnagul N, Page DH, Jurasek L (1992) Mechanism of hemicellulose-directed prebleaching of kraft pulps. Enzym Microb Technol 14:272–276
Parvez S, Kang M, Chung H, Cho C, Hong M, Shin M, Bae H (2006) Survey and mechanism of skin depigmenting and lightening agents. Phytother Res 20:921–934
Patel C, Yadav S, Rahi S, Dave A (2013) Studies on biodiversity of fungal endophytes of indigenous monocotaceous and dicotaceous plants and evaluation of their enzymatic potentialities. Int J Sci Res Publ 3:1
Patil MG, Pagare J, Patil SN, Sidhu AK (2015) Extracellular enzymatic activities of endophytic fungi isolated from various medicinal plants. Int J Curr Microbiol App Sci 4:1035–1042
Pavithra N, Sathish L, Ananda K (2012) Antimicrobial and enzyme activity of endophytic fungi isolated from Tulsi. J Pharmaceut Biomed Sci 16:2014
Peng X-W, Chen H-Z (2007) Microbial oil accumulation and cellulase secretion of the endophytic fungi from oleaginous plants. Ann Microbiol 57:239
Petrini O (1991) Fungal endophytes of tree leaves. In: Microbial ecology of leaves. Springer, Berlin, pp 179–197
Poling SM, Wicklow DT, Rogers KD, Gloer JB (2008) Acremonium zeae, a protective endophyte of maize, produces dihydroresorcylide and 7-hydroxydihydroresorcylides. J Agric Food Chem 56:3006–3009
Polizeli M, Rizzatti ACS, Monti R, Terenzi HF, Jorge JA, Amorim DS (2005) Xylanases from fungi: properties and industrial applications. Appl Microbiol Biotechnol 67:577–591
Prade RA (1996) Xylanases: from biology to biotechnology. Biotechnol Genet Eng Rev 13:101–132
Pradeep SM, Riaz M, Jagadeesh KS (2010) Screening and characterization of L-asparaginase producing microorganisms from tulsi (Ocimum sanctum L.). Karnataka J Agric Sci 23:660–661
Prathyusha P, Rajitha AB, Satya Prasad K (2015) Diversity and enzymatic activity of foliar endophytic fungi isolated from medicinal plants of Indian dry deciduous forest. Pharm Lett 7:244–251
Priebe S, Linde J, Albrecht D, Guthke R, Brakhage AA (2011) FungiFun: a web-based application for functional categorization of fungal genes and proteins. Fungal Genet Biol 48:353–358
Pritchard PE (1992) Studies on the bread-improving mechanism of fungal alpha-amylase. J Biol Educ 26:12–18
Rabha AJ, Naglot A, Sharma GD, Gogoi HK, Veer V (2014) In vitro evaluation of antagonism of endophytic Colletotrichum gloeosporioides against potent fungal pathogens of Camellia sinensis. Indian J Microbiol 54:302–309
Rajput K, Chanyal S, Agrawal PK (2016) Optimization of protease production by endophytic fungus, Alternaria alternata isolated from gymnosperm tree-Cupressus torulosa D Don. World J Pharm Pharmaceut Sci 5:1034–1054
Rajulu MBG, Thirunavukkarasu N, Suryanarayanan TS, Ravishankar JP, El Gueddari NE, Moerschbacher BM (2011) Chitinolytic enzymes from endophytic fungi. Fungal Divers 47:43–53

Rao MB, Tanksale AM, Ghatge MS, Deshpande VV (1998) Molecular and biotechnological aspects of microbial proteases. Microbiol Mol Biol Rev 62:597–635

Redecker D, Kodner R, Graham LE (2000) Glomalean fungi from the Ordovician. Science (80–) 289:1920–1921

Reddy PV, Lam CK, Belanger FC (1996) Mutualistic fungal endophytes express a proteinase that is homologous to proteases suspected to be important in fungal pathogenicity. Plant Physiol 111:1209–1218

Robl D, da Silva Delabona P, Mergel CM, Rojas JD, dos Santos Costa P, Pimentel IC, Vicente VA, da Cruz Pradella JG, Padilla G (2013) The capability of endophytic fungi for production of hemicellulases and related enzymes. BMC Biotechnol 13:94

Rubini MR, Silva-Ribeiro RT, Pomella AWV, Maki CS, Araújo WL, Dos Santos DR, Azevedo JL, Cruz R, Romero JV, Cortina Guerrero H (1980) Diversity of endophytic fungal community of cacao (Theobroma cacao L.) and biological control of Crinipellis perniciosa, causal agent of Witches' Broom disease. CATIE, Turrialba (Costa Rica)

Russell JR, Huang J, Anand P, Kucera K, Sandoval AG, Dantzler KW, Hickman D, Jee J, Kimovec FM, Koppstein D (2011) Biodegradation of polyester polyurethane by endophytic fungi. Appl Environ Microbiol 77:6076–6084

Saikkonen K, Mikola J, Helander M (2015) Endophytic phyllosphere fungi and nutrient cycling in terrestrial ecosystems. Curr Sci 109:121–126

Saran S, Isar J, Saxena RK (2007) A modified method for the detection of microbial proteases on agar plates using tannic acid. J Biochem Biophys Methods 70:697–699

Schulz B, Boyle C (2005) The endophytic continuum. Mycol Res 109:661–686

Sheik S, Chandrashekar KR, Swaroop K, Somashekarappa HM (2015) Biodegradation of gamma irradiated low density polyethylene and polypropylene by endophytic fungi. Int Biodeter Biodegr 105:21–29

Sherameti I, Shahollari B, Venus Y, Altschmied L, Varma A, Oelmüller R (2005) The endophytic fungus Piriformospora indica stimulates the expression of nitrate reductase and the starch-degrading enzyme glucan-water dikinase in tobacco and Arabidopsis roots through a homeodomain transcription factor that binds to a conserved motif in. J Biol Chem 280:26241–26247

Shubha J, Srinivas C (2017) Diversity and extracellular enzymes of endophytic fungi associated with Cymbidium aloifolium L. Afr J Biotechnol 16:2248–2258

Siddikee MA, Zereen MI, Li C-F, Dai C-C (2016) Endophytic fungus Phomopsis liquidambari and different doses of N-fertilizer alter microbial community structure and function in rhizosphere of rice. Sci Rep 6:32270

Smith DJ, Burnham MK, Bull JH, Hodgson JE, Ward JM, Browne P, Brown J, Barton B, Earl AJ, Turner G (1990) Beta-lactam antibiotic biosynthetic genes have been conserved in clusters in prokaryotes and eukaryotes. EMBO J 9:741–747

Soleimani M, Afyuni M, Hajabbasi MA, Nourbakhsh F, Sabzalian MR, Christensen JH (2010) Phytoremediation of an aged petroleum contaminated soil using endophyte infected and non-infected grasses. Chemosphere 81:1084–1090

Sorgatto M, Guimarães NCA, Zanoelo FF, Marques MR, Peixoto-Nogueira SC, Giannesi GG (2012) Purification and characterization of an extracellular xylanase produced by the endophytic fungus, Aspergillus terreus, grown in submerged fermentation. Afr J Biotechnol 11:8076–8084

Strobel GA (2002) Rainforest endophytes and bioactive products. Crit Rev Biotechnol 22:315–333

Sudha V, Govindaraj R, Baskar K, Al-Dhabi NA, Duraipandiyan V (2016) Biological properties of endophytic fungi. Braz Arch Biol Technol 59:e16150436

Sun X, Guo L-D, Hyde KD (2011) Community composition of endophytic fungi in Acer truncatum and their role in decomposition. Fungal Divers 47:85–95

Sundarram A, Murthy TPK (2014) α-amylase production and applications: a review. J Appl Environ Microbiol 2:166–175

Sunitha VH, Ramesha A, Savitha J, Srinivas C (2012) Amylase production by endophytic fungi Cylindrocephalum sp. isolated from medicinal plant Alpinia calcarata (Haw.) Roscoe. Braz J Microbiol 43:1213

Sunitha VH, Devi DN, Srinivas C (2013) Extracellular enzymatic activity of endophytic fungal strains isolated from medicinal plants. World J Agric Sci 9:1–9
Suryanarayanan TS, Thirunavukkarasu N, Govindarajulu MB, Gopalan V (2012) Fungal endophytes: an untapped source of biocatalysts. Fungal Divers 54:19–30
Suto M, Takebayashi M, Saito K, Tanaka M, Yokota A, Tomita F (2002) Endophytes as producers of xylanase. J Biosci Bioeng 93:88–90
Thangavel A, Krishnamoorthy G, Subramanian M, Maruthamuthu M (2013) Seaweed endophytic fungi: a potential source for Glutaminase free L-Asparaginase. Che Sci Rev Lett 2:354–384
Theantana T, Hyde KD, Lumyong S (2009) Asparaginase production by endophytic fungi from Thai medicinal plants: cytoxicity properties. Int J Integr Biol 7:1–8
Thirunavukkarasu N, Jahnes B, Broadstock A, Rajulu MBG, Murali TS, Gopalan V, Suryanarayanan TS (2015) Screening marine-derived endophytic fungi for xylan-degrading enzymes. Curr Sci 109:112–120
Tian X, Schaich KM (2013) Effects of molecular structure on kinetics and dynamics of the trolox equivalent antioxidant capacity assay with ABTS+•. J Agric Food Chem 61:5511–5519
Tiwari K (2015) The future products: endophytic fungal metabolites. J Biodivers Biopros Dev 2:214–2376
Toghueo RMK, Zabalgogeazcoa I, de Aldana BRV, Boyom FF (2017) Enzymatic activity of endophytic fungi from the medicinal plants Terminalia catappa, Terminalia mantaly and Cananga odorata. S Afr J Bot 109:146–153
Tuppad DS, Shishupala S (2014) Evaluation of endophytic fungi from Butea monosperma for antimicrobial and enzyme activity. J Med Plants Stud 2:38–45
Urairuj C, Khanongnuch C, Lumyong S (2003) Ligninolytic enzymes from tropical endophytic Xylariaceae. Fungal Divers 13:209–219
Uzma F, Konappa NM, Chowdappa S (2016) Diversity and extracellular enzyme activities of fungal endophytes isolated from medicinal plants of Western Ghats, Karnataka. Egypt J Basic Appl Sci 3:335–342
Uzma F, Hashem A, Murthy N, Mohan HD, Kamath PV, Singh BP, Venkataramana M, Gupta VK, Siddaiah CN, Chowdappa S, Alqaeawi AA, Abd Allah EF (2018) Endophytic fungi—alternative sources of cytotoxic compounds: a review. Front Pharmacol 9(309):1–37. doi: https://doi.org/10.3389/fphar.2018.00309.
Venkatachalam A, Rajulu G, Thirunavukkarasu N, Suryanarayanan TS (2015) Endophytic fungi of marine algae and seagrasses: a novel source of chitin modifying enzymes. Mycosphere 6:345–355
Venkatesagowda B, Ponugupaty E, Barbosa AM, Dekker RFH (2012) Diversity of plant oil seed-associated fungi isolated from seven oil-bearing seeds and their potential for the production of lipolytic enzymes. World J Microbiol Biotechnol 28:71–80
Wakelin SA, Warren RA, Harvey PR, Ryder MH (2004) Phosphate solubilization by Penicillium spp. closely associated with wheat roots. Biol Fertil Soils 40:36–43
Wang JW, Wu JH, Huang WY, Tan RX (2006) Laccase production by Monotospora sp., an endophytic fungus in Cynodon dactylon. Bioresour Technol 97:786–789
Wang Z, Wang Y, Zhang D, Li J, Hua Z, Du G, Chen J (2010) Enhancement of cell viability and alkaline polygalacturonate lyase production by sorbitol co-feeding with methanol in Pichia pastoris fermentation. Bioresour Technol 101:1318–1323
Wipusaree N, Sihanonth P, Piapukiew J, Sangvanich P, Karnchanatat A (2011) Purification and characterization of a xylanase from the endophytic fungus Alternaria alternata isolated from the Thai medicinal plant, Croton oblongifolius Roxb. Afr J Microbiol Res 5:5697–5712
Wong KK, Tan LUL, Saddler JN (1988) Multiplicity of beta-1, 4-xylanase in microorganisms: functions and applications. Microbiol Rev 52:305
Wu S, Chappell J (2008) Metabolic engineering of natural products in plants; tools of the trade and challenges for the future. Curr Opin Biotechnol 19:145–152
Wu B, Wu L, Chen D, Yang Z, Luo M (2009) Purification and characterization of a novel fibrinolytic protease from Fusarium sp. CPCC 480097. J Ind Microbiol Biotechnol 36:451–459

Wu W, Davis RW, Tran-Gyamfi MB, Kuo A, LaButti K, Mihaltcheva S, Hundley H, Chovatia M, Lindquist E, Barry K (2017) Characterization of four endophytic fungi as potential consolidated bioprocessing hosts for conversion of lignocellulose into advanced biofuels. Appl Microbiol Biotechnol 101:2603–2618

Yan JF, Broughton SJ, Yang SL, Gange AC (2015) Do endophytic fungi grow through their hosts systemically? Fungal Ecol 13:53–59

Yuan Y, Feng H, Wang L, Li Z, Shi Y, Zhao L, Feng Z, Zhu H (2017) Potential of endophytic fungi isolated from cotton roots for biological control against verticillium wilt disease. PLoS One 12:e0170557

Zaferanloo B, Virkar A, Mahon PJ, Palombo EA (2013) Endophytes from an Australian native plant are a promising source of industrially useful enzymes. World J Microbiol Biotechnol 29:335–345

Zaferanloo B, Bhattacharjee S, Ghorbani MM, Mahon PJ, Palombo EA (2014) Amylase production by Preussia minima, a fungus of endophytic origin: optimization of fermentation conditions and analysis of fungal secretome by LC-MS. BMC Microbiol 14:55

Zaidi KU, Ali AS, Ali SA (2014a) Purification and characterization of melanogenic enzyme tyrosinase from button mushroom. Enzyme Res 2014:120739

Zaidi KU, Ali AS, Ali SA, Naaz I (2014b) Microbial tyrosinases: promising enzymes for pharmaceutical, food bioprocessing, and environmental industry. Biochem Res Int 2014

Zanotto SP, Romano IP, Lisboa LUS, Duvoisin S Jr, Martins MK, Lima FA, Silva SF, Albuquerque PM (2009) Potential application in biocatalysis of mycelium-bound lipases from Amazonian fungi. J Braz Chem Soc 20:1046–1059

Chapter 15
Secondary Metabolites from Marine Endophytic Fungi: Emphasis on Recent Advances in Natural Product Research

P. V. Bramhachari, S. Anju, Ganugula Mohana Sheela, T. Raja Komaraiah, Devanaboyina Venkataiah, and A. M. V. N. Prathyusha

15.1 Introduction

An enormous biological and chemical diversity exists in the oceans, which make up the major portion of our planet's surface. Many diverse groups of microorganisms inhabit the marine habitat compared to terrestrial habitation. The discovery of marine natural products (MNPs) initially focused on the easily accessible macroorganisms—algae, corals, and sponges—of which a range of bioactive compounds have been described (Leal et al. 2012; Almeida et al. (2011); Aus et al. (2007)). Nonetheless, efforts have steadily turned to the smaller forms of life such as bacteria and fungi (Gerwick and Moore 2012), which constitute a large portion of the marine biomass (Sogin et al. 2006). In a recent chemoinformatics survey, it was determined that approximately 71% of the natural products obtained from marine organisms were not represented in terrestrial species, and among them 53% had been found barely once (Montaser and Luesch 2011). Harmonizing reports exploring the distribution of natural products in chemical space has shown clearly that MNPs have the broadest distribution, covering many drug-relevant areas (Tao et al. 2015). As such, the spotlight has now shifted to MNP bioprospecting, which has delivered extremely high hit rates (Blunt et al. 2015).

The marine environment is well known for its exceptional groups of organisms, which are the source of a broad collection of alluring structures. These marine

P. V. Bramhachari (✉) · A. M. V. N. Prathyusha
Department of Biotechnology, Krishna University, Machilipatnam, Andhra Pradesh, India

S. Anju
Bhavan's Vivekananda College of Science, Humanities and Commerce, Secunderabad, India

G. Mohana Sheela
Department of Biotechnology, Vignan University, Guntur, Andhra Pradesh, India

T. R. Komaraiah · D. Venkataiah
Department of Microbiology, Kakatiya University, Warangal, Telangana, India

B. P. Singh (ed.), *Advances in Endophytic Fungal Research*, Fungal Biology,
https://doi.org/10.1007/978-3-030-03589-1_15

microbes are without doubt attractive resources because they produce new natural products with antimicrobial activities. Marine-derived fungi are known to produce a surplus of novel bioactive secondary metabolites (Hasan et al. 2015). A few metabolites among them featured as new carbon frameworks, have been produced from exceptional microorganisms including bacteria, cyanobacteria, microalgae and fungi that become an important source of novel pharmacologically active metabolites. Many questions in marine natural product research are related to the biology and chemistry of various forms of associations between marine microbes: Whose partner produces what secondary metabolites? Thus far, no firm evidence of the biosynthetic source of most secondary metabolites isolated from marine invertebrates exists. The outcome of the few investigations at the genetic level must still be considered uncertain, in view of the fact that most information gained in this way assumes a putative function of the pertinent gene clusters or fragments thereof (König et al. 2006a).

Recent reviews make obvious the significance of these organisms as possible sources of pharmaceutical leads (Bugni and Ireland 2004; König et al. 2006b; Schulz 2001; Blunt et al. 2008; Ebel 2006). Marine-derived fungi are a profuse source of novel chemical diversity and have provided more than 1000 new natural products, a few of them with clinically significant pharmacological activity. These spiky rises in figures signify an intensifying curiosity in marine-derived fungi as sources of new bioactive metabolites. However, extensive information previously existed on the metabolic potential of marine endophytic fungi and stimulating potential for exploiting endophytic fungi for the manufacture of a surfeit of distinct and novel biologically active secondary metabolites (Aly et al. 2011; Mousa and Raizada 2013).

Metabolic compounds released by endophytes are expected to trigger plant defense systems and stimulate plant growth machinery below the diseased condition to maintain cellular loss caused by apoptosis (Alvin et al. 2014). In addition, endophytes are a source of numerous compounds having distinct properties. The large-scale production of endophyte-derived secondary metabolites may gain importance in agriculture and pharmaceuticals (Uzma et al. 2018; Mishra et al. 2017b; Golinska et al. 2015) due to their antibacterial, antifungal, anticancerous, immunosuppressive, and antioxidative nature (Mishra et al. 2017a; Zhang et al. 2006). Promising results on their pharmacological applications encourage a further look into this group of microorganisms (Firáková et al. 2007). In the early 1940s, following the discovery of penicillin in 1928, there was an incredible upsurge in the discovery of natural products from microbes that could be used as drugs. About 1500 fungal metabolites were added to the list of substances with antitumor and antibiotic activity, through 9 years from 1993 (Pelaez 2006; Bhimba et al. 2012). There have been several investigations of all habitats to isolate fungi that might be sources of metabolites of medical importance.

Initially, the most frequently isolated MNPs were from terrestrial fungi, but steadily, with investigations of microbial diversity in marine habitats, it became feasible to isolate marine fungi to produce a diversity of natural products. Nonetheless, a successful cultivation of a new fungal strain is often sufficient to identify new natural products (Marmann et al. 2014). Nevertheless the typical examples are isolation of arugosins from the marine-derived fungus *E. nidulans* var. *acristata* (Kralj et al. 2006) or the deliverance of deoxypodophyllotoxin, a prodrug of the anticancer drug podophyllotoxin, in the endophytic fungus *A. fumigatus*

Fresenius (Kusari et al. 2009). The most prominent groups of bioactive compounds obtained from marine-derived fungi are still the cephalosporines, with cephalosporine C first isolated in 1945 from *Acremonium chrysogenum* (Newton and Abraham 1955). There are also more recent promising examples, which include halovir and numerous naturally occurring analogs, which are potent inhibitors of herpes simplex viruses 1 and 2. Perhaps the most essential signature molecule is the diketopiperazine halimide, which was first discovered in 1990s and which acts as a tubulin depolymerizing molecule. However, this molecule served as a lead structure for closely related synthetic analog plinabulin (NPI-2358), which at the time underwent phase II clinical trials in patients with superior non-small-cell lung cancer (Aren et al. 2010). The new alkaloid sorbicillactone A isolated from the sponge-derived fungus *Penicillium chrysogenum* showed promising activity against leukemia cells without exhibiting noteworthy cytotoxicity (Bringmann et al. 2005). Other important classes of metabolites include, for example, the phomactins, peribysins, and tryprostatins, the marine-derived pain killer Prialt, and the antitumor drug Yondelis, previously available on the market. The SWOT analysis of the natural product chemistry of marine-derived fungi has received a increased attention in recent years, predominantly starting in the preceding decade, and there continues to be stable growth in the total number of novel structures reported in the chemical repertoire. In light of the pronounced disparity between the actual number of cultivated strains and the estimated biodiversity of fungi in marine habitats in general, it is yet to be likely that this trend will continue.

Studies have revealed that marine fungal endophytes are explored to identify their characteristic enzymes, vitamins, polysaccharides, pigments, and lipids of commercial importance. Future research should provide information to suggest methods and media for culturing these fungi and identifying the majority of their secondary metabolites (Damare et al. 2012). Endophytes in association with plants have an exceptionally appealing function as they can reduce harmful pathogens and nematodes. They are categorized as obligate endophytes and facultative endophytes Carroll (1988). Most of them are facultative and adjust to the rhizosphere region where they coexisted with other microbes (Brader et al. 2014). This review summarizes the exciting developments in the understanding of marine-associated endophytic fungi with a unique emphasis on secondary metabolite production and metagenomics for the exploitation of marine microbial diversity for its huge potential in biomedical applications.

15.2 Secondary Metabolite Potential of Marine Endophytic Fungi

Many new secondary metabolites have been identified in recent years that are products of marine fungi. Most secondary metabolites were recognized to have new chemical compositions that were very rarely similar to the ones found in terrestrial fungi. Fungal endophytes might dwell in a metabolically antagonistic milieu, steadily encountering host defense chemicals (Schulz et al. 2002, 2008). Such a hostile environment might account for the evolution of a potentially augmented

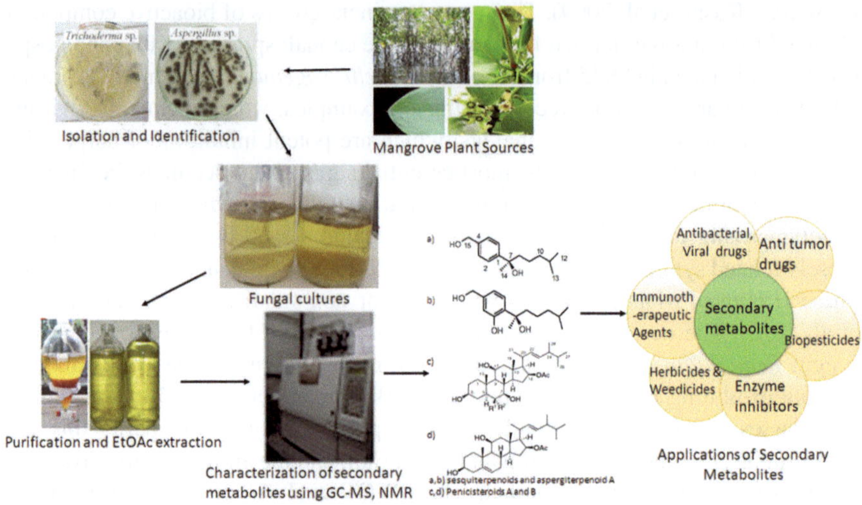

Fig. 15.1 Selected bioactive secondary metabolites isolated from marine endophytic fungi with medicinal importance

synthetic ability of endophytes. This is perhaps proven from the fact that an endophyte, when isolated from certain hosts, is found not to produce a metabolite that was otherwise expressed in certain hosts (Li et al. 1996). Marine fungi have potential significance in ecology, which is why they have caused many researchers to want to study them (Mayer and Hamann 2004; Bourguet-Kondracki and Kornprobst 2005). Developments in molecular separation techniques and spectroscopic analysis have spurred tremendous advances in configuring the chemical scaffolding and properties of the natural products obtained from marine endophytes (Kobayashi and Tsuda 2004). Nevertheless, the incidence of fungal endophytes related to many plant species is not yet understood. Hence, there is huge opportunity to identify novel bioactive molecules from microbes associated with therapeutically valuable plants (Nisa et al. 2015). One of the major factors in not being able to identify them is their inefficiency at producing in sufficient quantity for drug discovery. Many compounds that have antimicrobial (Maria et al 2005), antioxidant, and anticancer activities are found to have been produced by marine endophytic fungi (Fig. 15.1) (Chandra 2012; Nisa et al. 2015; Demain 2014).

15.3 Types of Secondary Metabolites Produced by Marine Fungi

Marine fungi are well known to produce a wide range of terpenes that have striking structural diversity with effective pharmacological properties. Mostly phomactins and peribysins have been identified as typical marine terpenes derived from fungi (Ebel 2010). *Apostichopus japonicus*, a fungus associated with sea cucumber, was identified as producing diaporthin B and three novel pimarane diterpenes (Xia et al. 2012). A bisabolane type of sesquiterpenoid and aspergiterpenoid A fermented by *Aspergillus* spp. was retrieved from a marine sponge *Xestospongia testudinaria* (Li et al. 2012). Interestingly, from an Ethyl acetate extract of *Aspergillus insuetus* (OY-207), three novel compounds, meroterpenoids, insuetolidesA-C, and (E)-6-(40-hydroxy-20-butenoyl)-strobilactone A, were derived. The fungus *Aspergillus insuetus* (OY-207) was found as an endophyte in the marine sponge *Psammocinia* sp.

Among the group of steroid-type secondary metabolites isolated that are of marine origin are penicisteroids A and B from *Penicillium chrysogenum* QEN-24S from red alga *Laurencia* (Gao et al. 2010). From an algal endophytic fungus, *Coniothyrium cereal*, seven phenalenone derivatives were obtained (Elsebai et al. 2010). Aspergillin A, isolated from *A. versicolor*, belongs to a group of aromatic polyketides (Li et al. 2011; Lee et al. 2010). Secondary metabolites belonging to a category of ethers, depsidones, aspergillusidones A-C, diaryl ether, and aspergillusether A, were isolated from a marine fungus of the sponge *Aunguis* (Sureram et al. 2012). An alcohol-containing type of secondary metabolite, chlorogentisyl alcohol, was derived from a marine species of *Aspergillus* in red alga *H. saidana* (Li et al. 2005). However, Kjer et al. (2010) gave a comprehensive explanation of isolation and cultivation methods for fungi associated with various marine organisms (e.g., sponges, algae, and mangroves).

15.4 Marine Endophytic Fungi of Sponges

The majority of marine endophytes were found to be associated with sponges. Their soft bodies and sedentary habitats facilitate effortless propagation of the fungal endophyte. The defense against the predators of sponges is the release of enzymes like phospholipases (Taylor et al. 2007; Selvin 2009; Ding et al. 2011). Metabolites released by fungi associated with sponges are known to have chemical structures different from that of other fungi (Gewonnenen et al. 2007). Sorbicillactone A is an alkaloid from the sponge-associated *Penicillium chrysogenum* that shows potential activity against leukemia cells. Polyketide synthase (PKS) and nonribosomal peptide synthase (NRPS) were identified in an endophytic fungus in sponges that showed significant antimicrobial activity on *P. fluorescens, S. aureus*, and *B. subtilis*. The microorganisms in invertebrate hosts establish themselves by means of a receptor binding. The sponges specifically produce lectins and surface glycans that allow ideal interactions (Höller et al. 2000). The host develops specific mechanisms like antibiosis that enables it to overcome the action of antibioticlike substances produced by endophytes. The symbiotic existence of the microbes inside the host may generate immunological reactions, but these antigens are recognized in the host as part of cross reactions that lead to a secure survival (Debbab et al. 2010).

A large number of endosymbionts that inhabit marine corals could be sites to isolate fungi that produce bioactive compounds. Characteristically, corals show tremendous adaptation to changes in the environment that may be greatly influenced by these endosymbionts (Reshef et al. 2006; van Oppen et al. 2009). *Emericella* sp.

filamentous marine fungus in association with *S. arenicola* was found to produce the novel compounds cyclic depsipeptides and emericellamides A and B isolated in sea corals. Some endophytic fungi are documented to be opportunistic pathogens in stressed sponges. An endophytic fungus in some species is believed to show mutualistic and pathogenic features. These feature have been distinctively identified only in marine fungi. It was reported that *A. syndowii* was identified as a pathogen of *S. obscura,* a sponge that has been isolated from the Bahamas (Debbab et al. 2011). Several other bioactive compounds, like 2,6-dimethoxy-1,4-benzoquinone from *D. salina*, Siccayne from *H. vilosa*, Trichoharzin from *T. harzianum*, and gymnastatins A, B, and C, were among the secondary metabolites identified from marine fungi (Gewonnenen et al. 2007).

15.5 Marine Endophytic Fungi of Algae

The majority of marine endophytic fungi are not obligate in nature. They represent a vast source of natural products (Bugni and Ireland 2004). Studies by Sarasan et al. (2017) reported the dominance of marine metabolites from endophytic fungi as brown algae (39%), red algae (28%), and green algae (23%) (Sarasan et al. 2017). Diverse types of pure compounds and crude extracts isolated from fungal endophytes of seaweed origin reveal different biological activities such as antifungal, insecticidal, antibacterial, antialgal, antiplasmodial, anticancer, acetylcholine esterase inhibitor, cytotoxicity, antiviral, antiangiogenic, and antioxidant activities (Ohkawa et al. 2010). Metabolites from endophytic fungi of marine algae are chemically characterized as alkaloids (Tsuda et al. 2004), aromatic polyketides (Pontius et al. 2008), terpenes (Lösgen et al. 2007), sesquiterpenes (Bugni and Ireland 2004), and steroids (Zhang et al. 2016). Thus, these fungi can be used for making drugs as they are a source of pharmacologically active substances and compounds for drug synthesis (Blunt et al. 2010).

Interestingly, prenylated polyketides were isolated from the endophytic fungus *E. nidulans* var. *a cristata* of a marine alga. The group of prenylated polyketides showed significant antitumor activity and immunostimulatory activity (Kralj et al. 2006). The endophytic fungus *C. globosum*, isolated from the tissues of the marine red alga *P. urceolata*, produced chaetopyranin, a novel benzaldehyde secondary metabolite (Wang et al. 2006). A marine endophytic fungus, *A. niger* EN-13, was isolated from the inner tissue of the marine brown alga *C. sinuosa* and was found to produce a new naphthoquinoneimine that displayed antifungal activity (Zhang et al. 2007). It is notable that the fungal endosymbiont *Chaetomium* species residing in marine algae is a source of novel polyketides, chaetocyclinone A, B, and C, as evidenced by Lösgen et al. (2007). It is interesting to note that asperamides A and B, a sphingolipid, and their corresponding glycosphingolipid were isolated from a culture extract of *A. niger* EN-13, an endophytic fungus isolated from the marine brown alga *C. sinuosa* (Zhang et al. 2007). However, Zhang et al. (2007) reported the isolation of an endophyte, *A. niger* EN-13, from the brown seaweed *C. sinuosa.* The endophyte was able to produce the antifungal compound naphthoquinoneimine derivative. In another report, the red alga *Ceramium* sp. collected from the North Sea, Busum, Germany, was shown to host the fungal endophyte *P. spartinae* as a potential source of spartinoxide, 4-hydroxy-3-prenyl-benzoic acid, and anofinic acid (Elsebai et al. 2010). *Penicillium chrysogenum* QEN-24S, a marine endophytic fungus isolated from an unidentified marine red algal species of the genus *Laurencia*, was identified with two polyketide derivatives, penicitides A and B, two glycerol monoterpene derivatives that were not reported previously among natural products (Gao et al. 2010). The green seaweed *U. pertusa* was reported as the host of the fungal endophyte *C. globosum* QEN-14, which was explored for the isolation of a cytochalasan derivative, cytoglobosins A-G (Cui et al. 2010a). This research group also isolated an endophyte, *A. ochraceus*, recovered from the brown seaweed *S. kjellmanianum*, and found a novel source of a 7-norergosteroid-bearing pentalactone B-ring (Cui et al. 2010b).

It is noteworthy that Pacific seaweed turned into an invasive species that had tremorgenic mycotoxin indoloditerpenes identified from *A. oryzae*, an endophyte of the marine red alga *H. japonica*. It is noteworthy that this metabolite, JBIR-03, exhibited strong insecticidal activity against *A. salina*, while asporyzin C exhibited potent activities against *E. coli* (Qiao et al. 2010). Strikingly, Wang (2012) reported the isolation and purification of seven different compounds from ethyl acetate extract of *Guignardia* spp. endophytic fungi inhabiting the edible marine alga *U. pinnatifida*. The compounds recognized were ergosterol, ergosterol epoxide, and cyclic dipeptides. Four new indole alkaloids, cristatumins A–D, together with six known congeners were identified and evaluated for antimicrobial and insecticidal activity from a culture extract of *E. cristatum* EN-220, an endophytic fungus isolated from the marine alga *S. thunbergii* (Du et al. 2012). Culture extract fractionation of *A. wentii* EN-48, an endophytic fungus isolated from a marine brown alga *Sargassum*, led to the isolation of three new seco-anthraquinone, benzamide derivative tetranorlabdane diterpenoids, and asperolides A–C evaluated for cytotoxic and antibacterial activities (Li et al. 2014). Hulikere et al. (2016) reported on the isolation of an endophyte, *C. cladosporoides*, that was isolated from the seaweed *S. wightii*. The extract from the fungus was evaluated for antioxidant properties, reducing power, total phenolic content, flavonoid content, and antiangiogenic activity. Li et al. (2017) isolated and evaluated the antimicrobial, antioxidant, and cytotoxic activity of five different polyhydroxylated hydroanthraquinone derivatives from a culture extract of the endophytic fungus *T. islandicus* EN-501 inhabiting the tissues of rhodophyte *L. okamurai*.

15.6 Biosynthetic Pathways for the Discovery of Fungal Secondary Metabolites

Studies on biosynthetic pathways implicated in secondary metabolites indicate that gene clusters involved in this are regulated by a chromatin-based mechanism along with the enzymes histone acetyltransferases, methyltransferases, and deacetylases. These chromatin-based clusters are repressed by primary metabolic product synthesis genes during their growth phase. Consequently, conditions should be transformed and signal molecules introduced in in vitro conditions that might activate secondary metabolite production (Blunt et al. 2013). Most secondary metabolites are cryptic and must be detected using certain methods that can instigate the expression of genes that might otherwise not be active (Brakhage and Schroeckh 2011). Among commonly produced natural metabolites, many cannot be detected in in vitro growth conditions, as they may not be triggered in the absence of certain signal molecules. Also, the concentrations at which they are produced are less likely to be detected. It is a well-understood fact that genes documented for secondary metabolism are more than the products that have been identified. To optimize the number of metabolites, a one strain many compounds (OSMAC) approach to cultivation was developed (Debbab et al. 2010).

Some exciting research has revealed that natural bioactive compounds from the large group of marine fungi can be produced by optimization of ribosomal peptide pathways using genome-based detection. This may be a pretty demanding process but there are exceptional ways to tap the rare products of marine fungal diversity (Donia et al. 2011). To explicate and classify the compounds, various molecular biological methods, like the regulation of transcription factors, so-called knockouts, and promoter exchange, have been used with success. Naturally occurring cryptic product evaluation has been attempted based on epigenetics. As a process to identify the bioactive compounds produced by *A. nidulans*, simulation strategies were followed. The results encourage the application of this method to open new avenues to discovering novel products (Brakhage and Schroeckh 2011).

15.7 Challenges Associated with Marine Natural Product Metagenomics

Though many potentially therapeutic substances have been isolated from marine organisms, few have been commercially exploited because of the inadequate quantities being produced. Also, many microorganisms are uncultivable, as their specific growth requirements are not known, and this must be a hindrance to the discovery of many more bioactive compounds. In this context, metagenomics would be an alternative and can play a vital role in the discovery of novel bioactive molecules (Osburne et al. 2000). In a metagenomic approach there is more scope to discover organisms, overcoming the drawbacks of culturing methods (Handelsman et al. 1998). Metagenomic approaches to MNP discovery can be applied using two methods: a sequence-based method and a function-based method. MNP detection using metagenomics faces a number of challenges and limitations when employing typical functional screening approaches (Kennedy et al. 2010; Li and Neubauer 2014; Reen et al. 2015). Analysis by metagenomic methods has revealed that the genes involved in MNP expression are mostly clustered and are highly conserved genes. The majority of MNPs are not successfully synthesized and cannot be optimized as these products are not expressed in in vitro conditions (Hertweck 2009). To accomplish heterologous expression of new bioactive compounds, development of marine-derived hosts such as actinomycete, cyanobacteria, and symbiotic fungi were developed (Rocha-Martin et al. 2014). Investigations of secondary metabolite production in marine endophytic fungi in the course of genetics have made only baseline contribution to current knowledge (Fig. 15.1). Consistent updating of the genetics and biochemistry of secondary metabolite production in marine fungi would facilitate development of MNP research.

15.8 Conclusions and Future Perspectives

Endophytic marine-derived fungi continue to be a productive resource of exceptional secondary metabolites with interesting structural features, a considerable number of which exhibit potential biological activities. Due to a current setback in natural product research in terrestrial habitats, there is rising concern in the exploration of marine microbes for novel metabolites using biological and chemical methodologies. The introduction of marine natural products to the market has highlighted that classical chemistry can yield sufficient quantities of such molecules if their therapeutic activity is adequately demonstrated. With regard to biochemistry, marine fungal endophytes produce several chemical classes of secondary metabolites, with terpenoids and polyketides being the most common, and flavonoids and lignans being the rarest. To understand the activities of marine fungal endophytes, details about their structure and function relationships, particularly with respect to configuring moieties to enhance or reduce the toxicity of compounds, remain insufficient. Despite clear developments, notable gaps remain in this field of research. Many biosynthetic pathways and enzymes remain unexplained. How an endophyte inside a host plant coordinates metabolic biosynthesis remains unidentified. Very few genes relevant to biosynthetic enzymes have been identified, and inadequate research has been conducted so far on the expression of these genes at the molecular level. Greater emphasis on marine genome and microbial biotechnology approaches is warranted to discover the optimum utilization of marine ecosystems for natural product drug discovery. Continuous genomic library generation, functional proteomics, and DNA microarray analysis of genes expressed in surrogate hosts will give qualitative and quantitative information on the secondary metabolites of uncultivable marine microbes. Finally, a more inclusive understanding of the endophyte–host relationship can lead to new prospects for developing commercial products for the benefit of humanity.

Acknowledgment The author gratefully acknowledges Krishna University for its encouragement and support.

Conflict of Interest We declare we don't have any conflict of interest.

References

Almeida C, Kehraus S, Prudêncio M, König GM (2011) Marilones AC, phthalides from the sponge-derived fungus Stachylidium sp. Beilstein J Org Chem 7:1636–1642

Alvin A, Miller KI, Neilan BA (2014) Exploring the potential of endophytes from medicinal plants as sources of antimycobacterial compounds. Microbiol Res 169(7):483–495

Aly AH, Debbab A, Proksch P (2011). Fungal endophytes: unique plant inhabitants with great promises. Applied microbiology and biotechnology, 90(6):1829–1845

AUS, IVNN, WON, SUAP, & MUSHROOMS, E. (2007). Isolation of new secondary metabolites from spongeassociated and plant-derived endophytic fungi

Berdy J (2005) Bioactive microbial metabolites. J Antibiot 58(1):1

Bhimba BV, Franco DAD, Mathew JM, Jose GM, Joel EL, Thangaraj M (2012) Anticancer and antimicrobial activity of mangrove derived fungi Hypocrea lixii VB1. Chin J Nat Med 10(1):77–80

Blunt JW et al (2008) Marine natural products. Nat Prod Rep 25:35–94

Blunt JW, Copp BR, Keyzers RA, Munro MH, Prinsep MR (2013) Marine natural products. Nat Prod Rep 30(2):237–323

Blunt J, Copp B, Keyzers R, Munro M, Prinsep M (2015) Marine natural products. Nat Prod Rep 32:116–211

Brader G et al (2014) Metabolic potential of endophytic bacteria. Curr Opin Biotechnol 27:30–37

Brakhage AA, Schroeckh V (2011) Fungal secondary metabolites–strategies to activate silent gene clusters. Fungal Genet Biol 48(1):15–22

Bringmann G, Lang G, Gulder TAM, Hideyuki H, Mühlbacher J, Maksimenka K, Steffens S, Schaumann K, Stöhr R, Wiese J, Imhoff JF, Perovi-Ottstadt S, Boreiko O, Müller WEG (2005) The first sorbicillinoid alkaloids, the antileukemic sorbicillactones A and B, from a sponge derived Penicillium chrysogenum strain. Tetrahedron 61:7252–7265

Bugni TS, Ireland CM (2004) Marine-derived fungi: a chemically and biologically diverse group of micro-organisms review. Nat Prod Rep 21:143–163

Carroll G (1988) Fungal endophytes in stems and leaves: from latent pathogen to mutualistic symbiont. Ecology 69(1):2–9

Cui CM, Li XM, Li CS, Proksch P, Wang BG (2010a) Cytoglobosins A–G, cytochalasans from a marine-derived endophytic fungus, Chaetomium globosum QEN-14. J Nat Prod 73(4):729–733

Cui CM, Li XM, Meng L, Li CS, Huang CG, Wang BG (2010b) 7-Nor-ergosterolide, a pentalactone-containing norsteroid and related steroids from the marine-derived endophytic Aspergillus ochraceus EN-31. J Nat Prod 73(11):1780–1784

Damare S, Singh P, Raghukumar S (2012) Biotechnology of marine fungi. Biology of marine fungi. Springer, Berlin, Heidelberg, pp 277–297

Debbab A, Aly AH, Lin WH, Proksch P (2010) Bioactive compounds from marine bacteria and fungi. Microb Biotechnol 3(5):544–563

Debbab A, Aly AH, Proksch P (2011) Bioactive secondary metabolites from endophytes and associated marine derived fungi. Fungal Divers 49(1):1

Donia MS et al (2011) Accessing the hidden majority of marine natural products through metagenomics. ChemBioChem 12(8):1230–1236

Du FY, Li XM, Li CS, Shang Z, Wang BG (2012) Cristatumins A–D, new indole alkaloids from the marine-derived endophytic fungus Eurotium cristatum EN-220. Bioorg Med Chem Lett 22(14):4650–4653

Ebel R (2006) Secondary metabolites from marine derived fungi. In: Proksch P, Müller WEG (eds) Frontiers in marine biotechnology. Horizon Scientific Press, Norwich, pp 73–143

Ebel R (2010) Terpenes from marine-derived fungi. Mar Drugs 8(8):2340–2368

Elsebai MF, Kehraus S, Gütschow M, Koenig GM (2010) Spartinoxide, a new enantiomer of A82775C with inhibitory activity toward HLE from the marine-derived Fungus Phaeosphaeria spartinae. Nat Prod Commun 5(7):1071–1076

Firáková S, Šturdíková M, Múčková M (2007) Bioactive secondary metabolites produced by microorganisms associated with plants. Biologia 62(3):251–257

Gao SS, Li XM, Du FY, Li CS, Proksch P, Wang BG (2010) Secondary metabolites from a marine-derived endophytic fungus Penicillium chrysogenum QEN-24S. Mar Drugs 9(1):59–70

Gerwick W, Moore B (2012) Lessons from the past and charting the future of marine natural products drug discovery and chemical biology. Chem Biol 19:85–98

Golinska P, Wypij M, Agarkar G, Rathod D, Dahm H, Rai M (2015) Endophytic actinobacteria of medicinal plants: diversity and bioactivity. Antonie Van Leeuwenhoek 108(2):267–289

Hasan S, Ansari MI, Ahmad A, Mishra M (2015) Major bioactive metabolites from marine fungi: a review. Bioinformation 11(4):176

Hertweck C (2009) Hidden biosynthetic treasures brought to light. Nat Chem Biol 5:450–452
Höller U, Wright AD, Matthee GF, Konig GM, Draeger S, Hans-Jürgen AUST, Schulz B (2000) Fungi from marine sponges: diversity, biological activity and secondary metabolites. Mycol Res 104(11):1354–1365
Hulikere MM, Joshi CG, Ananda D, Poyya J, Nivya T (2016) Antiangiogenic, wound healing and antioxidant activity of Cladosporium cladosporioides (Endophytic Fungus) isolated from seaweed (Sargassum wightii). Mycology 7(4):203–211
Kennedy J, Flemer B, Jackson S, Lejon D, Morrissey J, O'Gara F et al (2010) Marine metagenomics: new tools for the study and exploitation of marine microbial metabolism. Mar Drugs 8:608–628
Kjer J, Debbab A, Aly AH, Proksch P (2010) Methods for isolation of marine-derived endophytic fungi and their bioactive secondary products. Nat Protoc 5(3):479
König J, Seithel A, Gradhand U, Fromm MF (2006a) Pharmacogenomics of human OATP transporters. Naunyn-Schmiedeberg's archives of pharmacology, 372(6), 432–443
König GM, Kehraus S, Seibert SF, Abdel-Lateff A, Müller D (2006b) Natural products from marine organisms and their associated microbes. ChemBioChem 7(2):229–238
Kralj A, Kehraus S, Krick A, Eguereva E, Kelter G, Maurer M et al (2006) Arugosins G and H: prenylated polyketides from the marine-derived fungus Emericellanidulans var. acristata. J Nat Prod 69(7):995–1000
Kusari S, Lamshöft M, Spiteller M (2009) Aspergillus fumigatus Fresenius, an endophytic fungus from Juniperus communis L. Horstmann as a novel source of the anticancer pro-drug deoxypodophyllotoxin. J Appl Microbiol 107:1019–1030
Leal M, Puga J, Serôdio J, Gomes N, Calado R (2012) Trends in the discovery of new marine natural products from invertebrates over the last two decades–where and what are we bioprospecting? PLoS One 7:e30580
Lee YM, Li H, Hong J, Cho HY, Bae KS, Kim MA et al (2010) Bioactive metabolites from the sponge-derived fungus Aspergillus versicolor. Arch Pharm Res 33(2):231–235
Li Y, Li X, Son BW (2005) Antibacterial and radical scavenging epoxycyclohexenones and aromatic polyols from a marine isolate of the fungus Aspergillus. Nat Prod Sci 11:136–138
Li X, Li XM, Xu GM, Li CS, Wang BG (2014) Antioxidant metabolites from marine alga-derived fungus Aspergillus wentii EN-48. Phytochem Lett 7:120–123
Li HL, Li XM, Li X, Wang CY, Liu H, Kassack MU, Meng LH, Wang BG (2017) Antioxidant Hydroanthraquinones from the marine algal-derived endophytic fungus Talaromyces islandicus EN-501. J Nat Prod 80(1):162–168
Liu X et al (2010) Bioprospecting microbial natural product libraries from the marine environment for drug discovery. J Antibiot 63(8):415
Lösgen S, Schlörke O, Meindl K, Herbst-Irmer R, Zeeck A (2007) Structure and biosynthesis of chaetocyclinones, new polyketides produced by an endosymbiotic fungus. Eur J Org Chem 2007(13):2191–2196
Maria GL, Sridhar KR, & Raviraja NS (2005) Antimicrobial and enzyme activity of mangrove endophytic fungi of southwest coast of India. Journal of Agricultural technology, 1, 67–80
Marmann A, Aly AH, Lin W, Wang B, Proksch P (2014) Co-cultivation—a powerful emerging tool for enhancing the chemical diversity of microorganisms. Mar Drugs 12(2):1043–1065
Mishra VK, Passari AK, Chandra P, Leo VV, Kumar B, Gupta VK, Singh BP (2017a) Determination and production of antimicrobial compounds by *Aspergillus clavatonanicus* strain MJ31, an endophytic fungus from *Mirabilis jalapa* L. using UPLC-ESI-MS/MS and TD GC-MS. PLoS One 12(10):1–24. https://doi.org/10.1371/journal.pone.0186234
Mishra VK, Passari AK, Leo VV, Singh BP (2017b) Molecular diversity and detection of endophytic fungi based on their antimicrobial biosynthetic genes. In: Singh BP, Gupta VK (eds) Molecular markers in mycology, fungal biology. Springer International Publishing, Switzerland, pp 1–35. https://doi.org/10.1007/978-3-319-34106-4_1
Montaser R, Luesch H (2011) Marine natural products: a wave of new drugs? Future Med Chem 3:1475–1489
Mousa WK, Raizada MN (2013) The diversity of anti-microbial secondary metabolites produced by fungal endophytes: an interdisciplinary perspective. Frontiers in microbiology, 4:65

Newton GGF, Abraham EP (1955) Cephalosporine C, a new antibiotic containing sulphur and D-α-aminoadipic acid. Nature 175:548

Ohkawa Y, Miki K, Suzuki T, Nishio K, Sugita T, Kinoshita K, Takahashi K, Koyama K (2010) Antiangiogenic metabolites from a marine-derived fungus, Hypocrea vinosa. J Nat Prod 73(4):579–582

Osburne MS, Grossman TH, August PR, MacNeil IA (2000) Tapping into microbial diversity for natural products drug discovery. ASM News 66:411–417

Pelaez F (2006) The historical delivery of antibiotics from microbial natural products—can history repeat?. Biochemical pharmacology, 71(7):981–990

Pontius A, Krick A, Mesry R, Kehraus S, Foegen SE, Müller M, Klimo K, Gerhäuser C, König GM (2008) Monodictyochromes A and B, dimeric xanthone derivatives from the marine algicolous fungus Monodictys putredinis. J Nat Prod 71(11):1793–1799

Qiao MF, Ji NY, Liu XH, Li K, Zhu QM, Xue QZ (2010) Indoloditerpenes from an algicolous isolate of Aspergillus oryzae. Bioorg Med Chem Lett 20:5677–5680

Reen F, Gutiérrez-Barranquero J, Dobson A, Adams C, O'Gara F (2015) Emerging concepts promising new horizons for marine biodiscovery and synthetic biology. Mar Drugs 13:294–2954

Rocha-Martin J, Harrington C, Dobson A, O'Gara F (2014) Emerging strategies and integrated systems microbiology technologies for biodiscovery of marine bioactive compounds. Mar Drugs 12:3516–3559

Sarasan M, Puthumana J, Job N, Han J, Lee JS, Philip R (2017) Marine algicolous endophytic fungi—a promising drug resource of the era. J Microbiol Biotechnol 27:1039

Schulz B (2001) Bioactive fungal metabolites–impact and exploitation, 20. British Mycological Society. In: International symposium proceedings, University of Wales, Swansea, UK

Schulz B, Boyle C, Draeger S, Römmert AK, Krohn K (2002) Endophytic fungi: a source of novel biologically active secondary metabolites. Mycol Res 106(9):996–1004

Schulz B, Draeger S, dela Cruz TE, Rheinheimer J, Siems K, Loesgen S et al (2008) Screening strategies for obtaining novel, biologically active, fungal secondary metabolites from marine habitats. Bot Mar 51(3):219–234

Sogin M, Morrison H, Huber J, Mark Welch D, Huse S, Neal PR et al (2006) Microbial diversity in the deep sea and the underexplored "rare biosphere". Proc Natl Acad Sci U S A 103:12115–12120

Tao L, Zhu F, Qin C, Zhang C, Chen S, Zhang P et al (2015) Clustered distribution of natural product leads of drugs in the chemical space as influenced by the privileged target-sites. Sci Rep 5:9325

Tsuda M, Kasai Y, Komatsu K, Sone T, Tanaka M, Mikami Y, Kobayashi JI (2004) Citrinadin A, a novel pentacyclic alkaloid from marine-derived fungus Penicillium citrinum. Org Lett 6(18):3087–3089

Uzma F, Hashem A, Murthy N, Mohan HD, Kamath PV, Singh BP, Venkataramana M, Gupta VK, Siddaiah CN, Chowdappa S, Alqaeawi AA, Abd_Allah EF (2018) Endophytic fungi—alternative sources of cytotoxic compounds: a review. Front Pharmacol 9(309):1–37. https://doi.org/10.3389/fphar.2018.00309

Wang GY (2006) Diversity and biotechnological potential of the sponge–associated microbial consortia. J Ind Microbiol Biotechnol 33(7):545–551

Wang F-W (2012) Bioactive metabolites from Guignardia sp., an endophytic fungus residing in Undaria pinnatifida. Chin J Nat Med 10(1):72–76

Wang S, Li XM, Teuscher F, Li DL, Diesel A, Ebel R et al (2006) Chaetopyranin, a benzaldehyde derivative, and other related metabolites from Chaetomium globosum, an endophytic fungus derived from the marine red alga Polysiphonia urceolata. J Nat Prod 69(11):1622–1625

Zhang HW, Song YC, Tan RX (2006) Biology and chemistry of endophytes. Nat Prod Rep 23(5):753–771

Zhang Y, Wang S, Li XM, Cui CM, Feng C, Wang BG (2007) New sphingolipids with a previously unreported 9-methyl-C20-sphingosine moiety from a marine algous endophytic fungus Aspergillus niger EN-13. Lipids 42(8):759–764

Index

B. P. Singh (ed.), *Advances in Endophytic Fungal Research*, Fungal Biology,
https://doi.org/10.1007/978-3-030-03589-1